ALLE ZEIT WACH

1842

M. Samii (Ed.)

Surgery of the Sellar Region and Paranasal Sinuses

With 350 Figures

Springer-Verlag
Berlin Heidelberg New York
London Paris Tokyo
Hong Kong Barcelona Budapest

Chapter 2: Sellar Region

A. General Aspects

Chapter 3: Other Space-Occupying Lesions

List of Contributors

Abbassioun, K.: Arad General Hospital, Somaye Ave., Tehran, Iran

Aghzadi, A.: Service de Neurochirurgie, Hopital des Specialites O.N.O., BP 6220 Institits, Rabat, Morocco

Agnoli, A. L.: Department of Neurosurgery, University of Giessen, Klinikstr. 29, 6300 Giessen, FRG

Amedee, R. G.: Department of Otolaryngology-Head and Neck Surgery, Tulane University Medical Center, 1430 Tulane Avenue, New Orleans, LA 70112, USA

Amini, J.: Arad General Hospital, Somaye Ave., Tehran, Iran

Amirjamshidi, A.: Arad General Hospital, Somaye Ave., Tehran, Iran

Anile, C.: Istituto di Neurochirurgia, Policlinico Gemelli, Largo Gemelli 8, 00168 Roma, Italy

Anthofer, R.: Abteilung für Kiefer- und Gesichtschirurgie, Landeskrankenanstalten, Müllner Hauptstr. 48, 5020 Salzburg, Austria

Arnot, R.: Department of Medical Physics, Royal Postgraduate Medical School, Hammersmith Hospital, Du Cane Rd., London W12 ONS, UK

Ausman, J. I.: Department of Neurological Surgery, Henry Ford Neurosurgical Institute, 2799 West Grand Blvd., Detroit, MI 48202, USA

Bamberg, M.: Department of Radiotherapy West German Tumor Center, University Hospital Essen, Hufelandstr. 55, 4300 Essen 1, FRG

Basso, A.: Department of Neurosurgery, Hospital Santa Lucia, Buenos Aires, Argentina

Bebear, J. P.: Hôpital Pellegrin, Place Amélie Raba-Léon, 33076 Bordeaux Cedex, France

Behr, R.: Department of Neurosurgery, Justus-Liebig University, Klinikstr. 29, 6300 Giessen, FRG

Bellotti, C.: Divisione di Neurochirurgia, Ospedale Maggiore della Carità, 28100 Novara, Italy

Benedetti, D.: Av. La Plata 393 – 8° "C", 1235 Capital, Buenos Aires, Argentina

Berner, S.: Department of Neurosurgery, Hospital Santa Lucia, Buenos Aires, Argentina

Bini, W.: Neurosurgical Clinic, Nordstadt Hospital, Haltenhoffstr. 41, 3000 Hannover 1, FRG

Blitzer, A.: Department of Otolaryngology, College of Physicians and Surgeons, Columbia University, New York, NY 10032, USA

Bonneville, J. F.: Département de Neuroradiologie et de Radiologie Ostéo-Articulaire, Centre Hospitalier Universitaire, Hôpital Jean Minjoz, Blvd. A. Fleming, 25030 Besançon Cedex, France

Brück, K.: Department of Physiology, Justus-Liebig University, Klinikstr. 29, 6300 Giessen, FRG

Brunner, F. X.: Universitätsklinik für Hals-Nasen-Ohrenkranke, Josef-Schneider-Str. 11, 8700 Würzburg, FRG

Buchfelder, M.: Neurochirurgische Klinik mit Poliklinik der Universität Erlangen-Nürnberg, Schwabachanlage 6, 8520 Erlangen, FRG
Burrin, J. M.: Department of Medicine, Royal Postgraduate Medical School, Hammersmith Hospital, Du Cane Rd., London W12 ONS, UK
Buschmann, W.: Universitäts-Augenklinik, Josef-Schneider-Str. 11, 8700 Würzburg, FRG
Busse, H.: University Eye Hospital, Kardinal-von-Galen-Ring 10, 4400 Münster, FRG
Calbucci, F.: Divisione di Neurochirurgia, Ospedale Bellaria, Via Altura 3, 40139 Bologna, Italy
Caputi, E.: Department of Neurosurgery, Hospital Santa Lucia, Buenos Aires, Argentina
Caron, J. P.: Service de Neurochirurgie, Hôpital Henri-Mondor, 94010 Creteil, France
Carr, D. H.: Department of Radiology, Royal Postgraduate Medical School, Hammersmith Hospital, Du Cane Rd., London W12 ONS, UK
Cattin, F.: Département de Neuroradiologie et de Radiologie Ostéo-Articulaire, Centre Hospitalier Universitaire, Hôpital Jean Minjoz, Blvd. A. Fleming, 25030 Besançon Cedex, France
Chavez, V.: "Mariano R. Castex" Hospital, Neurological Service, Buenos Aires, Argentina
Chiaradio, J. C.: Virrey del Pino 1981, 1426-Buenos Aires, Argentina
Chiu, W.-T.: Division of Neurosurgery, Department of Surgery, Mackay Memorial Hospital, No. 92, North Chung San Rd., Sec. 2, Taipei 104, Taiwan, ROC
Christophis, P.: Department of Neurosurgery, University of Giessen, Klinikstr. 29, 6300 Giessen, FRG
Clar, H.-E.: Neurosurgical Clinic, Klinikum Ingolstadt, Krumenauerstr. 25, 8070 Ingolstadt, FRG
Collmann, H.: Abteilung für Pädiatrische Neurochirurgie der Klinik und Poliklinik für Neurochirurgie der Universität Würzburg, Josef-Schneider-Str. 11, 8700 Würzburg, FRG
Conesa, H.: Av. La Plata 393 – 8° "C", 1235 Capital, Buenos Aires, Argentina
Courtheoux, J.: C.H.U. de la Cote de Nacre, 14000 Caen, France
Daher, A.: Hôpital Neurologique et Neurochirurgical, 59, Blvd. Pinel, 69003 Lyon, France
Darrouzet, V.: Hôpital Pellegrin, Place Amélie Raba-Léon, 33076 Bordeaux Cedex, France
de Giacomi, B.: Universitätsspital Zürich, Kieferchirurgische Klinik und Poliklinik, Frauenklinikstr. 10, 8091 Zürich, Switzerland
Derkrikorian, J.: Av. La Plata 393 – 8° "C", 1235 Capital, Buenos Aires, Argentina
Deschepper, B.: C.H.U. Nord, Service de Radiologie, 80000 Amiens, France
de Silva, D.: Neuroradiologische Klinik, Alfried Krupp Krankenhaus, Alfried Krupp Str. 21, 4300 Essen 1, FRG
Diaz, F. G.: Department of Neurological Surgery, Henry Ford Neurosurgical Institute, 2799 West Grand Blvd., Detroit, MI 48202, USA
Dickob, M.: Rehabilitationskrankenhaus Ulm, Orthopädische Klinik, Lehr- und Forschungsbereich der Universität, Oberer Eselsberg 45, 7900 Ulm, FRG
Dietrich, C.: Department of Neurosurgery, Justus-Liebig University, Klinikstr. 29, 6300 Giessen, FRG
Dolan, E. J.: 4 Campus Way, Billings, MT 59102, USA
Draf, W.: Department of ENT Diseases, Head, Neck and Facial Plastic Surgery of Städtische Kliniken Fulda, Pacelliallee 4, 6400 Fulda, FRG

Dujovny, M.: Department of Neurological Surgery, Henry Ford Neurosurgical Institute, 2799 West Grand Blvd., Detroit, MI 48202, USA
Egan, P. J.: Department of Neuropathology, Medical School, University of Tübingen, Calwerstr. 7, 7400 Tübingen, FRG
Ehrenfeld, M.: Klinik für Kiefer- und Gesichtschirurgie der Universität Tübingen, Osianderstr. 2–8, 7400 Tübingen, FRG
El Khamlichi, A.: Service de Neurochirurgie, Hopital des Specialites O.N.O., BP 6220 Institits, Rabat, Morocco
El Ouahabi, A.: Service de Neurochirurgie, Hopital des Specialites O.N.O., BP 6220 Institits, Rabat, Morocco
Engel, W.: Neurosurgical Clinic, University Clinic of Essen, Hufelandstr. 55, 4300 Essen 1, FRG
Eskinja, N.: Dobrilina 3, 51410 Opatija, Yugoslavia
Etou, A.: Finkenstr. 12C, 6074 Rödermark, FRG
Fahlbusch, R.: Neurochirurgische Klinik mit Poliklinik der Universität Erlangen-Nürnberg, Schwabachanlage 6, 8520 Erlangen, FRG
Farmand, M.: Klinik und Poliklinik für Mund-Kiefer-Gesichtschirurgie der Friedrich-Alexander Universität Erlangen-Nürnberg, Glückstr. 11, 8520 Erlangen, FRG
Foerster, M.: Clinic of Ophthalmology, University Clinic of Steglitz, Hindenburgdamm 30, 1000 Berlin, FRG
Frank, G.: Divisione di Neurochirurgia, Ospedale Bellaria, Via Altura 3, 40139 Bologna, Italy
Fürst, G.: Institute of Diagnostic Radiology, Hospital of the University of Düsseldorf, Moorenstr. 5, 4000 Düsseldorf 1, FRG
Gaab, M. R.: Neurosurgical Department, Hanover Medical School, Konstanty-Gutschow-Str. 8, 3000 Hannover 61, FRG
Gádor, I.: National Institute of Neurosurgery, Amerikai út 57, Budapest, 1145, Hungary
Galanski, M.: Abteilung für Diagnostische Radiologie, Medizinische Hochschule Hannover, Konstanty-Gutschow-Str. 8, 3000 Hannover 61, FRG
Gamache, F. W. Jr.: Starr 6, Neurosurgery, The New York Hospital, Cornell Medical Center, 525 East 68th Street, New York, NY 10021, USA
Gandolfi, A.: Clinica O.R.L. II, Universita' degli Studi di Parma, Ospedale Maggiore, Via A. Gramsci 14, 43100 Parma, Italy
Gaston, A.: Service de Neuroradiologie, Hôpital Henri-Mondor, 94010 Creteil, France
Gentili, F.: Toronto General Hospital, 14 Eaton Wing, Rm 222, 200 Elizabeth Street, Toronto, Ontario, M5G 2C4, Canada
Gerhard, L.: Institute of Neuropathology, University Hospital Essen, Hufelandstr. 55, 4300 Essen 1, FRG
Geuna, E.: Divisione di Neurochirurgia, Ospedale Maggiore della Carità, 28100 Novara, Italy
Geyer, G.: Universitätsklinik für Hals-Nasen-Ohrenkranke, Josef-Schneider-Str. 11, 8700 Würzburg, FRG
Giers, U.: Eye Department, Military Hospital, Heldmanstr. 24, 4930 Detmold, FRG
Gilsbach, J.: Neurochirurgische Klinik der Medizinischen Falkultät der RWTH, Pauwelstr. 30, 5100 Aachen, FRG
Gnjidić, Ž.: Department of Neurosurgery, Clinical Hospital "Dr. M. Stojanović", Vinogradska 29, 41000 Zagreb, Yugoslavia
Grote, W.: Department of Neurosurgery, University of Essen, Medical Center, Hufelandstr. 55, 4300 Essen 1, FRG
Gullane, P.: Toronto General Hospital, 14 Eaton Wing, Rm 222, 200 Elizabeth Street, Toronto, Ontario, M5G 2C4, Canada

Haas, J. P.: The Institute of Radiology of Städtische Kliniken Fulda, Pacelliallee 4, 6400 Fulda, FRG
Habas, J.: Department of Endocrinology, College of Physicians and Surgeons, Columbia University, New York, NY 10032, USA
Hachleitner, J.: Abteilung für Kiefer- und Gesichtschirurgie der Landeskrankenanstalten Salzburg, Müllner Hauptstr. 48, 5020 Salzburg, Austria
Hajda, M.: National Institute of Neurosurgery, Amerikai út 57, Budapest, 1145, Hungary
Haßler, W.: Städtische Kliniken, Neurochirurgie, Zu den Rehwiesen/Kalkweg, 4100 Duisburg 1, FRG
Heider, W.: Neurochirurgische Abteilung, Klinikum der Wolfgang Goethe-Universität, Schleusenweg 2–16, 6000 Frankfurt/M 71, FRG
Heienbrok, W.-H.: Department of Neurosurgery, Alfried-Krupp-Krankenhaus, Alfried Krupp Str. 21, 4300 Essen 1, FRG
Heiss, E.: Department of Neurosurgery, Medical School, University of Tübingen, Calwerstr. 7, 7400 Tübingen, FRG
Helms, J.: Hals-, Nasen-, Ohrenklinik, Universitätsklinikum Würzburg, Joseph-Schneider-Str. 11, 8700 Würzburg, FRG
Heran, F.: Fondation Ophtalmologique A. de Rothschild, 25 a 29, rue Manin, 75940 Paris Cedex 19, France
Hildebrandt, G.: Universitätsklinik Köln, Neurochirurgie, Josef-Stelzmann-Str. 9, 5000 Köln 41, FRG
Holzer, F.: Department of Neurological Surgery, Henry Ford Neurosurgical Institute, 2799 West Grand Blvd., Detroit, MI 48202, USA
Howng, S.-L.: Division of Neurosurgery, Department of Surgery, Kaohsiung Medical College Hospital, Kaohsiung, Taiwan, ROC
Huk, W.: Neurochirurgische Klinik mit Poliklinik der Universität Erlangen-Nürnberg, Abteilung für Neuroradiologie, Schwabachanlage 6, 8520 Erlangen, FRG
Jadresic, A.: Department of Medicine, Royal Postgraduate Medical School, Hammersmith Hospital, Du Cane Rd., London W12 ONS, UK
Joplin, G. F.: Department of Medicine, Royal Postgraduate Medical School, Hammersmith Hospital, Du Cane Rd., London W12 ONS, UK
Kageyama, I.: Anatomical Institut, University of Würzburg, Koellikerstr. 6, 8700 Würzburg, FRG
Kahle, G.: The Institute of Radiology of Städtische Kliniken Fulda, Pacelliallee 4, 6400 Fulda, FRG
Kahn, T.: Institute of Diagnostic Radiology, Hospital of the University of Düsseldorf, Moorenstr. 5, 4000 Düsseldorf 1, FRG
Khamlichi, A. M.: Service de Neurochirurgie, Hopital des Specialites O.N.O. BP 6220 Institits, Rabat, Morocco
Kitz, K.: Department of Neurosurgery, Medical School of the University of Vienna, Währinger Gürtel 18–20, 1090 Vienna, Austria
Kiwit, J. C. W.: Department of Neurosurgery, Hospital of the University of Düsseldorf, Moorenstr. 5, 4000 Düsseldorf 1, FRG
Klein, H. J.: Universitätsklinik Köln, Neurochirurgie, Josef-Stelzmann-Str. 9, 5000 Köln 41, FRG
Klug, N.: Universitätsklinik Köln, Neurochirurgie, Josef-Stelzmann-Str. 9, 5000 Köln 41, FRG
Klun, B.: Department of Neurosurgery, University Medical Center, Ljubljana, Yugoslavia
Knosp, E.: Neurochirurgische Universitätsklinik, Langenbeckstr. 1, 6500 Mainz, FRG
Korn, A.: Department of Neurosurgery, Medical School of the University of Vienna, Währinger Gürtel 18–20, 1090 Vienna, Austria

Krenkel, C.: Abteilung für Kiefer- und Gesichtschirurgie, Landeskrankenanstalten, Müllner Hauptstr. 48, 5020 Salzburg, Austria
Kroll, P.: University Eye Hospital, Kardinal-von-Galen-Ring 10, 4400 Münster, FRG
Kühne, D.: Neuroradiologische Klinik, Alfried Krupp Krankenhaus, Alfried Krupp Str. 21, 4300 Essen 1, FRG
Künzler, A.: Universitätsspital Zürich, Kieferchirurgische Klinik und Poliklinik, Frauenklinikstr. 10, 8091 Zürich, Switzerland
Kuroiwa, A.: Department of Neurosurgery, University of Tokyo Hospital, Hongo, Bunkyo-ku, Tokyo, Japan
Lang, J.: Anatomical Institut, University of Würzburg, Koellikerstr. 6, 8700 Würzburg, FRG
Lányi, F.: National Institute of Neurosurgery, Amerikai út 57, Budapest, 1145, Hungary
Lasjaunias, P.: Departement de Neuroradiologie, Hopitaux de Paris, Rue du General Leclerc 78, 94270 Le Kremlin-Bicetre Cedex, France
Laws, E. R.: Department of Neurological Surgery, George Washington University Medical Center, 2150 Pennsylvania Ave., NW, Washington, DC 20037, USA
Lesoin, F.: Department of Neurosurgery B, Hôpital B, CHU, 59037 Lille Cedex, France
Lin, L.-S.: Division of Neurosurgery, Department of Surgery, Mackay Memorial Hospital, No. 92, North Chung San Rd., Sec. 2, Taipei 104, Taiwan, ROC
Lins, E.: Department of Neurosurgery, Hospital of the University of Düsseldorf, Moorenstr. 5, 4000 Düsseldorf 1, FRG
Lixl, G.: Medizintechniker der Abteilung für Kiefer- und Gesichtschirurgie, Landeskrankenanstalten, Müllner Hauptstr. 48, 5020 Salzburg, Austria
Loftus, C.: Department of Neurological Surgery, University of Iowa, Iowa City, Iowa 52242, USA
Loh, J.-K.: Division of Neurosurgery, Department of Surgery, Kaohsiung Medical College Hospital, Kaohsiung, Taiwan, ROC
Lorenz, M.: Neurosurgical Department, Hanover Medical School, Konstanty-Gutschow-Str. 8, 3000 Hannover 61, FRG
Lorenz, R.: Neurochirurgische Abteilung, Klinikum der Joh. Wolfgang Goethe-Universität, Schleusenweg 2–16, 6000 Frankfurt/M 71, FRG
Lorenz, R.G: Augenklinik des Marienkrankenhauses, Richard-Wagner-Str. 14, 6000 Frankfurt 1, FRG
Lovrić, D.: Department of Neurosurgery, Clinical Hospital "Dr. M. Stojanović", Vinogradska 29, 41000 Zagreb, Yugoslavia
Lozano, M. C.: "Mariano R. Castex" Hospital, Neurological Service, Buenos Aires, Argentina
Lupret, V.: Department of Neurosurgery, Clinical Hospital "Dr. M. Stojanovic", Vinogradska 29, 41000 Zagreb, Yugoslavia
Lustgarten, J.: Department of Neurological Surgery, College of Physicians and Surgeons, Columbia University, New York, NY 10032, USA
Ma, S.: Department of Neurological Surgery, Henry Ford Neurosurgical Institute, 2799 West Grand Blvd., Detroit, MI 48202, USA
Maira, G.: Istituto di Neurochirurgia, Policlinico Gemelli, Largo Gemelli 8, 00168 Roma, Italy
Mann, W.: Department of Otolaryngology, University of Mainz, Langenbeckstr. 1, 6500 Mainz, FRG
Matras, H.: Abteilung für Kiefer- und Gesichtschirurgie, Landeskrankenanstalten, Müllner Hauptstr. 48, 5020 Salzburg, Austria
McCormick, P.: Department of Neurological Surgery, College of Physicians and Surgeons, Columbia University, New York, NY 10032, USA

McLure, T.: ENT Department, The New York Hospital/Cornell Medical Center, 525 East 68th Street, New York, NY 10021, USA
Mehdorn, H. M.: Department of Neurosurgery, University of Kiel, Medical Center, Weimarer Str. 8, 2300 Kiel, FRG
Mirchandani, H. G.: Department of Neurological Surgery, Henry Ford Neurosurgical Institute, 2799 West Grand Blvd., Detroit, MI 48202, USA
Mohadjer, M.: Neurochirurgische Universitätsklinik Freiburg, Abteilung Stereotaxie und Neuronuklearmedizin, Hugstetter Str. 55, 7800 Freiburg, FRG
Mooij, J. J. A.: Department of Neurosurgery, University Hospital Groningen, Oostersingel 59, Groningen, The Netherlands
Moussaif, D.: Service de Neurochirurgie, Hopital des Specialites O.N.O., BP 6220 Institits, Rabat, Morocco
Mühling, J.: Klinik und Poliklinik für Mund-, Kiefer-, Gesichtschirurgie der Universität Würzburg, Pleicherwall 2, 8700 Würzburg, FRG
Müller, J.: Universitätsklinik für Hals-Nasen-Ohrenkranke, Josef-Schneider-Str. 11, 8700 Würzburg, FRG
Mundinger, F.: Abteilung für Neurochirurgie, St. Josefskrankenhaus, Hermann-Herder-Str. 1, 7800 Freiburg, FRG
Nau, H.-E. (†): Neurochirurgische Universitätsklinik Essen, Hufelandstr. 55, 4300 Essen 1, FRG
Negovetić, L.: Department of Neurosurgery, Clinical Hospital "Dr. M. Stojanović", Vinogradska 29, 41000 Zagreb, Yugoslavia
Nistor, R.: Neurochirurgische Klinik mit Poliklinik der Universität Erlangen-Nürnberg, Schwabachanlage 6, 8520 Erlangen, FRG
Nover, A.: Hebbelstraße 65, 6500 Mainz, FRG
Oeckler, R.: Neurochirurgische Klinik der Ludwig-Maximilians-Universität München, Klinikum Großhadern, Marchioninistr. 15, 8000 München 70, FRG
Ohaegbulam, S. C.: Neurosurgery Unit, University of Nigeria, Enugu Campus, Enugu, Nigeria
Pahnke, J.: Department of Anatomy, University of Würzburg, Koellikerstr. 6, 8700 Würzburg, FRG
Panzarasa, G.: Divisione di Neurochirurgia, Ospedale Maggiore della Carità, 28100 Novara, Italy
Pásztor, E.: National Institute of Neurosurgery, Amerikai út 57, Budapest, 1145, Hungary
Peirano, M.: Av. La Plata 393 – 8° "C", 1235 Capital, Buenos Aires, Argentina
Pellerin, P.: Department of Plastic Surgery, Hôpital Claude Huriez, CHU, 59037 Lille Cedex, France
Perneczky, A.: Neurochirurgische Universitätsklinik, Langenbeckstr. 1, 6500 Mainz, FRG
Piazza, F.: Clinica O.R.L. II, Universita' degli Studi di Parma, Ospedale Maggiore, Via A. Gramsci 14, 43100 Parma, Italy
Piffkó, P.: National Institute of Neurosurgery, Amerikai út 57, Budapest, 1145, Hungary
Post, K. D.: Department of Neurological Surgery, College of Physicians and Surgeons, Columbia University, New York, NY 10032, USA
Pruvost, P.: Hopital de Bicetre, Unité de neuroradiologie vasculaire diagnostique et therapeutique, 79, rue du Gal Leclerc, 94270 Kremlin Bicetre Cedex, France
Rabadán, A.: Av. La Plata 393 – 8° "C", 1235 Capital, Buenos Aires, Argentina
Ramamurthi, B.: V.H.S. Medical Centre, Madras-600 113, India

Ramina, R.: Sociedade Brasileira de Cirurgia de Base do Cranio, R. Goncalves Dias, 713, Seminario, 80420 Curitiba Paraná, Brasil
Rath, S. A.: Neurochirurgische Abtlg. der Universität Ulm, Bezirkskrankenhaus Günzburg, Ludwig-Heilmeyer-Str. 2, 8870 Günzburg, FRG
Rauhut, F.: Neurosurgical Clinic, University Clinic of Essen, Hufelandstr. 55, 4300 Essen 1, FRG
Raulo, Y.: Service de Chirurgie maxillar faciale, Hôpital Henri-Mondor, 94010 Creteil, France
Regalia, E.: Divisione di Neurochirurgia, Ospedale Maggiore della Carità, 28100 Novara, Italy
Reuther, J.: Klinik und Poliklinik für Mund-, Kiefer-, Gesichtschirurgie der Universität Würzburg, Pleicherwall 2, 8700 Würzburg, FRG
Richter, H.-P.: Former Department of Neurosurgery of Städtische Kliniken Fulda, Pacelliallee 4, 6400 Fulda, FRG
Riechelmann, H.: Department of Otolaryngology, University of Mainz, Langenbeckstr. 1, 6500 Mainz, FRG
Riediger, D.: Klinik für Kiefer- und Gesichtschirurgie, Plastische Chirurgie, Katharinenhospital Stuttgart, Kriegsbergstr. 60, 7000 Stuttgart 1, FRG
Rimbach, F.: Neurochirurgische Universitätsklinik Essen, Hufelandstr. 55, 4300 Essen 1, FRG
Rochels, R.: University Eye Clinic, Hegewischstr. 2, 2300 Kiel, FRG
Rodesch, G.: Hopital de Bicetre, Unité de neuroradiologie vasculaire diagnostique et thérapeutique, 79, rue du Gal Leclerc, 94270 Kremlin Bicetre Cedex, France
Roosen, N.: Brain Tumor Research Center, Department of Neurosurgery, University of California, San Fransisco, CA 94143, USA
Rutka, J.: Toronto General Hospital, 14 Eaton Wing, Rm 222, 200 Elizabeth Street, Toronto, Ontario, M5G 2C4, Canada
Sadasivan, B.: Department of Neurological Surgery, Henry Ford Neurosurgical Institute, 2799 West Grand Blvd., Detroit, MI 48202, USA
Samii, M.: Medical School Hanover, Neurosurgical Clinic, Nordstadt Hospital, Haltenhoffstr. 41, 3000 Hannover 1, FRG
Sandler, L. M.: Wycombe General Hospital, High Wycombe, Bucks, HP11 2TT, U.K.
Schatz, H.: Department of Neurosurgery, University of Giessen, Klinikstr. 29, 6300 Giessen, FRG
Scheunemann, H.: Klinikum der Johannes Gutenberg-Universität, Klinik für Mund-, Kiefer- und Gesichtschirurgie, Augustusplatz 2, 6500 Mainz, FRG
Schmidt, K.: Neurochirurgische Abteilung der Universität Ulm, Bezirkskrankenhaus Günzburg, Ludwig-Heilmeyer-Str. 2, 8870 Günzburg, FRG
Schreck, R.: "Mariano R. Castex" Hospital, Neurological Service, Buenos Aires, Argentina
Schroth, G.: Department of Neuroradiology, Medical School, University of Tübingen, Calwerstr. 7, 7400 Tübingen, FRG
Schütte, E.: Ophthalmological Department of the Military Hospital in Ulm, Oberer Eselsberg 40, 7900 Ulm, FRG
Schwenzer, N.: Klinik für Kiefer- und Gesichtschirurgie der Eberhard-Karls-Universität Tübingen, Osianderstr. 2–8, 7400 Tübingen, FRG
Sepehrnia, A.: Neurosurgical Clinic. Nordstadt Hospital, Haltenhoffstr. 41, 3000 Hannover 1, FRG
Shih, C.-J.: Department of Health, The Executive Yuan, P.O. Box 91103, Taipei, Taiwan, ROC
Shitara, N.: Department of Neurosurgery, University of Tokyo Hospital, Hongo, Bunkyo-ku, Tokyo, Japan

Sindou, M.: Hôpital Neurologique et Neurochirurgical, 59, Blvd. Pinel, 69003 Lyon, France
Sörensen, N.: Abteilung für Pädiatrische Neurochirurgie der Klinik und Poliklinik für Neurochirurgie der Universität Würzburg, Josef-Schneider-Str. 11, 8700 Würzburg, FRG
Šteňo, J.: Department of Neurosurgery of Comenius University, Limbová 5, 833 05 Bratislava, Czechoslovakia
Steudel, W.: Neurochirurgische Abteilung, Klinikum der Joh. Wolfgang Goethe-Universität, Schleusenweg 2–16, 6000 Frankfurt/M 71, FRG
Stoll, D.: Hôpital Pellegrin, Place Amélie Raba-Léon, 33076 Bordeaux Cedex, France
Stoll, W.: University ENT Hospital, Kardinal-von-Galen-Ring 10, 4400 Münster, FRG
Stracke, H.: Department of Neurosurgery, University of Giessen, Klinikstr. 29, 6300 Giessen, FRG
Sweet, W. H.: Harvard Medical School, Neurological Service, Massachusetts General Hospital, 15 Parkman Street, Boston, MA 02114, USA
Symon, L.: Gough-Cooper Department of Neurological Surgery, Institute of Neurology, The National Hospital, Queen Square, London WC1N 3BG, UK
Takakura, K.: Department of Neurosurgery, University of Tokyo Hospital, Hongo, Bunkyo-ku, Tokyo, Japan
Teramoto, A.: Department of Neurosurgery, University of Tokyo Hospital, Hongo, Bunkyo-ku, Tokyo, Japan
Terbrugge, K.: Toronto Western Hospital, Department of Radiology, 399 Bathurst Street, Toronto, Ontario, Canada
Thaller-Antlanger, H.: Abteilung für Orthoptik und Pleoptik der Landeskrankenanstalten Salzburg, Müllner Hauptstr. 48, 5020 Salzburg, Austria
Thomas, C. E.: Department of Neurosurgery B, Hôpital B, CHU, 59037 Lille Cedex, France
Tognetti, F.: Divisione di Neurochirurgia, Ospedale Bellaria, Via Altura 3, 40139 Bologna, Italy
Tsirimbas, J. P.: 401 GMH, 13, Argostoliou Street, 113 62 Athens, Greece
Turel, K. E.: 628, Turel Terrace, Parsee Colony, Dadar, Bombay-400014, India
van Overbeeke, J. J.: Department of Neurosurgery, University Hospital, P.O. Box 85.500, 3508 GA Utrecht, The Netherlands
Vidović, D.: Department of Neurosurgery, Clinical Hospital "Dr. M. Stojanović", Vinogradska 29, 41000 Zagreb, Yugoslavia
Vignaud, J.: Foundation Ophtalmologique A. de Rothschild, 25 a 29, rue Manin, 75940 Paris Cedex 19, France
Voci, A.: Divisione di Neurochirurgia, Ospedale Maggiore della Carità, 28100 Novara, Italy
von Wild, K.: Neurochirurgische Abteilung des Clemenshospitals, Duesbergweg 124, 4400 Münster, FRG
Vuletić, S.: Department of Neurosurgery, Clinical Hospital "Dr. M. Stojanovic", Vinogradska 29, 41000 Zagreb, Yugoslavia
Wagner, W.: Klinikum der Johannes-Gutenberg-Universität, Klinik für Mund-, Kiefer- und Gesichtschirurgie, Augustusplatz 2, 6500 Mainz, FRG
Wahlmann, U. W.: Klinikum der Johannes-Gutenberg-Universität, Klinik für Mund-, Kiefer- und Gesichtschirurgie, Augustusplatz 2, 6500 Mainz, FRG
Wittkowski, W.: Institut für Anatomie der Westfälischen Wilhelms-Universität Münster, 4400 Münster, FRG
Yampolsky, C.: Department of Neurosurgery, Hospital Santa Lucia, Buenos Aires, Argentina

Zamorano, L. J.: Department of Neurological Surgery, Henry Ford Neurosurgical Institute, 2799 West Grand Blvd., Detroit, MI 48202, USA
Zentner, J.: Department of Neurosurgery, Medical School, University of Bonn, Sigmund-Freud-Str. 25, 5300 Bonn 1, FRG
Zini, C.: Clinica O.R.L. II, Universita'degli Studi di Parma, Ospedale Maggiore, Via A. Gramsci 14, 43100 Parma, Italy
Zohdi, A.: Neurosurgical Department, Cairo-University-Clinic, Kasr el Aini Hospital, El Manya 1, Cairo, Egypt

Chapter 1: Paranasal Sinuses and Orbit

Paranasal Sinuses: Anatomical Considerations

J. LANG, Würzburg/FRG

In a short overview of the paranasal sinuses it is only possible to point out some anatomical facts and their clinical importance. (For further details see Lang and Oehmann 1976; Lang and Schlehahn 1978; Lang and Haas 1979; Lang and Schäfer 1979; Lang and Papke 1980, 1984; Lang et al. 1979; Lang 1985a, b, 1988a, b; Lang 1989).

Paranasal Sinuses: Development

In fetuses with a crown-rump length of 8 cm a nasal capsule of cartilage is developed (Fig. 1). On its outer side there are desmal ossifications representing separate courses of the lacrimal, nasal, and premaxillary bones. Also seen in this stage are the anlagen of the sphenoidal concha and the vomer (Fig. 2). In the 4th month of fetal life one sees connective tissue buds in the area of the side

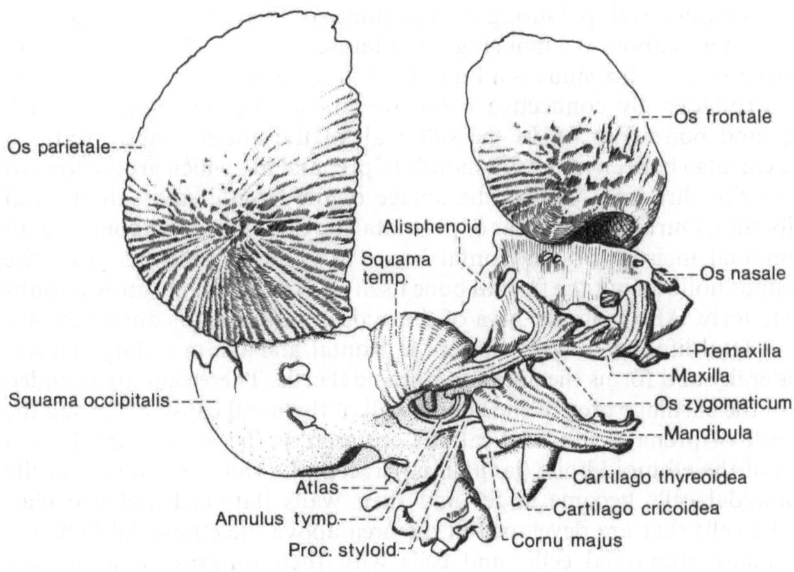

Fig. 1. Skull and upper cervical column, cartilage parts and membranous bone parts in a fetus of the 3rd month. (From Broman 1911)

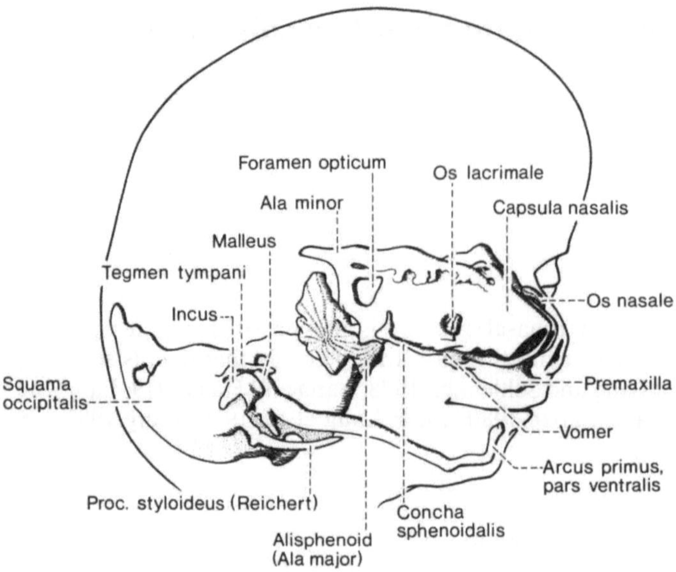

Fig. 2. Cartilage parts of the head of a 3-month-old fetus. Also seen are parts composed of membranous bones (lacrimal bone, nasal bone, sphenoidal concha and maxilla-premaxilla). (From Hertwig and Kollmann, cited in Broman 1911)

wall and the posterior superior part of the nasal cavity from which the paranasal sinuses develop. During the 4th month of fetal life, for example, the maxillary sinus is about 0.5 mm (Fig. 3) in fetuses of 25 cm length and 5 mm deep. The anlage of the sinus is a bud of the mucous membrane of the nasal cavity surrounded by connective tissue which has the potential to resorb cartilage and bone (Fig. 4). In the side wall of the middle nasal meatus a recessus can also be seen in the 4th month of prenatal life which grows upward in an anterior direction. This is the anlage of the frontal sinus, the frontal infundibulum. During the 1st year of postnatal life this anlage is in contact with the ethmoidal incisure of the frontal bone. In children aged 3.5 years the frontal sinus hollows out the frontal bone itself and can normally grow upward and posteriorly. Also from the area of the middle nasal meatus during the 4th month of fetal life, an outgrowth (as the frontal and the maxillary sinuses) bores laterally and forms the anterior ethmoidal cells. These cells are rounded and lie in the cartilage area of the lateral wall of the nasal capsule. During the 7th month of prenatal life ossifications are seen in the area of the lamina orbitalis of the ethmoid bone (lamina papyracea). During the 3rd year of life the ethmoidal cells become larger and their walls flattened and irregular. Ethmoidal cells that are developed in the area above the ethmoidal bulla are called middle ethmoidal cells, and cells with their outgrowths above the middle nasal concha are called posterior ethmoidal cells. It should be mentoined that ethmoidal cells partly communicate with neighboring cells or

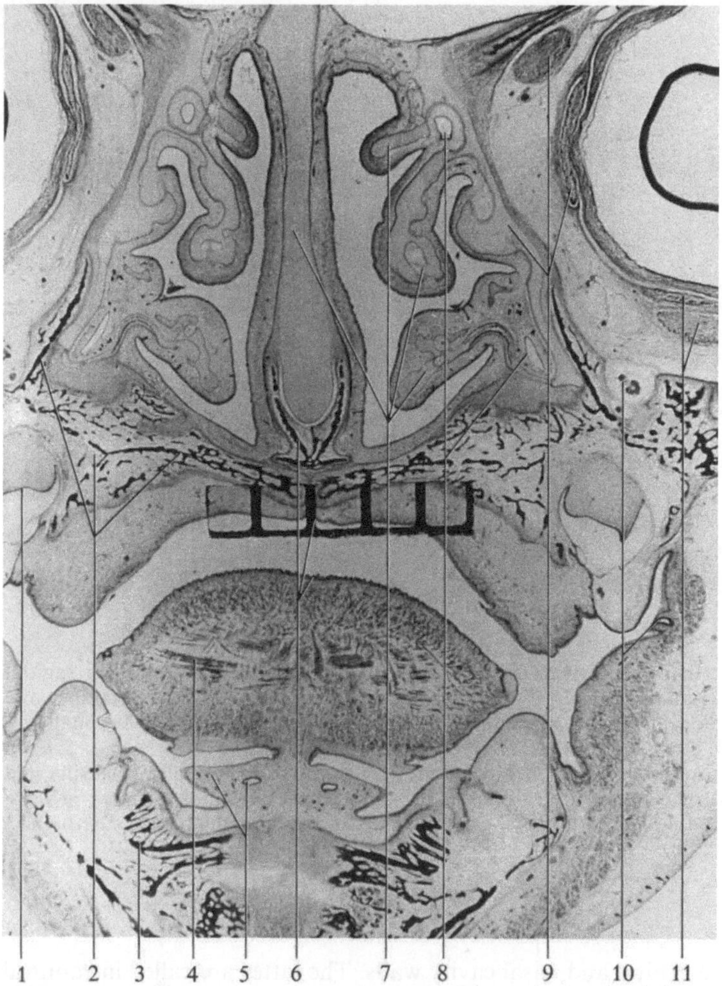

Fig. 3. Development of the paranasal sinuses (coronal section through the head of a 18-cm-long fetus). *1*, Dens molaris I, anlage; *2*, different parts of the maxilla (membranous bone); *3*, mandibula (on the outer side of Meckel's cartilage), membranous bone part; *4*, tongue, transversal muscle system; *5*, sublingual gland and submandibular duct; *6*, vomer (membranous bone), mm paper, and dorsum linguae; *7*, cartilage of septum nasi, superior, middle and inferior nasal concha; *8*, anterior ethmoidal cell and maxillary sinus, anlage; *9*, superior oblique and medial rectus muscles and capsula nasi (still cartilage); *10*, infraorbital nerve and duct of the parotid gland; *11*, inferior rectus and inferior oblique muscles

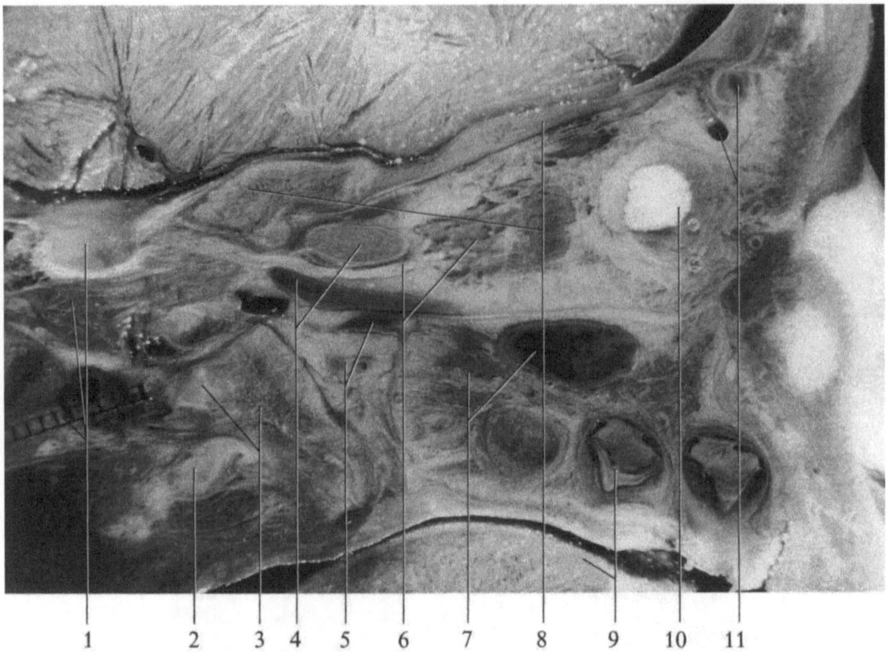

Fig. 4. Maxillary sinus in a newborn und surrounding structures. Paramedian sagittal section, viewed from the lateral. *1*, Upper part of the side wall of the cavernous sinus and trigeminal ganglion, mm paper; *2*, cartilage of the auditory tube; *3*, sphenooccipital synchondrosis and pterygoid process; *4*, inferior rectus muscle and optic nerve; *5*, fossa pterygopalatina and m. orbitalis; *6*, third nerve, twig to m. rectus medialis; *7*, maxilla and maxillary sinus (thick mocous membrane); *8*, frontal bone and anterior clinoid process; *9*, dens caninus and tongue; *10*, medial part of sclera; *11*, superior oblique muscle (trochlea and ophthalmic vein)

lie between the orbital and nasal cavity walls. The latter are called intramural cells.

The sphenoidal sinus anlage can also be seen in the 4th month of prenatal life. This outgrowth is in the posterior end of the nasal cavity and has been termed the paleosinus sphenoidalis (van Gilse 1926). This bud grows downward and medial to the sphenoidal concha (Bertin's ossicles). In newborns there is a growth of this sinus in the cartilage of the later body of the sphenoid bone. Between the 7th and 12th year the sphenoidal concha becomes connected to the sphenoidal body and forms the anterior wall of the neosinus sphenoidalis (see Lang and Papke 1984; Lang 1985, 1987, 1988). It was earlier thought that at the age of 20 years the sinuses have their final shapes and outlines (Koch 1930). Other researchers (Stern 1939) have maintained that the final form is reached only in the 40th year of life. Finby and Kraft (1972) showed that after the 40th year, for example, the frontal sinus has a clearly recognizable growth in height and depth.

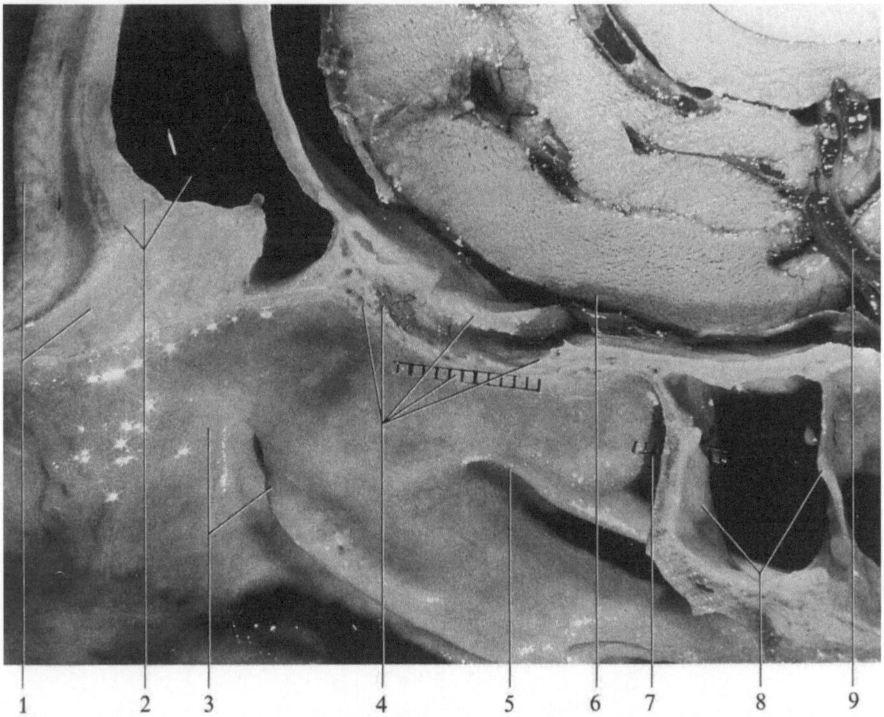

Fig. 5. Frontal sinus and surrounding structures in a median sagittal section. *1*, Skin and os nasale; *2*, nasion, frontal bone, and frontal sinus; *3*, agger nasi and middle nasal concha; *4*, lamina cribrosa, olfactory bulb, and olfactory fossa, mm paper; *5*, superior nasal concha; *6*, gyrus rectus; *7*, mm paper in aperture of the sphenoid sinus; *8*, sphenoid sinus, anterior wall and oblique septum; *9*, anterior cerebral artery (A_2 segment)

Frontal Sinus

In our material ($n = 205$) the height of the frontal sinus (Fig. 5) in the anterior wall of the anterior cranial fossa averaged 24.3 mm (5–66 mm). Its width was 29 mm (17–49 mm) in developed sinuses; here the anterior posterior extension averaged 20.5 mm (10–46.5 mm), and the roof of the orbit was doubled.

Aplastic or hypoplastic sinuses have been reported in 2%–38.8% by various authors (with racial differences). The thickness of the anterior wall of the frontal sinus in our material averaged 4.0 mm (0.45–12.5 mm). Its posterior wall in the frontal squama measured 1.9 mm (0.1–4.8 mm). In the median sagittal plane but beside the septum, the anterior posterior depth was 10.49 mm (0–17.7 mm); 2 cm lateral to the median plane the sinus was 9 mm (0–27.5 mm). The distance between the posterior side of the anterior wall and the cribrous lamina was 12.6 mm (4.7–21.25 mm). One sometimes finds defects of the walls of the frontal sinus to the orbit and to the anterior cranial fossa area (see Fig. 38 in Lang 1983).

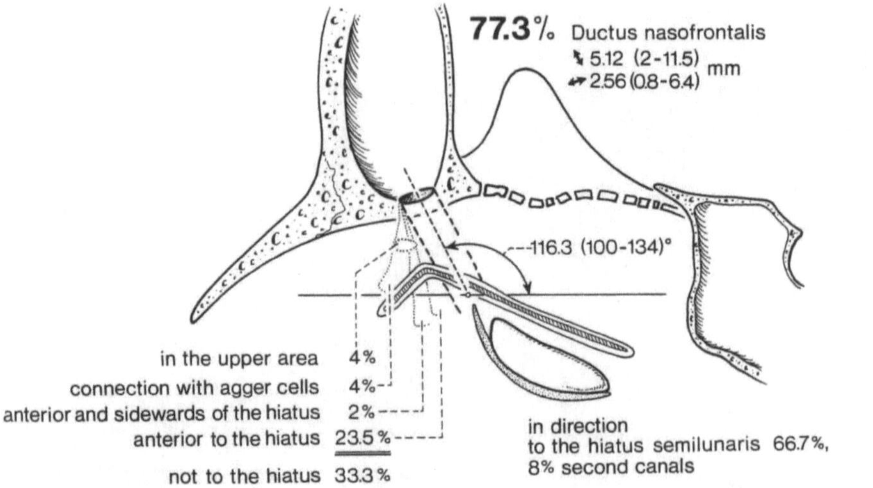

Fig. 6. Nasofrontal duct. In about 77% of cases we found ducts with a length of more than 3 mm. Seen are the diameters of these ducts, the angle of these ducts with the Frankfurt horizontal plane, and its direction to the anterior part of the hiatus semilunaris or to other areas

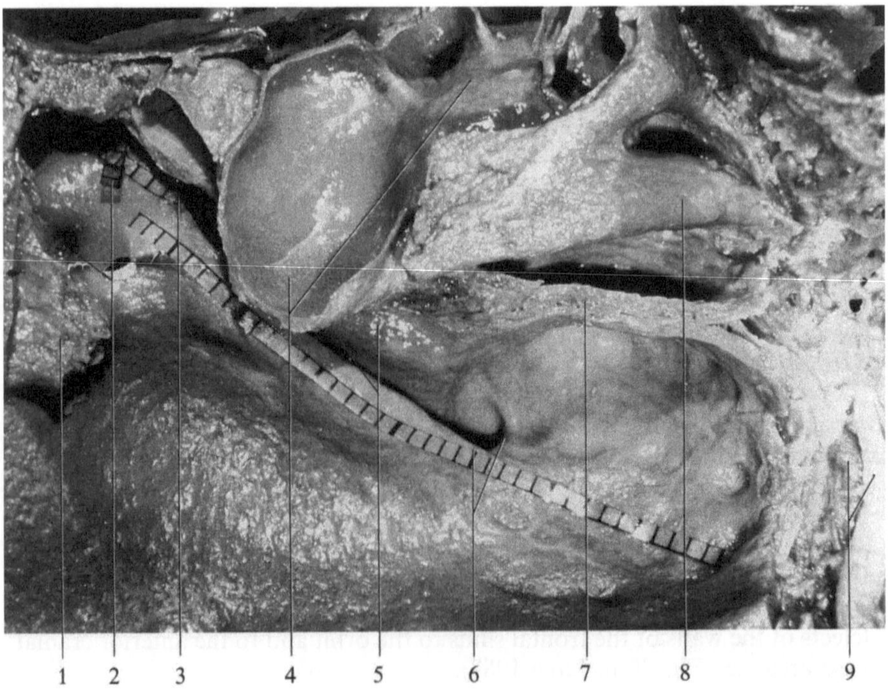

Nasofrontal Duct. In 77% of our cases the communication of the frontal sinus with the middle nasal meatus was longer than 3 mm; this we term a nasofrontal duct (Fig. 6). Its width averaged 5.12 mm and its depth 2.56 mm. In about 20% of cases the narrow zone between the frontal sinus and the nasal cavity was very short; we term this the ostium frontale. In cases with ducts its direction to the Frankfurt horizontal plane measured 116° (100°–134°). In 67% of cases the duct was found in the direction to the anterior end of the hiatus semilunaris.

Ethmoidal Infundibulum. In the upper part of the middle nasal meatus there is a niched area called the ethmoidal infundibulum (Fig. 7). This is anterior to the ethmoidal bulla. Its medial wall is the lateral wall of the middle nasal concha. In cases of pneumatized conchae (bullous conchae) the infundibulum is bordered medially by these cells. The lateral wall of the infundibulum is the area between the anterior and upper end of the uncinate process and the ethmoidal bulla.

Ethmoidal Cells

Anterior Ethmoidal Cells. The openings of anterior ethmoidal cells are below the anterior and upper part of the middle nasal meatus, in most cases in the area of the ethmoidal infundibulum (see Fig. 8). They grow in different directions. In the area of the agger nasi we found cells in 77%. In our material 30% showed connection to the lacrimal bone (lacrimal cells) and 39% to the nasal bone area (nasal cells). Cells extending in the bottom area of the anterior cranial fossa were developed in 17% of cases. It should be noted that the bottom area of the frontal sinus reached the area of the nasal bones in 23% and was very deep in these cases.

Ethmoidal Bulla and Middle Ethmoidal Cells. In about 70% of cases we found a pneumatized ethmoidal bulla; in 30% it was only developed as a so-called torus lateralis (see Fig. 8). The length of the ethmoidal bulla measured 18 mm (9–28 mm) and its height 5.4 mm (2–13 mm). In cases of pneumatized bullae ethmoidales the opening was usually found in the middle area of the upper wall, in others in the anterior or posterior area. Sometimes the bulla contained two openings on its upper surface. The cells with openings above the

◄

Fig. 7. Anterior and middle ethmoidal cells, viewed from the medial after resection of the middle and a part of the superior nasal concha. *1*, Resection area of the middle nasal concha; *2*, infundibular cells and mm graph in a lacrimal cell and agger cell; *3*, mm graph in the nasofrontal duct; *4*, a large middle ethmoidal cell; *5*, hiatus semilunaris and bulla ethmoidalis; *6*, mm graph and posterior border of hiatus semilunaris; *7*, posterior cut area of the middle nasal concha; *8*, superior nasal concha; *9*, n. palatinus major and a. palatina major, dissected

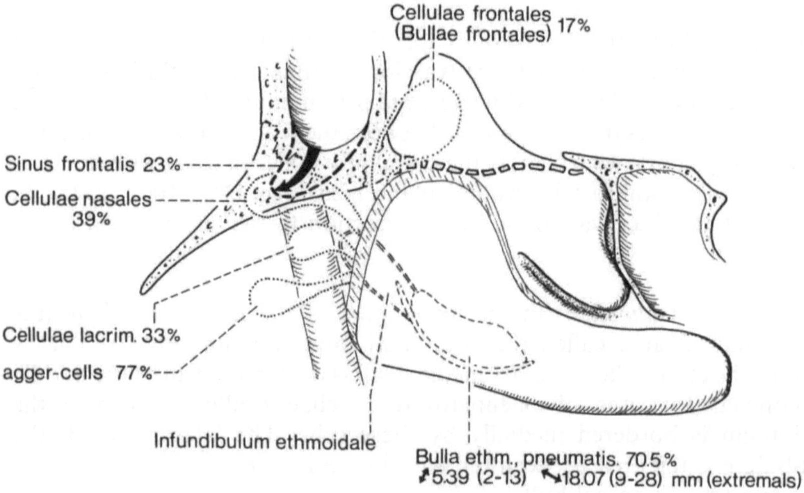

Fig. 8. Various kinds of infundibular cells, from our material. Also seen is a deeply situated bottom area of the frontal sinus, which we found in 23% of cases. It should be noted that the crista galli was found in about 10% to be pneumatized from the frontal sinus, sometimes also the anterior part of the nasal septum

bulla are called middle ethmoidal cells. We found that one to three of these cells may extend posteriorly far into the ethmoidal labyrinth.

Haller's Cells. In 1743 Haller described middle ethmoidal cells that reach above the maxillary sinus laterally far into the bottom area of the orbit. We also found such cells with their lateral walls lateral to the inferior rectus muscle of the bulb (see Fig. 98 in Lang 1988). Sometimes these cells have their openings above the middle nasal concha.

Posterior Ethmoidal Cells. The openings of posterior ethmoidal cells are openings above and behind the middle nasal concha (Fig. 9). In our material we found one posterior ethmoidal cell in 21% of cases, two cells in 55%, three in 21% and four, five, or six in about 2% each. Cells with openings in the meatus nasi supremus are called postreme ethmoidal cells.

Intramural Cells. Ethmoidal cells embedded in other ethmoidal cells are called intramural cells. It should be noted that one ethmoidal cell may connect with another inside the ethmoidal labyrinth.

Onodi's Cells. Onodi's, or Grünwald's, cells border the medial wall of the optic canal. We found such cells in about 12% of cases. The medial wall may be comprised of one Onodi's cell, by two Onodi's cells, or partly by the sphenoidal sinus and one Onodi's cell. The upper wall and a part of the medial wall of the optic canal is composed of a posterior superior ethmoidal cell in

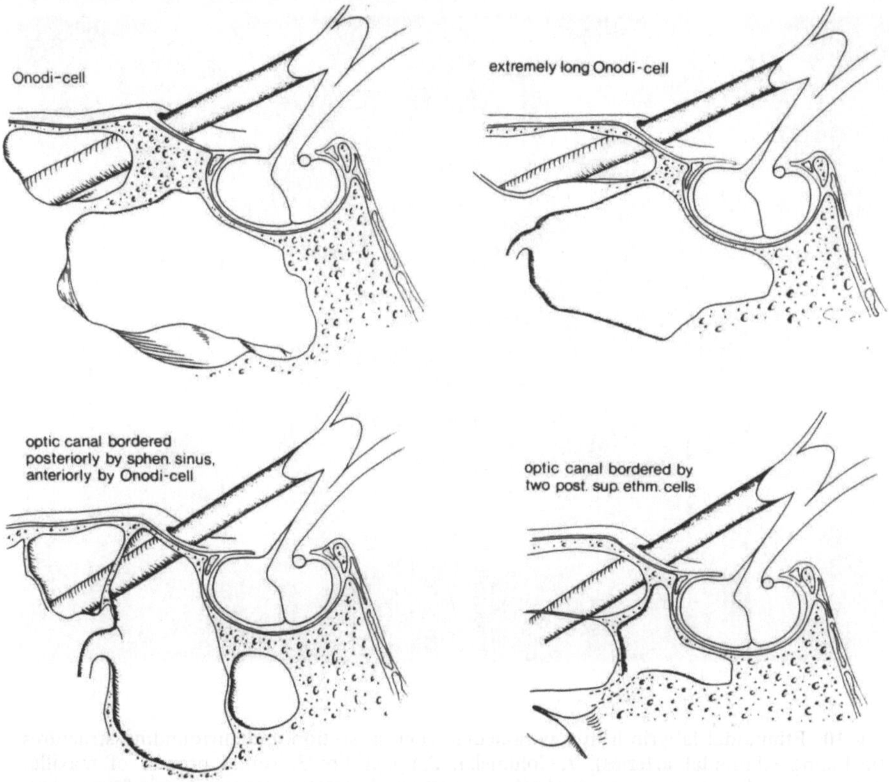

Fig. 9. Posterior superior ethmoidal cells surrounding the medial wall of the optic canal and their neighborhood to the sphenoid sinus. Cells of this kind were found in about 12% of our cases (Lang 1988)

25%. The wall thickness in these cells was 0.9 mm (0.5–1.5 mm) in our material.

Dangerous Areas in the Ethmoidal Labyrinth. Well-known are the anterior and posterior ethmoidal canals (Figs. 10, 11). Their length, width, and directions and their width were measured in our material (1979). In about one-third of cases we also found a third ethmoidal canal and an ethmoidal artery; here the height of the anterior ethmoidal canal measured 2 mm (−1 to +4 mm) above the cribrous lamina. The third ethmoidal canal was 1 mm (−1.1 to 3.25 mm) above the cribrous lamina, and the posterior ethmoidal canal was 1.5 mm (0–3.1 mm) above it. The olfactory groove has a different size and a different depth. In the anterior area we found a depth of 4.8 mm (0.6–11.7 mm). In the middle area the depth measured 5 mm (0–15.6 mm), and in the posterior area of the cribrous lamina it was 3.2 mm (0–10 mm; for details see Figs. 114, 115 in Lang 1988). We found dehiscences of the anterior ethmoidal canal in 92%, of the third ethmoidal canal in 39%, and of the

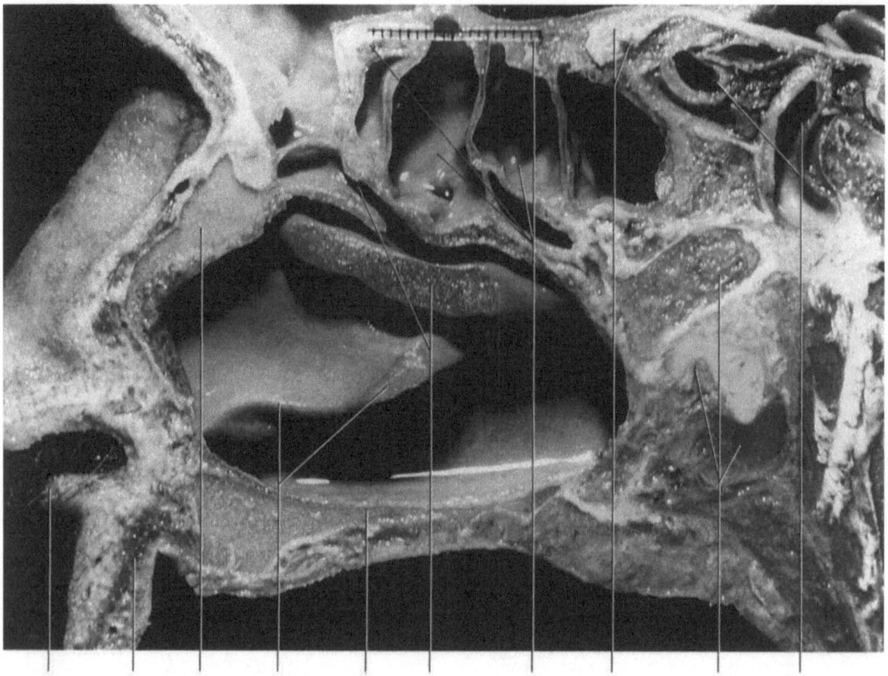

Fig. 10. Ethmoidal labyrinth in a paramedian sagittal section and surrounding structures (including ethmoidal arteries). *1*, Columella; *2*, upper lip; *3*, frontal process of maxilla; *4*, Deviation of the septum nasi, in the dorsal part cut; *5*, palatum durum; *6*, nasofrontal duct and middle nasal concha, cut; *7*, anterior and posterior ethmoidal arteries, mm graph and cellulae ethmoidales mediae; *8*, optic nerve and ophthalmic artery; *9*, pterygoid process, auditive tube and m. levator veli palatini; *10*, internal carotid artery inside the cavernous sinus

posterior ethmoidal canal in 59%. The anterior part of the middle nasal concha is inserted on the lateral border of the cribrous lamina. This part is termed the lamina conchalis. Its posterior insertion is found in the posterior third of the crista galli. We sometimes found the insertion slightly downward on the posterior end of the ethmoidal bulla. Small arachnoid-lined pouches of dural extensions accompany the olfactory fila through the cribrous lamina. The olfactory fila run on the side wall and the medial wall of the middle nasal meatus. Therefore the area is very dangerous in all operations (and traumata): Liquorrhoea.

Maxillary Sinus

In adults the height of the maxillary sinus (Figs. 12, 13) measured 40.4 mm (29–57 mm) in our material. There are some differences between the right and left sides (see Lang and Papke 1984). The width of the maxillary sinus in

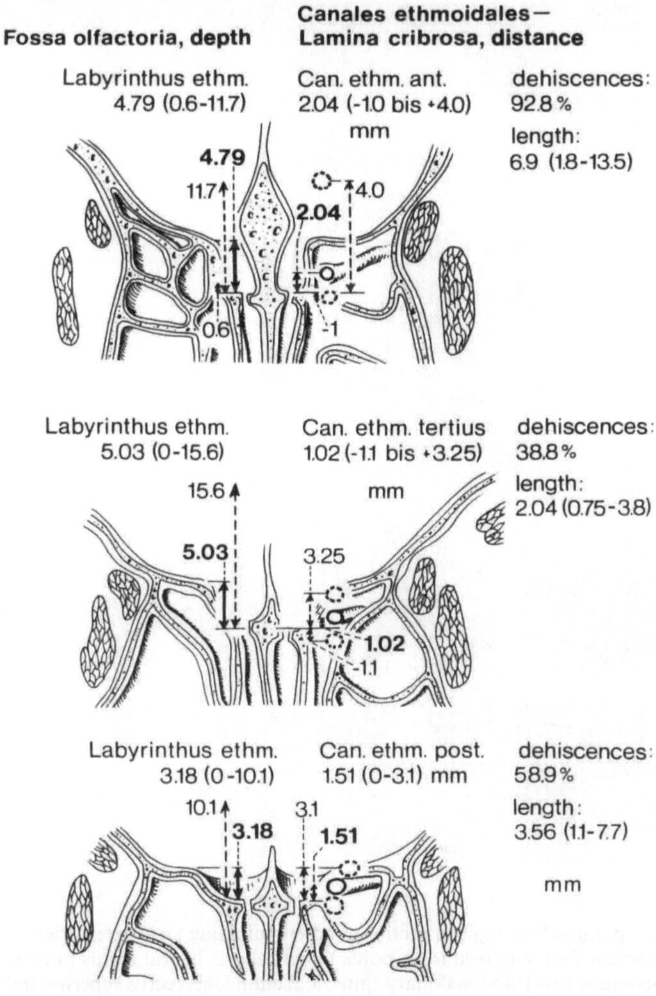

Fig. 11. The depth of the olfactory fossa and the anterior, middle, and posterior parts of the lamina cribrosa, the course of the anterior, third, and posterior ethmoidal canals to the lamina cribrosa and their dehiscences (after Lang and Haas 1990). It should be noted that the width of the anterior ethmoidal artery in the upper area is 0.55 mm, in the third ethmoidal artery 0.23 mm, and in the posterior ethmoidal artery 0.38 mm (mean values according to Lang and Schäfer 1979)

the roof area was 26.5 mm (16.1–39.8 mm) and the length 38.7 mm (30.1–49.2 mm). The recessus that have been described are the palatine recess, infraorbital medial recess (prelacrimal recess), zygomatic recess, and alveolar recess. The maxillary sinus was doubled in 6% of our cases with an ostium to the middle nasal meatus and in 6% with an ostium above the middle nasal meatus (see Lang 1988). In over 50% haustra are found in the bottom area of the maxillary sinus. The bottom area of the maxillary sinus was found in about

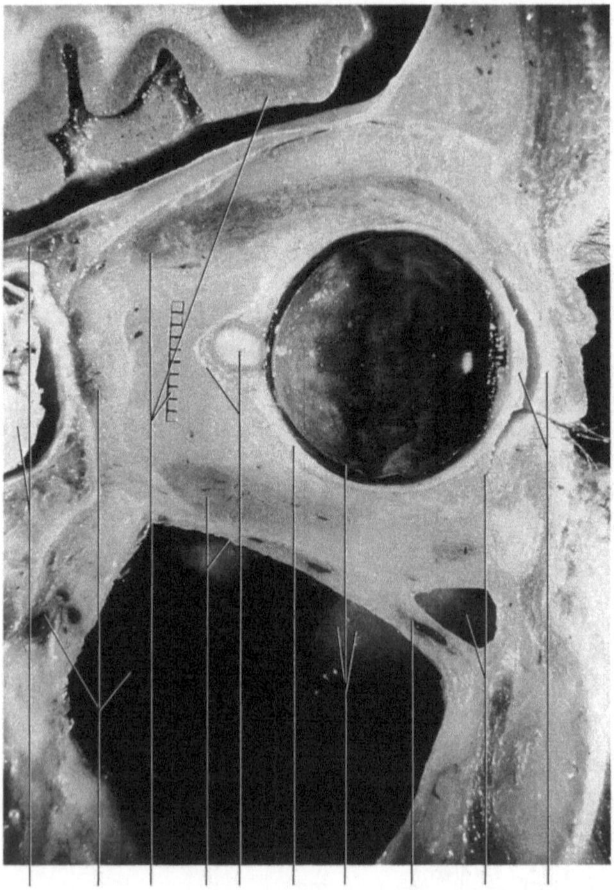

Fig. 12. Maxillary sinus, paramedian sagittal section, and surrounding structures. Viewed from the lateral. *1*, Ala minor and temporal lobe, polus temporalis; *2*, lateral rectus muscle, a. maxillaris in pterygopalatine fossa, and maxillary sinus; *3*, frontal lobe, rectus superior and levator palpebrae muscles and mm graph in orbital fatty tissue; *4*, inferior rectus muscle and thin roof of maxillary sinud; *5*, optic nerve, cut and dura mater of optic nerve; *6*, sclera; *7*, retina and protuberance of nasolacrimal canal in maxillary sinus; *8*, infraorbital canal and infraorbital artery, cut; *9*, inferior fornix of conjunctival sac and prelacrimal recess of maxillary sinus; *10*, cornea and superior eyelid (tarsus)

25% of cases to be below the bottom area of the nasal cavity. The third tooth (dens caninus) has contact with the maxillary sinus in nearly all cases between the 6th and 11th year of postnatal life. After this time its route is in the anterior wall of the sinus (with an average length of 7.1 mm; Bonnsdorff 1925). The bottom area of the maxillary sinus has dehiscences against the roots of the first, second, and third upper molars in about 2% each (Harrison 1961). The anterior, posterior, superior, and medial walls of the maxillary sinus are very

Fig. 13. Coronal section through the maxillary sinus and surrounding structures. *1*, Dura mater of the anterior cranial fossa, connective tissue strands in the area behind the bulb and zygomatic bone (lateral wall of the orbit); *2*, lateral rectus muscle, mm paper, and corpus adiposum buccae; *3*, inferior connective tissue strands of the orbital fatty tissue; *4*, infraorbital nerve and artery and periorbita (periost of the orbit); *5*, inferior rectus muscle and thin upper wall of the maxillary sinus, and maxillary sinus; *6*, m. levator palpebrae superioris, m. rectus superior, and optic nerve; *7*, ophthalmic artery, superior oblique muscle, and medial rectus muscle; *8*, ethmoidal cell, oblique medial wall of the orbit, wall of a Haller's cell, hiatus semilunaris, and inferior nasal concha; *9*, middle nasal meatus and middle nasal concha; *10*, forebrain, olfactory tract, and superior nasal concha

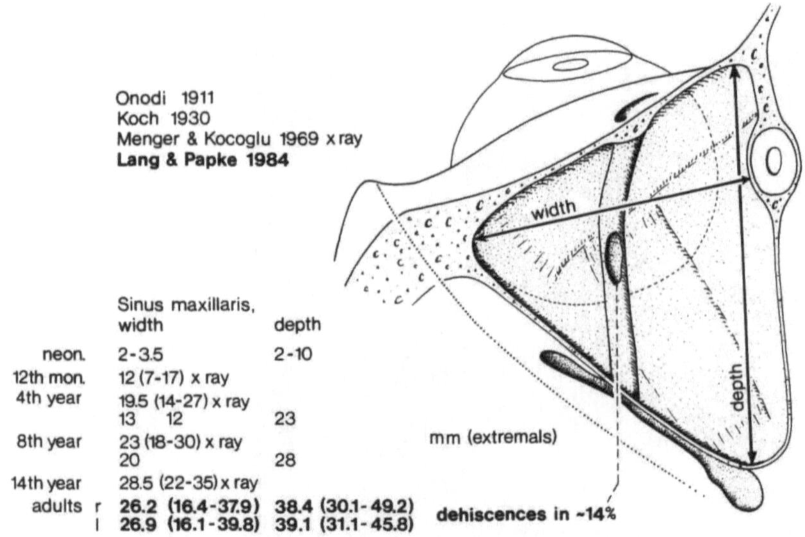

Fig. 14. Maxillary sinus, roof area, course of the infraorbital sulcus and canal, and inferior rectus and inferior oblique muscles. (Data from Onodi 1911 and Koch 1930; X-ray diagram from Menger and Kocoglu 1969; reprinted from Lang and Papke 1984; further details see Lang 1989)

thin and sometimes have dehiscences (see Lang 1988). The bony dehiscences in the area of the medial wall of the maxillary sinus are called fonticuli. In their areas the mucous membranes of the maxillary sinus and the nasal cavity are found together, and accessory ostia of the maxillary sinus sometimes develop in these regions. These were found in our material in 28% (see, for example, Fig. 78 in Lang 1988). The ostium of the maxillary sinus is situated in the hiatus semilunaris area of the middle nasal meatus; its upper border was 1.3 mm (0–9.5 mm) below the roof area of the sinus. The nasolacrimal duct was 4.35 mm (1.3–11.5 mm) in front of the ostium. In our material the thickness of the roof area of the maxillary sinus medial to the infraorbital groove and canal was 0.46 mm (0.05–4.0 mm); lateral to this area we measured a thickness of 0.56 mm (0.06–2.0 mm). Dehiscences in the area of the bottom of the groove or canal were found on the right side in 12% and on the left in 15.4% (Fig. 14).

Blow-out fractures were described by W. Lang in 1889. In such cases it may be impossible to rotate the bulb upward – double vision – because the inferior rectus muscle, the periorbit, or the fascia of the inferior rectus muscle are between the borders of the broken area of the maxillary sinus. In addition the medial and anterior walls in the area of the fossa canina may be very thin. The medial wall of the orbit is also very thin but in most cases is supported by the walls of the ethmoidal cells. Especially below the mucous membrane of the side wall and the anterior wall of the maxillary sinus are running twigs of Meckel's ganglion and infraorbital nerve and vessels (see Fig. 110 in Lang 1988). If the mucous membrane of the sinus is removed these nerves may be disturbed. The clinical anatomy of the sphenoidal sinus is discussed by Lang and Pahnke (this volume).

Clinical Conclusions

We may summarize our findings in the study of 200 cases as follows:

1. Paranasal sinuses vary in size; the left are usually larger than the right (except in the posterior ethmoidal labyrinth, which is broader on the right).
2. The walls are composed not only of the bones of its name.
3. The bony walls sometimes have deficiencies.
4. In the ethmoid and frontal sinuses run arteries, veins (and nerves).
5. The roof of the maxillary sinus is very thin and can have dehiscences (blow-out fractures).
6. Duplications are sometimes found in the maxillary and frontal sinuses (different outlines).

References

Bonnsdorff P von (1925) Untersuchungen über Maßverhältnisse des Oberkiefers mit spezieller Berücksichtigung der Lagebeziehungen zwischen den Zahnwurzeln und der Kieferhöhle. Akad Abh Helsingfors

Bressel S (1989) Klinisch-anatomische Befunde zur Nase und den Nasennebenhöhlen unter besonderer Berücksichtigung des Os ethmoidale. Inaug Diss Würzburg

Broman I (1911) Die knorpeligen Nasenwände. In: Normale und abnormale Entwicklung des Menschen. Bergmann, Wiesbaden

Finby N, Kraft E (1972) The aging skull: comparative roentgen study 25 to 34 year interval. Clin Radiol 23:410–414

Haas A (1989) Erweiterte Daten zur topographischen Anatomie von Lamina cribrosa, Sinus frontalis und Labyrinthus ethmoidalis. Inaug Diss Würzburg

Haas RJ (1980) Neue Befunde zur Bodenregion der Fossa cranialis anterior. Inaug Diss Würzburg

Harrison DFN (1961) Oro-antral fistulae. Br J Clin Pract 15:169–174

Hertwig O (1907) Handbuch der vergleichenden und experimentellen Entwicklungslehre des Menschen und der Wirbeltiere 3. Aufl. Fischer, Jena

Koch J (1930) Die normale und pathologische Pneumatisation der Stirnhöhlen. Ref Zentralbl Hals Heilkd 15:620

Kollmann J (1907) Handatlas der Entwicklungsgeschichte des Menschen. Fischer, Jena

Lang J (1983) Clinical anatomy of the head. Neurocranium – orbit – craniocervical regions. Translated by Wilson RR, Winstanley DP. Springer, Berlin Heidelberg New York Tokyo

Lang J (1985a) Hypophyseal region anatomy of the operative approaches. Neurosurg Rev 8:93–124

Lang J (1985b) Praktische Anatomie: Ein Lehr- u. Hilfsbuch der anatomischen Grundlagen ärztlichen Handelns. Begr. von T. von Lanz, W. Wachsmuth. Fortgef. u. hrsg. von J. Lang, W Wachsmuth. Springer, Berlin Heidelberg New York Tokyo

Lang J (1987) Fossa cranialis anterior, mediale Bodenregion. Roundtable-Konferenz: das frontobasale Trauma – Diagnostik und Behandlungsablauf. In: Schuchardt K, Schwenzer N (Hrsg) Fortschritte der Kiefer- und Gesichts-Chirurgie, Bd. XXXII: Bildgebende Untersuchungsverfahren in der Mund-, Kiefer- und Gesichts-Chirurgie. Thieme, Stuttgart, S 210–218

Lang J (1988 a) Klinische Anatomie der Nase, Nasenhöhle und Nebenhöhlen. Grundlagen für Diagnostik und Operation. Thieme, Stuttgart

Lang J (1989) Clinical Anatomy of the Nose, Nasal Cavity and Paranasal Sinuses. Thieme, Stuttgart; Thieme Medical Publishers, Inc New York

Lang J (1988 b) Über die Cellulae ethmoidales posteriores und ihre Beziehungen zum Canalis opticus. HNO 36:5–32

Lang J, Haas A (1990) Über die Sagittalausdehnung des Sinus frontalis, dessen Wanddicke, Abstände zur Lamina cribrosa, die Tiefe der sogenannten Olfactorius-Rinne und die Canales ethmoidales. Gegenbaurs Morph Jb 134:459–469

Lang J, Haas R (1979) Neue Befunde zur Bodenregion der Fossa cranialis anterior. Verh Anat Ges 73:77–86

Lang J, Oehmann G (1976) Formentwicklung des Canalis opticus, seine Maße und Einstellung zu den Schädelebenen. Verh Anat Ges 70:567–574

Lang J, Papke J (1980) Über anatomische Grundlagen der blow out-fractures der Orbita. Second workshop of the Anatomische Gesellschaft Würzburg, 8–10 Oct. 1980

Lang J, Papke J (1984) Über die klinische Anatomie des Paries inferior orbitae und dessen Nachbarstrukturen. Gegenbaurs Morphol Jahrb 130:1–47

Lang J, Schäfer K (1979) Arteriae ethmoidales: Ursprung, Verlauf, Versorgungsgebiete und Anastomosen. Acta Anat 104:183–197

Lang J, Schlehahn F (1978) Foramina ethmoidalia und Canales ethmoidales. Verh Anat Ges 72:433–435

Lang J, Schlehahn F, Schäfer K (1979) Über den Inhalt der Canales ethmoidales. Verh Anat Ges 73:87–94

Lang J, Väth H (1989) Über die Abstände zwischen dem Ductus nasolacrimalis und dem Canalis palatinus major an der seitlichen Nasenwand sowie Distanzen zwischen Ostium pharyngeum tubae auditivae und Canalis palatinus major sowie Oberseite des Palatum molle. Anat Anz 168:315–320

Lang W (1889) Injuries and diseases of the orbit. I. Traumatic enophthalmos with retention of perfect acuity of vision. Trans Ophthalmol Soc UK 9:41–45

Menger W, Kocoglu V (1969) Paranasale Sinusitiden im Säuglings- und Kindesalter und ihre Beziehung zu den Bronchitiden. Med Welt 31:1686–1693

Onodi A (1911) Die Nebenhöhlen der Nase beim Kinde: 102 Präparate in natürlicher Größe nach pathologischen Aufnahmen dargestellt. Kabitzsch, Würzburg

Stern L (1939) Röntgenologische Betrachtung der Entwicklung und Ausdehnung der Nasennebenhöhlen. HNO 30:169–199

Van Gilse PHG (1926) Über die Entwicklung der Keilbeinhöhle des Menschen. Beitrag zur Kenntnis der Pneumatisierung des Schädels von der Nase aus. Z Hals Nasen Ohrenheilkd 16:202–298

The Growth of the Paranasal Sinuses in Craniostenosis

J. MÜHLING, J. REUTHER, N. SÖRENSEN, and H. COLLMANN, Würzburg/FRG

The paranasal sinuses are a functional and anatomic unit. They stand in close relationship to the brain, the orbits, and the teeth. Functionally they participate in breathing, phonation, and smelling (Wagemann 1964a). The development of the paranasal sinuses starts in the 3rd month of embryogenesis from the epithelium of the nose (Plenk and Tschabitscher 1986). The primary small scions extend into the surrounding bone, reaching final size in adulthood. There are two concepts explaining the mechanism of pneumatization. The first defines development of the paranasal sinuses as an active process; the osteolytic potency of the epithelium leads to the retraction of surrounding bone structure and thereby to the inflation of the sinuses (Schwarz 1934). On the other hand, Weidenreich (1924) explained the growth as a passive process; by the growth of the midface the neurocranium withdraws from the alveolar process. The structurally superfluous spongiosa is replaced by sinuses. There is a large variety in form and size of paranasal sinuses as well as in their asymmetric extension. This is due to mechanical and structural features caused by the function of the surrounding muscles and soft tissues (Wagemann 1964b).

This knowledge and the related clinical implications lead to the conclusion that craniostenoses also affect the development of the paranasal sinuses. In craniostenoses the growth of the neuro- and viscerocranium is impaired by premature synostoses of the sutures. Depending on the localization of the affected sutures, typical skull deformities may be observed restricting the intracranial volume (Virchow 1851/1852). By further extension of the brain the intracranial pressure increases. This leads to a thinning of the skull and in extreme cases to bone erosions. In X-rays the digital impressions are increased. Cephalometric investigations on 24 patients have shown that increased intracranial pressure leads to restricted development of the frontal sinuses. Craniostenosis also limits the forward-downward rotation of the midface. This results in hypoplastic maxilla with small maxillary sinuses. Dentition in these cases is not restricted, but since the alveolar process is only small, misplacements of the teeth may occur (Fig. 1).

In surgical treatment of craniostenoses we use a standardized fronto-orbital advancement (Mühling et al. 1984) which includes osteotomy, removal and shaping, and advanced reconstruction of the fronto-orbital bone segments. By osteotomy of the glabella the region of the frontal sinuses is separated (Fig. 2). Postoperative investigations show that small frontal sinuses may develop after reossification in spite of the advancement. Several months after fronto-orbital

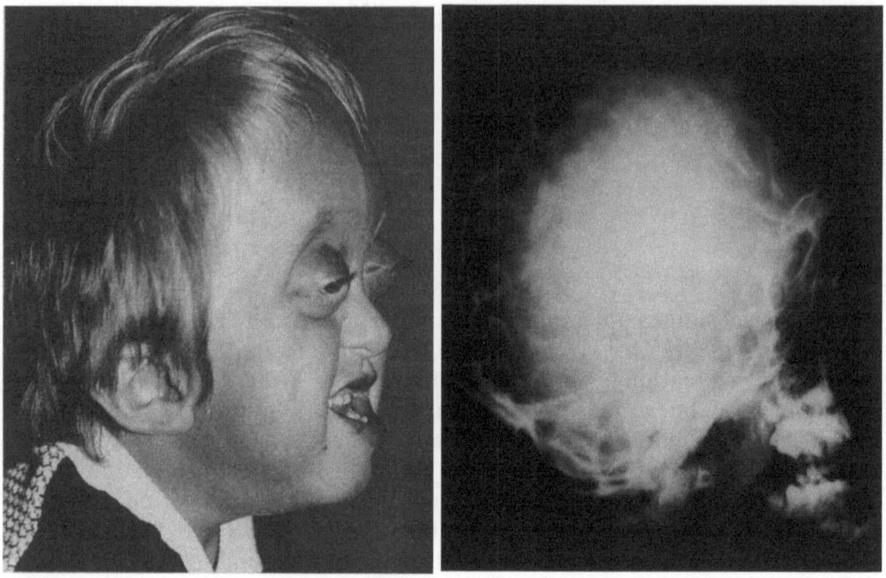

Fig. 1. A 7-year-old untreated patient with extreme brachycephalus, midface hypoplasia, and restricted pneumatization of the sinuses

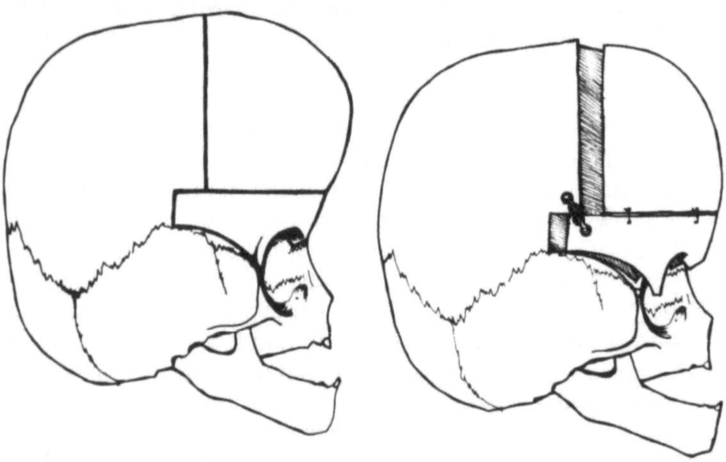

Fig. 2. Fronto-orbital advancement for surgical treatment of craniostenoses

advancement with reduction of intracranial pressure frontal pneumatization may become visible (Fig. 3). If a frontal sinus is opened by the operation the epithelium must be removed to prevent infection. Additionally, the ostium to the nose must be closed. In spite of this procedure the further pneumatization of the sinuses is not impaired. There is re-epithelization and further expansion of the sinuses, which has been confirmed in cephalometric investigations and in reoperations. The hypoplasia of the midface is corrected with a Le Fort III

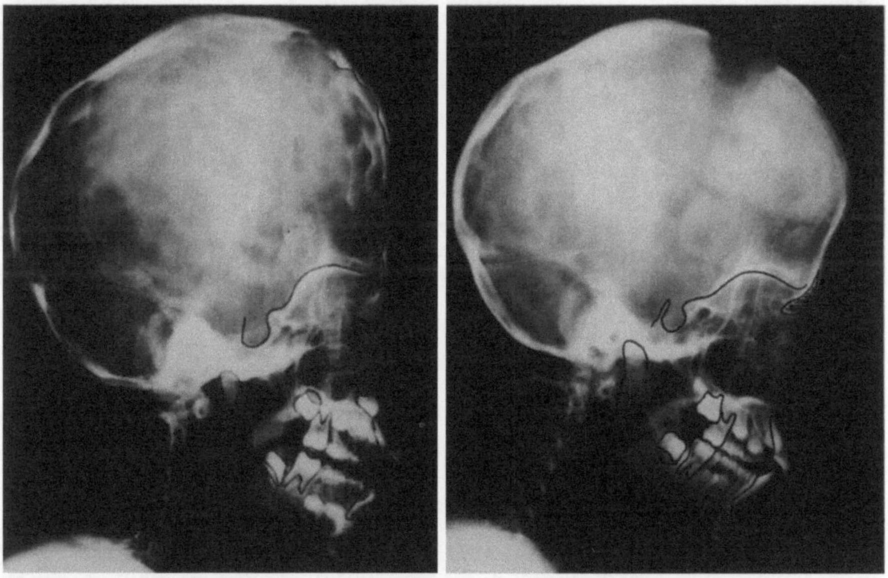

Fig. 3. Pre- and postoperative X-ray of an 8-year-old patient with oxycephalus; the digital impressions disappear, and frontal pneumatization becomes visible

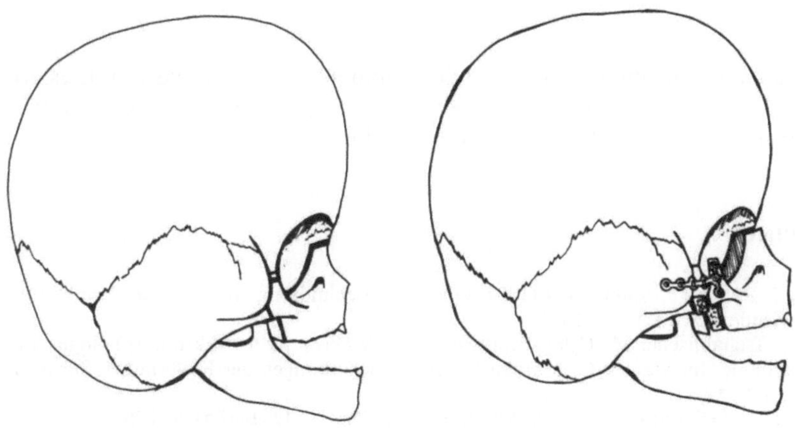

Fig. 4. Le Fort III osteotomy

osteotomy. The complete midface including the lower half of the orbits is advanced. Thereby the maxillary sinuses are opened (Fig. 4).

Even after these surgical injuries the maxillary sinuses develop unrestricted (Fig. 5). After a primary X-ray opacity caused by the blood, the sinuses keep growing. In spite of close relationship to the developing teeth we could not find any restriction of dentition.

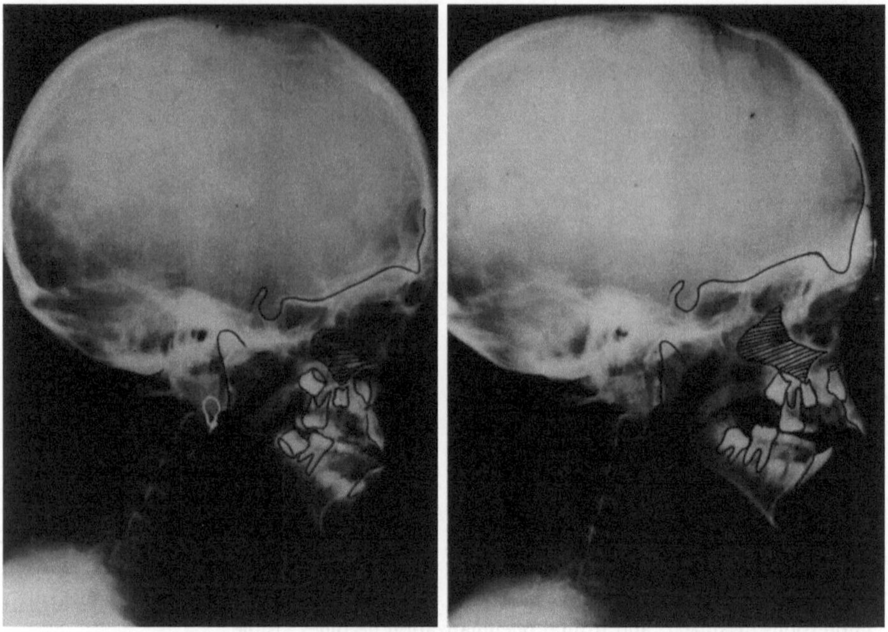

Fig. 5. Pre- and postoperative development of the maxillary sinuses. After midface advancement, further pneumatization is unrestricted; dentition is not impaired

In summary, pneumatization and development of the paranasal sinuses are narrowed by craniofacial stenoses. Neither their growth nor the dentition is affected by the fronto-orbital advancement in early childhood.

References

Mühling J, Reuther J, Sörensen N (1984) Operative Behandlung craniofacialer Fehlbildungen. Kinderarzt 15:1022–1023

Plenk H, Tschabitscher M (1986) Entwicklung, Makro- und Mikromorphologie der Kieferhöhle. In: Watzek G, Matyka M (eds) Erkrankungen der Kieferhöhle. Springer, Wien, p 1–12

Schwarz M (1934) Entwicklung der Nasennebenhöhlen. HNO (Berlin) 36:296

Virchow R (1851/52) Über den Cretinismus, namentlich in Franken, und über pathologische Schädelformen. Verh Phys-Med Ges Würzburg 2:230–271

Wagemann NKW (1964a) Funktionen von Nase und Nebenhöhlen. In: Berendes J, Link R, Zöllner F (eds) HNO-Heilkunde, vol 1. Thieme, Stuttgart, p 30–38

Wagemann NKW (1964b) Entwicklungsgeschichte, Wachstum. In: Berendes J, Link R, Zöllner F (eds) HNO-Heilkunde, vol 1. Thieme, Stuttgart, p 26

Weidenreich F (1924) Über die pneumatischen Nebenräume des Kopfes. Ein Beitrag zur Kenntnis des Bauprinzips der Knochen des Schädels und des Körpers. Anat Embryol 72:55–93

Clinical Anatomy of the Sphenoidal Sinus

J. Lang and J. Pahnke, Würzburg/FRG

The sphenoidal sinus is an important space for surgical procedures. It is opened in inflammatory diseases of the paranasal sinuses, and the neurosurgeon passes through this sinus in order to reach the hypophyseal region. We report here some characteristics and new measurements of the sphenoidal sinus and hypophyseal region.

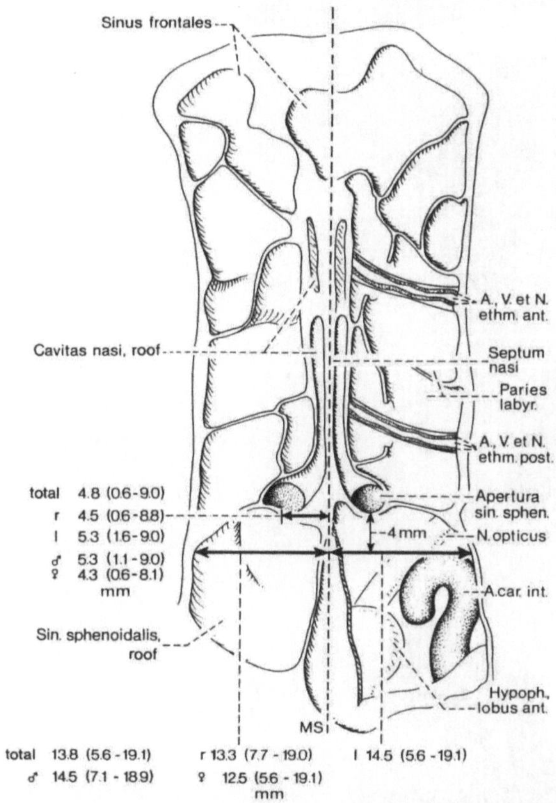

Fig. 1. Drawing of a transverse section, seen from below. The paramedian position of the aperture of the sphenoidal sinus and the paramedian distance of the side wall of the sphenoidal sinus in the level of the aperture are shown. *Right*, the upper wall of the sphenoid sinus is removed. Measurements in millimetres; means and ranges (From Lang et al. 1988)

Sphenoidal Sinus

For the transnasal approach the aperture of the sphenoidal sinus is an important landmark. According to Lang and Baumeister (1982) its distance to the nasospinal point averages 53.7 mm in adult skulls, varying between 43 and 64 mm. In our cadavers we measured distances of 61.5 mm (43–69 mm). Lateral to the posterosuperior part of the nasal septum the sphenoethmoidal recess was well developed in 48.3% of cases (Lang and Sakals 1981). The aperture leads into this small space. The distance between the midpoint of the aperture and the median-sagittal plane averaged 4.8 mm–5.3 mm on the left and 4.5 mm on the right (varying between 0.6 mm and 9.0 mm). The aperture is usually oval shaped. In 20 specimens the long axis measured 3.7 mm (2.0–

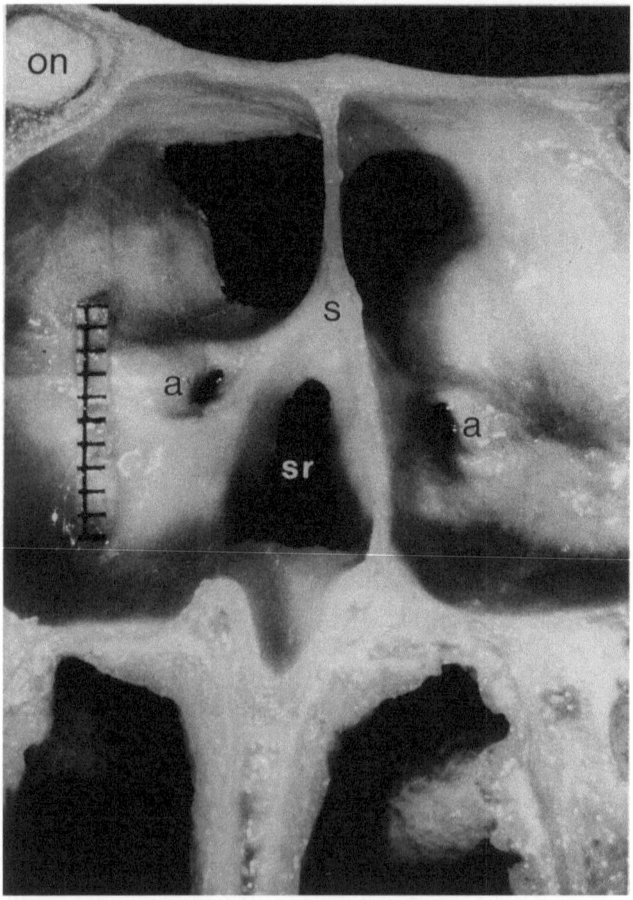

Fig. 2. Frontal section through the sphenoidal sinus area, dorsal view to the anterior of the sphenoidal sinus. *a*, Aperture of the sphenoidal sinus; *s*, septum of the sphenoidal sinus; *sr*, septal recess; *on*, optic nerve

9.1 mm). The transverse diameter averaged 1.7 mm (0.7 mm–3.6 mm). The long axis of the aperture normally converges on the median-sagittal plane in superior direction (see Figs. 2, 3), but in 10% of our cases it diverged. The angle between the long axis of the aperture and the median-sagittal plane varied from −11° (diverging) to +51° (converging). Therefore an aperture cannot be assigned with certainty to the left or right side by observing its form and the direction of the long diameter. In some cases a septal recess can be found in the anterior wall between the apertures (Fig. 2). At the level of the sphenoidal aperture and 4 mm behind the paramedian distance of the side wall of the sphenoidal sinus (Fig. 1) averaged 13.8 mm (5.6–19.1 mm) – 14.5 mm in men and 12.5 mm in women. In the side wall various prominences are visible These are caused by the optic nerve, internal carotic artery, abducent nerve,

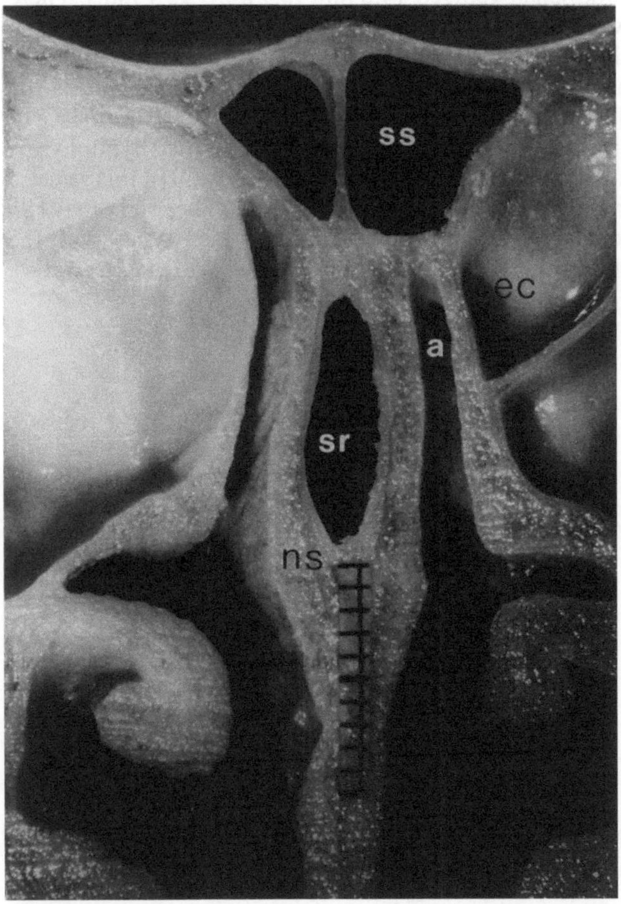

Fig. 3. Frontal section, anterior view (same specimen as in Fig. 2). *ss*, Sphenoidal sinus; *ec*, ethmoidal cell; *sr*, septal recess; *ns*, nasal septum; *a*, aperture of the sphenoidal sinus in the sphenoethmoidal recess

and maxillary nerve. According to Lang et al. (1988), the prominence of the optic canal was found in 51.8% of cases. In the anterior part the internal carotid artery caused a bulge of the side wall in 47.6%. Lateral and inferior to the hypophyseal fossa the prominence of the carotid artery was found in 14.3% and in the lower posterior part in 23.8%. The abducent nerve bulged the side wall in only 4.8% of cases. The prominence of the maxillary nerve was seen in 42%. In the floor of the sphenoidal sinus the pterygoid canal was prominent in about 29% and was sometimes dehiscent.

Hypophyseal Region

The distance from the midpoint of the sphenoidal aperture to the nearest point of the anterior wall of the sella averaged 14.6 mm (8.9 mm–23.1 mm). There was no correlation between this distance and age (46–92 years). The thickness of the sella floor in the anterior-inferior region was 0.7 mm (0.2–1.8 mm; $n = 44$). Bergland et al. (1968) stressed that in 72% of their specimens the anterior floor of the sella was 1 mm or less in thickness. On the lower surface of the pituitary gland the medial of the main branches of the inferior hypophyseal artery is usually found. Its diameter measured 0.32 mm. Very often this vessel is situated between the anterior and posterior pituitary lobes. The artery may run within the inferior intercavernous sinus and normally anastomoses with the corresponding vessel of the other side. These findings agree with the description

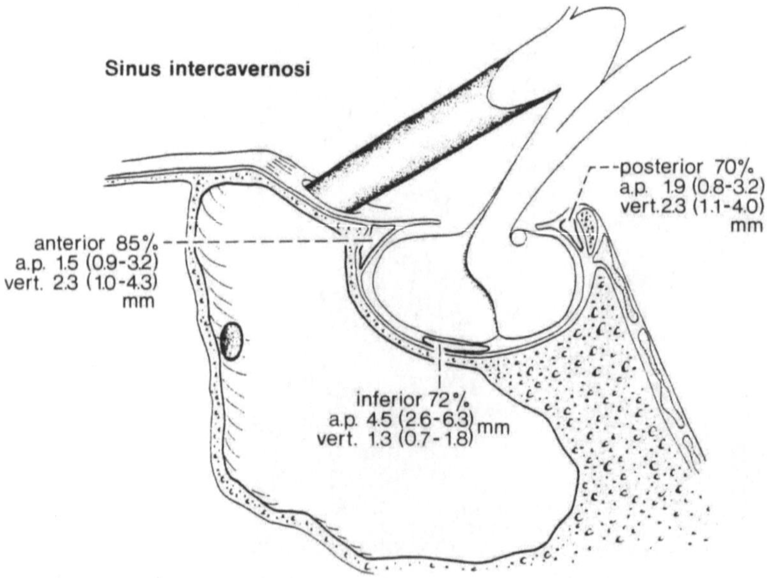

Fig. 4. Drawing of a medial view of a median-sagittal section. Frequency, anterior-posterior diameter, and vertical diameter of the intercavernous sinuses are shown

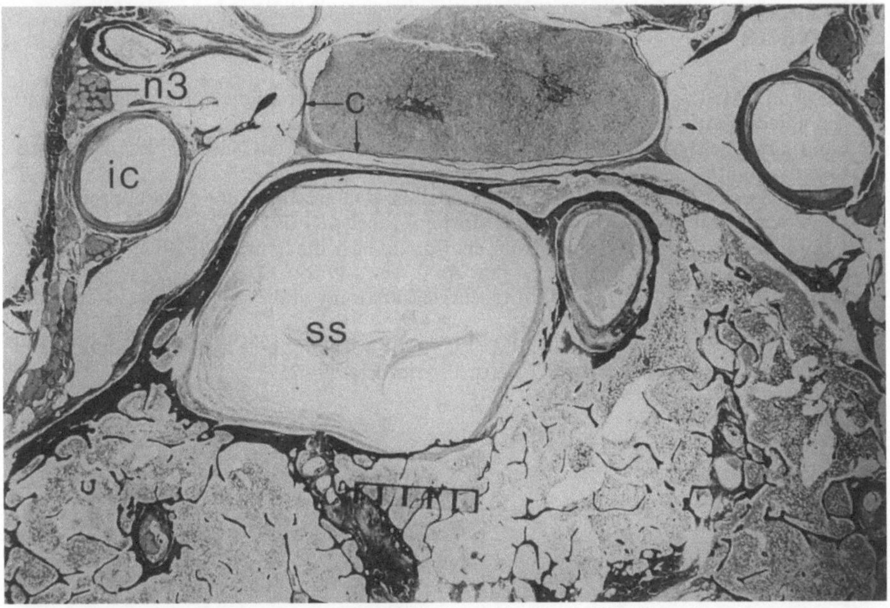

Fig. 5. Coronal section, 50-year-old man (Goldner elastica staining). *ic*, Intracavernous part of the internal carotid artery; *n3*, oculomotor nerve; *ss*, sphenoidal sinus; *c*, capsule of the pituitary gland with artificial shrinking cleft

of Xuereb et al. (1954). In sagittal sections, performed with a paramedian distance of up to 3 mm, the anterior intercavernous sinus was developed in about 85%. Its anterior-posterior extension was 1.5 mm (0.9–3.2 mm). The inferior intercavernous sinus was found in 72%, the anterior-posterior diameter averaging 4.5 mm (2.6–6.3 mm). The inferior branch of the inferior hypophyseal artery may be situated within this sinus. In 70% of cases the posterior intercavernous sinus was seen (Fig. 4). Its anterior-posterior extension was 1.9 mm (0.8–3.2 mm). In 65% of cases the posterior sinuses have a connection to the basilar venous sinus. Renn and Rhoton (1975) observed the anterior intercavernous sinus in 76% and the posterior in 32%. The diaphragm of sella may be a barrier during transsphenoidal hypophysectomy (Renn and Rhoton 1975). Its foramen often has no sharp edge when it is intermingling with arachnoid tissue. The anterior-posterior diameter averaged 6.5 mm (4.1–9.2 mm) in 14 specimens with well-delineated foramina. Renn and Rhoton found this opening greater than 5 mm in 56% of cases. Sometimes the bottom area of the hypophyseal fossa is extremely thin; this is illustrated in Fig. 5, which also shows the hypophyseal capsule and nerves and blood spaces of the cavernous sinus.

References

1. Bergland RM, Ray BS, Torack RM (1968) Anatomical variations in the pituitary gland and adjacent structures in 225 human autopsy cases. J Neurosurg 28:93–99
2. Lang J, Baumeister R (1982) Über das postnatale Wachstum der Nasenhöhle. Gegenbaurs Morphol Jahrb 128:354–393
3. Lang J, Bressel S, Pahnke J (1988) Sphenoidal sinus, clinical anatomy of the approaches to the hypophyseal region. Gegenbaurs Morphol Jahrb 134:291–307
4. Lang J, Sakals E (1981) Über die Höhe der Cavitas nasi, die Länge ihres Bodens und Maße sowie Anordnung der Conchae nasales. Anat Anz 149:297–318
5. Renn WH, Rhoton AL Jr (1975) Microsurgical anatomy of the sellar region. J Neurosurg 43:288–298
6. Xuereb GP, Prichard ML, Daniel PM (1954) The arterial supply and venous drainage of the human hypophysis cerebri. Q J Exp Physiol 39:199–217

Space-Occupying Processes of Paranasal Sinuses

I. GÁDOR, P. PIFFKÓ, M. HAJDA, F. LÁNYI, and E. PÁSZTOR,
Budapest/Hungary

According to Hommerich (1964) malignant tumors of paranasal sinuses account for 10%–13% of the otorhinolaryngologic malignancies and for 0.2% of malignancies overall. These tumors are usually diagnosed late, since in view of the mild symptoms patients tend to not see the physician early. Vollrath and Lasch (1988) detected stage III or IV tumors in 75% of his 61 patients. Since precise diagnosis of these tumors is not always easy, even given the case history and symptoms, a team including an ophthalmologist, a radiologist and a neurologist is needed.

Over the past 10 years at the National Institute of Neurosurgery in Budapest, we have found space-occupying lesions of the paranasal sinuses or basal tumors showing paranasal infiltration in 91 patients. We observed frontoethmoidal mucoceles in 28 patients, in whom frontoethmoidectomy was performed with nasofrontal drainage by silicon tube. The study revealed tumors in a further 63 patients, 37 of them malignant, representing 0.1% of the total patient population. The 25 primary benign tumors included 9 osteomas, 7 juvenile angiofibromas, 4 dermoid cysts, 3 eosinophilic granulomas, and 2 fibromas. The 19 primary malignant tumors included 10 carcinomas, 3 carcomas, 2 melanomas, and 4 other types of tumor (neuroblastoma, malignant angiofibroma, and hemangiopericytoma).

Due to great extent and infiltration of the tumors in some patients it is often impossible to localize precisely the site of tumor origin. If the tumor invades the sinuses from the base, or vice versa, the term secondary neoplasm can be used. Such processes were observed in 19 patients and included 8 carcinomas, 5 chordomas, 3 sarcomas, and one case each of osteoclastoma, chondroma, and melanoma. In 63 patients suffering from tumor invading paranasal sinuses 71 operations have been carried out. The surgical methods included the following: orbitomy with exenteration (14), modified Moure's incision (13), frontoethmoidectomy (11), transsphenoidal approach (9), endonasal biopsy (7), op. sec. Caldwell-Luc operation (7), transoral approach (5), and epipharyngeal biopsy (5). In vascularized tumors surgery may be hampered by bleeding. Sellars (1980) reports a 3% intraoperative mortality rate in such cases. To diminish the risk due to extensive bleeding we performed surgery on six patients 4–5 days after free embolization; this allowed intraoperative bleeding to be much more easily controlled.

This can be illustrated by the case of a *20-year-old woman*. She presented with auricular complaints of 2 years' history, conductive hearing loss, and progressive trismus. We performed adenotomy, septal resection, and extraction of the upper right 8th tooth, but no long-lasting improvement was achieved. Thorough reexamination followed, and computed tomography verified a contrast-enhancing tumor that infiltrated the right maxillary sinus and orbital base, partially filled the sphenoidal sinus, and extended both para- and retropharyngeally. Right percutaneous carotid angiography also revealed pathological vascularization in the tumor localization by the internal maxillary and the facial and lingual arteries. Intracranial propagation of the tumor was not detected. A biopsy was performed through the fossa canina, and histology presented evidence of ameloblastoma. On September 24, 1987 the internal maxillary artery, the middle meningeal artery, and the pharyngeal ascendent artery were obliterated. Four days later, the malignant tumor involving also the retropharyngeal space was macroscopically totally removed using a modified Moure's incision. The final histology showed ameloblastic carcinoma. Due to signs of buccal infiltration the patient was given postoperative cytostatic treatment, and prosthetic management at the Clinic of Stomatology in Budapest. The patient was without a sign of recurrency 3 months after surgery. This case demonstrates that space-occupying processes of the paranasal sinuses warrant thorough investigation and close interdisciplinary team work.

References

Hommerich KW (1964) Die Geschwülste der Nase und Nasennebenhöhlen. In: Berendes J, Link R, Zöllner F (Hrsg) HNO-Heilkunde, Bd 1. Thieme, Stuttgart

Sellars SL (1980) Juvenile nasopharyngeal angiofibroma. South Afr Med J 58:961–964

Vollrath M, Lasch V (1988) Klinik und Prognose der Nasennebenhöhlenmalignome unter besonderer Berücksichtigung ihrer Ätiologie. HNO 36:22–27

Clinical Manifestation of Sinus Disorders – ENT

J. HELMS, Würzburg/FRG

The clinical manifestation of diseases of the paranasal sinuses is mainly otorhinolaryngological. The majority of symptoms arise first as problems of the nose or face. In a single year the number of inpatients and outpatients showing typical clinical manifestations were: malformations 8, infections, allergy 2043, trauma 141, and tumors 10.

A clinical example of malformation is extended hemangioma of the nose with possible involvement of the paranasal sinuses. Malformations may be subdivided clinically into aplasia; clefts, medial and paramedial; dysmorphic syndromes; Meningoceles (and meningoencephaloceles); Dermoids; and others. The major clinical importance of a unilateral aplasia of the maxillary or frontal sinus is in cases of misinterpretation of X-rays. A Sturge-Weber syndrome with nevus flammeus of the skin along one branch of the trigeminal nerve and with ophthalmological and intracranial involvement leads to extended asymmetry of the frontal sinus. An oculo-oto vertebral syndrome may involve the sinuses even more. It is a complex dysmorphic dysplasia involving the areas mentioned in its name.

The category of inflammation and allergy are the most frequent clinical manifestations of sinus disorders. These may be subdivided as follows: acute, unspecific (e.g., virus); specific (e.g., tuberculosis); chronic manifestations; polyps, mucoceles, pyoceles, etc.; and complications (orbital, endocranial, osteomyelitis). Fluid levels are best seen in classical X-ray projections. Mucoceles and complications of a sinusitis are analysed in more detail by more sophisticated radiological methods.

Traumas may be distinguished into the following categories: fractures; Le Fort etc.; soft tissue damages; superficial and deep structures; orbit, brain, others; different kinds of forces; penetrating foreign bodies; burns; and others. In contrast to the vast majority of infections, the number of patients with serious impairment by trauma is high and reaches 50 % of clinical cases. Traffic accidents are the most frequent causes of facial and sinus lacerations. Gun bullets and other foreign bodies, such as splinters from small handicraft machines, are less frequently causes of sinus trauma. In a psychopathic patient two bronze needles had to be removed from the orbit; these had been placed close to the bulb by the patient himself in the attempt to treat pain.

Tumors may be benign or malignant. Benign tumors originate in soft tissues, cartilage, bone, and teeth and include hemangioma, fibroma, and osteoma. Malignant tumors may be epithelial, mesenchymal, or metastatic; these include carcinoma, sarcoma, and melanoma. Tumors of the paranasal sinuses are still a clinical challenge. These are frequently diagnosed only after advanced growth, and the treatment can be complicated and disappointing for the patient. This is the case for extended benign and the majority of malignant neoplasias.

A thorough preoperative diagnosis made in cooperation with neuroradiology collegues is mandatory to minimize mistakes. The subtraction angiogram of a juvenile angiofibroma and its location may be typical and the embolization is in most cases a great success. In spite of this, histology may then confirm the doubts of the ENT surgeon which develop during surgery.

Ophthalmological Manifestations of Paranasal Sinus Diseases

R. Rochels, Kiel/FRG

Introduction

Paranasal sinus diseases manifest themselves in a high percentage of cases by striking and threatening ocular and orbital signs and symptoms (Ducrey 1985; Duke-Elder 1972, 1974; Ganz 1977; Herrmann 1958; Jakobiec and Jones 1986; Jakobiec et al. 1986; Seiferth and Wustrow 1977; Siegert 1975; Smith et al. 1986; Wustrow 1965). This explains why the ophthalmologist is often the first to see and to examine such patients. The anatomic basis of ophthalmic complications in paranasal sinus diseases is in the following (Jakobiec and Jones 1986):

1. The orbit is nearly completely surrounded by the sinuses, except laterally.
2. The separating bones are very thin.
3. There are numerous suture lines.
4. There is a direct communication between the ethmoidal cells and the orbit by the anterior and posterior ethmoidal vessels.
5. The orbital contents such as the extraocular muscles and the optic nerve are in close relationship to the sinuses.
6. The orbital veins lack valves, allowing for a free communication between the facial vein and the cavernous sinus.

Ocular and orbital complications of inflammatory, traumatic, and tumorous paranasal sinus diseases often imitate affections of the eye or the orbit itself. The differential diagnosis is rendered even more difficult by the fact that there are no pathognomonic symptoms allowing for a clear-cut distinction between these entities. The ophthalmologist should therefore always bear in mind that an apparent ocular or orbital symptom may be merely an ophthalmic manifestation of a paranasal sinus disease.

Ophthalmic Signs of Inflammatory Paranasal Sinus Diseases

Ophthalmic complications of inflammatory paranasal sinus diseases can be classified as orbital edema, periostitis, subperiosteal abscess, orbital cellulitis, retrobulbar neuritis, orbital apex syndrome, cavernous sinus thrombosis, and muco(pyo)cele (Ducrey 1985; Duke-Elder 1972, 1974; Ganz 1977; Herrmann 1958; Jakobiec and Jones 1986; Siegert 1975). Orbital edema either shows up as a collateral preseptal edema of the lid(s) (Fig. 1), or it is a real inflammatory

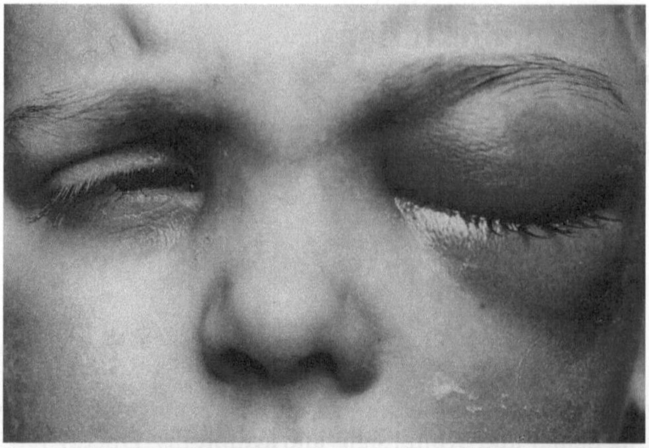

1

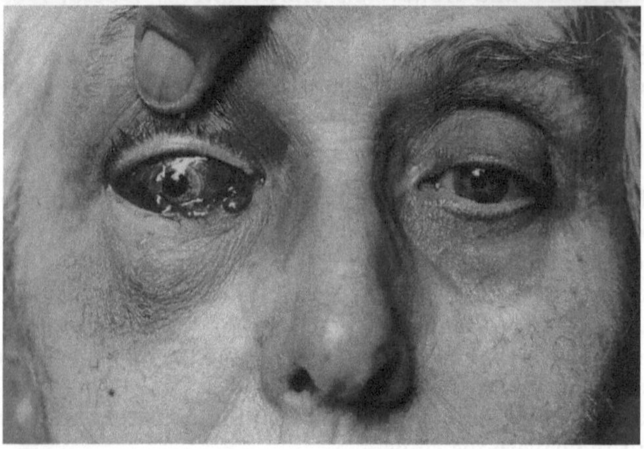

2

Fig. 1. Collateral preseptal edema of the lids in anterior ethmoidal sinusitis

Fig. 2. Orbital cellulitis in ethmoidal sinusitis with proptosis and dislocation of the globe, swelling of the conjunctiva and lids, and motility disorders

edema involving the lids and the conjunctiva. Proptosis, displacement of the globe, and motility disorders are usually missing. In orbital periostitis a circumscribed pain is induced by touch, in addition. Dislocation and protrusion of the globe along with minimal restriction of motility are observed in a subperiosteal abscess. Orbital cellulitis is characterized by massive (bilateral) swelling of the lids and conjunctiva, and an extremely painful exophthalmos with gross motility disorders (Fig. 2). Ophthalmoscopy sometimes reveals a papilledema and a dilation of the retinal veins caused by pressure or vascular involvement of the optic nerve, followed by an impairment of visual acuity. Rarely, an involvement of the posterior ciliary vessels and the ophthalmic artery ends up in an infarction of the uvea and the retina.

A collateral rhinogenic retrobulbar neuritis can develop in inflammation of the sphenoidal sinus and the posterior ethmoid as well as in orbital cellulitis. Ophthalmic symptoms are a gross deterioration of visual acuity, visual field defects (central scotoma), impairment of color vision, and retrobulbar pain on ocular movements. An optic atrophy with temporal pallor of the optic nerve head is usually the consequence.

In inflammatory orbital apex syndrome an involvement of the second to sixth cranial nerves presents along with the symptoms of retrobulbar neuritis, oculomotor, trochlear, and abducent palsy. Diplopia is not a common sign because of the concomitant ptosis and retrobular neuritis. Marked temporoparietal pain and massive protrusion of the globe are leading symptoms.

Frontal sinusitis is often complicated by a preseptal cellulitis, a subperiosteal abscess of the orbital roof and an intralid abscess, whereas maxillary sinusitis without a history of injury or surgery rarely affects the orbit. Inflammation of the ethmoidal cells can lead to preseptal edema, lid abscess, subperiosteal abscess of the medial orbital wall, and orbital cellulitis.

The life-threatening cavernous sinus thrombosis is, among other things, a complication of sphenoethmoidal sinusitis. Ophthalmic signs are those of a fulminant inflammatory orbital apex syndrome that most often develops bilaterally and ends up in blindness by massive papilledema (Fig. 3), acute obstruction of the retinal vessels, and consecutive optic atrophy. High fever, shivering fit, dizziness, and vomiting are general symptoms.

Orbital mucoceles with a pathognomonic dislocation of the globe are the result of the expansion of a sinus, induced by obstruction of its ostium, with consecutive thinning and erosion of the bone. Causes include paranasal inflammation, scarring after trauma or surgery, and benign and malignant sinus tumors. In frontal mucoceles the globe is displaced downward and

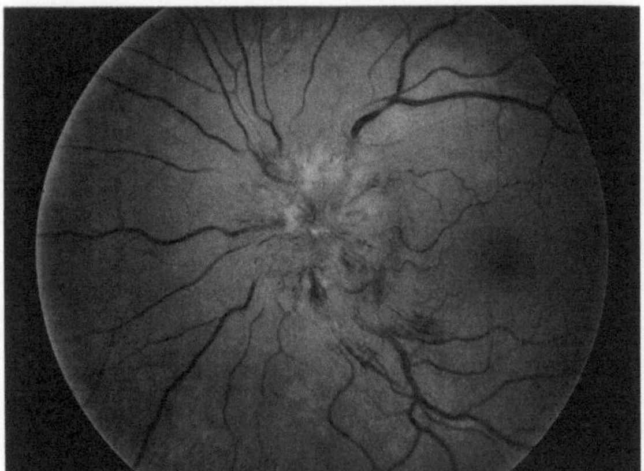

Fig. 3. Papilledema, striate retinal hemorrhages and engorgement of the veins in cavernous sinus thrombosis

outward. On palpation a crepitant feeling is present. Mucoceles of the anterior ethmoid displace the globe laterally and sometimes cause epiphora or chronic dacryocystitis. Despite the dislocation of the globe, diplopia is not a common sign in either of these types. Mucoceles of the sphenoidal sinus affect the third cranial nerve with a consecutive external (!) oculomotor palsy; an exophthalmos is observed only if the posterior ethmoid is also involved. Large sphenoidal mucoceles can furthermore compress the optic nerve and, after destruction of the sellar region, the chiasm. Along with additional disturbances of the pituitary gland, the clinical picture resembles that of a craniopharyngioma.

As a complication, frontal, ethmoidal, and sphenoidal mucoceles – induced by paranasal inflammation, trauma, surgery, or tumor – can become superinfected, leading to the above mentioned symptoms of orbital edema, and even cellulitis. Mucoceles of the maxillary sinus with orbital involvement are extremely rare.

Ophthalmic Signs of Traumatic Paranasal Sinus Diseases

In about 10% of cases the orbit is involved in paranasal injuries; depending on the fracture lines, injuries of the orbital walls, the orbital margin, or a combination of them can be distinguished (Duke-Elder 1972, 1974; Seiferth and Wustrow 1977; Siegert 1975; Smith et al. 1986).

In blow-out fractures of the medial orbital wall, ophthalmic complications, such as intraorbital, subconjunctival and intralid hemorrhage, and an emphysema with a consecutive proptosis, can be expected. If the perimuscular tissue is entrapped into the fracture line, a restriction of motility on abduction and adduction ensues. A muscle hematoma must be considered in differential diagnosis. Blow-out fractures of the orbital floor cause slight enophthalmos, restriction of motility in vertical directions (Fig. 4), sometimes an orbital

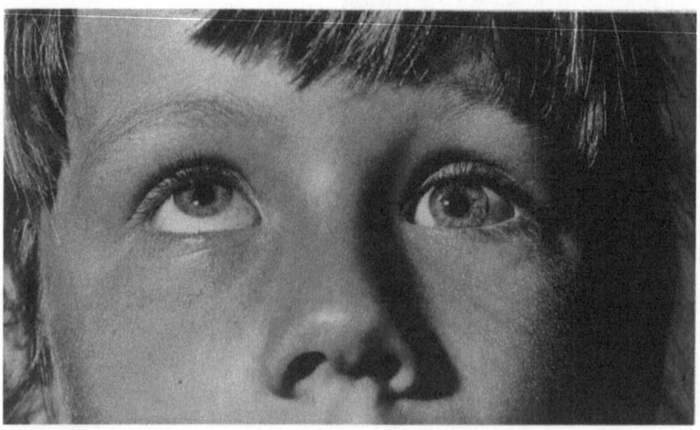

Fig. 4. Enophthalmos on the left with restriction of motility on upgaze in blow-out fracture of the orbital floor

hematoma, and often an infraorbital paresthesia. Concomitant injuries of the globe, the lids, and the lacrimal drainage system by bony fragments must be ruled out carefully. Unilateral or bilateral hematoma of the lids without other ocular signs must prompt the diagnosis of a skull base fracture until the contrary is proved.

In all orbito-paranasal fractures a secondary posttraumatic orbital inflammation can develop.

Further ophthalmic complications of paranasal injuries are (a) motility disorders with diplopia by avulsion of the trochlea, palsy of the third (Fig. 5), fourth, or sixth cranial nerve and (b) a compression of the optic nerve by a subperiosteal hematoma of the medial or upper orbital wall, an optic sheath hematoma, or by fractures in the optic channel with laceration of the nerve. Visual acuity decreases dramatically, the pupil becomes enlarged and unresponsive to light stimuli. Ophthalmoscopy shows either a normal fundus picture or an edema and hemorrhages of the optic nerve head, an engorgement of the retinal veins, and sometimes even an occlusion of the central retinal artery. Without rapid decompression a total optic atrophy with blindness is the consequence.

The posttraumatic carotid cavernous sinus fistula is characterized by a bilateral pulsating exophthalmos (Fig. 6) with a systolic noise in the lid and

5 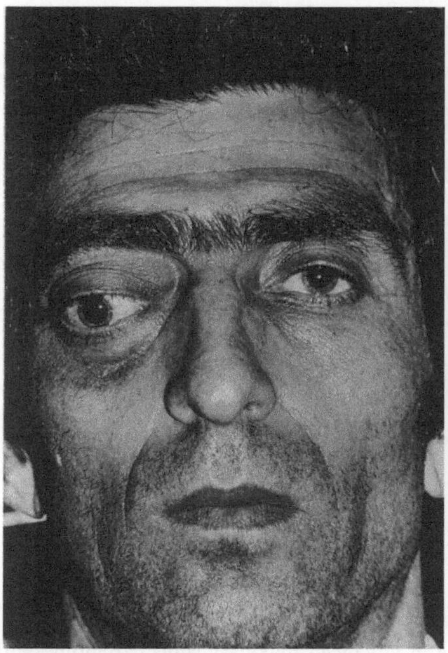

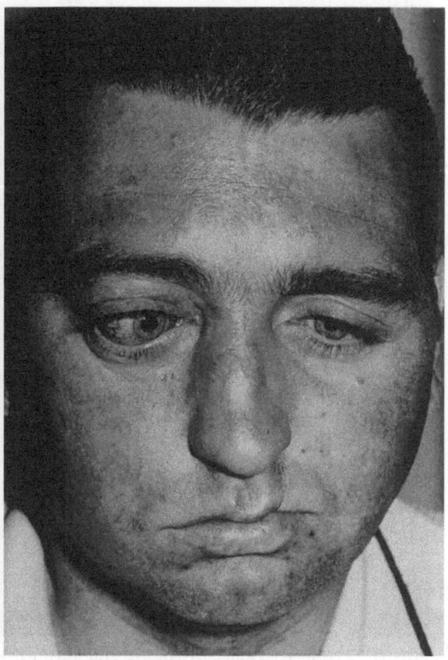 6

Fig. 5. Partial palsy of the third cranial nerve in LeFort III fracture

Fig. 6. Bilateral pulsating exophthalmos with engorgement of the episcleral vessels and paresis of extraocular muscels in traumatic carotid cavernous sinus fistula

temporal region. Retinal, conjunctival, and episcleral veins are engorged, the lids are edematous. Further complications of this fistula are paresis of extraocular muscles, optic atrophy, and secondary glaucoma.

Ophthalmic Signs of Tumorous Paranasal Sinus Diseases

Paranasal sinus tumors involve the orbit in up to 65% of cases by destruction of its bone walls. Squamous and transitional cell, adenoid cystic, and mucoepidermoid carcinomas account for about 80%, while sarcomas are rare (Ducrey 1985; Duke-Elder 1972, 1974; Herrmann 1958; Jakobiec et al. 1986; Wustrow 1965).

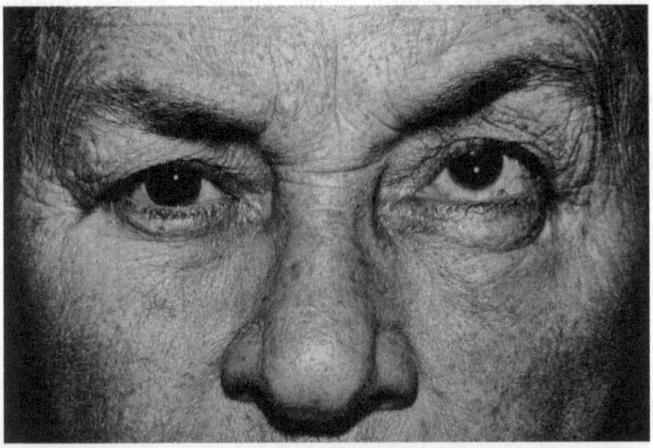

7

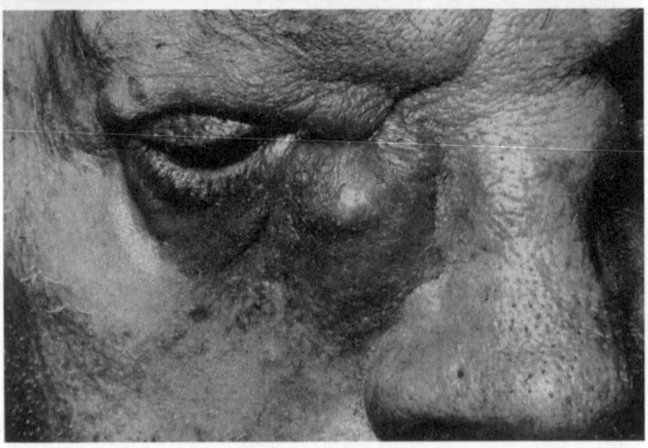

8

Fig. 7. Upward dislocation of the left globe in advanced maxillary sinus carcinoma with orbital involvement

Fig. 8. Chronic bilateral dacryocystitis in advanced carcinoma of the anterior ethmoid with infiltration of the lacrimal drainage system

Table 1. Differential diagnosis of ophthalmological signs and symptoms in paranasal sinus diseases

Periorbital swelling	
Inflammation	Collateral edema
	Periostitis
	Subperiosteal abscess
	Orbital cellulitis
	Lid abscess
	Osteomyelitis
	Cavernous sinus thrombosis
	Pyocele
Trauma	Hematoma (uni-, bilateral)
	Emphysema
	Carotid cavernous sinus fistula
Tumor	Congestive edema
	Mass lesion
Exophthalmos	
Unilateral	
Inflammation	Subperiosteal abscess
	Orbital cellulitis, myositis
	Cavernous sinus thrombosis
	Pyocele
Trauma	Orbital hemorrhage, emphysema
	Carotid cavernous sinus fistula
	Mucocele
Tumor	
Bilateral	
Inflammation	Cavernous sinus thrombosis
Trauma	Carotid cavernous fistula
Tumor	
With dislocation	
Downward	Subperiosteal abscess of roof
	Frontal mucocele
	Subperiosteal hematoma of roof
	Frontal sinus tumor
Outward	Subperiosteal abscess
	Sphenoethmoidal mucocele
	Subperiosteal hematoma
	Sphenoethmoidal tumor
Upward	Maxillary sinus tumor
Pulsating	Carotid cavernous sinus fistula
	Traumatic defect of roof
Painless	Benign paranasal tumor
Painful	
Inflammation	Subperiosteal abscess
	Orbital cellulitis, myositis
	Cavernous sinus thrombosis
	Pyocele
Trauma	Orbital, subperiosteal hematoma
Tumor	

Table 1 (continued)

Motility Disorders		
	Inflammation	Subperiosteal abscess
		Orbital cellulitis, myositis
		Sphenoidal sinus mucocele
		Cavernous sinus thrombosis
	Trauma	Blow-out fracture
		III, IV, VI nerve injury
		Avulsion of the trochlea
		Hematoma of muscle
		Carotid cavernous sinus fistula
	Tumor	Infiltration of muscle
Visual impairment		
	Inflammation	Subperiosteal abscess [b, c]
		Orbital cellulitis [b, c, e]
		Sphenoethmoidal sinusitis [c]
		Cavernous sinus thrombosis [b, e]
		Sphenoidal mucocele [b]
	Trauma	Subperiosteal, orbital, optic nerve sheath hematoma [b, d, f]
		Fracture in optic channel [b, d]
		Carotid cavernous sinus fistula [b]
		Avulsion of optic nerve [d]
		Retinal edema, hemorrhage [a]
	Tumor	Large sinus tumor [b, f]
Ptosis		
	Inflammation	Pseudoptosis
		Frontal, sphenoethmoidal mucocele
	Trauma	Eyelid laceration
		Fracture of orbital roof
		Hematoma of lid
		III nerve palsy
	Tumor	Frontal sinus tumor
		Sphenoethmoidal tumor
Enophthalmos		Blow-out fracture
		Herniation of orbital contents
		Atrophy of orbital fat
Epiphora – dacryocystitis		
	Inflammation	Paranasal sinusitis
		Mucocele
	Trauma	Injury of drainage system
	Tumor	Obstruction, invasion of drainage system
Infraorbital paresthesia		
	Inflammation	Orbital apex syndrome
	Trauma	Blow-out fracture
	Tumor	Maxillary, sphenoethmoidal

[a] Cause: ocular trauma.
[b] Cause: compression of optic nerve (vessels), papilledema, hemorrhage, optic atrophy.
[c] Cause: retrobulbar neuritis.
[d] Cause: injury of optic nerve.
[e] Cause: retinochoroidal infarction.
[f] Cause: retinochoroidal folds.

Tumors of the maxillary sinus cause an upward displacement of the globe (Fig. 7) with or without motility disorders and consecutive diplopia. The visual acuity can decrease. Leading symptoms are infraorbital paresthesia, swelling of the lower lid and the cheek, and epiphora and chronic dacryocystitis, which most often is misdiagnosed as a primary inflammation of the lacrimal drainage system.

Tumors of the anterior ethmoid induce swelling of the medial canthal region and a dislocation of the globe laterally without gross restriction of motility, unless – in advanced stages – the medial rectus muscle is infiltrated. Epiphora and chronic dacryocystitis are, again, common signs (Fig. 8).

Sphenoethmoidal tumors early cause a proptosis and by compression of the optic nerve a deterioration of visual acuity. Ophthalmoscopy reveals a papilledema and later an optic atrophy. Motility disorders are the result of a compressive orbital apex syndrome.

In frontal sinus tumors with orbital involvement a dislocation of the globe downward and outward is characteristic.

Differential Diagnosis in Ophthalmic Signs in Paranasal Sinus Diseases

The variety of inflammatory, traumatic, and tumorous diseases of the frontal, ethmoidal, sphenoidal, and maxillary sinuses is accompanied by some major ophthalmic signs and symptoms, such as periorbital swelling, exophthalmos with or without dislocation of the globe, motility disorders, visual impairment, ptosis, enophthalmos, epiphora, chronic dacryocystitis, and infraorbital paresthesia (Pau 1986; Rochels 1987; Roy 1984). But it must be stressed that none of these is at all pathognomonic for one specific orbital or ocular complication of any paranasal sinus disease. They are merely local manifestations of an ocular, orbital, or sinus disease. There are, of course, some combinations of symptoms that might verify a specific diagnosis. Nevertheless, each of the above mentioned ophthalmic signs can have a variety of causes (see Table 1), and the ophthalmologist must consider and check each one systematically and thoroughly. An interdisciplinary diagnostic approach is almost always mandatory to shorten the sometimes long route from ophthalmic signs to a prompt diagnosis and thus an appropriate therapy.

References

Ducrey N (1985) Les affections orbitaires non traumatiques. Masson, Paris, pp 164–182

Duke-Elder SS (1972, 1974) System of ophthalmology. Kimpton, London, vol 13, part 2, pp 811–825, 866–889, 1167–1204; vol 14, part 1, pp 212–310

Ganz H (1977) Komplikationen der unspezifischen Nasen- und Nebenhöhlenentzündungen. In: Berendes J, Link R, Zöllner F (Hrsg) Hals-Nasen-Ohrenheilkunde in Praxis und Klinik, 2. Aufl., Bd 1. Thieme, Stuttgart, S 1–16

Herrmann R (1958) Die rhinogenen Erkrankungen der Orbita. Thieme, Stuttgart

Jakobiec FA, Jones AS (1986) Orbital inflammations. In: Duane TD, Jaeger EA (eds) Clinical ophthalmology, vol 2. Harper & Row, Philadelphia, pp 50–56

Jakobiec FA, Rootman J, Jones IS (1986) Secondary and metastatic tumors of the orbit. In: Duane TD, Jaeger EA (eds) Clinical ophthalmology, vol 2. Harper & Row, Philadelphia, pp 35–43

Pau H (1986) Differentialdiagnose der Augenkrankheiten, 2. Aufl. Thieme, Stuttgart

Rochels R (1987) Rationelle Diagnostik beim Leitsymptom "Exophthalmus". In: Lund OE, Waubke TN (Hrsg) Okuläre Symptome – Strategien der Untersuchung. Enke, Stuttgart, S 221–234

Roy FH (1984) Ocular differential diagnosis, 3rd ed. Lea & Febiger, Philadelphia

Seiferth LB, Wustrow F (1977) Verletzungen im Bereich der Nase, des Mittelgesichts und seiner Nebenhöhlen sowie frontobasale Verletzungen. In: Berendes J, Link R, Zöllner F (Hrsg) Hals-Nasen-Ohrenheilkunde in Praxis und Klinik, 2. Aufl, Bd 1. Thieme, Stuttgart, S 7–41

Siegert P (1975) Erkrankungen der Orbita. In: Velhagen K (Hrsg) Der Augenarzt, 2. Aufl, Bd 3. Thieme, Leipzig, S 422–449, 527–543

Smith B, Grove A, Guibor P (1986) Fractures of the orbit. In: Duane TD, Jaeger EA (eds) Clinical ophthalmology, vol 2. Harper & Row, pp 1–10

Wustrow F (1965) Die Tumoren des Gesichtsschädels. Topographie, Pathologie und Klinik. Urban & Schwarzenberg, München, S 7–32, 201–229

Imaging of the Paranasal Sinuses

B. DESCHEPPER, Amiens/France and J. VIGNAUD, Paris/France

Over the past decade new techniques of imaging have progressively modified the strategy of imaging of the paranasal sinuses. Computed tomography (CT) allows a very sophisticated analysis of all the bony components and the adjacent soft tissue due to its very good spatial and density resolution. More recently, magnetic resonance imaging has complemented CT, allowing better tissue characterization of a lesion, and in most of the cases allowing the differentiation of tumoral lesion from inflammatory process. CT definitely replaces conventional tomography and may totally replace the usual plain film study, which even with four projections is often very difficult to interpret, mainly due to the varying shape and thickness of the sinuses. CT is much more precise, more informative, easy to perform, and easy to interpret.

Computed Tomography

Direct sections require two orthogonal planes: the axial plane (orbitomeatal (OM) $= -10°$; Fig. 1) and the coronal plane (Fig. 2), theoretically perpendicular to the axial one but actually depending on the presence of teeth prothesis (which must be a avoided, to prevent artifacts). The sections are scanned directly on a sagittal "plane scan". A plane of section at OM $= -40°$ is sometimes necessary when the frontonasal canal is to be visualized. The sections must be thin (2–5 mm) and performed at every 8 mm or less, depending on the size of the lesion. Two sets of windows are necessary for each section: one for the bone and the other one for the soft tissue. Injection is indicated in cases of tumors and vascular malformations.

Multiplanar reconstruction (Fig. 3) allows sections according any plane, including sagittal and oblique ones. The resolution depends on the thickness of the sections. At least 3 mm thickness and joined sections are requested.

Three-dimensional images (Fig. 4) allow visualization of the head and face with a pseudorelief due to a shadowing of the structures. It is possible, with a sophisticated program, to have the head turned in any direction. Using a gray or color scale, it is possible to visualize the normal or abnormal soft tissue and to add or remove it from the skeleton. This requires thin (2 mm) and joined sections of the whole head. The result is fascinating and is useful in planning surgery of facial reconstruction. However, many pitfalls exist, mainly false fractures and defects.

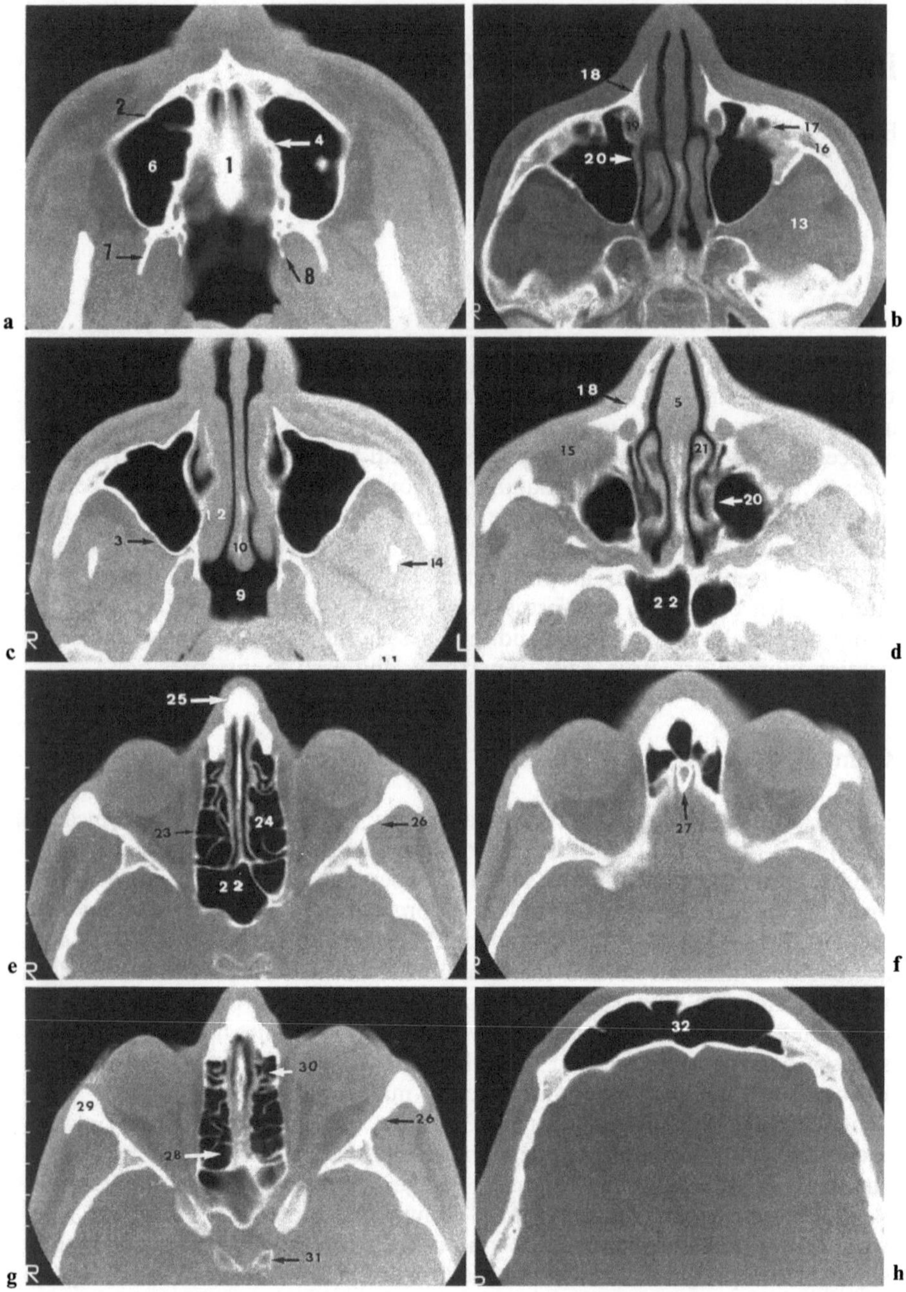

Fig. 1 a–h. Axial view of paranasal sinuses. *1*, Hard palate; *2*, anterior wall maxillary antrum; *3*, posterior wall maxillary antrum; *4*, medial wall maxillary antrum; *5*, septal cartilage; *6*, maxillary antrum; *7*, lateral pterygoid plate; *8*, medial pterygoid plate; *9*, nasopharynx; *10*, bony septum; *11*, mandibular head; *12*, inferior concha; *13*, pterygopalatine fossa; *14*, coronoid process of the mandible; *15*, orbit; *16*, zygomatic process; *17*, infra-orbital nerve canal; *18*, frontal process of maxillary; *19*, lacrymonasal duct; *20*, maxillary ostium; *21*, middle concha; *22*, sphenoid sinus; *23*, lamina papyracea; *24*, ethmoid sinus; *25*, nasal bone; *26*, lateral wall of orbit; *27*, crista galli; *28*, posterior cells of ethmoid sinus; *29*, zygomatic bone; *30*, anterior cells of ethmoid sinus; *31*, dorsum sellae; *32*, frontal sinus

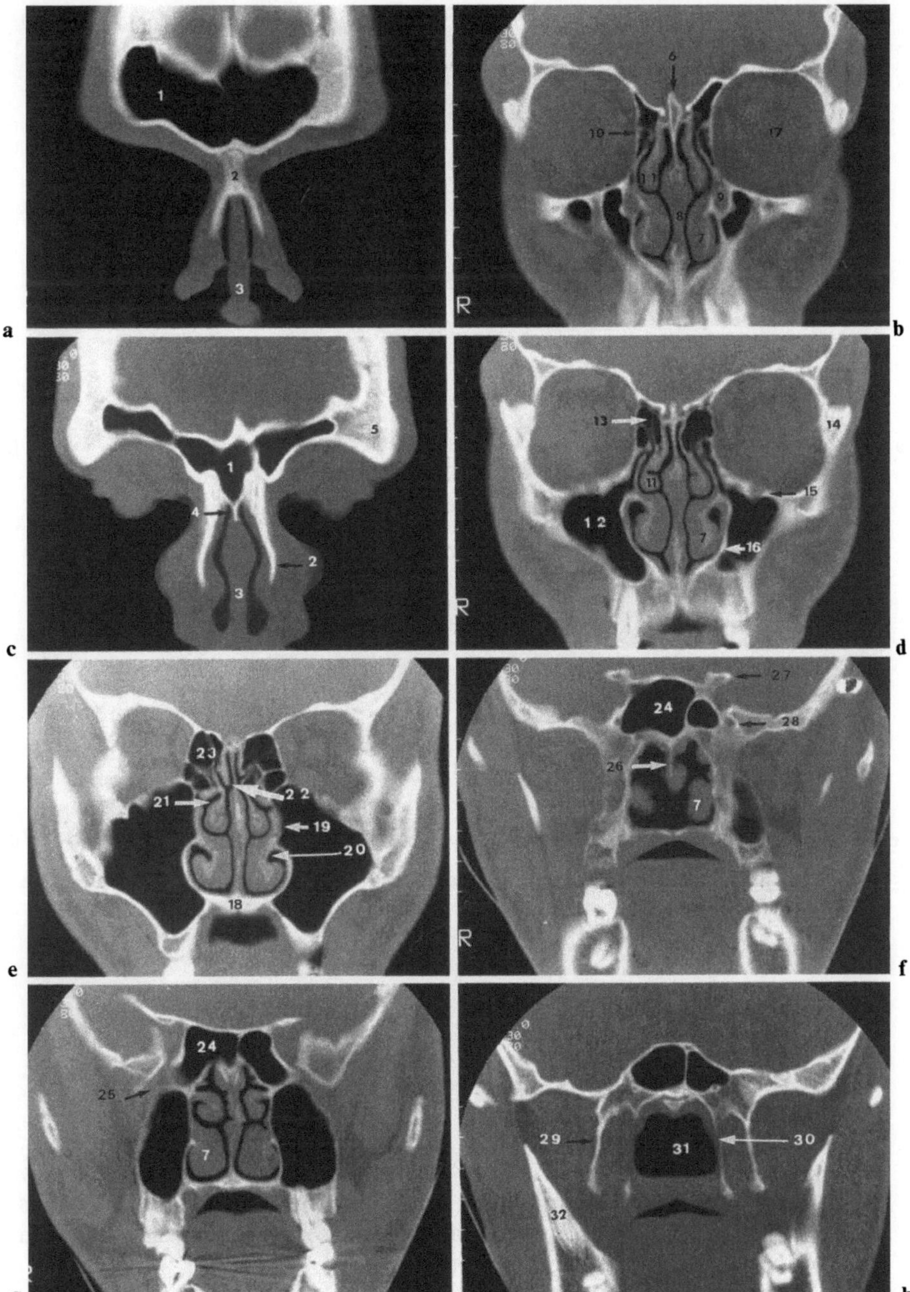

Fig. 2a–h. Coronal view of paranasal sinuses. *1*, Frontal sinus; *2*, nasal bone; *3*, nasal cartilage; *4*, nasal spine of frontal bone; *5*, frontal sinus; *6*, crista galli apophysis; *7*, inferior concha; *8*, nasal septum; *9*, lacrymonasal duct; *10*, lacrymal bone; *11*, middle concha; *12*, maxillary antrum; *13*, anterior ethmoid cells; *14*, zygomatic process of frontal nerve; *15*, infra-orbital nerve canal; *16*, medial maxillary bone; *17*, orbit; *18*, hard palate; *19*, uncinatus process; *20*, inferior meatus; *21*, middle meatus; *22*, superior concha; *23*, middle ethmoid cells; *24*, sphenoid sinus; *25*, inferior orbital fissure; *26*, vomer; *27*, anterior clinoid; *28*, superior orbital fissure; *29*, lateral pterygoid process; *30*, medial pterygoid process; *31*, nasopharynx; *32*, mandible

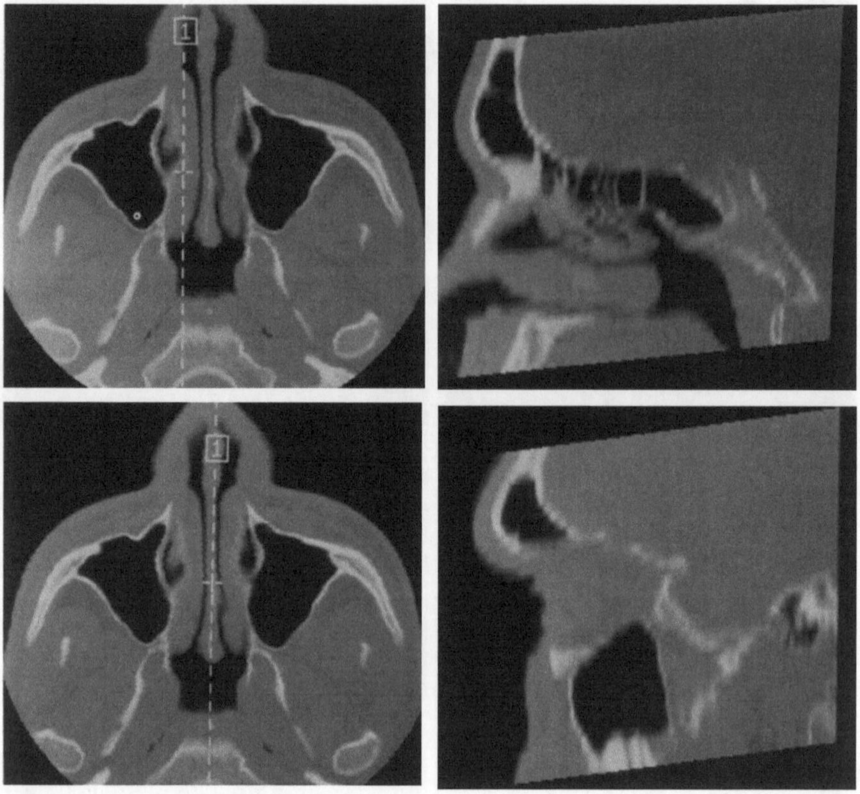

a b c d

Fig. 3a–d. Reformated images of a normal face in sagittal (**b**) and parasagittal plane (**d**), from a set of axial sections

Magnetic Resonance Imaging

This technique complements the CT scan in cases of pathological findings. T_1W sequences are performed in axial and frontal plane. Lateral sections may be performed in addition, depending on the location of the lesion. T_2Wi is performed in axial or frontal plane. Special blood flow sequences are useful in cases of hypervascularized abnormalities.

It is not the purpose of this chapter to give a survey of the pathology. With some examples of tumoral (Figs. 5–7) and traumatic (Fig. 8) cases, we should like to emphasize the interest of modern imaging.

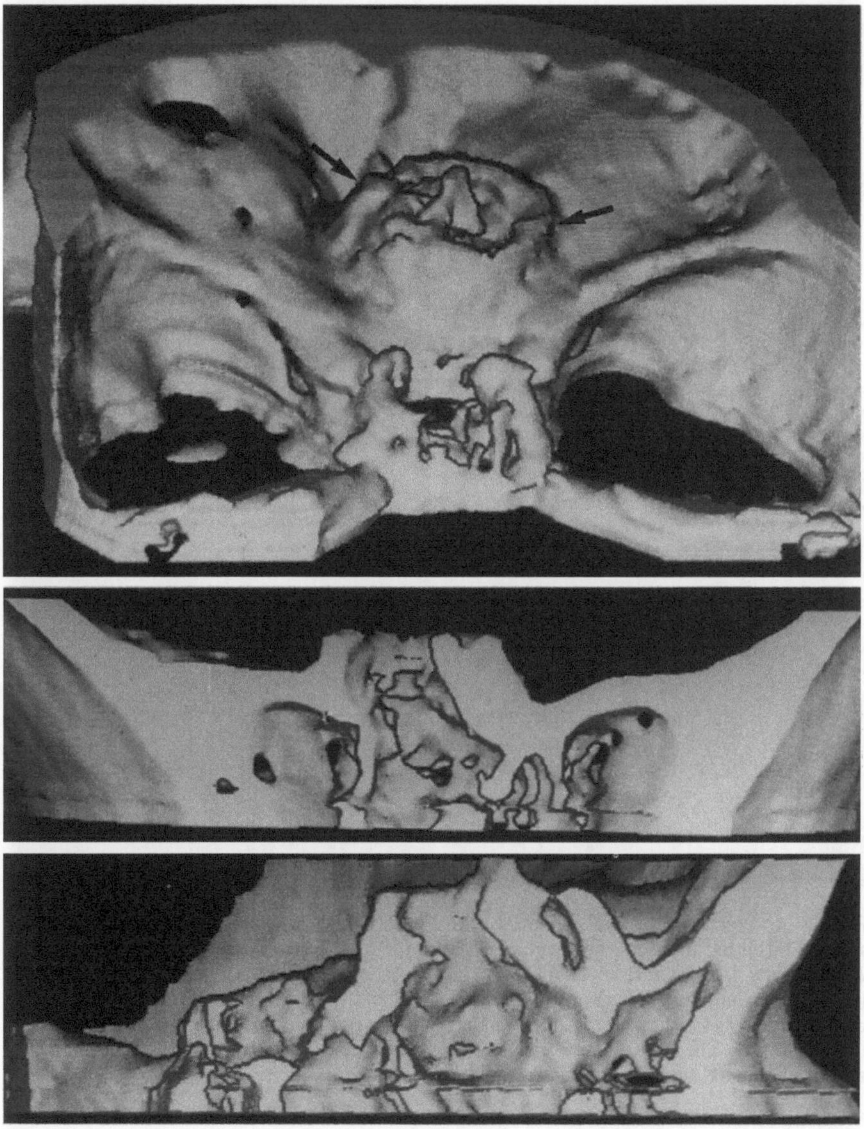

a

b

c

Fig. 4a–c. Three-dimensional images. Calcified ethmoid tumor, seen from above (**a, b**) and from the interior after "removal" of the anterior part of the face (courtesy of P. Meriot C.H.U. Service Radiologie, Brest, France)

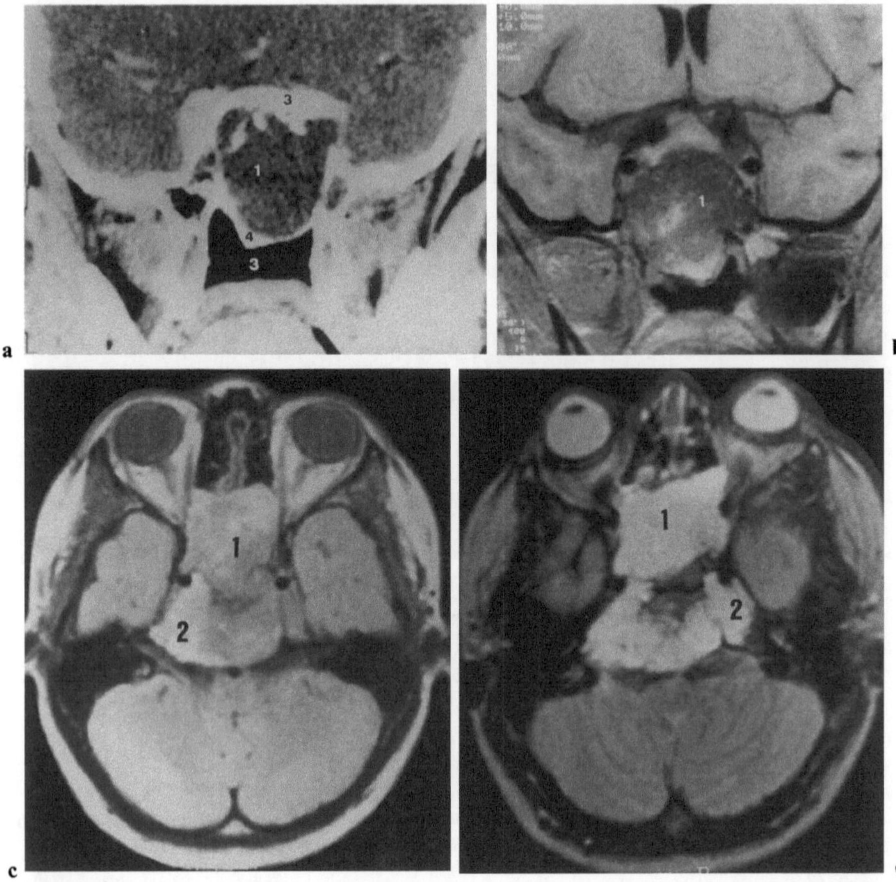

Fig. 5 a–d. Mucocele of the sphenoid sinus. **a** CT + IV injection, coronal view. Huge sphenoidal mass thinning and displacing the walls of the sinuses (*3*, *4*). No enhancement. *1*, Mucocele; *2*, nasopharynx. **b–d** MRI. **b** T_1 Wi, coronal section. The mass is well outlined, less intense than parenchyma. Relationship with the compressed cavernous sinus is well demonstrated as well as upward displacement of the hypophysis. **c, d** Proton density and T_2Wi in axial section. The mass (*1*) is very well outlined. There is an invasion of both petrous apex (*2*) and a high signal intensity in T_2Wi

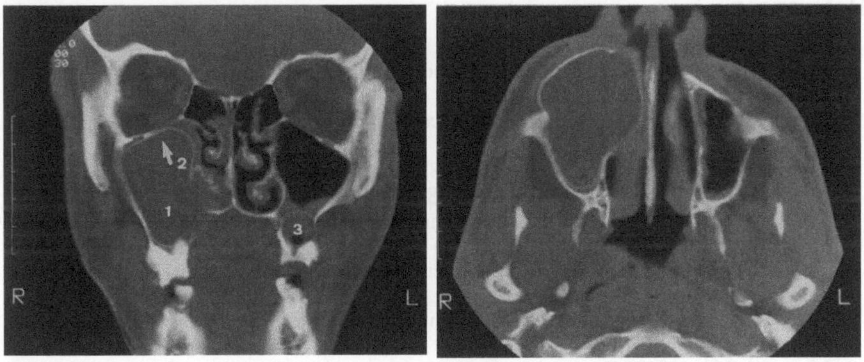

a b

Fig. 6. C.T. coronal (**a**) and axial (**b**) section. Bilateral periradicular dental cyst. Inflammatory granuloma induces a cystic process (*1*) with invagination of the floor of the maxillary sinus within its own cavity (*2*). On the left there is a smaller cyst (*3*)

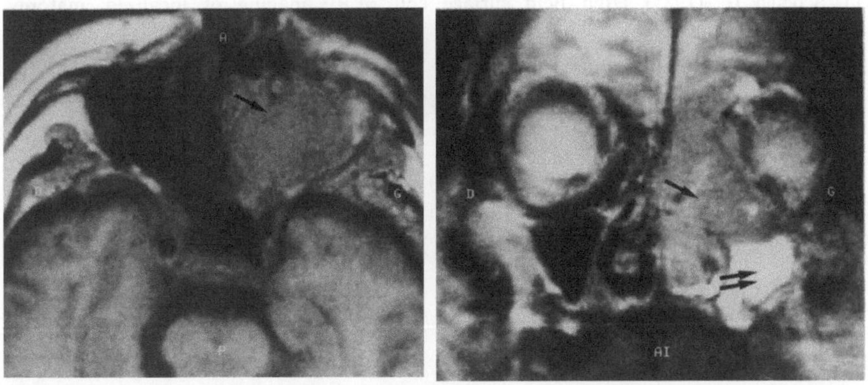

a b

Fig. 7 a, b. Adenocarcinoma of the ethmoid, maxillary, and sphenoid sinuses. **a** T_1Wi (axial view); the mass is homogenous (→). **b** T_2Wi (AP view) in the maxillary sinus. There are two components: the tumor (low signal; →) and retention inflammatory lesion (high signal; ⇉)

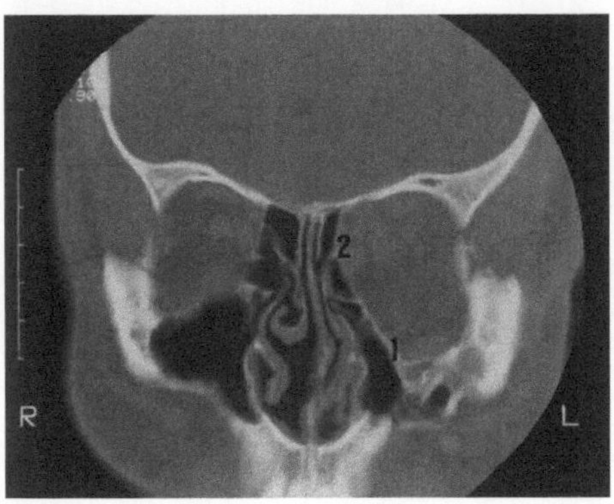

Fig. 8. CT coronal section. Blow-out fracture: fracture – displacement of the floor (*1*) and inner wall (*2*) of the left orbit

References

1. Hasso AN (1986) Radioanatomie tomodensitométrique normale du massif facial. Radiologie J CEPUR 6:215–228
2. Boulin A, Vignaud J (1984) Anatomie scanographique du massif facial. Feuillets de Radiologie, 24, n°6, 369–380
3. Bourjat P, Kahn JL, Roy C, Veillon F (1988) Tomodensitométrie cervicofaciale. Masson
4. Bouchet A, Cuilleret J (1985) Anatomie. SIMEP, Tome 1
5. Daireaux P, Garcia-Torres E, Doyon D, Piekarski JD, Halimi P (1987) I.R.M. du massif facial. Radiologie J CEPUR
6. Doyon D, Halimi P, Desprez A, Iba-Zizen MT, Cabanis EA (1984) I.R.M. du massif facial. J Biophys Med Nucl 2–3:51–52
7. Halimi P, Doyon D, Madoule P, Iba-Zizen MT, Lopez A (1984) I.R.M. du massif facial. J Neuroradiol 11:274–284
8. Margules AR (1983) L'Imagerie par résonance magnétique nucléaire en 83. J Imag Méd 2:149–153
9. Meschan I (1975) An atlas of anatomy basic to radiology. Saunders
10. Piekarski JD, Masselot J (1984) Les tumeurs malignes du massif facial. Encyclopédie médico-chirurgicale Paris, Radiodiagnostic II, 31675, A50, 4
11. Schenck JF (1985) High field surface coil M.R. imaging of localized anatomy. AJNR 6:181–186

Influence of Sphenoidal Sinus Size on the Sella Angle in Lateral Cephalograms

R. Anthofer, C. Krenkel, and H. Matras, Salzburg/Austria

Orthodontic and surgical management of malocclusion requires a detailed and careful analysis of the underlying dental and facial deformities. Lateral cephalograms provide necessary information on three relationships important for planning orthodontic-surgical treatment: that between the dental axes and the alveolar processes in the jaws, that between the maxillary and mandibular bases, and that between the jaw base and the skull base. As the facial profile is determined mainly by the position of the maxilla and mandible relative to the skull base, a full understanding of the malposition is imperative for restoring normal anatomical relations.

In general, the position of the maxilla and mandible relative to the skull base is defined by two angles: SNA from the mid-sella point through the bony nasion to the A point and SNB from the mid-sella point through the bony nasion to the B point (Fig. 1).

However, cephalometric analyses have repeatedly shown that the SNA and SNB angles poorly or incorrectly reflect the type of malocclusion identified by all other parameters evaluated. On detailed analysis of these cephalometric films the sphenoid sinus was found to be either excessively or poorly pneumatized and the sella angle to be either too large or too small. These observations suggest that the position of the sella point relative to the skull base varies as a function of the pneumatization of the sphenoid, just as the orbital lamellae may be separated by a large orbital recess in the frontal sinus.

To test this hypothesis, 100 cephalometric films from our maxillofacial surgery patients were reviewed. Of these, 95 were reanalyzed. To establish the relationship between the sella point and the skull base, a baseline connecting the nasion with the basion was defined and the sella-nasion-basion angle and the perpendicular distance of the sella point from the baseline were determined (Fig. 2). In the cephalograms studied the SNBa angle averaged 19.1°, while the mean sella-baseline distance was 24.3 mm (Figs. 3, 4). Allowing for the standard deviations of 2.1 and 2.9, an SNBa angle of $19° \pm 1°$ and a sella-base distance of 24 ± 2 mm can thus be regarded as normal. There was no relationship to age or sex. The cephalograms reevaluated were all taken from patients aged over 15 years.

On the basis of these data, all cephalometric films showing sella angles that deviated from normal were reexamined for the relation between the sella point and the baseline. Of 33 films with abnormal sella angles, 23 were found to show SNBa angles that were either too large or too small. In addition, the distance of

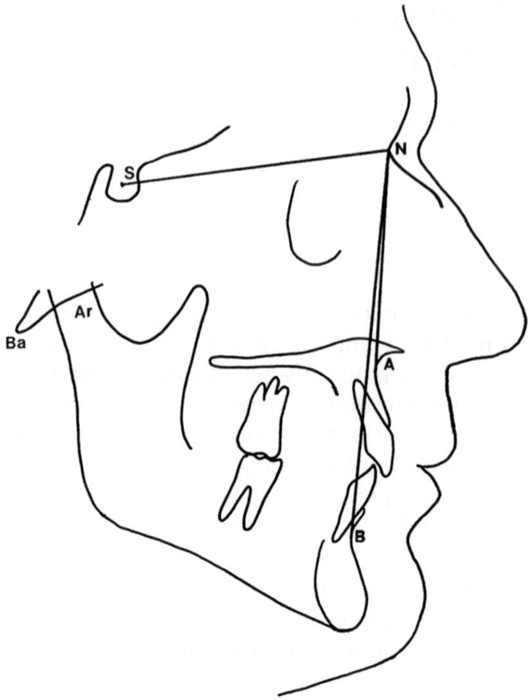

Fig. 1. Template taken from a lateral cephalogram showing the S-N-A angle from the mid-sella point (*S*) through the bony nasion (*N*) to the A point (*A*) and the S-N-B angle from the mid-sella point through the bony nasion to the B point (*B*), as well as the sella angle from the articular (*Ar*) through the mid-sella point to the bony nasion. *Ba*, basion

the sella point from the baseline was abnormal on 21 films. All except one of these patients had malocclusion angle class II or III with abnormal dentoalveolar or maxillomandibular base relationships. A comparison of the SNA and SNB angles found with the amount of surgical correction required to restore an average profile and to produce normal occlusion showed that the angles found incorrectly reflected the relation of the jaw bases to the base of the skull in seven patients. Three of these had maxillary retrusion with normal SNA angles. In two the SNA angles were normal inspite of maxillary protrusion. Of two patients with orthognathic relations, one had a too small and the other a too large SNA angle. When the position of the sella point was corrected for a normal sella-nasion-basion angle of 19° and the new sella point plotted in the vertex of the 123° angle between the articular and the retraced sella-nasion line, the resulting SNA and SNB angles correctly reflected the underlying malposition of the jaw bases in all cases (Fig. 5). This enabled us to correct the relationship between the maxillary base and the mandibular base to

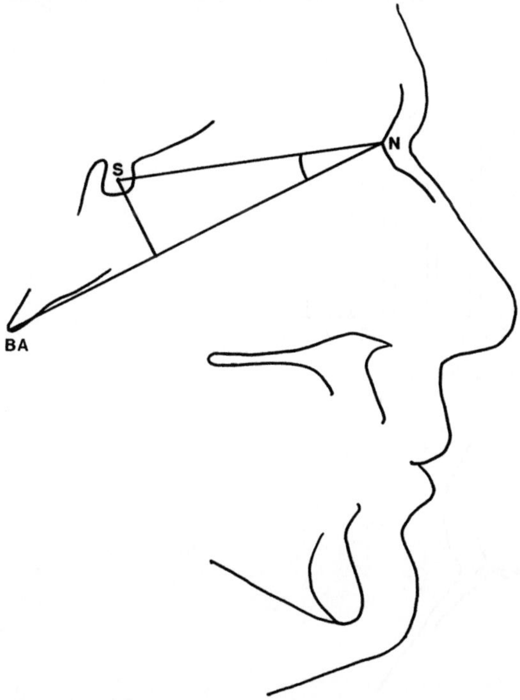

Fig. 2. The S-N-Ba angle from the mid-sella point (*S*) through the bony nasion (*N*) to the basion (*Ba*) and the perpendicular distance of the S point to the base line N-Ba

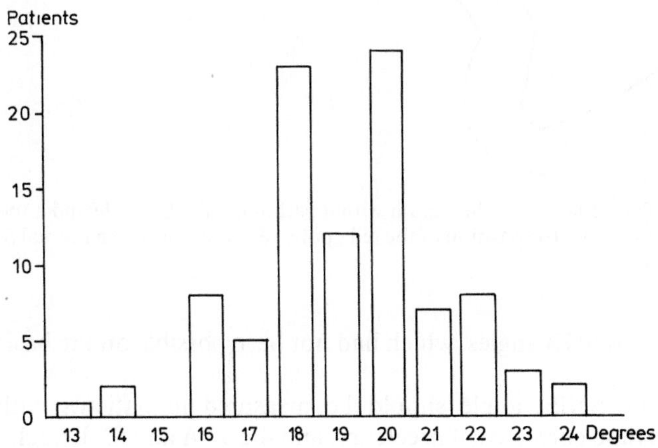

Fig. 3. The distribution of S-N-Ba angles in the 95 cases of our investigation

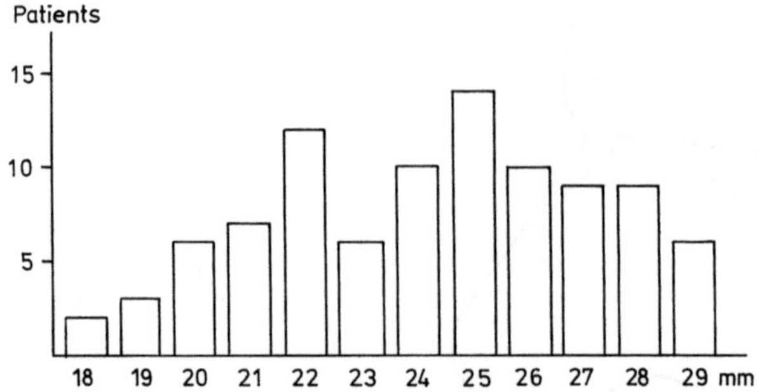

Fig. 4. The distribution of sella-base-line distance

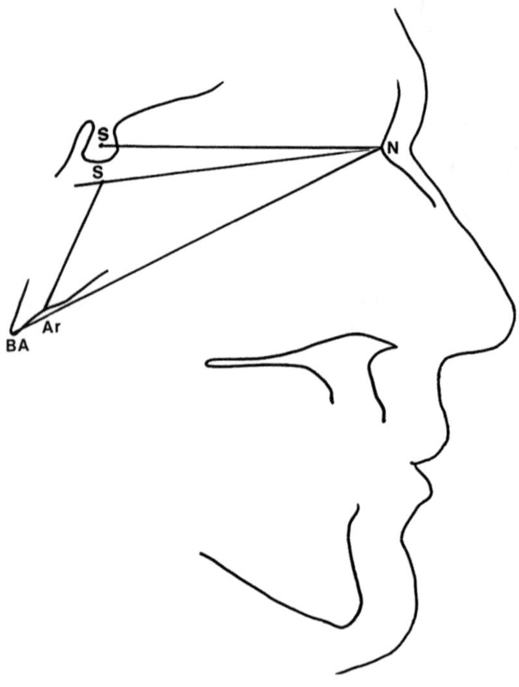

Fig. 5. Template taken from a lateral cephalogram with a pathological sella angle and a too large S-N-Ba angle and the corrected position of the sella point (*S'*) after forming a normal S-N-Ba angle of 19° (*S'-N-Ba*)

within normal SNA and SNB angles, which had not been possible on the basis of the original SN line.

In conclusion, the SNBa angle should be measured in patients with abnormal sella angles and corrected, if necessary, and the SNA and SNB angles should be redetermined on the basis of the corrected position of the sella point.

The Sphenoid Sinus: Typical and Variable Morphologic Pattern Demonstrated by High-Resolution and Multiplanar CT

J.P. Haas, G. Kahle, and W. Draf, Fulda/FRG

Pneumatization of the sphenoid bone starts about the 4th year of life. The body of the sphenoid bone of a 4-year-old girl (not illustrated) shows the beginning of a lowering in density – a sphenoid sinus in statu nascendi; a 10-year-old boy (not illustrated) presents an extensive pneumatization of sphenoid bone, extending to the spheno-occipital synchondrosis. Extensive pneumatization of the sphenoid bone may extend to the clivus and aerate the dorsum sellae. Spreading of pneumatization into the pterygoid process forms a pterygoid recess, and pneumatization of the sphenoidal crest leads to a recess of vomer. Configuration and craniocaudal extension of the sphenoid sinus are illustrated in coronal and sagittal reconstructions. The coronal reconstruction shown in Fig. 1 shows a well-pneumatized sphenoid sinus. The section runs through anterior clinoid processes, foramen rotundum, vomer, and pterygoid processes.

Drainage of the sphenoid sinus is to the sphenoethmoid recess of the nasal cavity. This recess can be demonstrated in axial scans, semicoronal scans, and coronal reconstructions. The important relationships between the nasopharynx border below the sphenoid sinus and the hypophysis within the sella above are best shown in multiplanar reconstructions. Coronal reconstructions

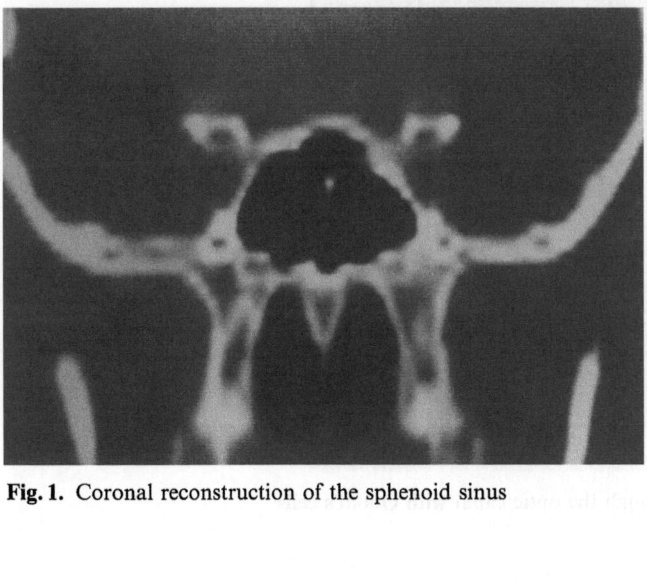

Fig. 1. Coronal reconstruction of the sphenoid sinus

demonstrate the roof of the sphenoid sinus, forming the sellar bottom, and the posterior clinoid process and the pterygoid-vidian canals.

The relationship to the nasopharynx becomes evident in complementary coronal and sagittal (Fig. 2) reconstructions. Posteriorly the clivus forms the border to the posterior cranial fossa. Rostrally the sphenoid sinus is linked with the ethmoid cell system. In this border range the optic nerve runs through the optic canal to the orbit. The medial wall of the optic canal is formed by the sphenoid sinus in 90% of cases. This is demonstrated in transverse cuts and in paraxial reconstructions.

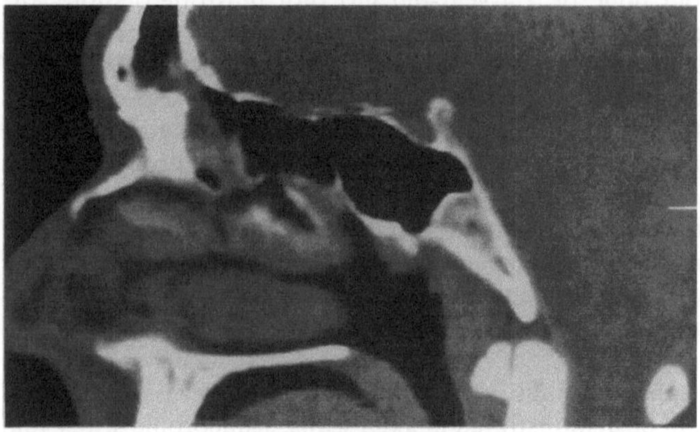

Fig. 2. Sagittal reconstruction of the sphenoid sinus

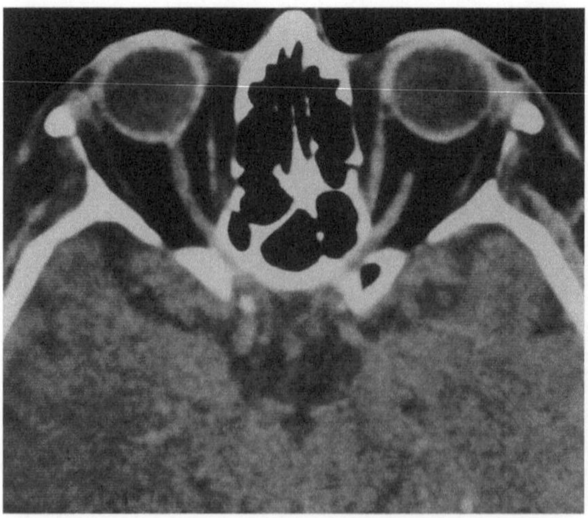

Fig. 3. Axial scan through the optic canal with Onodi's cells

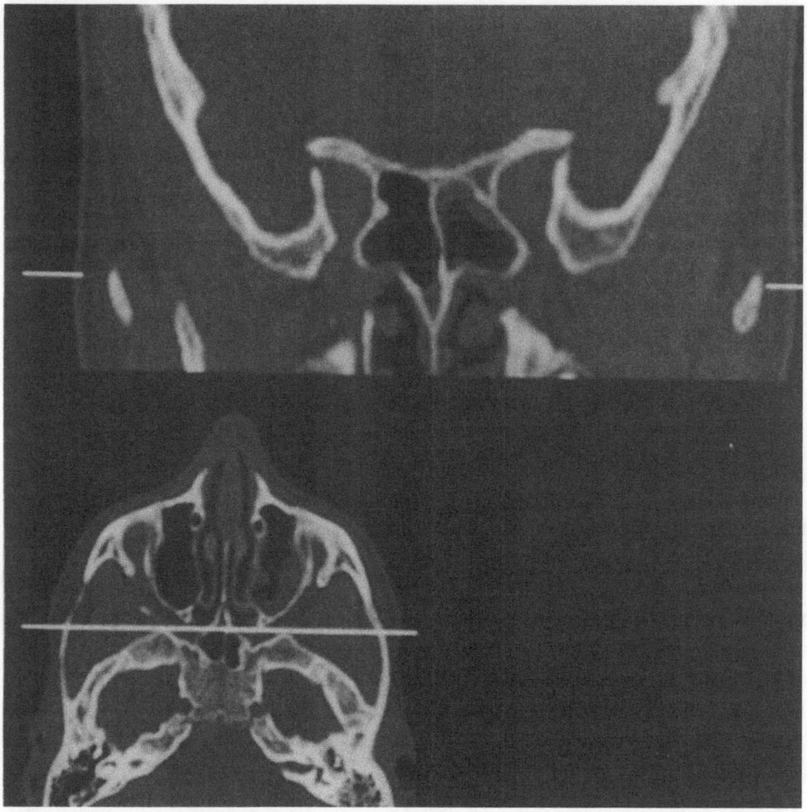

Fig. 4. Coronal reconstruction of the sphenoid sinus with the pterygopalatine fossa

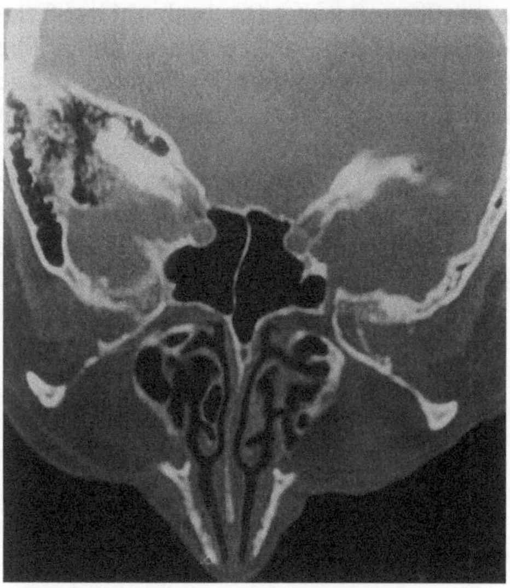

Fig. 5. Axial scan through the sphenoid sinus with the bulge of the internal carotid artery

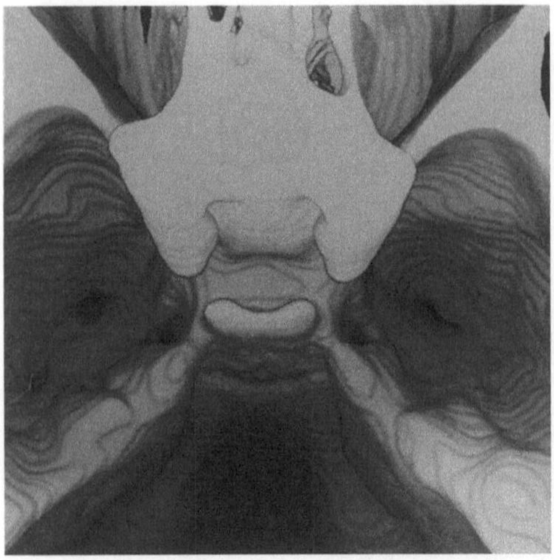

Fig. 6. Three-dimensional reconstruction of the sphenoid bone, view of the sellar bottom in the center and middle cranial fossa bilaterally, posterior fossa, and orbits

In cases of small sphenoid sinuses, posterior ethmoid cells grow into the medial wall of the optic canal (Fig. 3), the so-called Onodi's or Grünwald's cells. As shown in coronal reconstructions, the superior orbital fissure is also confined by the sphenoid sinus. The pterygopalatine fossa is of great importance, securing neural and vasal connections between skull and face. Its medial and cranial boundaries are formed by the sphenoid sinus. Axial scans together with sagittal and coronal (Fig. 4) reconstructions show the pathways: to the orbit via the inferior orbital fissure, to the middle cranial fossa via the foramen rotundum, to the nasal cavity via the sphenopalatine foramen, and to the temporal fossa via the pterygomaxillary fissure. On both sides of the sphenoid sinus runs the internal carotid artery within the cavernous sinus. The first postpetrosal part of the carotid artery ascends in close contact to the sphenoid sinus and thus bulging into it (Fig. 5). This large vessel is often covered by only a thin bony layer, sometimes without any bony cover, the so-called unprotected carotid. The carotid knee can also protrude into the sphenoid sinus. Due to the small sagittal diameter of the sphenoid sinus the bulge of the carotid artery may come into close contact with the anterior wall of the sinus. Three-dimensional reconstructions underline the central position of the sphenoid sinus in the skull base, depicting the inner and outer (Fig. 6) relief of the sinus.

Interventional Angiography

D. KÜHNE and D. DE SILVA, Essen/FRG

Neuroradiological intervention in the region of the paranasal sinuses is basically similar to intervention in other areas of the head and neck. The vessels involved are the external and internal carotid arteries. Great strides forward have been made in the past few years due to the development of new embolic materials and new catheter systems, with new microcatheters as fine as 1 F in diameter and combined with detachable balloons (Debrun 1983). The superselective catheterization enables the operator to place the catheter tip very near the pathological lesion and to deliver the emboli accurately without compromising adjacent healthy tissues. The indications for intervention have thus been widened and the complication rate reduced.

The basic types of embolic materials are: particles (Lyodura and Gelfoam, polyvinyl alcohol particles, angiostat collagen), coils and springs, fluids (bucrylate, Ethibloc), and releasable balloons. When only a temporary occlusion is required, Gelfoam (Bank and Kerber 1979; Horton et al. 1983) or lyophilized dura may be used. Recently the use of polyvinyl alcohol granules (Ivalon; Horton et al. 1983; Latchaw and Gold 1979) ranging in size from 125 to 1000 µm has become increasingly popular due to the ease of its handling. The semisolid and liquid embolic materials, which are satisfactorily tolerated by the soft tissues are Ethibloc, an alcohol-based maize protein, and Collagen Angiostat. When large-caliber emboli are required, releasable balloons of various systems may be used.

An essential requirement for interventional neuroradiology is digital substraction angiography (DSA; Zeumer 1987). This system enables vessels as small as 1 mm in diameter to be visualized with low-concentration contrast materials. During embolization, DSA enables the progress to be evaluated without loss of time and using only small volumes and low concentrations of contrast agents. This is especially so in the region of the skull base and in that of the paranasal sinuses, where elimination of disturbing bony structures can be accomplished. Technical decisions can thus be made quickly during the examination, increasing the safety and accuracy of the intervention. Embolization of lesions in the region of supply of the external carotid artery (ECA) requires precise angiographic analysis. Special attention must be paid to the presence of collaterals not only between the branches of the ECA and the internal carotid artery (ICA) but also between the ECA branches and the vertebral artery branches (Lasjaunias and Berenstein 1987 a, b).

Preoperative embolization in the ECA region is indicated in highly vascular postnasal tumors, especially juvenile angiofibroma (Fig. 1), in the total therapeutic embolization of arteriovenous malformations, in severe recurrent epistaxis, and less commonly as a palliative, such as in inoperable vascular malignant tumors. Juvenile angiofibromas can develop into massive tumors, grow out of the postnasal space and into the paranasal sinuses, and invade the skull base (Fig. 1). The blood supply is mainly from branches of the internal maxillary artery, but in the larger ones dural branches of the ICA branches can also be involved. These tumors are very favorably embolizable preoperatively with Ivalon particles (Horton et al. 1983; Latchaw and Gold 1979). These particles pack in like logs in a river occluding the vessels. To cap the procedure and precipitate thrombosis the stem of the internal maxillary artery is embolized with a large piece of resorbable gelfoam (Bank and Kerber 1979). After embolization by this regime it is usually possible to resect the tumor without major loss of blood.

The indications for and planning of embolization procedures depend on individual features and on the requirements of the surgeons. Their management requires cooperation between the interventionalist and the surgeon. The operation must take place not later than 4 days after embolization due to the possibility of collateral revascularization.

Severe and recurrent epistaxis is an indication for immediate angiography. Thereafter, when indicated, intervention may follow. We encountered one case with an arteriovenous fistula associated with the Osler-Rendu-Weber syndrome and another with a massive angiomatous malformation involving the

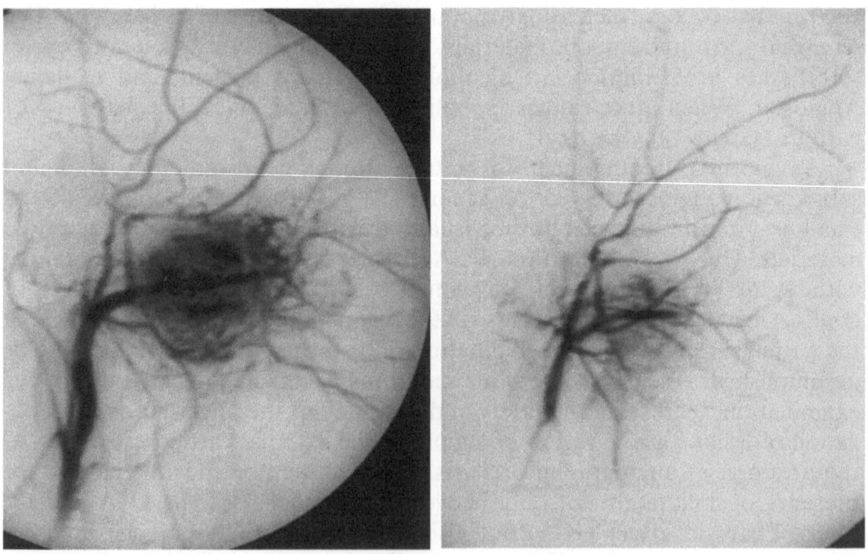

a b

Fig. 1 a, b. Juvenile angiofibroma in a 12-year-old boy. Internal maxillary artery before (**a**) and after (**b**) embolization (tumor almost totally devascularized)

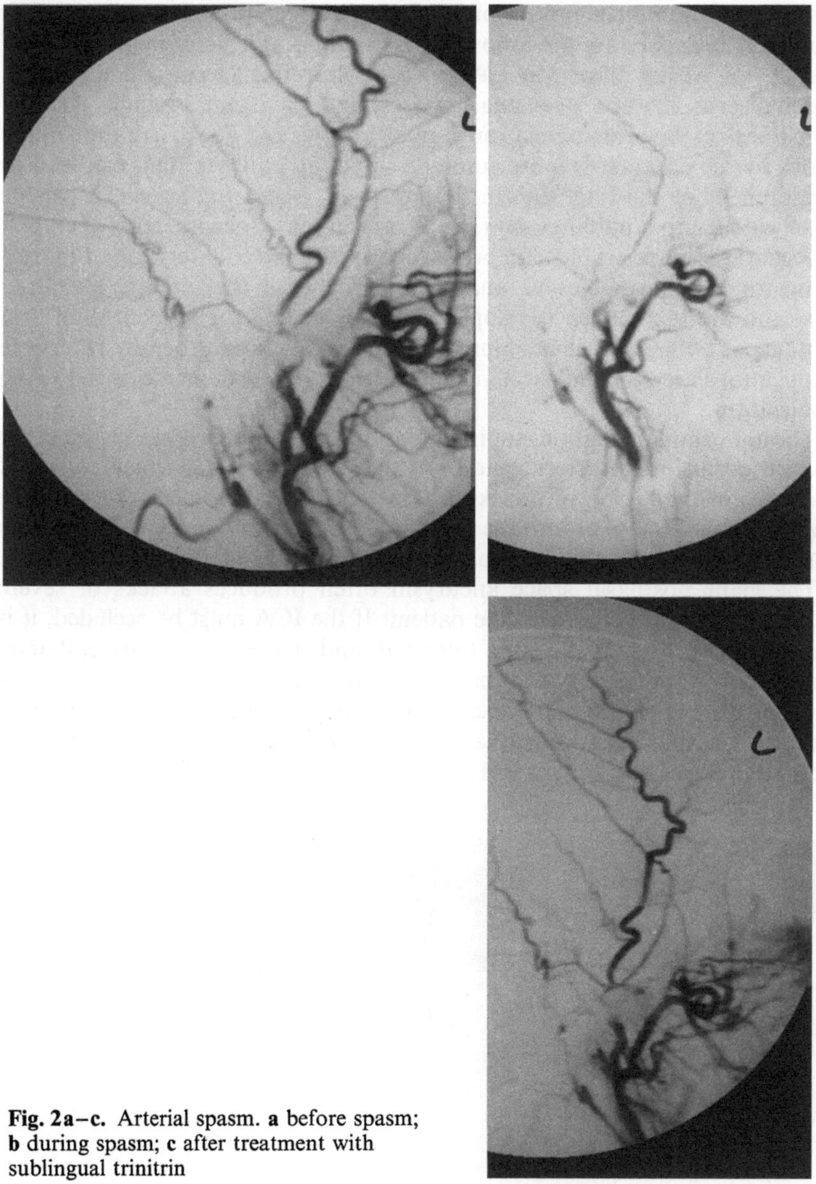

a b c

Fig. 2a–c. Arterial spasm. **a** before spasm; **b** during spasm; **c** after treatment with sublingual trinitrin

upper and lower jaws, associated with the Weyborn-Mason syndrome, in which severe nasal bleeding occurred. The latter case was satisfactorily managed by repeated embolization of the internal maxillary and mandibular artery branches.

The most common complications and dangers of embolization in the region of the ECA branches are the following (Lasjaunias and Berenstein 1987a, b): vessel spasm, vessel dissection, catheter emboli, reflux of embolic materials, accidental embolization over anastamoses, and pain and swelling. Arterial spasm due to catheter manipulation in small arteries and due to irritation of the arteries by the contrast is more common in young patients. This can lead to abandonment of the intervention. If the spasm is not too severe, it can be treated with trinitrin sublingually (Fig. 2). When spasm occurs reflux of emboli may occur into undesirable territories, such as into the ICA territory. Figure 3 demonstrates such dangerous anastomoses between the internal maxillary artery and the ICA which developed suddenly during the embolization of a spontaneous soft tissue hemangioma. Less dramatic and generally reversible complications are pain and swelling and stiffness of jaw movements following embolization.

The important indications for interventional procedures in the region of the ICA are occlusion of arteriovenous fistulas and the management of giant aneurysms in the region of the skull base. The latter may penetrate into the sphenoid sinus (Fig. 4) or into the postnasal space. Giant aneurysms are often not selectively embolizable, and many need to be trapped along with the ICA.

The giant postnasal space aneurysm often produces attacks of severe epistaxis leading to collapse of the patient. If the ICA must be occluded, it is necessary to verify that it can be tolerated, and that a satisfactory collateral blood supply is present. A trial balloon occlusion of the ICA over 20 min is therefore done under neurological monitoring. If the collateral supply is inadequate, as may occur with aplasia of the anterior communicating artery, an external/internal carotid bypass (EC/IC bypass) must be performed before the

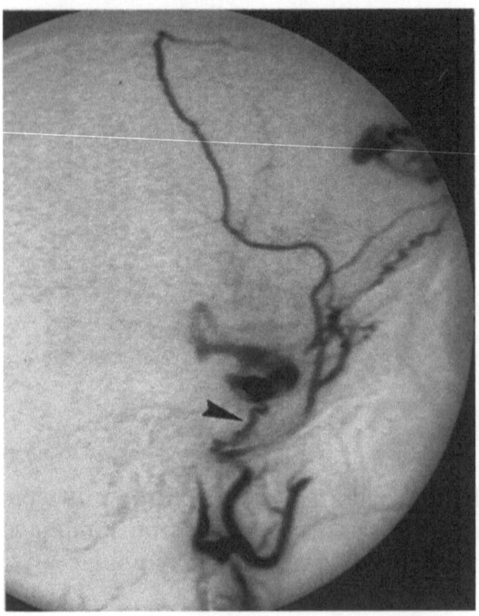

Fig. 3. Dangerous anastomoses between the internal maxillary and the internal carotid

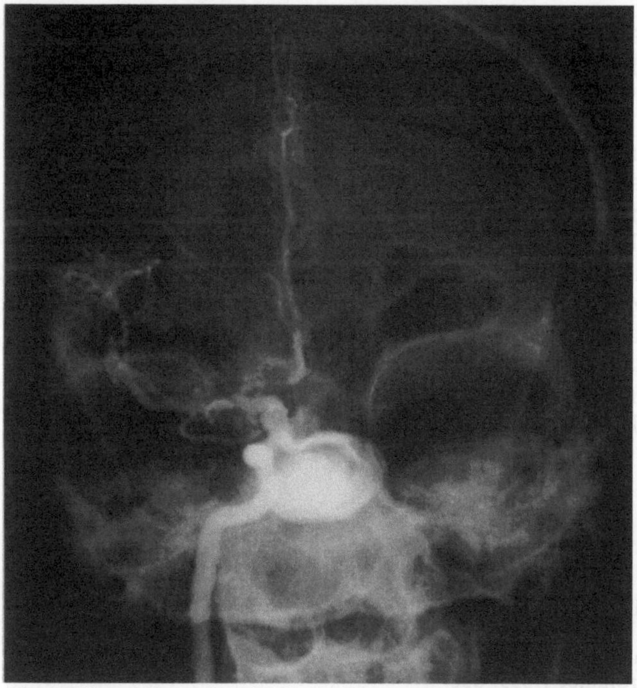

Fig. 4. Giant posttraumatic aneurysm occupying the sphenoid sinus

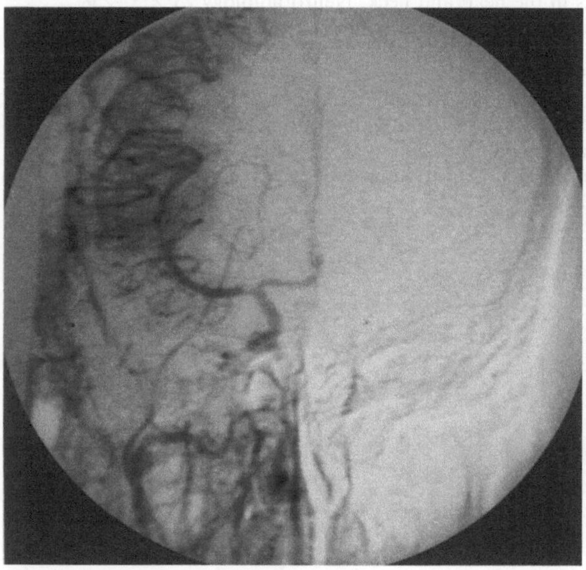

Fig. 5. EC/IC bypass. The left hemisphere is totally supplied by the bypass immediately after the trapping occlusion of the ICA

embolization. Figure 5 demonstrates good blood supply of the hemisphere following EC/IC bypass. Nuclear magnetic resonance imaging is used to observe the thrombosis and shrinking of aneurysms after embolization. Iatrogenic injuries to the ICA during transnasal transphenoid pituitary surgery may produce hemorrhage, pseudoaneurysms, or carotid cavernous fistulas. These may also be satisfactorily treated by balloon occlusion.

In conclusions, interventional neuroradiology plays an important role in the management of highly vascular tumors and vascular malformations, in arteriovenous fistulas, and in the giant aneurysms that penetrate into the paranasal sinuses. Most important are adequate experience, judgement of correct indications, careful analysis of the presenting lesions, knowledge of and appreciation of the dangers associated with each case, a search for dangerous anastomoses between the external and internal carotid territories, careful angiographic control during and after embolization, and the availability of DSA.

References

Bank WO, Kerber CW (1979) Gelfoam embolization: a simplified technique. Am J Röntgenol 132:299–301

Debrun G (1983) Treatment of traumatic carotid-cavernous fistula using detachable balloon catheters. Am J Neuroradiol 4:355–356

Horton JA et al. (1983) Polyvinyl alcohol foam-gelfoam for therapeutic embolization: a synergistic mixture. Am J Neuroradiol 4:143–147

Kühne D, Helmke K (1982) Embolization with "Ethibloc" of vascular tumors and arteriovenous malformations in the head and neck. Neuroradiology 23:253–258

Lasjaunias P, Berenstein A (1987a) Surgical neuroangiography. I. Functional anatomy of craniofacial arteries. Springer, Berlin Heidelberg New York Tokyo

Lasjaunias P, Berenstein A (1987b) Surgical neuroangiography. II. Endovascular treatment of craniofacial lesions. Springer, Berlin Heidelberg New York Tokyo

Latchaw RE, Gold LHA (1979) Polyvinyl foam embolization of vascular and neoplastic lesions of the head, neck and spine. Radiology 131:669–679

Zeumer H (1987) Intraarterielle digitale Subtraktionsangiographie in der interventionellen Neuroradiologie. Electro-Medica 55:1–18

Surgical Management of Maxillary Sinus Pathology

N. Schwenzer, Tübingen/FRG

The maxillary sinuses are without doubt the paranasal sinuses most often involved by pathology of dentogenic origin. In addition, benign and malignant tumors originating in the oral cavity can involve the maxillary sinuses. The most important diseases of the maxillary sinuses and their surgical therapy are presented here briefly from the maxillofacial point of view.

Operative Management of Dentogenic Cysts of the Maxillary Sinus

Two types of odontogenic cysts can generally be distinguished: radicular and follicular cysts. Radicular cysts, also known as root cysts, develop from chronic apical parodontitis (dental granulomas) of nonvital teeth. Their slow growth can result in expansion toward and into the maxillary sinus. Two cavities can thus be identified in this region: a maxillary sinus which is reduced in size and lined with respiratory epithelium, and the cystic cavity, lined with a squamous epithelium; a separating osseous wall may not be present. In extreme cases, these cysts may replace the entire maxillary sinus. If the tooth that was the cause of this pathology is already missing, this cyst is usually referred to as a residual cyst.

The follicular cyst develops from remnants of the dental lamina and outer enamel epithelium of developing teeth, which undergoes proliferative and cystic changes. Typically, the follicular cyst envelops the crown of an unerupted, often displaced tooth (Figs. 1, 2). In the maxilla, third molars and occasionally canines are the most common source of follicular cysts. Often, the lateral wall of the maxillary sinus has already been eroded, resulting in a wide access to the cyst and the maxillary sinus. If the antral walls are still present, good exposure can be obtained by creating an osseous window in the fossa canina, which can subsequently be closed by replantation of the bone lid (Figs. 3, 4). Treatment of such large cysts consists of extraction of the responsible tooth in all cases, as well as transmaxillary cystectomy with retention of the antral mucosa if it shows no pathological changes. This treatment was suggested by Wassmund as early as 1935. Fenestration toward the inferior meatus of the nose as described by Caldwell and Luc is always performed as part of this operation.

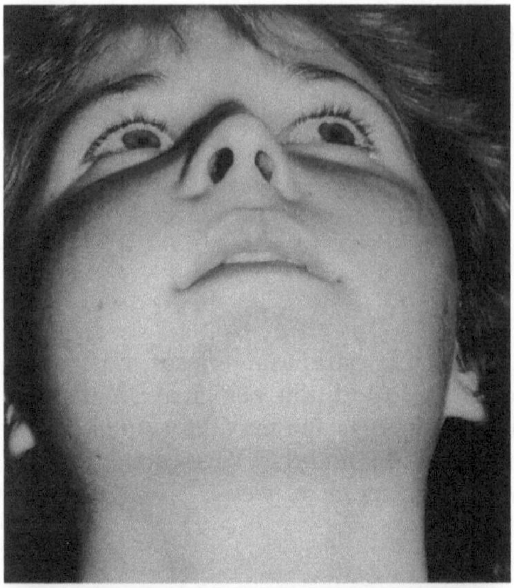

Fig. 1. Swelling of the right cheek due to an odontogenic cyst. This large follicular cyst of the maxilla was caused by an upper third molar

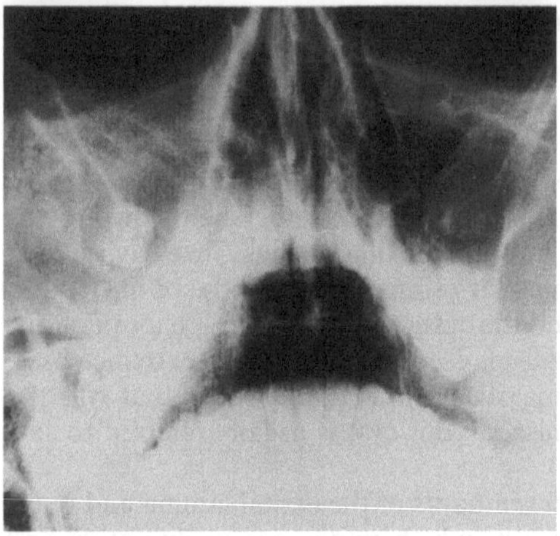

Fig. 2. The entire maxillary sinus has been invaded by the cyst. The crown of the third molar is enveloped by the cyst

The incision must be made so as to ensure adequate closure of the defect caused by the tooth extraction. To reach this goal, Wassmund recommended placing the incision in the marginal gingiva. For reconstruction and/or reinforcement of the lateral sinus wall, Lyodura may be used.

Cyst Development Following Surgery

A less well-known type of cyst is the so-called occlusion cyst, which may occur in the maxillary sinus following surgery. Following radical operations, septations may develop in the maxillary sinus, which can be lined with an epithelium derived from remnant cell islands showing autonomic growth. In the cases that

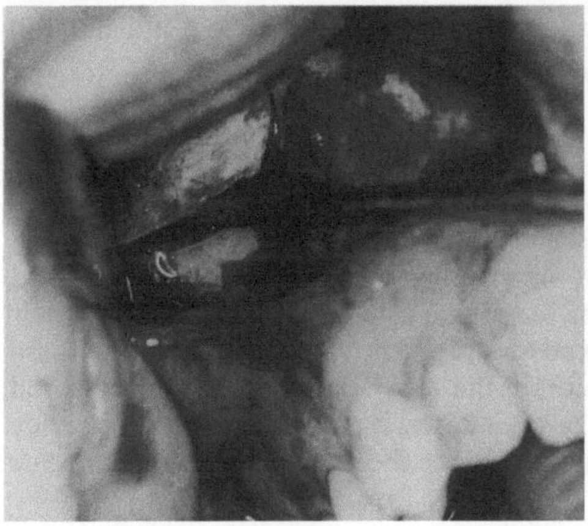
3

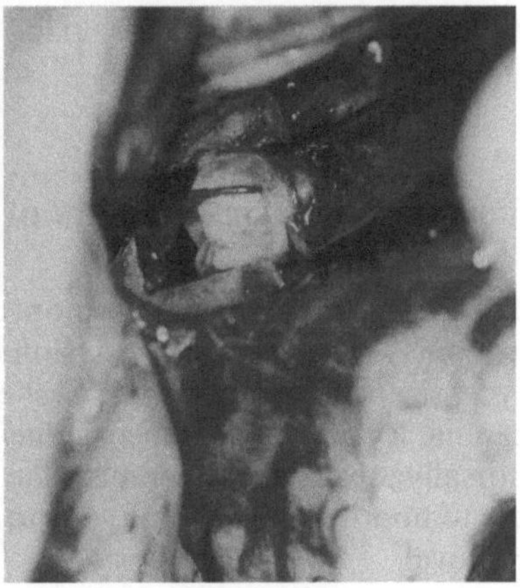
4

Fig. 3. Creation of a bone lid in the fossa canina. The bone lid remains attached cranially to the soft tissues

Fig. 4. The third molar is removed through the bone window

we have observed the operations had been 5–7 years earlier, and the cysts presented clinically as bulging, fluctuating tumors in the vestibulum, which could be confused clinically with odontogenic residual cysts. An infection of these cysts via the maxillary sinus or hematogenously is possible. The treatment of occlusion cysts is similar to that of odontogenic cysts and consists of excision and a wide nasoantral window. Large cysts may also extend toward the orbit, destroy the orbital floor, and raise the globe. During surgery the orbital floor should be reconstructed to prevent the development of an enophthalmos.

Infections of these cysts may spread toward the orbit and can result in a life-threatening phlegmonous infection of the orbit (Figs. 5, 6). Aggressive treat-

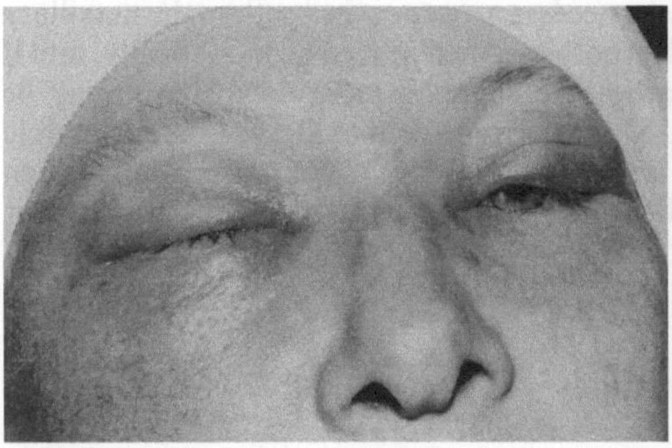
5

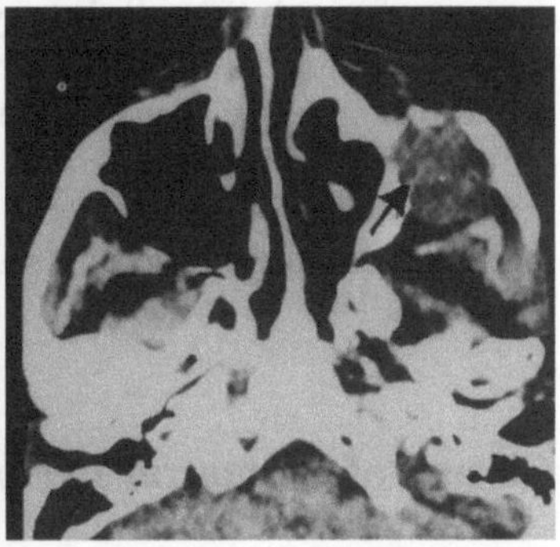
6

Fig. 5. Phlegmonous infection of the right orbit (following an operation of a residual cyst of the maxillary sinus). The cyst had destroyed the orbital floor, allowing a sinusitis to spread into the orbit

Fig. 6. The cyst is clearly visible on this computed tomogram (*arrow*)

ment, including external drainage and intravenous antibiotics, is indicated in these cases.

Therapy of Benign Tumors of the Maxillary Sinus

The most common benign tumors of the maxillary sinus are hemangiomas and osteomas. The access for excision of these rare tumors is determined by their type, size, and location. The best exposure is provided by access via the fossa canina. If needed, the mucosal incision in the vestibular area may be extended dorsally toward the tuber maxillae. In some cases, it may be necessary to divide the tumor (i.e., an osteoma) due to its size inside the maxillary sinus prior to its removal.

Transantral Tooth Removal

The removal of a tooth through the antrum is indicated if an approach via the alveolar ridge would result in injury of neighboring teeth or a significant defect of the alveolar ridge that would impair a later prosthetic rehabilitation. A further indication is the removal of a tooth or roots that have been incidently dislocated into the antrum during extraction or osteotomy. Occasionally, upper third molars can be located in the floor of the maxillary sinus. Displaced canines and premolars may be found as well in the antral walls in very rare instances.

Following roentgenological diagnosis in two planes, access to the maxillary sinus is obtained as well through the fossa canina. The necessary exposure can be obtained by creating the largest possible bone window with a replantable bone lid. Hence the window must be large enough that the crown of the tooth can pass through (Fig. 4). It may be necessary to turn the tooth in the maxillary sinus and grasp it at its root. If the tooth can be removed atraumatically, and if the antral mucosa shows no pathologic changes, nasal antrostomy may be omitted. Transalveolar removal of roots from the maxillary sinus is possible only if they are located in the immediate vicinity of the perforation. When making the incision, the possibility of creating a gingival flap for plastic repair should be kept in mind.

Closure of Oroantral Perforations

Closure of oroantral perforations can be secured in various ways. A trapezoid-shaped mucoperiosteal flap is most commonly used, especially following recent perforations. This flap, first described by Rehrmann, is based laterally and consists of the gingiva, the periostium, and the buccal mucosa; length can be obtained by periosteal incisions to secure a closure without any tension. Horizontal mattress sutures may be used, or the flap may be slid under the palatinal mucosa once its medial margin has been deepithelialized.

To protect the wound the incorporation of a custom-made plastic splint is quite successful. In inflammatory changes of the antral mucosa the plastic closure is combined with a maxillary sinus revision, whereby the vestibular incision may be combined with the trapezoid flap. Particularly in localized polypoid changes (usually in the vicinity of the perforation) only the areas with pathologic changes are removed, while the remnant normal mucosa is spared. We have treated 97% of all patients with oroantral fistulas following this method. If the patient is edentulous, the incision should be on the alveolar ridge.

To cover perforations of the alveolar ridge dorsally or ventrally based buccal flaps may be used which must be rotated 90° and then placed in the defect. For large perforations, especially those following unsuccessful plastic closures with vestibular mucosa, a palatinal island flap may be used. This was our method of choice in 0.8% of all oroantral fistulas. Double coverage following the principle of Lautenschläger may be necessary for larger perforations in some cases. This technique is helpful mainly in covering vestibular perforations. The procedure is not as helpful for perforations of the alveolar ridge, and was utilized in 2% of our cases.

In the edentulous patient, vestibularly and palatinally based bridge flaps may be used. These flaps can be shifted over the wound, both from the dorsal (Kazanjian 1959) or the ventral (Schuchardt 1953). A basic principle for all successful plastic closures remains that no tension be applied.

Even though closure of defects with soft tissue is adequate in most patients, additional osteoplasties are indicated in some instances. In patients with severely atrophic alveolar ridges, the floor of the maxillary sinus can be filled with cancellous bone to permit a dental allograft in the future. For this purpose, an incision is placed on the alveolar crest, then the alveolar ridge is divided, and the bone graft is placed between the buccal and palatinal bone segments.

If the soft tissues are missing as well, and no local flaps can be formed, (microvascular) free flaps, tube grafts, or (vascular and myocutaneous) island flaps may be used. This situation usually follows resection of malignant tumors. The reconstruction of the maxillary sinus is the procedure of choice if the patient does not wish a strictly prosthetic solution. An example is an osteoplasty following soft tissue coverage with a tube graft from the upper arm for reconstruction of the defect resulting from radical resection of an osteosarcoma of the alveolar ridge (Fig. 7).

In summary, it can be noted that odontogenic pathologic changes of the maxillary sinuses are an indication for surgery. Incision and access must be planned carefully and require some experience. Quite important from the maxillofacial point of view are also oroantral fistulas and their closure since the principles of reconstruction of form and function are also valid in this area.

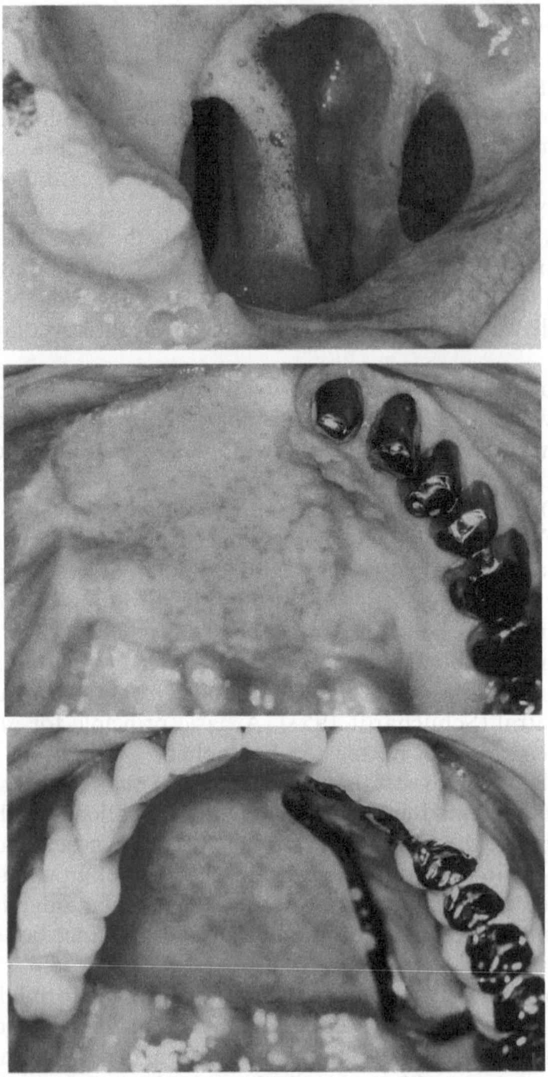

a

b

c

Fig. 7. a Defect of the maxilla, 5 years after radical resection of an osteosarcoma of the alveolar ridge. **b** After plastic reconstruction with a rib graft and a tube graft from the arm. **c** With prosthesis

References

Kazanjian VH (1959) In: Kazanjian VH, Converse JM (eds) The surgical treatment of facial injuries 3.934. Williams & Wilkins, Baltimore

Schuchardt K (1953) Zur Methodik des Verschlusses von Defekten im Alveolarfortsatz zahnloser Oberkiefer. Dtsch Zahn Mund Kieferheilkd 17:366

Wassmund M (1935) Lehrbuch der praktischen Chirurgie des Mundes und der Kiefer, Bd. 1. Meusser, Leipzig

Neurosurgical/ENT Management of Paranasal Sinus Lesions Extending Through the Skull Base

E. J. Dolan, Billings/USA, P. Gullane, F. Gentili, and J. Rutka, Toronto/Canada

The present work illustrates our experience at Toronto General Hospital with extensive anterior skull base lesions. This involves an interdisciplinary approach combining neurosurgical and otolaryngological expertise to manage neoplasms in this complex area, especially for malignant lesions, where we believe an en bloc resection of the tumor is necessary. This interdisciplinary approach is most helpful in enabling single-stage resection and reconstruction of the skull base with minimal morbidity and mortality.

We divide skull base lesions into three groups: group I comprises lesions of the paranasal sinuses and usually requires some form of midface approach in addition to an anterior skull base exposure; group II consists of lesions located primarily in the middle fossa but extending below the skull and into the orbit and requiring some form of infratemporal fossa approach; group III lesions are located primarily in the temporal bone. As lesions become more extensive, sometimes involving the entire skull base, there is some overlapping of these groups, but we have found the classification useful for planning surgical strategies to manage skull base neoplasms.

Materials and Methods

Patients. The present review comprises eight cases from group I and two cases from group II, with lesions extending through the skull base. Table 1 summarizes the demographic data on the ten patients. There were eight men and two women. Ages ranged from 17 to 85 years at time of surgical resection; most patients were over 40 years of age. Eight cases were recurrences of the primary disease, and most patients had undergone numerous previous surgeries. Four of these cases had also had prior radiation therapy of 50–60 Gy to the area of the lesion. The length of time from initial onset of disease until basal skull surgery varied in this group from 2 to 14 years. The two cases seen primarily had symptoms for 1 month and 7 months, respectively. Symptoms at presentation consisted of facial pain in three cases and epistaxis in two. Six patients had a visible mass or proptosis. Two complained of decreased vision, and facial numbness was mentioned by three. Cranial nerve palsies were present in four patients.

Imaging. All patients underwent preoperative axial and coronal computed tomography (CT) with contrast. More recently, patients are examined using

Table 1. Patient data

Patient	Group	Sex	Age	Location	Pathology	Previous surgery	Previous irradiation	Recurrence time (months)	Follow-up time (months)	Outcome
IJ	I	M	66		Chondrosarcoma	Yes			3	NED
RM	I	F	22	Frontal bone	Ossifying fibroma	Yes	No		9	NED
DI	I	M	48	Ethmoids, maxilla, orbit	Squamous cell carcinoma	Yes	Yes	4	6	Died
SS	I	M	17	Ethmoid	Squamous cell carcinoma	No	No		12	NED
HS	I	M	69	Ethmoids, maxilla, orbit	Squamous cell carcinoma	Yes	Yes	1	7	Died
HB	I	M	85	Ethmoids, orbit	Squamous cell carcinoma	Yes	Yes		16	NED
JH	I	M	48	Ethmoids, orbit	Basal cell carcinoma	Yes	Yes		15	NED
CD	I	M	45	Clivus (extensive)	Chordoma	Yes	?			AWD a
SH	II	F	50	Ethmoids, orbit, infratemporal fossa	Meningioma	Yes	No	11	42	NED
JS	II	M	32	Middle fossa, infratemporal fossa	Osteosarcoma	No	No	6	13	Died

NED, No evidence of disease; AWD, alive with disease.
[a] Incomplete resection.

magnetic resonance imaging (MRI) to further delineate the extent of lesions. We believe both investigations are necessary as they provide complementary information. The MRI readily demonstrates the lesion and its intradural or intraparenchymal infiltration but does not show bony boundaries or bony infiltration well. The CT scan has problems with skull base lesions because of the numerous bone:tissue or tissue:air interfaces, but it superbly demonstrates bone margins and bone changes due to lesions at the skull base. Difficulty is still encountered in deciding between infiltration and compression of the cavernous sinus by parasellar lesions. Selected cases have preoperative angiography to delineate the intracranial carotid artery or to embolize a vascular tumor preoperatively.

Perioperative Management. Diphenylhydantoin (Dilantin) was used in all cases and maintained for 3 months postoperatively. Dexamethasone 16 mg/day in four doses was started 48 h prior to operation and then tapered over 5 days postoperatively in most cases. Antibiotics cefazolin (Ancef) 1 g was administered every 8 h and metronidazole (Flagyl) 500 mg every 8 h on the day of surgery and for 48 h thereafter. All patients had arterial catheters and central venous pressure lines in place. Monitoring of evoked potentials was used in operations around the intracranial carotid artery. During the operation, lumbar spinal fluid drainage was used in only one case. Adequate relaxation of the brain was achieved with hyperventilation to pCO_2 levels of 25–30 mmHg, mannitol (20%), and direct drainage of CSF from the basal cisterns.

Operative Procedures. The three cases of recurrent ethmoidal squamous cell carcinoma and the recurrent basal cell carcinoma of the orbit were treated with a bifrontal craniotomy to expose the anterior fossa floor, combined with a modified Weber-Ferguson incision for exposure of the midface (Johns and Kaplan 1986). An en bloc unilateral orbitectomy was combined with maxillectomy and bilateral ethmoidectomy. In one case the nasal bones were also resected, and reconstructed later. The de novo case of ethmoid squamous cell carcinoma was treated with a bifrontal exposure and en bloc resection of the ethmoids (Schramm 1984). Bifrontal craniotomy was also used for complete resection of the ossifying fibroma of the frontal bone. A combination transsphenoidal/transmaxillary/transethmoidal approach through a modified Weber-Ferguson exposure gave excellent exposure to the clivus for resection of an extensive recurrent chordoma. Unfortunately, the lesion could be only incompletely removed. Two cases were approached via an extended pterional exposure as described by Mickey et al. (1988). One case had only the horizontal portion of the zygoma resected (Fujitsu and Kuwabara 1985), while the other had the complete tripod of the zygoma resected and subsequently replaced. This allowed excellent exposure of the infratemporal area. In these two cases, tumor was removed piecemeal, but in the other malignant lesions tumor was removed en bloc.

Reconstruction. The aims of reconstruction are to provide a complete dural seal to prevent CSF leak, to fill in any dead space and provide support for neural structures, to provide bony coverage where necessary, and to provide epithelial coverage where necessary. For the dura a pericranial flap was elevated in all unilateral or bilateral frontal exposures. The pericranial flap forms the outermost layer of our dural closure and lines the aerodigestive tract. This was tacked down posteriorly over the sphenoid wing and anterior clinoid areas. Any defects or holes in the dura were repaired separately with lyodura, fascia lata, or in one case temporalis fascia. Two layers of tissue therefore always separated the brain from the nasal cavity. For the skull we hesitated to use methymethacrylate in these potentially infected areas, preferring reconstruction with autologous materials. Reconstruction with split calvarial bone was used to repair one anterior fossa defect and two calvarial defects after tumor resection, while in one case split rib grafts were used for repair of a calvarial defect. For the skin split thickness skin grafts were applied to the pericranium in all cases of orbital exenteration. In one case a scalp rotation flap provided additional coverage, while a pectoralis major flap was necessary in another case. A subsequent local recurrence of basal cell carcinoma required a free radial forearm flap for repair.

Results

As expected, the outcome varied according to the primary pathology of the lesion resected, and analysis of the effectiveness of basal skull surgery must take this into account. Of the four patients with squamous cell carcinoma, two are alive with no evidence of disease at 12 and 16 months, while two died 6 and 7 months postoperatively. The latter two had recurrence of disease at 4 months and 1 month, respectively. The patient with osteosarcoma died at 13 months after developing lung metastases 8 months and cervical spine metastases 10 months after operation. The primary cranial site was clear of tumor. The clival chordoma recurred very rapidly within 4 months of resection, and the patient died 7 months following resection. The other four cases had no evidence of disease when last seen – at 9 months (ossifying fibroma), 14 months (meningioma), 36 months (chondrosarcoma), and 31 months (basal cell carcinoma).

Conclusions

The improved diagnostic imaging techniques that are available today provide accurate delineation of basal skull tumors, although determination of cavernous sinus invasion still remains a problem. An interdisciplinary approach with improvement in surgical techniques, reconstructive options, and intra- and peri-operative monitoring has reduced the morbidity and mortality of extensive skull base resections to an acceptable level. These lesions are uncommon, and each center performing skull base surgery sees only a limited number of cases from each pathologic entity. Multicenter database collection is needed to accrue sufficient cases, and prospective multicenter studies are necessary to evaluate the long-term benefits of this extensive and often disfiguring surgery.

References

Fujitsu K, Kuwabara T (1985) Zygomatic approach for lesions in the interpeduncular cistern. J Neurosurg 62:340–343

Johns ME, Kaplan MJ (1986) Surgical approach to the anterior skull base. Ear Nose Throat J 65:117–124

Mickey B, Close L, Schaefer S, Samson D (1988) A combined frontotemporal and lateral infratemporal fossa approach to the skull base. J Neurosurg 68:678–683

Schramm VL (1984) Craniofacial resection. In: Sasaki CT, McCabe BF, Kirchener JA (eds) Surgery of skull base. Lippincott, Philadelphia

Therapy of Tumors Affecting the Paranasal Sinuses and Sellar Region

D. Riediger, Stuttgart/FRG, W. Hassler, Duisburg/FRG, and
N. Schwenzer, Tübingen/FRG

Introduction

Due to the close topographic relationship between maxillary sinuses, paranasal sinuses, and skull base the therapy of tumors affecting these regions often requires an interdisciplinary, joint surgical procedure by maxillofacial surgeons, ENT surgeons, and neurosurgeons (Riediger and Jahnke 1988). From the maxillofacial surgeons' point of view this is especially true in neoplasms expanding after destructive growth from the maxilla via the maxillary sinus towards the orbit and skull base. Surgical concepts are determined, especially regarding the degree of radicality, by the technical possibility of defect closure and avoidance of liquor fistulas. In extensive resections in the midface region, including maxillary segments, special attention must be paid to functional masticatory rehabilitation. Whereas minor defects can often be cared for according to the principles of local grafting, extensive defects including parts of the maxilla require distant grafts. Pedicled myocutaneous grafts and especially microsurgical techniques have decisively contributed to the scope of modern free tissue transfer (Watson et al. 1979; Riediger and Schmelzle 1986). Their employment provides us with means for secure coverage of the skull base as well as a bony midface reconstruction resembling the normal anatomical structures as much as possible, especially inmaxillary segments carrying masticatory functions.

In this contex revascularized iliac crest grafts in combination with enosseous implants have proven especially valuable, being applicable also following radiotherapy (Riediger 1988). Therapeutic concepts are presented below by clinical cases with special consideration given to reconstructive procedures.

Case Reports

Case Report 1. A hemimaxillectomy had been performed elsewhere in a 47-year-old man because of a left maxillary adenoid-cystic carcinoma. Four years later he was transferred to us because of an extensive relapse. The areously spreading tumor had invaded the ascending mandibular ramus, the left orbit, and the skull base (Fig. 1). An extensive soft tissue and bone resection of the midface region was required. Additionally, the sphenoid wing including the

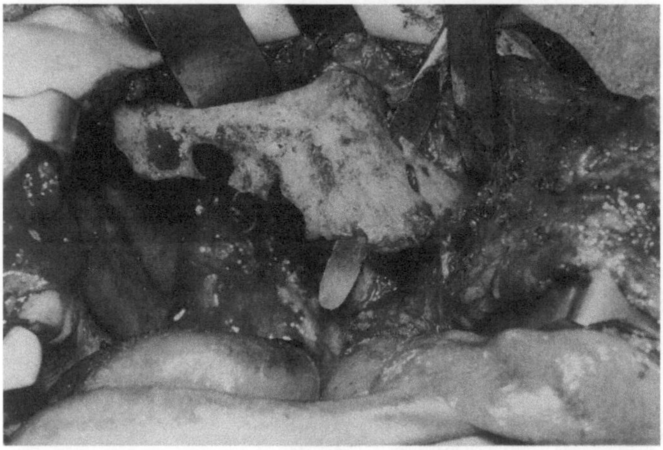

Fig. 1. Destruction of bony orbit and skull base by an extended adenoid-cystic carcinoma's relapse in a 47-year-old man

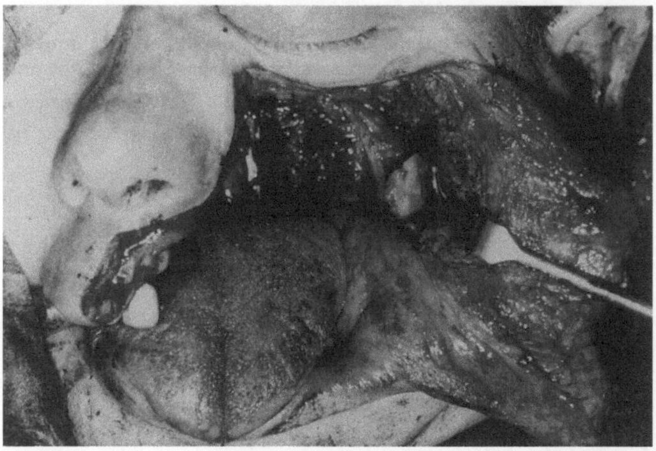

Fig. 2. Large defect following tumor resection including resection of the floor of the medial cranial fossa

superior and inferior orbital fissure as well as the ethmoid cells and the base of the medial cranial fossa including the round and spineous foramen were resected (Fig. 2). With no local material available for reconstruction a microsurgically transplanted modified latissimus dorsi flap was decided upon. This was folded in a sandwichlike manner with its cranial muscular portion covering the osseous skull base defect, especially the temporal lobe, after the skin layer had been removed. The lower parts served as closure of the oral cavity (Figs. 3, 4). A postoperative irradiation with a total tumor dose of 70 Gy was administered. Computed tomographic controls revealed no evidence of

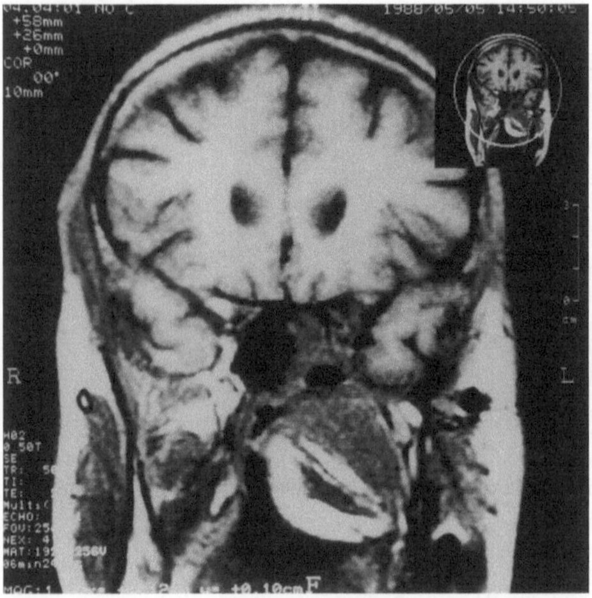

Fig. 3. Postoperative NMR tomography after microsurgical transplantation of a doubled latissimus dorsi flap with closure of the skull base defect and simultaneous closure of the oral cavity

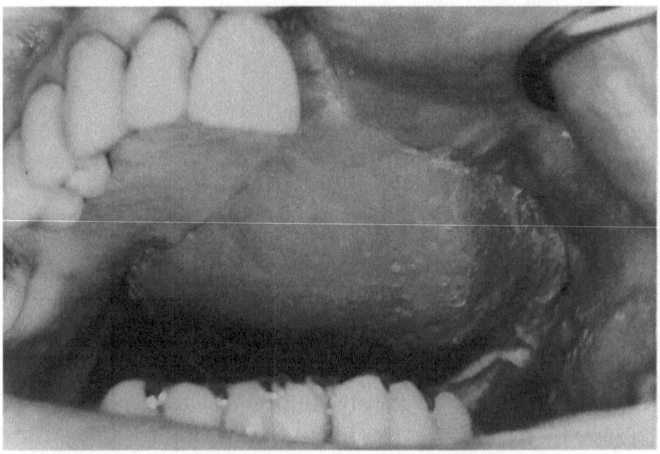

Fig. 4. Closure of the oral cavity defect by the latissimus dorsi myocutaneous flap. Situation following secondary grafting of a revascularized iliac crest for maxillary reconstruction

Fig. 5. Vascular pedicled iliac crest graft raised from the right groin for reconstruction of the maxilla

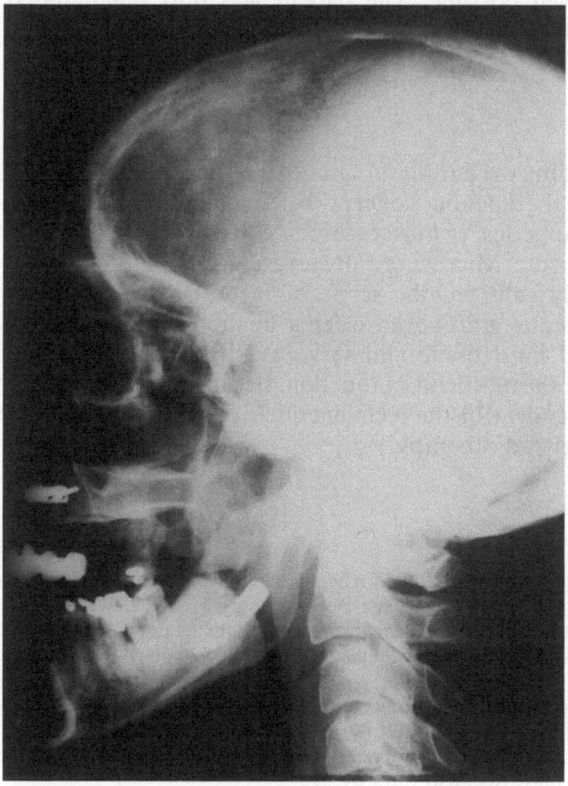

Fig. 6. Lateral radiograph of the skull with microsurgically revascularized iliac crest graft fixed by a miniaturized metal plate

recurrence. In the course of masticatory rehabilitation the left maxilla was reconstructed by a revascularized iliac crest graft (Taylor et al. 1979) connected to the left superficial temporal vessels. A free bone graft was used for reconstruction of the orbital floor at the same time. Figure 5 shows the iliac crest graft with its vascular pedicle of the deep circumflex ilium artery and vein, which allows reconstruction of the alveolar process and later the integration of enosseus implants. The iliac crest graft was fixed by a miniplate and showed uneventful healing, which is clearly visible in a lateral radiograph 3 weeks later. It forms a reconstructed maxillary base able to bear functional loads (Fig. 6).

Case Report 2. A similar reconstructive concept was chosen in a 67-year-old man who required partial maxillectomy due to an extensive maxillary ameloblastoma reaching the skull base. The large defect resulting in grave functional disturbances was primarily closed by a myocutaneous latissimus dorsi flap, attached microsurgically to the facial artery and vein. For reconstruction of the masticatory function in a second stage a revascularized iliac crest graft was transplanted and connected to the superficial temporal vessels. It served as a recipient site for titanium enosseous implants. Functional denture integration was then possible. The patient was thus rehabilitated esthetically and functionally.

Discussion

Tumors of the maxilla invading the paranasal sinuses and the skull base require a joint, interdisciplinary therapeutic concept by maxillofacial surgeons, neurosurgeons, and ENT surgeons. In most cases this concept is determined by the possibilities of defect closure. Microsurgically revascularized soft tissue and bone grafts have decisively enlarged the scope of reconstructive methods. Combined osteomyocutaneous grafts even offer a method of simultaneous reconstruction of soft and hard tissue and serve as a basis for functional rehabilitation, especially of the masticatory function. Because of their vascular attachment they are independent of the recipient site's tissue quality and are therefore also suitable after radiotherapy.

References

Riediger D (1988) Restoration of masticatory function by microsurgically revascularized iliac crest bone grafts using enosseous implants. Plast Reconstr Surg 81:857–861

Riediger D, Schmelzle R (1986) Modifizierte Anwendung des Latissimus dorsi-Lappens zur Defektdeckung im Mund-Kiefer-Gesichtsbereich. Dtsch Z Mund Kiefer Gesichts Chir 10:364–374

Riediger D, Jahnke K (1988) Mikrochirurgisch revaskularisierte Transplantate zur Defektdeckung im Mittelgesicht. Laryngol Rhinol Otol 67:7–12

Taylor G, Townsend R, Corlett R (1979) Superiority of the deep circumflex iliac vessels as the supply for free groin flaps. Plast Reconstr Surg 64:595

Watson JS, Craig ROP, Orton CE (1979) The free latissimus dorsi myocutaneous flap. Plast Reconstr Surg 64:299

Microscopic Endonasal Surgery of the Paranasal Sinuses and the Parasellar Region *

R. G. AMEDEE, New Orleans/USA, W. MANN, Mainz/FRG, and
J. GILSBACH, Aachen/FRG

Introduction

Microscopic endonasal surgery (MES) of the paranasal sinuses has been performed in adults since the late 1950s and was first reported for sinus surgery in children in 1976 [1, 3]. It was initially used to create an antrostomy and has recently been used for more difficult intranasal ethmoidectomy and transethmoido-sphenoidectomy [2, 4]. The development of special self-retaining nasal speculum systems has greatly aided this form of surgery both in adults and children with a variety of regional pathologies, including chronic sinus infection, benign polyps, orbital abscess, and tumors of the orbit, parasellar, and hypophyseal regions. The speculum holder is attached to the operating table enabling the surgeon to work bimanually. The surgeon may stand or be seated at the patient's side with the operating hands resting on the patient, thus affording excellent control of the instruments. Binocular, stereoscopic vision with excellent illumination creates additional precision and control in delicate ethmoid and sphenoid sinus surgery. The Zeiss operating microscope with a 330-mm objective lens and halogen illumination is recommended, while mounting an observer arm or videocamera has provided an outstanding opportunity for resident teaching (Figs. 1–3).

Preoperative shrinkage of the nasal mucous membranes is achieved with topical vasoconstrictors; however, injections with local anesthetics plus adrenaline have proven unnecessary. MES should be performed using hypotensive anesthesia (systolic blood pressure of 70–90 mmHg) to decrease intraoperative bleeding, and the patient should be placed in a semi-Fowler's position enabling blood and mucous to drain from the skull base and away from the operative field. The eyelids are left uncovered and should always remain in the peripheral field of vision. Bilateral surgery attempted at the same sitting is not contraindicated and is usually completed within 60–90 min.

The complex anatomy of the paranasal sinuses, the parasellar region, and the surrounding vital structures has become more familiar to our specialty with the advent of endoscopic endonasal sinus surgery [5–7, 9]. Currently, fewer Caldwell-Luc and external approaches are deemed necessary, except in cases where malignancies are suspected [4]. High-resolution axial and coronal

* This study was supported by Deutscher Akademischer Austauschdienst Grant 315-402-021-8.

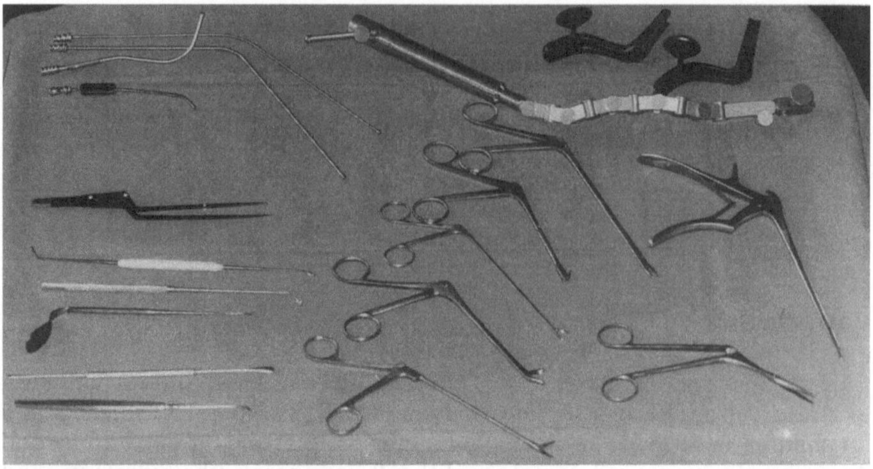

Fig. 1. Set of useful microscopic endonasal instruments

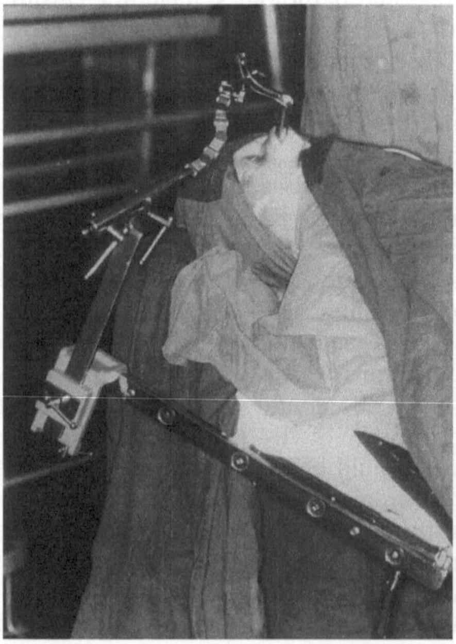

2

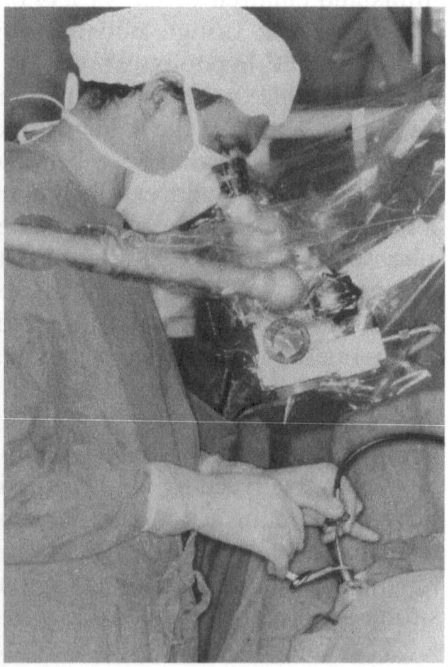

3

Fig. 2. Self-retaining speculum system inserted into the nose and fixed to the operating table

Fig. 3. Bimanual microscopic endonasal surgery with opportunity for resident teaching

computed tomography (CT) offer a precise way to evaluate the presence of sinus disease and further aid in determining when surgery is indicated [10].

Many patients with chronic sinusitis and/or polyposis require specific allergy testing and treatment. We advocate that definitive surgery, when indicated, be performed first [8]. If a rare postoperative infection develops, allergy testing should be deferred until all signs of infection are thoroughly cleared to avoid an inaccurate response. Postoperative antibiotics were routinely used in the initial 150 cases, while in the remaining cases none were given unless a purulent sinusitis or abscess was found at the time of surgery. The use of postoperative antibiotics in MES for chronic sinusitis (with or without polyposis) is the source of an ongoing study in our clinic.

Ethmoidal and Sphenoidal Sinus Surgery

The middle turbinate is displaced medially under microscopic vision, and the self-retaining speculum is then introduced into the nose and placed within the proximal middle meatus, holding the middle turbinate toward the septum (Fig. 4). An infundibulotomy is performed with a sharp double-ended Freer elevator by fracturing the ethmoidal bulla, and the large anterior ethmoidal cells are then entered. The bony labyrinth including mucosa, polyps, and/or secretions is removed with either a Weil ethmoid forceps or a Caspar bone forceps. The middle turbinate is used to guide the dissection directly posterior until the anterior wall of the sphenoid sinus is exposed and the roof of the ethmoid is identified. Resection then follows the ethmoidal roof from posterior

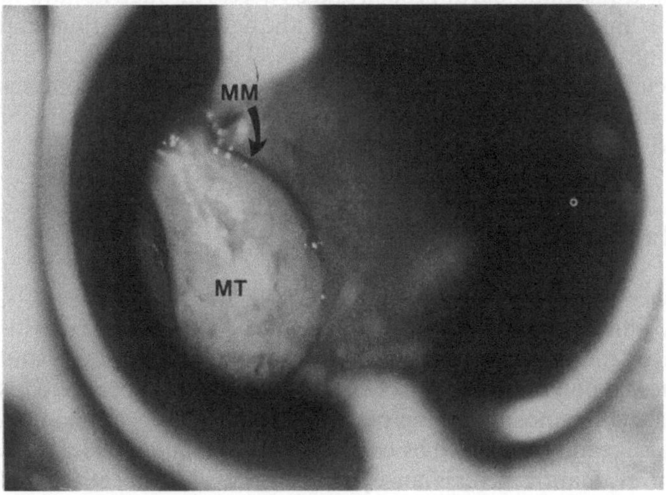

Fig. 4. Microscopic view through the speculum demonstrating the middle turbinate (*MT*) and the middle meatus (*MM*, *arrow*)

to anterior, and a 30° backward cutting curette is gently used to expose the entire roof and lateral sinus wall which is the lamina papyracea of the orbit. This lateral wall may be opened, when indicated, to allow for intranasal drainage of an orbital abscess. Bleeding is usually not a problem and is adequately controlled with selective bipolar cautery using long and delicate forceps.

Dissection should not be carried out medial to the superior attachment of the middle turbinate since the bone of the cribriform plate is extremely thin and delicate, and injury to this region is likely to result in CSF rhinorrhea. During the operations the anterior portion of the middle turbinate is carefully preserved as a useful anatomic landmark in case secondary procedures are indicated. The posterior half of the middle turbinate is routinely resected since in earlier cases we found it to obstruct proper cleansing of the posterior cavity, especially in the region of the sphenoidotomy. In addition to resecting this portion of the middle turbinate we cauterize the posterior end of the inferior turbinate. If a concha bullosa is present we recommend resection of the lateral half of the remaining middle turbinate with the associated air cell(s).

The incidence of sphenoidal sinus disease is greatly reduced compared with ethmoidal disease. If sphenoidotomy is indicated based on preoperative radiologic evaluation and intraoperative findings, this may be accomplished by gently pushing through the anterior sphenoid wall with the Weil ethmoid forceps or a curette. This opening is then enlarged with a fine 30° Stanze forceps. The diseased contents of the sinus must be cautiously removed since in several of our cases the lateral bony wall of the sinus was partially dehiscent, and careless curettage may have resulted in injury to the optic nerve or the internal carotid artery. We recommend blunt palpation of the lateral wall prior to removal of the sinus contents with forceps or suction. The sphenoid sinus has been entered by this transethmoidal approach and by access through its natural ostia in the sphenoethmoidal recess to repair traumatic defects of the ethmoidal and sphenoidal roof with concomitant CSF leakage. The dural dehiscence is repaired using fascia obtained from the anterior layer of the rectus abdominus sheath, and the sinus is subsequently packed with alternating layers of abdominal fat and pressed collagen sheets, all held in place by fibrin glue.

Frontral and Maxillary Sinus Surgery

External approaches to the frontal sinus are reserved for traumatic defects, tumors, and bony obliteration of the nasofrontal ducts. The frontal recess is usually identified in the anterior portion of the roof of the ethmoid, and occasionally ethmoidal polyps hinder direct visualization of this opening. This obstruction represents the most common cause of frontal sinus pathology in this study, a situation readily improved by complete polypectomy. In rare cases when preoperative evaluation reveals an opacified sinus, but no ethmoidal polyps are found intraoperatively, a check for patency of the duct is made using a Ritter bougie or a 70° backward cutting curette. If the preoperative CT shows a normal frontal sinus, and the patient is asymptomatic, this portion of the surgery may be omitted.

Similar problems within the maxillary sinus most commonly result from obstruction of the natural ostium in the fontanelle. Once this obstruction is removed, the natural ostium is gently probed to determine whether the 70° curette may be introduced. If this maneuver is not possible, the ostium is enlarged with a backward cutting antral punch avoiding the region of the nasolacrimal duct. A large nasoantral window is then created beneath the inferior turbinate.

Postoperative Care

A continuous nasal tamponade consisting of 2-cm-wide surgical gauze impregnated with sulfur gel is loosely packed into the sphenoidal and ethmoidal sinus cavities. In the event that a nasoantral window was created, a separate tamponade is placed into the window and brought out through the nares. One-half of the entire tamponade is removed from each side 24 h after surgery, and that remaining on the 2nd postoperative day. Gentle suction of the cavities including the internal nasal airways is then performed and the patient is instructed to begin nasal irrigations three times per day with lukewarm saline. Additionally, a decongestant nasal spray followed by a topical nasal steroid inhaler are both applied three times per day. Finally, a menthol nasal savle is applied to the nostrils as needed to help prevent postoperative drying and crusting.

It should be emphasized that careful débridement of the cavities must be performed at 2- to 3-day intervals during the 1st postoperative week. The

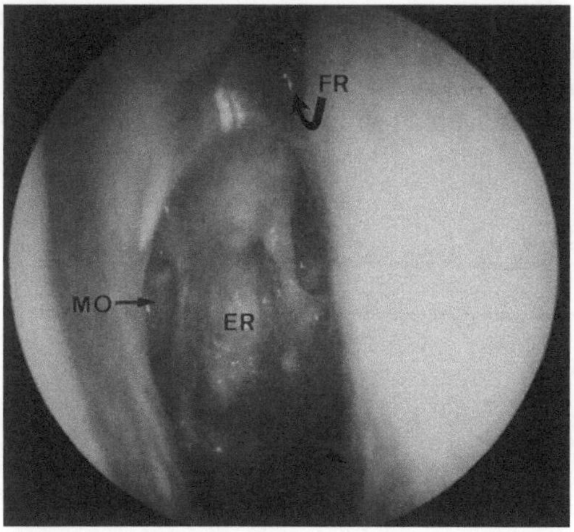

Fig. 5. Postoperative view of epithelialized ethmoidal cavity 2 years after surgery. *FR*, Frontal recess; *MO*, maxillary ostium; *ER*, ethmoidal roof

previously described regimen of irrigations and sprays has significantly improved patient comfort and reduced the occurrence of secondary infections. The steroid nasal inhaler is continued after the 1st week for periods of 3–12 months depending on the sinus pathology and postoperative evaluations, while the irrigations and decongestant spray are omitted. Occasionally the gel is continued for a longer period of time in an effort to enhance nasal moisturization. Allergy testing and immunotherapy, when indicated, are essential in preventing the early and undesirable recurrence of intranasal polyps (Fig. 5).

Discussion

The most common indication for MES in this series was chronic sinusitis frequently complicated by polyposis nasi, which accounted for 75% of all cases (Table 1). For each of these indications the results were quite satisfying, not only to the surgeon but also to uniformly grateful patients. All situations were successfully handled during the initial MES, including the biopsy of ten paranasal sinus tumors and the permanent control of 19 cases of traumatic ethmoidal and/or sphenoidal CSF rhinorrhea. No external approaches or secondary MES procedures were necessary to complete the diagnosis and/or treatment in these 219 patients. The patients with chronic sinusitis and polyposis are predictably well controlled by this procedure and experience a dramatic improvement in their overall condition and nasal respiration.

The rate of severe complications was markedly less with MES compared with the traditional external approaches. Bleeding which required transfusion, visual disturbances as a result of nerve or muscle injury, CSF fistula with abscess or meningitis, alterations of facial or dental sensation, or impairment of olfaction are all theoretically possible but have not occurred to date. Septal perforations have not been attributed to usage of the self-retaining speculum system or to septoplasty performed in conjunction with MES. The most common complications following MES have been limited to minor epistaxis, headache, and mild to moderate midfacial pain with swelling. The epistaxis has

Table 1. Indications for MES ($n = 219$)

	Number of patients
Choanal atresia	2
Chronic sinusitis and/or polyposis nasi	165
Maxillary sinus mucocele	5
Concha bullosa	4
Traumatic sphenoid CSF rhinorrhea	19
Orbital abscess/decompression	7
Sphenoidal/ethmoidal tumor biopsy	10
Inverted papilloma lateral nasal wall	1
Hypophysectomy	6

not required further packing to control, and the pain with swelling is improved tremendously when the intranasal packs are completely removed.

Improved diagnostic techniques, especially high-resolution CT scanning, now enable precise localization of paranasal sinus and parasellar diseases. Combined with major advances in operative instrumentation and a better understanding of the regional pathology, these improvements in diagnostic ability have led to this concept of MES. This technique results in decreased morbidity, allows preservation of normal tissue, and in the majority of cases avoids the necessity for external procedures.

Conclusions

MES using newly designed instruments with the self-retaining speculum system allows for bimanual surgery with excellent control of instruments, while the binocular vision ensures excellent surgical precision and illumination. MES of the paranasal sinuses and parasellar region affords adequate resection or biopsy of diseased tissues with less fear of serious complications. Troublesome bleeding is easily controlled by selective bipolar cautery, hypotensive general anesthesia, and use of the semi-Fowler's position which allows blood and mucous to flow away from the operative field. The use of an observer arm or videocamera mounted on the microscope provides an outstanding opportunity for resident teaching.

References

1. Dixon H (1976) Microscopic antrostomies in children: a review of the literature in chronic sinusitis and a plan of medical and surgical treatment. Laryngoscope 86:1796–1814
2. Dixon H (1983) Microscopic sinus surgery, transnasal ethmoidectomy and sphenoidectomy. Laryngoscope 93:440–444
3. Heermann H (1958) Endonasal surgery with the use of the binocular microscope. Arch Ohren Nasen Kehl Kopfheilkd 171:295–297
4. Heermann H, Neues D (1986) Intranasal microsurgery of all paranasal sinuses, the septum, and the lacrimal sac with hypotensive anesthesia. Ann Otol Rhinol Laryngol 95:631–637
5. Kennedy D (1985) Functional endoscopic sinus surgery-technique. Arch Otolaryngol 3:643–649
6. Kennedy D, Zinreich S, Rosenbaum A, Johns M (1985) Functional endoscopic sinus surgery-theory and diagnostic evaluations. Arch Otolaryngol 3:576–582
7. Messerklinger W (1978) Endoscopy of the nose. Urban & Schwarzenberg, Baltimore
8. Schlenter W, Mann W (1983) Operative Therapie der chronischen Sinusitis. Erfolg bei allergischen und nicht allergischen Patienten. Laryngol Rhinol Otol (Stuttg) 62:284–288
9. Stammberger H (1986) Endoscopic endonasal surgery-concepts in treatment of recurring rhinosinusitis. I. Anatomic and pathophysiologic considerations. Otolaryngol Head Neck Surg 94:143–149
10. Zinreich S, Kennedy D, Rosenbaum A, Stammberger H (1987) Paranasal sinuses: CT imaging requirements for endoscopic surgery. Radiology 163:769–774

Exeresis and Recontruction Techniques in the Surgical Treatment of Malignant Tumors of the Ethmoid

F. LESOIN, P. PELLERIN, and C. E. THOMAS, Lille Cedex/France

Malignant tumors which invade the ethmoid are usually associated with a very reserved or poor prognosis. The severity of the prognosis depends on the extent of infiltration of the tumor into the base of the skull [1, 5, 7]. When operating on these patients, it is therefore logical to choose a surgical approach which provides maximum exposure of the anterior base of the skull for inspection [4, 6, 8, 9, 10]. This can be done by using a monoblock orbitofrontral bone flap approach in conjunction with the standard sublabial transmaxillary approach.

Technique

Monoblock ethmoidectomy is performed simultaneously by two surgical teams using two separate approaches. The subfrontal approach with removal of a monoblock fronto-orbital bone flap exposes the superior portion of the tumor and allows inspection of both the width and depth or tumoral infiltration into the base of the skull. The sublabial transmaxillary gingivojugal approach provides access to the lower portion of the tumor.

With the patient lying on his back, the surgical teams must have free and ready access to both the frontal and buccal regions, as well as to one leg and the iliac crest. The first step of the frontal approach consists of making a bilateral Cairns-Unterberger incision which runs 2 cm above the frontal hairline, and an epicranial incision which forms a large flap medial to the inferior pedicule, using Derome's technique (Fig. 1). This is followed by suborbital dissection of the periosetium of the nasal bones, thus providing wide exposure of the frontal area. Four trepanations form the four corners of the frontal bone flap. The two lower holes are drilled 10–15 mm above the external third of the superior crest of the orbits (Fig. 2). Through these holes the fronto-naso-maxillary bone is cut free with an oscillating saw, and the flap is removed (Fig. 3).

After this step is completed, the subfrontal approach is begun. The dura mater is opened for inspection to look for possible tumoral infiltration and/or intradural tumoral budding. Any infiltrated tissue is removed, and gaps left in the dura mater are patched with fascia lata, which was removed from the leg during removal of the frontal bone flap. The maxillary sinus is then opened, and the maxillary and sphenopalatine arteries are inspected.

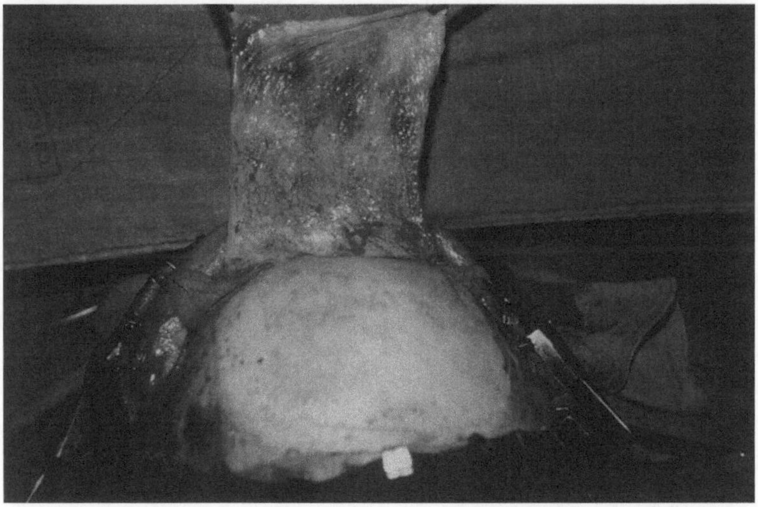

Fig. 1. Incision of the epicranium to form a large median flap with an inferior pedicle according to Derome's technique

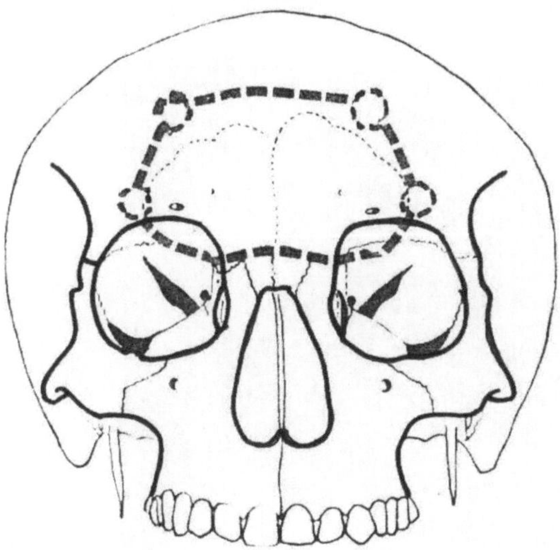

Fig. 2. The limits of the flap are defined by four trephine holes

The frontal approach permits visualization of the lateral spread of the tumor toward both the sphenoid sinuses and the nasal bone, while the submaxillary approach exposes the underside of the tumor (Fig. 4). Once the boundaries of the tumor are determined, it is freed and removed via the sublabial transmaxillary opening. Once the tumor is removed, the frontal dura mater is repaired with fascia lata, as previously described, and the epicranial flap is replaced and anchored to the jugum. In addition, a strip of fascia lata or

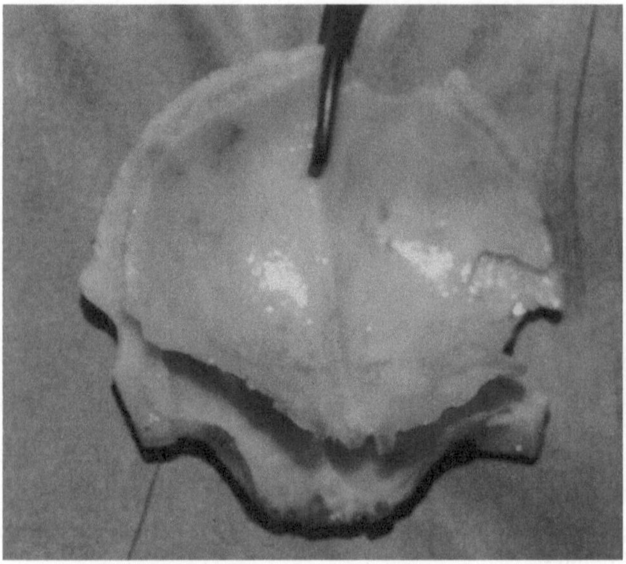

Fig. 3. The monobloc flap

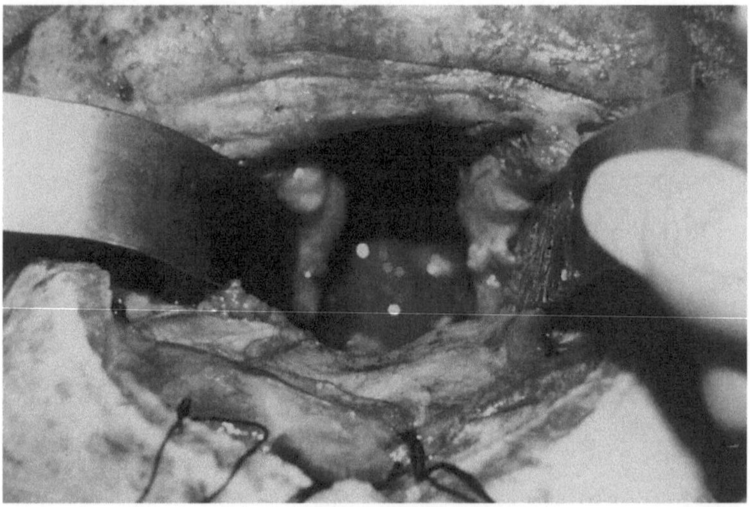

Fig. 4. The field of vision is increased by decreasing the retraction of the frontal lobes

of abdominal fat is inserted between the epicranium and the dura mater, and the frontal bone flap is either put back into place or replaced with a bone graft taken from the iliac crest. In either case, wire sutures are used to secure the flap.

Results

Over a 3-year period we have used this technique to operate on 28 patients suffering from malignant tumors of the ethmoid with tumoral invasion of the base of the skull. Eight of these patients suffered from intracranial infiltration; preoperative computed tomography revealed the extension in all eight cases. Total macroscopic exeresis of the tumor was possible in 24 of the cases; in the remaining four there was invasion of the cavernous sinuses, and complete removal was impossible.

Discussion

This technique is indicated in the surgical treatment of malignant tumors of the ethmoid with extension to the base of the skull. Frontal and axial computed tomography provides valuable information concerning the presence and extent of tumoral infiltration of the anterior dura mater, the orbits, and the cavernous sinuses [13]. In these types of cases, the double approach offers three distinct advantages. First, with the large frontal aperture it is possible to inspect the roof of the ethmoid, both extra- and intra-dural, the lamina cribrosia, the roof of the orbits, and the frontonasal junction. If necessary, the optic canals can be opened and inspected. With the frontal approach, the extent of lateral and posterior dissemination of the tumor is readily evident. Second, whereas postoperative leaking of CSF is an ever-present problem in the surgical treatment of these types of tumors [2, 4, 10, 11, 12], with this technique complete closure is possible, thereby avoiding this problem. Finally, the sublabial transmaxillary approach gives full access to the lower portion of the tumor, it leaves no visible facial scars, and it permits ligature of the internal maxillary artery and its branches, which reduces postoperative bleeding.

References

1. Bebear JP, Bagot d'Arc M (1982) L'ethmoidectomie totale par voie mixte: frontale extra-durale et paralatéronasale. Rev. Laryngol 3:179–188
2. Cophignon J, George B, Marchac D, Roux F (1983) Voie transversale élargie par mobilisation du bandeau fronto-orbitaire médian. Neurochirurgie 29:407–410
3. Derome P et al. (1972) Les tumeurs sphéno-ethmoidales. Possibilités d'exérèse et de réparations chirurgicales. Neurochirurgie 18 [Suppl 1]
4. Derome P (1983) Chirurgie des tumeurs osseuses de la base du crâne. Rev Laryngol 3:283–285

5. Hadjean E, Klap P, Thurel A, Bicot H, Tran Ba Huy P (1985) Le volet transfacial dans la chirurgie des tumeurs du tiers moyen de la face et de la base du crâne. Ann Otolaryng (Paris) 102:479–485
6. Johns M, Kaplan M, Jane J, Park TS, Cantrell R (1984) Supraorbital rim approach to the anterior skull base. Laryngoscope 94:1137–1139
7. Ketcham AS, Van Buren JM (1985) Tumors of the paranasal sinuses: a therapeutic challenge. Am J Surg 150:406–413
8. Lesoin F, Pellerin P, Villette L, Autricque A, Jomin M (1986) Mobilisation du bandeau fronto-orbitaire médian par volet monobloc Neurochirurgie 32:161–165
9. Mariel P, Barthelme A, Czorny A, Perrin C, Simon C (1986) Les cancers de l'ethmoide étendus à l'étage antérieur cérébral. Rev Laryngol 107:27–29
10. Portmann M, Guerrier Y (1983) Traité de Technique chirugicale, ORL, tome II. Nez et face. Masson, Paris
11. Schuller D, Goodmann J, Miller C (1984) Reconstruction of the skull base Laryngoscope 94:1359–1364
12. Thomassin JM, Pech A, Cannoni M et al. (1984) Eesthesioneuromes olfactifs. Compte-rendus Congrès Français ORL. Arnette, Paris, pp 359–362
13. Tiwari R, Gerritsen G, Balm A, Snow G (1986) A critical evaluation of the role of C.T. scanning in ethmoidal cancer. J Laryngol 100:412–428

Total Ethmoidectomy for Malignant Tumors of the Anterior Skull Base: 14 Years' Experience on 62 Cases

J. P. Bebear, D. Stoll, and V. Darrouzet, Bordeaux/France

Introduction

In the treatment of ethmoidal cancers by combined rhino-neurosurgical total ethmoidectomy we have 14 years of experience, having treated 62 patients in this way since 1974. This retrospective study describes the evolution of our surgical technique and considers the results obtained according to the histological nature of the tumor.

Technique

Following Ketcham et al. in 1963 [5] and Van Buren et al. in 1968 [8], the idea of a subfrontal approach for excision of ethmoidosphenoidal tumors was developed by Derome in 1972 [4]. With our experience in rhinology and knowing the frequent endocranial recurrence of ethmoidal cancers, since 1974 we have pursued the idea of a combined paralateronasal and subfrontal extradural excision of ethmoidal cancers [6]. This combined approach is based on the concept of ethmoidal cancer quickly overflowing the border of the sinus. Excision of the tumor according to the conventions of carcinology requires the control at the same time of all possible extensions, especially invasion of the roof of the ethmoid and of the cribriform plate, which are weak areas. Only a double approach satisfies this need and allows a monobloc excision of the ethmoid in its anatomic integrity. The surgical technique consists of five steps: (a) a single or double cutaneous incision; (b) a nasomaxillary flap at the end of the operation; (c) a frontal flap; (d) the monobloc excision of the tumorous ethmoid; and (e) the bony reconstruction of the anterior floor after dura mater plasty.

Technical Evolutions

Since 1974, radiologic imaging and its diagnostic possibilities have evolved along with surgical technique. As with all tumors capable of entering the orbit, but especially reaching the brain, the study of ethmoidal cancers benefits substantially from nuclear magnetic resonance imaging. Sagittal sections that show the anterior floor of frontal sections without the need for positioning or reconstruction allows one to visualize with great precision the invasion of the dura mater and the brain of the orbit. Such information is essential at the time of surgery.

Several modifications in our surgical technique seek to decrease postoperative morbidity by improving the fit of the bony assembly and, if possible, the conditions of the operation and its sequelae. For cutaneous incisions we now use a combined approach: a Cairns-Unterberger bifrontal scalp flap and a Moure paralateronasal incision. The cosmetic advantage is obvious, with the absence of a disfiguring frontal scar, and the preservation of the frontal branch of the facial nerve. A single incision is still indicated in the excision of large tumors invading the frontonasal bloc. In this case the necessary sacrifice of bone is made easier by the single incision which uncovers the frontonasal border. It is, however, necessary to be careful regarding the viability of the cutaneous flap at the internal canthus, especially after radiotherapy. In craniotomies we make the incision as low as possible to improve the view of the anterior floor and to avoid a bothersome bony edge. The frontal sinuses must be widely cranialized. Postoperative results are a consequence of the quality of the reconstruction. For the past year we have used a piece of cryopreserved iliac crest homograft instead of antologous iliac bone. The bone is taken in aseptic surgical conditions typical in organ transplants from a brain-dead patient free

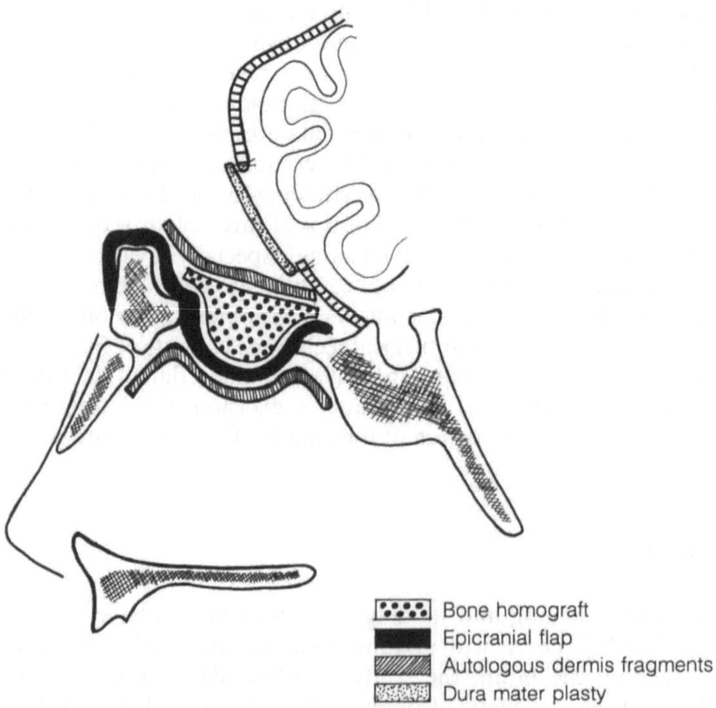

Fig. 1. Reconstruction of the anterior skull base

of any viral pathology. The iliac crest fragment is preserved until its use at −80 °C. At this temperature it loses all antigenic properties and so can be used as a free graft without risk of rejection. The piece of bone is shaped as a wedge, its cortical endocranial side being wider than its spongious endonasal side. This method eliminates the substantial bleeding due to the taking of the graft, as well as the frequent impotence and painful sequelae. Since 1984, we have cut a large rectangular epicranial flap with an inferior bridge before the craniotomy. This is flipped over backward after the excision and covers the anterior floor over the bony hiatus. The bone graft is secondarily interlocked, the epicranial flap used as a living joint for the assembly can be seen at its inferior endonasal side. It thus protects the bone graft from the septic conditions of the nasal cavity. Two fragments of autologous dermis, endocranial and endonasal, maintained with Tissucol, complete the assembly (Fig. 1).

Postoperative Complications

Infections and neurological problems predominate in postoperative morbidity. In a previous study involving 30 patients we noted the following complications [2]: CSF leaks with meningitis (three cases), orbit pus formation (three), frontal osteitis (one), deep hematomas of the iliac region (two), and intracranial hematoma (one). We have since operated on 32 patients without noticeable complications. The modifications in our technique therefore seem to carry a real benefit. As for postoperative morbidity, only four deaths were attributable to the intervention: one due to intracerebral hematoma, two to meningoencephalitis, and one to cardiac failure.

Results

All patients who were operated on presented with a large ethmoidal tumor, classified as T3 or T4. Only 55 patients had a long enough postoperative follow-up to be included in a survival study (Fig. 2). A total of 28 had adenocarcinomas. Of these, 16 received complementary irradiations. The corrected 5-year survival rate in this group was 80% (Fig. 3). Deaths were due to T4 tumors (with dura mater invasion). This was more often a matter of local recurrence (two cases) than of a lymph node recurrence or of an osseous metastasis (one case) or pulmonary metastasis (one case). There were 18 cases of epidermoid carcinoma. All of these patients were irradiated postoperatively. The corrected 5-year survival rate among these was only 35%. Recurrences were local (two cases) and metastatic (four cases). Olfactory bulb tumors (esthesioneurocytomas and esthesioneuroblastomas) were found in seven patients. All these were irradiated. They are all alive except for one who died of a local recurrence; this was the case of a second operation after incomplete excision. A further two patients had cylindromes.

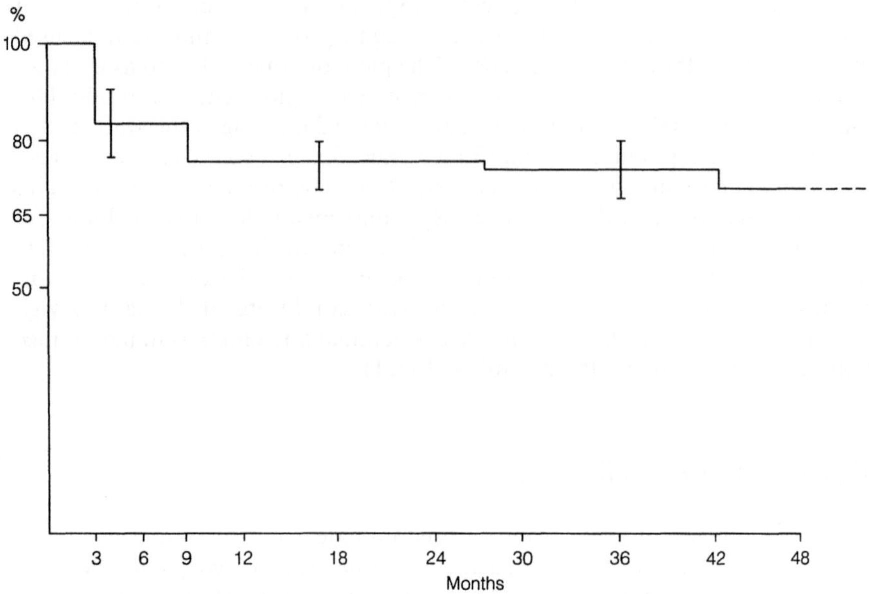

Fig. 2. Corrected survival rate in carcinomas (all types; $n = 55$)

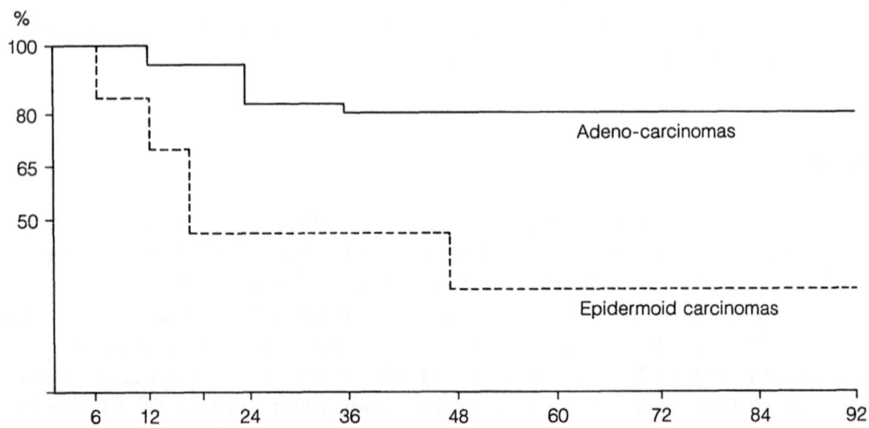

Fig. 3. Comparison of corrected survival rates in epidermoid carcinomas and adenocarcinomas

Indications and Contraindications

Analysis of the results demonstrates the benefits of this surgical approach in treating large ethmoidal tumors. This is particularly true for adenocarcinomas and olfactory bulb tumors. The latter, which are anterior floor tumors, are the typical indication for a surgical technique, which does seem to change their prognosis. The paralateronasal, purely rhinologic approach, which is not as difficult and possible whatever the grounds, remains acceptable for small tumors limited to one lateral mass of the ethmoid. The precision of anatomical figures offered by computed tomography and nuclear magnetic resonance imaging is very useful to the choice of treatment.

Contraindications stem from the tumor (massive brain as sphenoidal invasion) or from the grounds (elderly patient or one bearing severe visceral defects).

Conclusion

Total ethmoidectomy by paralateronasal and subfrontal approach, in extensive cancers of the ethmoid, allows a sensible improvement of survival. This is particularly true for adenocarcinomas and olfactory tumors. Recent improvements in the reconstruction technique of the anterior floor noticeably decrease morbidity secondary to a difficult surgical procedure.

References

1. Bebear JP, Stoll D (1982a) Epidémiologie des cancers de l'ethmoïde. Rev Laryngol 103:173–176
2. Bebear JP, Bagot d'Arc M (1982b) L'ethmoïdectomie totale par voie mixte: frontale extradurale et para-latéro-nasale. Rev Laryngol 103:179–188
3. Bebear JP, Stoll D (1986) A propos de 47 cas d'ethmoïdectomie totale par voie mixte. Congrès de la Société Française d'O.R.L et de pathologie cervico-faciale, Paris
4. Derome P (1972) Les tumeurs sphéno-ethmoïdales. Possibilités d'exérèse et de réparation chirurgicale. Neurochirurgie 18 [Suppl 1]
5. Ketcham AS, Wickins R, Van Buren JM, Smith RR (1963) A combined intracranial approach for the paranasal sinuses. Am J Surg 106:698
6. Portmann M, Bebear JP, Patris D (1977) Ethmoïdectomié totale par voie combinée: endocranienne et para-latéro-nasale. Technique, indications et limites. Communication 74ème Congrès Français d'O.R.L et de Pathologic cervico-faciale, Paris
7. Stoll D, Bebear JP (1984) The facial nerve in surgical approach to the subtemporal region and the jugular foramen. In: Portmann M (ed) Facial nerve. Masson, Paris
8. Van Buren JM, Ommaya AK, Ketcham AS (1968) Ten years experience with radical combined craniofacial resection of malignant tumors of the paranasal sinuses. J Neurosurg (Chicago) 28:341–350

The Status of the Frontal Sinus After Craniotomy

W. MANN, H. RIECHELMANN, Mainz/FRG, and J. GILSBACH, Aachen/FRG

Introduction

During a frontal, bifrontal, or pterional craniotomy for various neurosurgical procedures the frontal sinus is purposely or inadvertently opened. In case of a pterional craniotomy only lateral parts of the sinus are entered. If the frontal sinus is divided horizontally, the apical parts are removed with the osteoplastic flap, and the basal parts remain in place (Fig. 1). Ablation of the apical parts in the bone flap is best achieved by cranialization. The posterior wall and all sinus mucosa is carefully burred away. A galeaperiostal flap is split from the forehead flap and used to cover the basal parts of the frontal sinus. The residual basal sinus itself can be treated in three ways: (a) by obliteration with autogenous or alloplastic materials, (b) by performing a draining procedure, or (c) by preserving the structures of the basal parts of the sinus, thus creating a residual mini-sinus. In any event secondary diseases of the frontal sinus, such as chronic infection or mucocele formation that might follow frontal sinus surgery, must be prevented [2, 17].

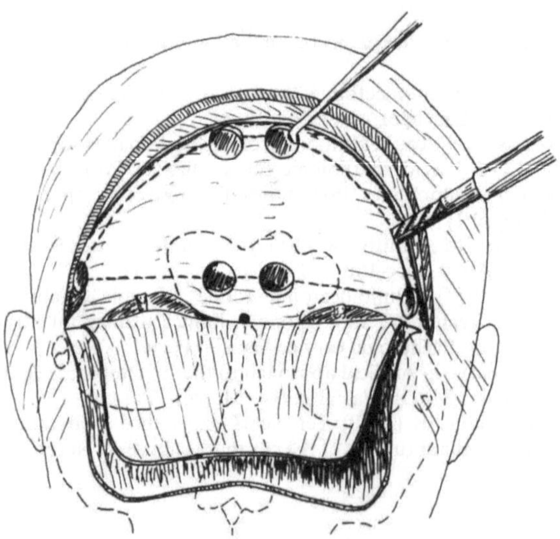

Fig. 1. Bifrontal craniotomy with burr holes over the frontal sinus. The sinus is bisected. The apical part is cranialized, and the basal part is covered with a galeaperiosteal flap

In the case of frontal craniotomy, obliteration of the residual sinus is recommended by various authors [7, 15, 18]. This is thought reliably to prevent secondary diseases of the frontal sinus. To accomplish this goal, all sinus mucosa must be carefully removed, the inner cortical layer of bone drilled away, the nasofrontal duct obliterated, and the sinus cavity filled with abdominal wall fat, skeletal muscle, or Gelfoam [4, 5, 14, 21]. Complications after obliteration are implant infection, re-aeration or mucocele formation [3, 17, 21]. The aim of this study was to evaluate the fate of the frontal sinus after various treatment modalities, concentrating on patients with attempted obliteration or creation of a residual mini-sinus.

Methods and Materials

The condition of the frontal sinus after previous craniotomy was evaluated in 39 patients (15 women and 24 men). The mean age of the patients was 38.9 years, ranging from 17 to 71 years. Patients were asked to return to the hospital after an average observation time of 24.3 months (7–48). In addition to the clinical examination, X-rays of the paranasal sinuses (Caldwell and Waters view), an ultrasound examination [9], and computed tomograms were taken with wide and narrow window settings [19]. In selected patients a nasal endoscopy was performed. In 25 patients an attempt had been made to obliterate the frontal sinus, mostly using free grafted muscle; in 11 a horizontal splitting of the frontal sinus had been performed, maintaining part of the original sinus lumen. In three patients, the frontal sinus had been entered laterally, and its lumen had been preserved.

Results

Follow-up examination revealed that sinus obliteration attempted after craniotomy was successful and complete in only 4 out of 25 patients. On the radiograms and the computed tomograms the former sinus lumen was completely replaced by dense bone, and ultrasound revealed no remaining sinus lumen or fluid retention. In 16 patients the obliteration attempt was less successful. The former sinus lumen was partly or completely re-aerated. Five patients had signs of sinus infection such as headache, discharge from the nasofrontal duct, and/or fluid retention or mucosal swelling in the ultrasonogram, the X-rays, and the tomogram. In patients in whom the frontal sinus had been split horizontally (Fig. 2), covering the basal part with an inferior based galeaperiosteal flap, a residual mini-sinus was created. During the control examination this mini-sinus was aerated and drained in ten patients. No one complained about recurrent episodes of sinusitis. In one patient, the frontal sinus was infected and filled with swollen mucosa (Table 1). In three patients the frontal sinus had been opened laterally and covered at the end of the procedure. Two patients showed ventilated and drained sinuses, where fascia

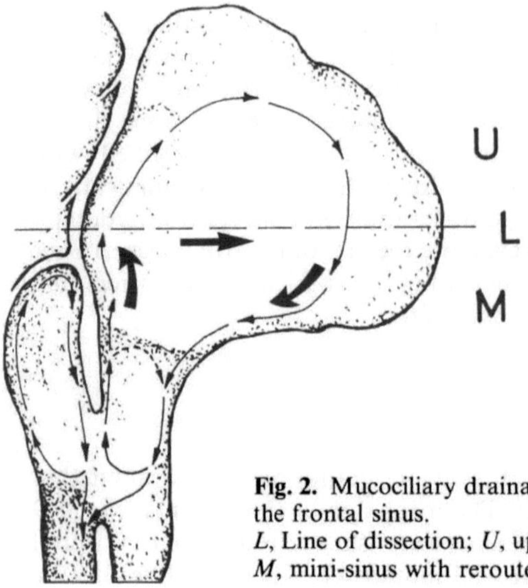

Fig. 2. Mucociliary drainage pattern after horizontal bisection of the frontal sinus.
L, Line of dissection; *U*, upper part is cranialized;
M, mini-sinus with rerouted mucociliary drainage pattern (*arrows*)

Table 1. Status of the frontal sinus 24 months after craniotomy and various surgical procedures

	Aerated	Re-aerated	Obliterated	Infected
Obliteration ($n = 25$)	–	16	4	5
Horizontal splitting ($n = 11$)	10	–	–	1
Lateral splitting ($n = 3$)	2	–	–	1

had been used to cover the sinus defect. One patient had a purulent frontal sinusitis. In his case, bone wax had been used to cover the lateral defect of the frontal sinus. The patient was admitted to surgery, the frontal sinus was opened and drained, and the thickened mucosa was removed together with fragments of bone wax.

Discussion

During frontal craniotomy severe trauma may be inflicted on the bony walls or the mucosa of the frontal sinus. This leads to the interruption of the normal mucociliary drainage pattern by simple loss of functional epithelium. If mucosa is removed from the sinus lumen, reepitheliazation with a functional mucosal layer takes up to 4 months [8]. In case of complete defects of the frontal sinus, including bone, periosteum, and mucosa, reepitheliazation occurs by migration of epithelium from the orders of the defect. A prerequisite for a successful epithelial migration is a vascularized and functional matrix which fills the defect. In case of frontal craniotomy with creation of a basal mini-sinus, the mucociliary clearance pattern, as described by Messerklinger [13], must adapt to the new anatomic situation (Fig. 2). But mucosal reepitheliazation, ventilation and drainage can be achieved only if the nasofrontal duct remains patent. Human observations and animal experiments have shown that even after severe impairment the frontal sinus is able to regenerate, provided the nasofrontal duct is not altered [4, 5, 6, 20, 21].

In our opinion there are some restrictions for preserving a residual sinus lumen. One is the case of a nasofrontal duct altered by a previous frontobasal trauma. May et al. [10] were able to demonstrate that impairment of the nasofrontal duct occurs in most frontobasal fractures. This results in chronic frontal sinusitis or mucocele formation [16]. A second restriction is, in our opinion, a history of chronic sinus infection with macroscopic mucosal changes, once the sinus is opened. And finally, there is no chance to preserve the sinus lumen without danger of later complication if the nasofrontal duct is damaged during craniotomy.

In each one of these conditions obliteration of the frontal sinus is mandatory. Removal of the inner cortical layer of the frontal sinus bone is part of the procedure to achieve safe obliteration [3]. This is not only a guarantee for complete removal of all mucosal remnants but opens up the Haversian canals for vascularization of implanted material. This technical detail, unfortunately, had not been observed initially in our patients, leading to a high percentage of re-aeration. A second reason for unsuccessful obliteration was the choice of implant material. Although free muscle grafts were used successfully in animal experiments to obliterate the frontal sinus [11, 17], this was not the case in our patients. Free fat grafts remain virtually unchanged when implanted [1, 14]. Muscle grafts of Gelfoam retract or are extruded via the nasofrontal duct. Replacement by fibrous tissue occurs with shrinkage and secondary epithelial ingrowth [11, 12].

In conclusion, on a long-term basis, the creation of a mini-sinus is an acceptable method to handle the frontal sinus during craniotomy, provided the nasofrontal duct is intact. Free muscle grafts implanted without removing the inner cortical layer of the frontal sinus were unreliable for obliteration. Bone wax, used to seal up frontal sinus defects, may initiate inflammation of the remaining mucosa.

Summary

In 39 patients (24 male, 15 female), the frontal sinus was opened in the course of a frontal, a bifrontal, or a pterional trephination for neurosurgical treatment of trauma, tumors, or aneurysms. In 25/39, the frontal sinus was subsequently obliterated; in 11/39, the apical half or two-thirds of the sinus was ablated and the remaining caudal part covered with a galea periosteal flap, leaving the nasofrontal duct intact. In 3/39 patients, the periphery of the sinus was ablated and the remaining part covered. The patients were followed up for a mean of 2 years and had a final evaluation using CT, X-rays, ultrasonography, and a complete ENT examination. In 4/25, obliteration was successful and complete; in 16/25, there was a partial or complete re-aeration of the residual sinus, and in 5/25 the sinus was infected. In one out of 11 patients in whom the frontal sinus was bisected with preservation of the lower half, the basal part of the sinus was infected; in 10/11, the residual sinus was aerated and well drained. In one out of three patients in whom the periphery of the sinus was ablated, there was infection; two of the three sinuses were aerated. Evaluating the experience in these 39 patients and our recent experience, obliteration of the sinus with muscle is very unreliable and the use of bone wax is obsolete. Covering a bisected or peripherally opened sinus with fascia or a galea periosteal flap yields a well aerated and drained sinus, provided the nasofrontal duct is intact.

References

1. Bergara AR, Itoiz AO (1955) Present state of the surgical treatment of chronic frontal sinusitis. Arch Otolaryngol 61:616–628
2. Bordley JE, Bosley WR (1973) Mucoceles of the frontal sinus: causes and treatment. Ann Otol 82:696–702
3. Donald PJ (1979) The tenacity of frontal sinus mucosa. Otolaryngol Head Neck Surg 87:557–566
4. Goodale RL (1965) Obliteration of the frontal sinus. Ann Otol 74:831–839
5. Goodale RL (1965) The rationale of frontal sinus surgery. Laryngoscope 75:981–987
6. Hybels RL, Newman MH (1977) Posterior table fractures of the frontal sinus. I. An experimental study. Laryngoscope 87:71–179
7. Malis LI (1985) Surgical resection of tumors of the skull base. In: Wilkins RH, Rengachary SS (eds) Neurosurgery. McGraw-Hill, New York, p. 1011
8. Mann W (1977) Neue diagnostische Möglichkeiten bei Entzündungen der Nasennebenhöhlen. Habilitationsschrift Med. Fakultät, Freiburg
9. Mann W (1984) Ultraschall im Kopf-Hals-Bereich. Springer, Berlin Heidelberg New York Tokyo
10. May M, Ogura JH, Schramm V (1970) Nasofrontal duct in frontal sinus fractures. Arch Otolaryngol 92:534–539
11. McCoy G (1963) Cerebrospinal rhinorrhea. Laryngoscope 73:1125–1156
12. McNeill RA (1967) Surgical obliteration of the maxillary sinus. Laryngoscope 77:202–217
13. Messerklinger W (1967) Über die Drainage der menschlichen Nasennebenhöhlen unter normalen und pathologischen Bedingungen. 2. Mitteilung: Die Stirnhöhle und ihr Ausführungsgangsystem. Mschr Ohrenheilkd 101:313–326
14. Montgomery WW, Pierce DL (1963) Anterior osteoplastic fat obliteration for frontal sinus: clinical experience and animal studies. Trans Am Acad Ophthalmol Otolaryngol 67:46–57
15. Ojemann RG (1982) Basal frontal meningeomas. In: Schmidek HH, Sweet WH (eds) Operative neurosurgical techniques. Grune & Stratton, New York London Paris, p. 549
16. Schenk NL, Rauchbach E, Ogura JH (1974) Frontal sinus disease. II. Development of the frontal sinus model: occlusion of the nasofrontal duct. Laryngoscope 84:1233–1247
17. Schenk NL (1975) Frontal sinus disease. III. Experimental and clinical factors in failure of the frontal osteoplastic operation. Laryngoscope 85:76–92

18. Seeger W (1983) Microsurgery of the cranial-base, 3rd vol. Frontobasal operations. Springer, Berlin Heidelberg New York Tokyo, figures 100ff
19. Som PM (1985) CT of paranasal sinuses. Neuroradiology 27:189–201
20. Walsh TE (1943) Experimental surgery of the frontal sinus: the role of the ostium and the nasofrontal duct in postoperative healing. Laryngoscope 84:75–92
21. Zonis RD, Montgomery WW, Goodale RL (1966) Frontal sinus disease: 100 cases treated by osteoplastic operation. Laryngoscope 76:1816–1476

Juvenile Nasopharyngeal Angiofibroma: The Update Concept of Diagnosis and Therapy

W. DRAF, J.-P. HAAS, and H.-P. RICHTER, Fulda/FRG

Juvenile nasopharyngeal angiofibromas frequently require surgery of the paranasal sinuses as well as of the sellar region. This demands a close interdisciplinary cooperation before, during, and after the intervention (Samii and Draf 1989). The symptoms depend on the size of the tumor. Nasal obstruction, nasal speech, cheek swelling, dislocation of orbital contents, recurrent middle ear effusions, and headache have been the most frequent ones in our seven fairly extended cases, operated on over the past 5 years. Recurrent bleeding occurred less commonly, as one might expect. In only one case was a cranial nerve deficit, namely a deterioration of vision, observed. This recovered completely following optic nerve decompression after microsurgical tumor removal. In three cases intracranial extension was obvious.

For therapy we prefer surgical removal of angiofibroma rather than irradiation, using the latter in cases of recurrence (Table 1). A successful operative strategy for eradication of this highly vascularized tumor requires a precise definition of the tumor's location, its vascular supply, and its relation to neighboring structures. High-resolution computed tomography with multiplanar reconstruction and magnetic resonance imaging provide precise information about the extent and possible intracranial invasion. The intracranial part is often firmly attached to the dura, thinning it out or penetrating it (Figs. 1, 2). Superselective digital substraction angiography is most valuable for determining the tumor extent and its blood supply. Following diagnostic

Table 1. Juvenile angiofibroma, strategy of treatment

A.	Complete knowledge of extensiveness (CT, MRI, DSA)
B.	Management of blood loss (Autotransfusion with cell saver, embolization)
C.	Adequate and atraumatic exposure Macro-microsurgical dissection Osteoplastic technique Preservation: infraorbital nerve Nasolacrimal duct
D.	Combination of anterior and lateral approach
E.	Reconstruction of facial sceleton
F.	Radiotherapy second choice

1 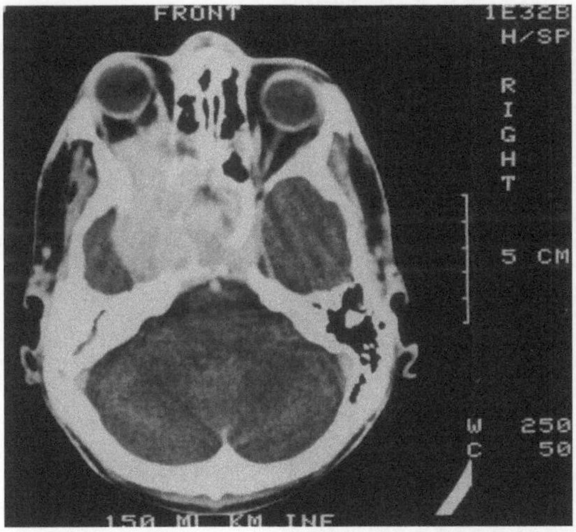

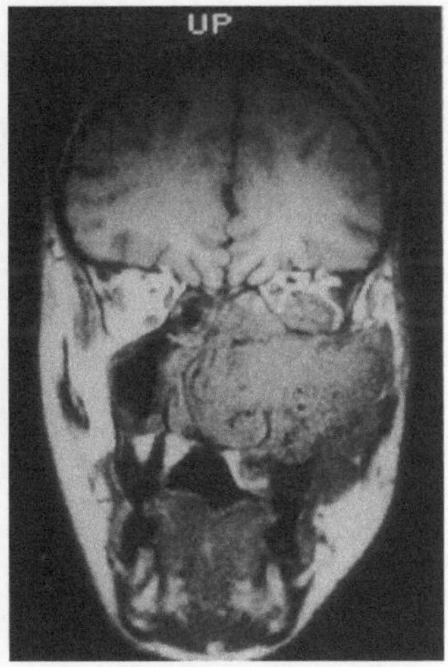 2

Fig. 1. Extensive juvenile angiofibroma invading the anterior and middle cranial fossae, the orbit, the pterygopalatine, pterygoid, and infratemporal fossae. High-resolution computed tomography, axial section

Fig. 2. Extensive juvenile angiofibroma (same case as Fig. 1), magnetic resonance imaging, coronal section. (Kindly provided by Dr. Keil and Dr. Hetzel, Würzburg, FRG)

angiography, therapeutic embolization of tumor feeders from the external and even the internal carotid arteries helps greatly to reduce the blood loss during subsequent surgery (Lasjaunias and Berenstein 1987).

Intraoperative hemorrhage increases postoperative morbidity and mortality (Panje and Gross 1987). If major blood loss during surgery can be expected, as in cases of vascular supply coming from external *and* internal carotid arteries, we now borrow routinely the cell saver from our cardiac surgeons (T. Stegmann, personal communication 1987; Fig. 3). This allows immediate autotransfusion of the patient's blood after its filtration and cleaning and enables us to continue tumor dissection in spite of major bleeding even out of the cavernous sinus. Cell saver and embolization are important in easing complete removal and reducing surgical trauma.

The adequate and least traumatic exposure possible using macro- and microsurgical techniques includes temporary removal of the anterior wall of the maxillary sinus and the ethmoidal cells as osteoplastic lateral rhinotomy. The infraorbital nerve should be preserved, mobilizing and displacing it laterally (Fig. 4). Since 1980 we also routinely save the nasolacrimal duct and its orifice after shifting it temporarily out of the surgical field. At the end of the operation it is sutured back into the cheek flap (Draf 1980).

In extensive angiofibromas such as the case demonstrated in Fig. 1, 2, 4, 5, 6, we combine the direct transzygomatic approach to the infratemporal and pterygoid fossa with the lateral rhinotomy. We thus avoid major weakening

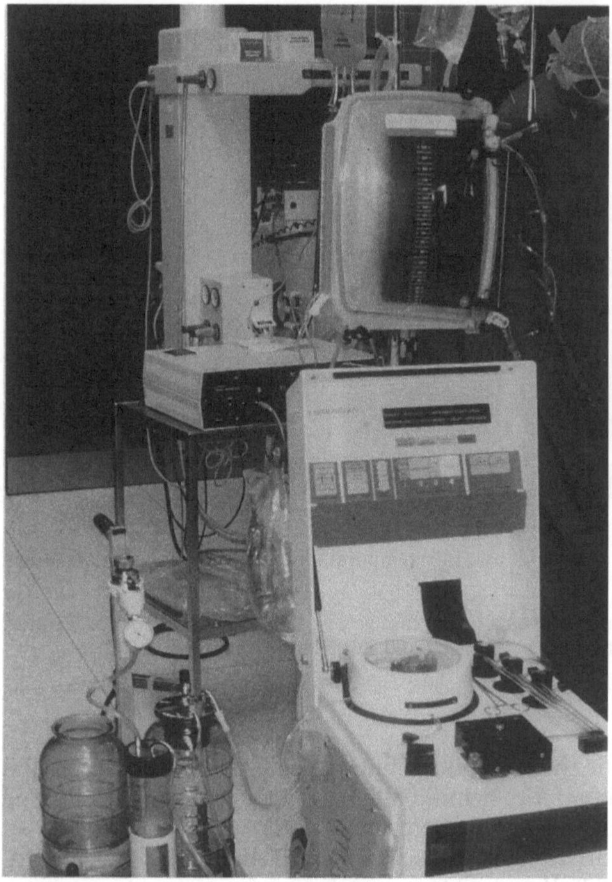

Fig. 3. So-called cell-saver for immediate blood retransfusion. (The valuable support of Prof. Dr. T. Stegmann, Department of Cardiac, Thoracic and Vascular Surgery, Städtische Kliniken, Fulda, FRG, is gratefully acknowledged)

and resection of the facial skeleton. In our series we never considered the transpalatal approach necessary.

After total removal of the tumor the facial skeleton is reconstructed to avoid early or late distortion of the face. Bone autografts and sometimes in addition stored cartilage are fixed with miniplates, and finally the cheek flap is replaced meticulously. Figure 6 shows the patient 2 months after surgery.

In summary, surgery of juvenile nasopharyngeal angiofibroma has been practiced for centuries; in fact it goes back to the time of Hippocrates, who described removal of a "hard nasal polyp." We think that surgery of this highly vascularized benign tumor has been improved by the strategy described here, not only in terms of total extirpation but also regarding reduced morbidity and mortality.

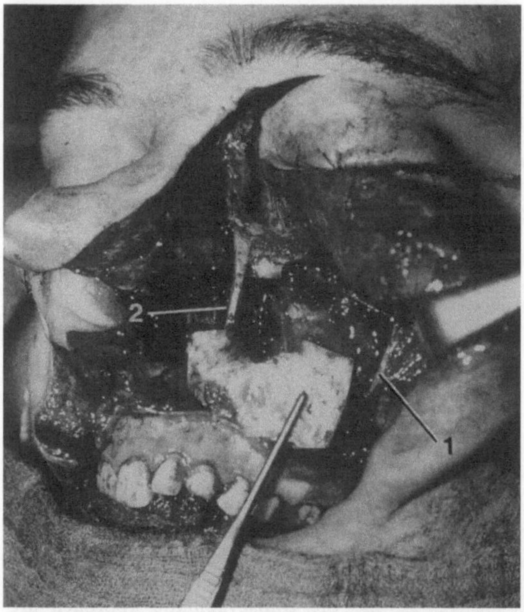

Fig. 4. Osteoplastic lateral rhinotomy. The infraorbital nerve (*1*) as well as the lacrimal sac and nasolacrimal duct (*2*) are preserved

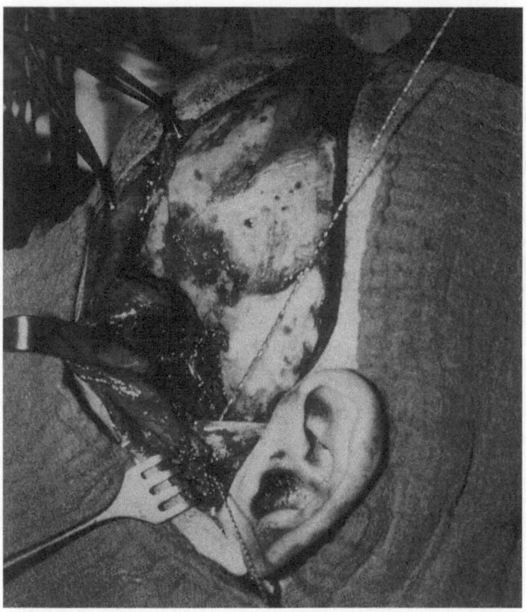

Fig. 5. Direct lateral, transzygomatic approach to the middle cranial fossa in combination with lateral rhinotomy helps to preserve as much as possible of the facial skeleton

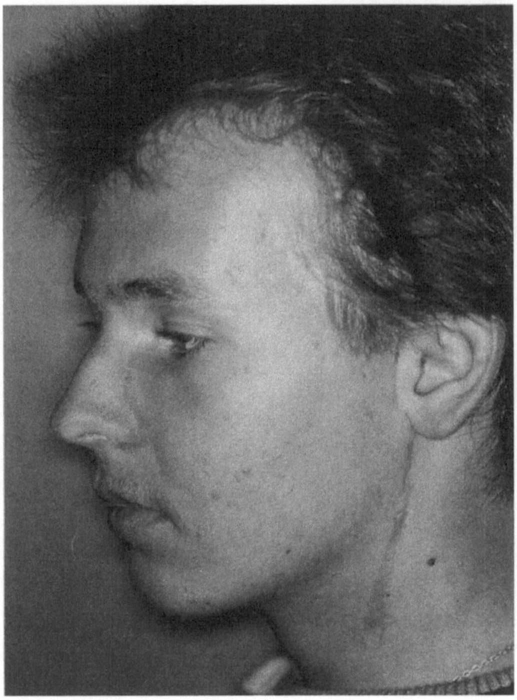

Fig. 6. Patient 2 months after surgery (same case as in Figs. 1, 2). The scars of the lateral rhinotomy and the transzygomatic approach are relatively discrete. The scar of the neck incision for control of the internal carotid artery is still obvious.

References

Draf W (1980) Tumors of the anterior skull base. Lecture at the Course on Surgery of the anterior skull base. Hannover

Lasjaunias P, Berenstein A (1987) Surgical Neuroangiography. Springer, Berlin Heidelberg New York Tokyo

Panje WR, Gross CE (1987) Surgical therapy of tumors of the pharynx. In: Thawley, Panje, Batsakis, Lindberg (eds) Comprehensive management of head and neck tumors, vol 1. Saunders, Philadelphia, pp 662ff

Samii M, Draf W (1989) Surgery of the skull base. An interdisciplinary approach. Springer, Berlin Heidelberg New York Tokyo

Computed Tomography-Stereotactic Curie Therapy of Recurrent Nasopharyngeal Carcinoma Invading the Skull Base

A. Etou, Rödermark/FRG, M. Mohadjer, and F. Mundinger, Freiburg i. Br./FRG

Introduction

Malignant nasopharyngeal tumors are frequently discovered late by unspecific symptoms such as headache, by infiltration of the paranasal sinuses or the skull base, or by swelling of the lateral neck by involvement of the related lymph nodes. In such an advanced stage effective surgical treatment can scarcely be achieved and is confined to tumor reduction or preradiotherapy neck dissection which may be indicated in certain cases (Mann et al. 1984; Scheunemann et al. 1986).

The primary treatment of the malignant nasopharyngeal tumors is the domain of radiotherapy (Wannenmacher 1984), especially since histologically they are mostly lymphoepithelial or undifferentiated carcinomas with a high radiosensitivity. However, neither by combined radiosurgical therapy nor by adjuvant chemotherapy (Yamashita et al. 1985) have the rates of cure and recurrence-free survival been improved satisfactorily. The major reason for the still rather poor prognosis is local tumor infiltration extending to the skull base and invading the intracranial space after the treatment. Generally, neurosurgical treatment is not reasonable in the case of tumor relapse extending to the middle cranial fossa or parasellar region.

In these cases as a possible further treatment with relatively low risk and burden for the patient the stereotactic operation guided by computed tomography (CT) or magnetic resonance imaging should be considered. First, by stereotactic biopsy the tumor recurrence must be confirmed histologically and differentiated from radionecrosis. Provided the tumor is tolerably delimited during the same operation, radionuclides for interstitial Curie therapy can be implanted.

Recently in several cases with reoccurring nasopharyngeal carcinoma after percutaneous irradiation we used the stereotactic device of Riechert and Mundinger (Birg and Mundinger 1973; Riechert and Mundinger 1956; Fig. 1) to perform stereotactic biopsy of the lesion in the skull base and intracranially and implanted radionuclides for interstitial radiotherapy. To date no tumor relapse has occurred. A description of the method and technique of the operative procedure, which can be performed in local anesthesia, has been published before (Mundinger 1970, 1979, 1982, 1988; Mundinger and Birg 1988; Mundinger and Weigel 1988).

a

b

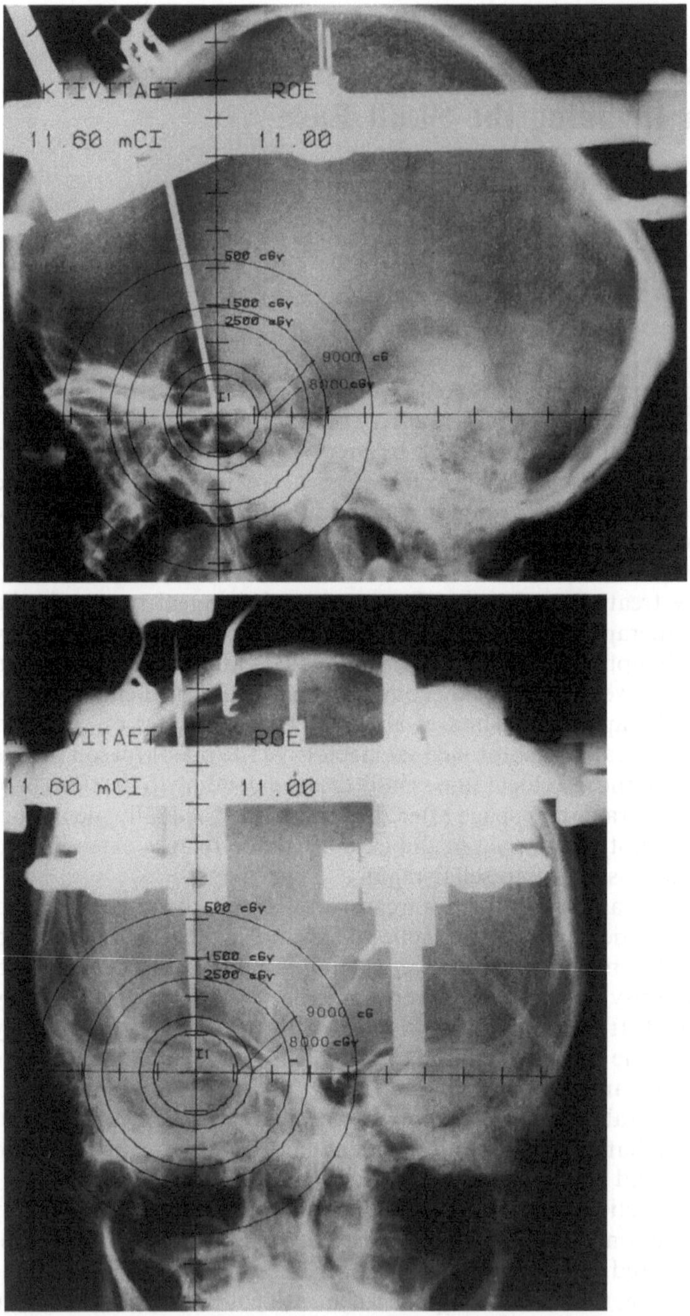

Fig. 1 a, b. The stereotactic device of Riechert and Mundinger in position with the probe situated and a superimposed isodosogram (**a**: lateral, **b**: a.-p. view)

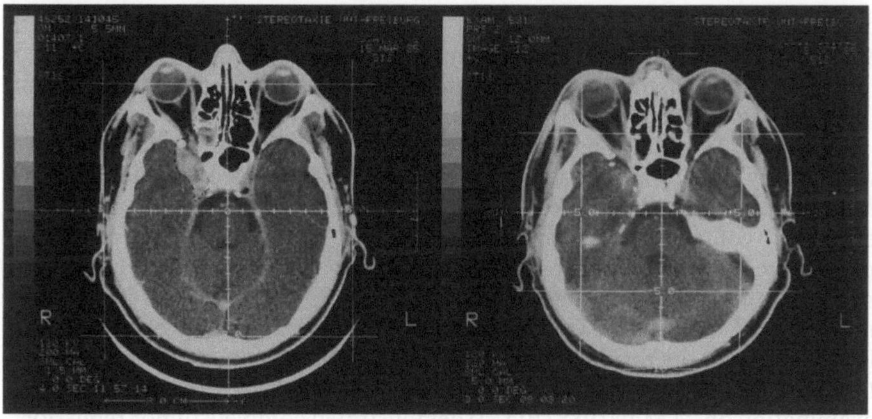

Fig. 2. Preoperative CT scan showing intracranial tumor extension (*left*). CT control, 7 months after interstitial Curie therapy

Case Report

A now 43-year-old man noticed a swelling at the right side of the neck in November 1982. Oto-rhino-laryngological examination and biopsy revealed a nasopharyngeal carcinoma in an early stage. After percutaneous radiation of the epipharynx including the regional lymph nodes with 70 Gy, tumor remission was observed. Two years later in November 1984 the patient complained of double vision and progressive vision disorder in the right eye. Abducens nerve palsy and optic nerve atrophy were found. The CT examination showed a tumor relapse extending to the parasellar area and to the right

3

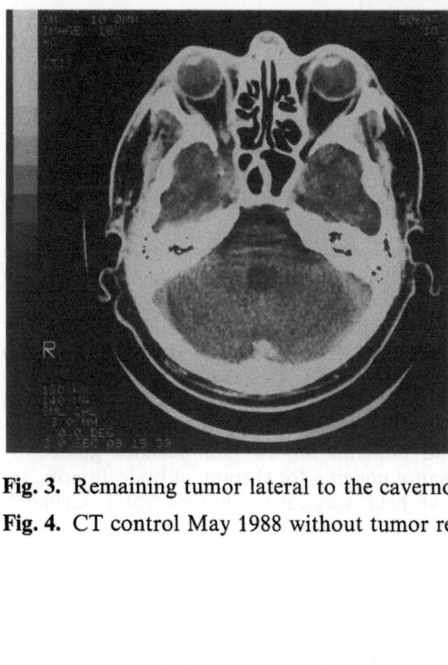

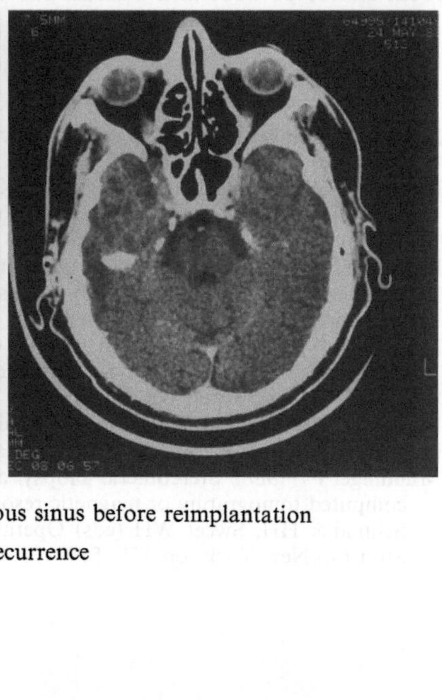

4

Fig. 3. Remaining tumor lateral to the cavernous sinus before reimplantation

Fig. 4. CT control May 1988 without tumor recurrence

orbital apex (Fig. 2). In April 1985, by stereotactic biopsy, an undifferentiated epithelial carcinoma with lymphatic infiltrates was confirmed. We implanted iodine-125 for interstitial radiation with 90 Gy calculated for a radius of 15 mm (see also Fig. 1).

In May 1985 additional focal percutaneous irradiation of the skull base with 50 Gy was applied, which had not been included at the primary radiotherapy. Tumor volume then reduced to approximately 40%, as shown on a CT control in January 1986 (Fig. 3). After verification of tumor persistence by stereotactic biopsy the remaining tumor lateral of the cavernous sinus was then again implanted with iodine-125 with a dose of 80 Gy at a 10-mm radius. Clinically the patient's visual disturbance had improved after the first stereotactic operation. Paraesthesias now occurring on the right side of the face were relieved by carbamazepine. The patient is now well. A CT control examination on 24 May 1988 (Fig. 4) demonstrated complete tumor remission.

Discussion

The therapy of malignant epipharyngeal tumors is a special problem when tumor relapse occurs after external radiation therapy, and the reoccurring tumor infiltrates the skull base and extends intracranially as a space-occupying lesion. In these cases, as an additional further treatment, stereotactic interstitial radiation should be considered. This indication is not yet common or routine, and to administer this treatment successfully an established stereotactic unit is required.

Although the cases that we have treated up to now show promising results, and the patient illustrated above has now been surviving in a good state for 6 years, further observation and experience is necessary to evaluate the effect on local tumor control and overall survival time.

References

Birg W, Mundinger F (1973) Computer calculations of target parameters for a stereotactic apparatus. Acta Neurochir 29:123–129

Mann WJ, Wannenmacher MR, Beck C, Kigele I (1984) Kombinierte radio-chirurgische Therapie der Nasopharynxtumoren – Freiburger Erfahrungen. Verh Dtsch Krebs Ges 5:319–323

Mundinger F (1970) The treatment of brain tumors with interstitially applied radioactive isotopes. In: Yen Wang, Paoletti P (eds) Radionuclide applications in neurology and neurosurgery. Thomas, Springfield/I11, pp 199–265

Mundinger F (1979) Rationale and methods of interstitial iridium-192-Brachy-Curie-therapy and iridium-192- or iodine-125 protracted long term irradiation. In: Szikla G (ed) INSERM Symposium No. 12. Elsevier/ North-Holland, Amsterdam, pp 101–115

Mundinger F (1982) Implantation of radioisotopes (Curie therapy). In: Schaltenbrand G, Walker AE (eds) Stereotaxy of the human brain. Thieme, Stuttgart, pp 410–435

Mundinger F (1988) Stereotactic biopsy and implantation of radionuclides guided by computed tomography or magnetic resonance imaging for therapy of brain tumors. In: Schmidek HH, Sweet WH (eds) Operative neurosurgical techniques, vol 1. Grune & Stratton, New York, pp 491–514

Mundinger F, Birg W (1988) The imaging-compatible Riechert-Mundinger-system. In: Lunsford LD (ed) Modern stereotactic surgery. Nijhoff, Boston, pp 13–25
Mundinger F, Weigel K (1988) Considerations in the usage and results of Curie therapy. In: Lunsford LD (ed) Modern stereotactic surgery. Nijhoff, Boston, pp 245–258
Riechert T, Mundinger F (1956) Beschreibung und Anwendung eines Zielgerätes für stereotaktische Hirnoperationen (2. Modell). Acta neurochir 3:308–337
Scheunemann H, Schürmann K, Helms J (eds) (1986) Tumors of the skull base. De Gruyter, Berlin New York
Wannenmacher MR (ed) (1984) Nasopharynxtumoren. Urban & Schwarzenberg, München Wien Baltimore
Yamashita S, Konto M, Hashimoto S (1985) Squamous cell carcinoma of the nasopharynx. Acta Radiol Oncol 24:4

Isolated Sphenoid Sinus Aspergillosis with Intracranial Extension: Report on Three Cases

L.-S. Lin, S.-L. Howng, W.-T. Chiu, J.-K. Loh,
and C.-J. Shih, Taipei and Kaohsiung, Taiwan/ROC

Introduction

Aspergillus is a fungus of the class Ascomycetes which is of worldwide distribution. It is commonly found on decaying debris and in soil. Man comes into contact with a variety of *Aspergillus* species, however only seven species have been implicated in human infections. *Aspergillus fumigatus* has accounted for approximately 90% of *Aspergillus* infections [9], and the common *Aspergillus flavus* accounts for the rest. Aspergillosis of the paranasal sinus is rare, except in the Sudan, where various publications have indicated the disease to be endemic [7, 10]. The maxillary sinus is most frequently involved. Peterson et al., in a review of world literature in 1982, reported fewer than 120 cases of aspergillosis of the maxillary sinus [9]. Infection of the sphenoid sinus is extremely rare. Comoretto et al. in 1986 reported only 14 cases, in a review of world literature [1].

We present here three additional cases of isolated sphenoid sinus aspergillosis, all of which involved intracranial extension and were encountered in the past 3 years.

Case Reports

Case 1. A 60-year-old man had a history of diabetes mellitus for 3 years without adequate treatment. He was admitted with the chief complaint of right temporal headache and blurring of the right eye for about 2 months. Neurological examination revealed bitemporal hemianopsia. A lateral view skull X-ray showed haziness of the sphenoid sinus and bony destruction of the floor of the sella turcica. Computed tomography disclosed a lesion in the sphenoid sinus with extension into the sella turcica; as seen in Fig. 1, the close view at the level of the sella turcica showed haziness of the sphenoid sinus and bony destruction of the floor of the sella turcica.

The sphenoid sinus was approached transseptally under general anesthesia. Removal of the anterior wall of the sphenoid sinus revealed the cavity to be filled with a cheesy, grayish-green material. Histological examination disclosed *Aspergillus* infection (Fig. 2), and Amphotericin-B therapy was given for the prevention of systemic spread of the disease. Following surgery, the patient noted a great improvement in his vision and cessation of headache. The patient has remained asymptomatic.

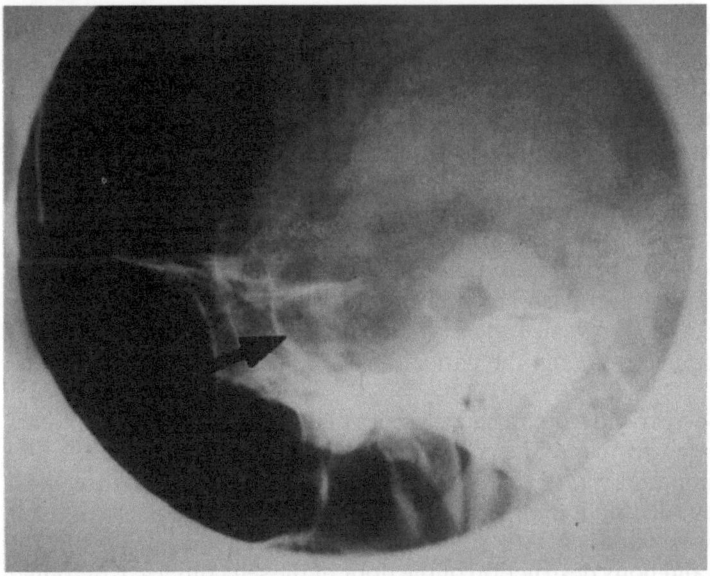

Fig. 1. Lateral view of the skull showing opacity of the sphenoid sinus and bone destruction of the floor of the sella turcica (*arrow*)

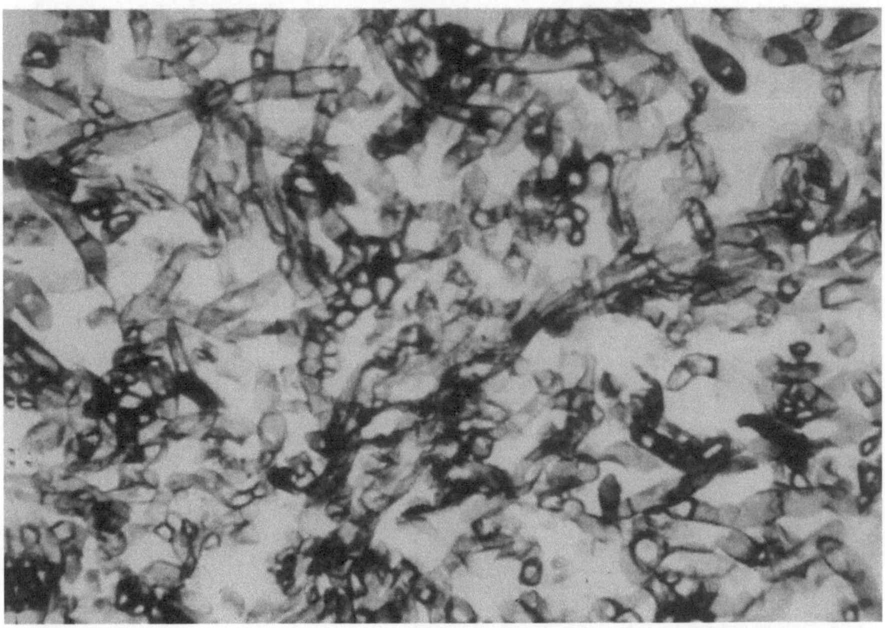

Fig. 2. Photomicrograph showing large septate hyphae compatible with *Aspergillus* infection of the sphenoid sinus

Case 2. A 57-year-old man had a history of diabetes mellitus for 12 years without adequate treatment. He complained of left temporo-occipital headache for about 3 months and left eyelid droop for 1 week. Neurological examination revealed left 3rd, 4th, and 6th cranial nerve palsy. A lateral view skull X-ray showed haziness in the sphenoid sinus and bony destruction at the floor of the sella turcica, as in case 1. Computed tomography, the coronal view of the sphenoid sinus, disclosed a lesion in the left sphenoid sinus with extension into the sella turcica and left cavernous sinus, as shown in Fig. 3. Similarly as above, we performed sublabial transseptal sphenoidotomy. Some necrotic material with grayish-green cheesy consistency was obtained. Histological examination revealed *Aspergillus* species. Again Amphotericin-B therapy was administered, and the patient has since remained asymptomatic.

Case 3. A 58-year-old woman had a history of diabetes mellitus for 10 years without adequate treatment. She complained of right temporal headache and blurring of vision and right eyelid droop for about 3 months. Neurological examination revealed right 2nd, 3rd, 4th, and 6th cranial nerve palsy, and left abducens nerve paralysis. A lateral view skull X-ray showed haziness in the sphenoid sinus and bony destruction of the floor of the sella turcica. Computed tomography disclosed a lesion in the sphenoid sinus with extension into the sella turcica and both superior orbital fissures, more on the right side (Fig. 4). A lateral view carotid angiogram showed elevation of the carotid siphon (Fig. 5). Again, we performed sublabial transseptal sphenoidotomy and discovered the same as in the previous two cases. Amphotericin-B was administered immediately after histological proof of the *Aspergillus* infection. The patient has since remained asymptomatic.

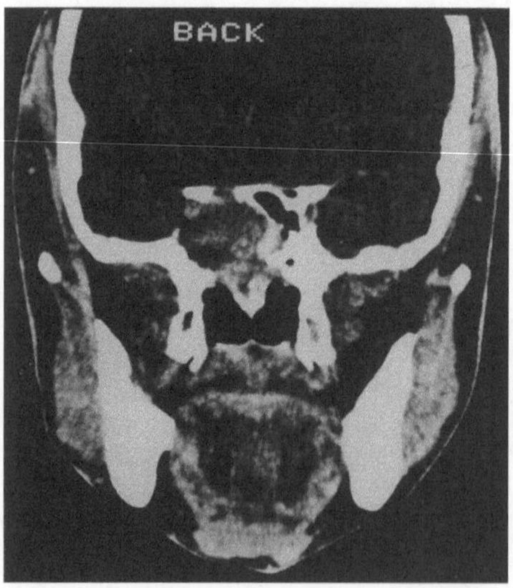

Fig. 3. CT scan: the coronal view of the sphenoid sinus shows a lesion in the left sphenoid sinus with extension into the sella turcica and left cavernous sinus

4 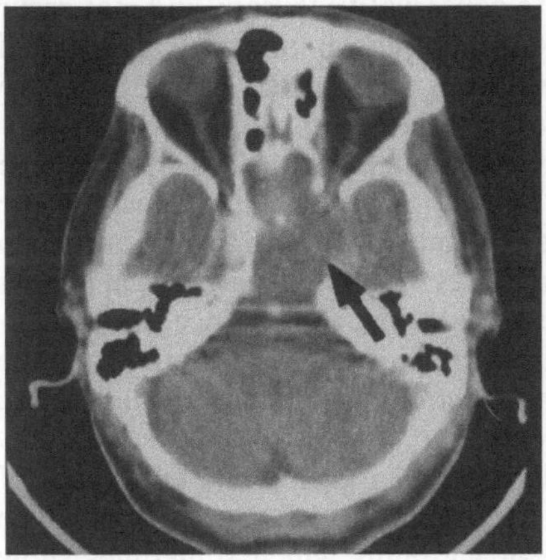 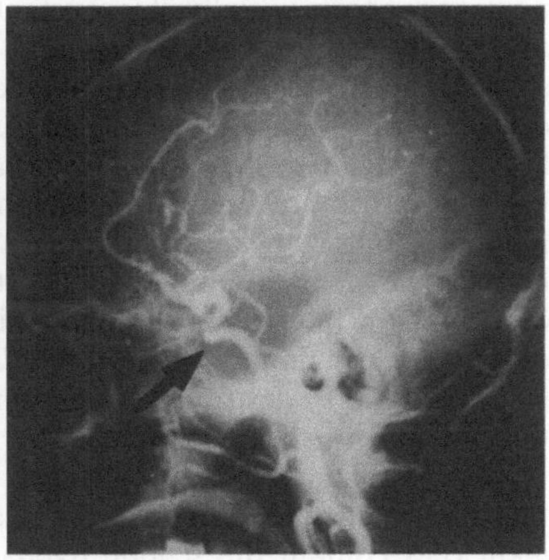 5

Fig. 4. CT scan of the sphenoid sinus showing a mass lesion with extension into the sella turcica and both superior orbital fissures, more on the right side (*arrow*)

Fig. 5. Lateral view of the carotid angiogram showing elevation of the carotid siphon (*arrow*)

Discussion

Aspergillus most commonly occurs as a saprophytic infection of the lining of the aerodigestive tract. It can also be a rapidly fatal disease with local invasion and hematogenous spread. Although *Aspergillus* spores are ubiquitous, and exposure is common, clinically apparent aspergillosis is rare. *Aspergillus fumigatus* and *flavus* are the most common species reported. The fungus becomes pathogenic under anaerobic condition [7].

It has been speculated that the warm moist climate of agricultural regions predisposes the population to a high rate of allergic, hypertropic, vasomotor, and infectious rhinosinusitis and provides the conditions for *Aspergillus* infection due to altered physiology of the upper airway [4]. Increasing use of immunosuppressive and oncological drugs further increase the incidence of aspergillosis due to lowered host resistance [4, 7, 11]. Metabolic diseases such as diabetes mellitus, are known to increase the susceptibility of patients to *Aspergillus* infections [5, 7, 12]. The presence of a foreign body in the sphenoid sinus has been reported to favor aspergillus infection [1].

In the few documented cases of isolated sphenoid sinus aspergillosis, headache was the most frequent symptom. Other symptoms included postnasal bleeding,facial pain, and nasal obstruction. Ophthalmological symptoms have been limited to blurred vision, but in two of the three cases that we describe the patients presented with drooping of eyelid on the same side as the headache.

Clinically, Hora [3] described two discrete forms of paranasal sinus aspergillosis, the noninvasive and invasive forms. The noninvasive form, which occurs more frequently, responds to removal of the aspergilloma and drainage of the involved sinus or sinuses. However, the invasive form may resemble the noninvasive form initially; it is resistant to conservative therapy and may recur after combined radical excision and Amphotericin-B therapy [6]. Isolated sphenoid sinus aspergillosis and paranasal sinus aspergillosis appear to be two distinct clinical entities [6]. Only 14 cases of isolated sphenoid sinus aspergillosis have been reported. All of these have been the noninvasive form but have behaved in a more aggressive fashion than noninvasive aspergillosis of the paranasal sinuses. This greater severity is explained by the intimate relationship between the sphenoid sinus and other structures of the skull base. Local factors disrupting the integrity of this anatomic relationship, for instance transsphenoid surgery, can predispose to *Aspergillus* meningitis [1]. The bony walls of the sinus need not be destroyed, intracranial infection may spread through septic thrombophlebitis via communicating veins.

The histological picture is essentially that of a granuloma composed of multinucleated giant cells scattered in a fibrotic connective tissue stroma. The giant cells contain the septate hyphae of a fungus which sometimes show branching, which are best visualized with PAS stain. Lymphocytes and plasma cells are often seen, and blood vessels may show endarteritis obliterans. The center of some granuloma may be presented as microabscess, in which free fungal hyphae are found (Fig. 5). In some cases, the center of the granuloma is composed of eosinophilic necrotic material surrounded by fungus containing giant cells and palisading histiocytes [7].

Radiographic studies show opacities of sphenoid sinus with bone destruction (Figs. 1–3), thus resembling a malignant neoplasm. It is therefore necessary to make differential diagnosis with the following sphenoid lesions: infections (polypoid sinusitis, mucocele or pyocele, etc.), primary tumors (squamous cell carcinoma, adenocarcinoma, cylindroma, giant cell tumors, sarcoma, etc.), and secondary invaders (craniopharyngioma, chordoma, pituitary tumor, meningioma, etc.). In differentiating among these, we find computed tomography to be very useful, and cerebral angiography can differentiate new growth from vascular lesions.

A review of documented reports reveals that surgical treatment has been the treatment of choice [9]. Sphenoid packing is not necessary, and it is in fact inadvisable. Amphotericin-B therapy should be reserved for those rare instances in which histological evidence of tissue invasion by fungus is present. The key to successful surgical treatment is the removal of diseased mucosa and aeration and drainage of the involved sinus. For invasive disease involving the tissue, conservative excision of all visible granuloma with concomitant sinus aeration is recommended. Radical surgery has not proven necessary. In fact, residual granuloma next to vital structures such as dura and sclera has been seen to regress and heal by fibrosis following excision of the gross disease with concomitant drainage and aeration of the involved sinus [4].

Summary

It must be emphasized that the clinical diagnosis of aspergillosis of the sphenoid sinus is difficult. From our experience of these three cases and a review of the world literature, it seems that the patients with isolated sphenoid sinus aspergillosis with intracranial extension commonly present with headache and ocular problems. One should suspect fungal infection of the sinus if, (a) the suspected sinusitis does not respond to routine treatment, (b) there is a localized mass, and (c) radiographic studies reveal bony destruction of the floor of the sella turcica. In addition, one should be aware of patients with depressed immune response such as those treated with steroids, antibiotics, cytotoxic, or other immunosuppressive agents. Also, in patients with general debilitating disease, such as tuberculosis, hepatitis, alcoholism, drug addictions, diabetes mellitus, leukemia, or neoplastic diseases. Transseptal sphenoidotomy and Amphotericin-B therapy are probably the best treatment for the isolated sphenoid sinus aspergillosis with intracranial extension. The use of proper preoperative serological tests and postoperative cultures on fungal medium will aid in the conclusive identification of the organism involved.

References

1. Comoretto R, Carbone A et al. (1986) Isolated sphenoid sinus aspergillosis: a case report. Rhinology 24:219–222
2. Feely M, Steinberg M (1977) Aspergillosis infection complicating transsphenoid yttrium-90 Pituitary implant. Report of two cases. J Neurosurg 46:530–532
3. Hora JF (1965) Primary aspergillosis of the paranasal sinuses and associated areas. Laryngoscope 75:769–773
4. McGuirt WF, Harrill JA et al. (1979) Paranasal sinus aspergillosis. Laryngoscope 89:1563–1568
5. Meyer RD et al. (1973) Aspergillosis complicating neoplastic disease. Am J Med 54:6–15
6. Miglets AW et al. (1978) Aspergillosis of the sphenoid sinus. Arch Otolaryngol 104:47–50
7. Milosev B, Mahgoub ES, Aal OA et al. (1969). Primary aspergilloma of paranasal sinuses in the Sudan. Br J Surg 56:132–137
8. Nielsen EW et al. (1983) Aspergillosis of the sphenoid sinus presenting as orbital pseudotumor. Otolaryngol Head Neck Surg 91:699–703
9. Peterson JM, Baldone SC et al. (1982) Paranasal sinuses aspergillosis. A report of two cases and review of the literature. J Am Osteopath Assoc 81:549–553
10. Veress B, Malik OA, E1 Tayeb AA et al. (1973) Further observations on the primary paranasal aspergillus granuloma in the Sudan: a morphological study of 46 cases. Am J Trop Med Hyg 22:6:765–772
11. Weinstein M, Theron J, Newton TH (1976) Aspergillosis involving the sphenoid sinus. Neuroradiology 11:137–139
12. Young RC et al. (1970) Aspergillosis. The spectrum of the disease in 98 patients. Medicine 49:147–173

Cranial Complications Following Dental Infection

W. WAGNER, U. W. WAHLMANN, and H. SCHEUNEMANN, Mainz/FRG

Introduction

Intracranial spread of infection deriving from the teeth is a rare but life-threatening complication, as described by Partsch (1900) and others at the beginning of this century. Nowadays, due to early antibiotic treatment, we observe central nervous complications of dental infection in the form of pyogenic meningitis or brain abscess only seldom. In recent years six representative cases (see Table 1) have been observed among the patients of our clinic.

Case Reports

Cases 1 and 2. Acute osteomyelitis of the maxilla in young children does not necessarily lead to an intracranial spread of the infection (Wunderer and Watzke 1984), but we saw two children at the age of 1 year having maxillary osteomyelitis and showing protrusion of the eyes and severe clinical symptoms of meningitis (Scheunemann 1964).

Case 3. A 5-year-old boy first showed a slight swelling near the left mandibular ramus, then a protrusion of the left eye with a slight paralysis of the fourth cranial nerve (Kempkes et al. 1984). A severe mandibular osteomyelitis developed, with a serious pyogenic menigitis and loss of the temporomandibular joint. Two weeks before being referred to us, the boy had been hospitalized in another town under the diagnosis of mumps meningitis. The

Table 1. Age, sex, and origin of infection and cranial complications

Age (years)	Sex	Dental infection	Cranial complication
1	M	Maxillary osteomyelitis	Meningismus
1	F	Maxillary osteomyelitis	Meningismus
5	M	Mandibular osteomyelitis	Meningitis
67	M	Mandibular radio-osteomyelitis	Meningitis
71	F	Maxillary sinusitis	Brain abscess
18	F	Maxillary sinusitis/dental abscess	Brain abscess

infection was caused by an untreated, destroyed deciduous tooth leading to osteomyelitis of the left mandible.

Case 4. In another patient we found a pyogenic meningitis resulting from infectious spread of mandibular radioosteomyelitis into the skull base. The patient had received 70 Gy radiation during the postoperative radiotherapy because of a squamous cell carcinoma of the floor of the mouth.

Case 5. Protrusion of the eye and palsy of the abducens nerve in this patient were caused by a brain abscess deriving from an infratemporal abscess of dental origin. After immediate incision and draining of the dental and cerebral abscesses the function of the abducens nerve returned completely.

Case 6. Two weeks after endonasal surgical treatment of an ethmoidal-maxillary sinusitis in another hospital, a young woman was found unconscious and referred to the nearest hospital. Computed axial tomography (CAT) detected a brain abscess in the frontotemporal region (Fig. 1). An initial conservative treatment with antibiotics and corticosteroids (Wallenfang 1984) was started by the neurosurgeons to precede the planned definitive neurosurgical therapy. Looking for the cause of the infection, we found a maxillary empyema initiated by an infected first upper molar, which was subsequently extracted along with an operative revision of the maxillary sinus. Finally, the young woman felt so well that she refused surgical treatment of the brain abscess and even failed to attend follow-up examinations. After 4 years she reported without any clinical neuropathological symptoms. On the control CAT-scan (Fig. 2) there was no evidence of the former brain infection. The

1 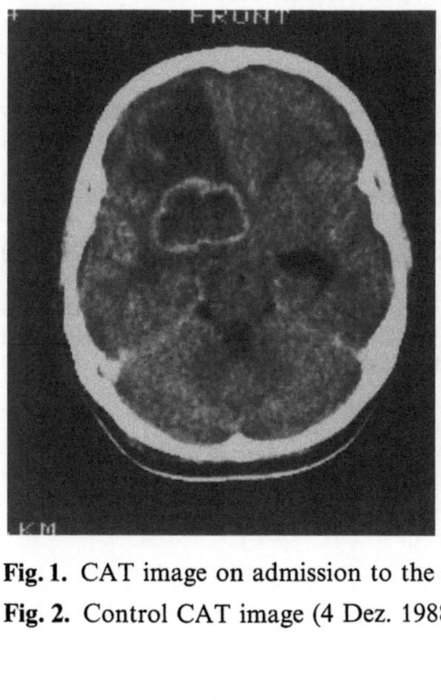

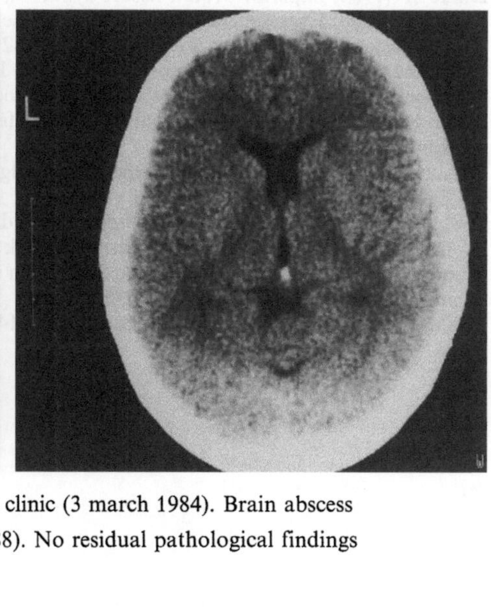 2

Fig. 1. CAT image on admission to the clinic (3 march 1984). Brain abscess

Fig. 2. Control CAT image (4 Dez. 1988). No residual pathological findings

patient refused further diagnostic measures to trace the remainders of the infection (Wahlmann and Hey 1990).

Discussion

A swelling of the temporal region with a protrusion of the eyes is an important sign of possible cranial spread of dental infection. Neurological deficits of cranial nerves can be an early warning sign, especially paralysis of the abducens nerve, which was found in five of our six cases and seems to be an important symptom. These symptoms seem to be a part of the superior orbital fissure syndrome, which was described by Lakke (1962) and Progrel (1980) as being caused by a spread of infection. Severe courses with fulminant spread of infection are seen mostly in small children (Scheunemann 1964) due to anatomic differences, such as missing pneumatization of the facial bones.

Conclusions. These cases with intracranial spread of infection deriving from the teeth and jaws show the importance of thoroughly investigating for any possible dental cause of infection in order to eliminate it as early as possible. Close cooperation between maxillofacial, ENT, and neurosurgeons facilitates the complex management of these severe complications.

References

Kempkes K, Kuffner H-D, Krämer G (1984) Intracranielle entzündliche Komplikationen bei odontogener Osteomyelitis des Unterkiefers. In: Pfeifer G, Schwenzer N (Hrsg) Fortschritte der Kiefer-Gesichtschirurgie, Bd 24. Thieme, Stuttgart, S 61–62

Lakke SP (1962) Superior orbital fissure syndrome. Arch Neurol 7:289–300

Partsch C (1900) Über seltene Verbreitungswege der von den Zähnen ausgehenden Eiterungen. Dtsch Monatsschr Zahnheilkd 6:242–252

Progrel MA (1980) The superior orbit syndrome. J Oral Surg 38:115–117

Scheunemann H (1964) Metastatische Eiterungen bei der Kieferosteomyelitis Neugeborener. In: Schuchardt K (Hrsg) Fortschritte der Kiefer- und Gesichtschirurgie, Bd 9. Thieme, Stuttgart, S 171

Wahlmann UW, Hey O (1988) Hirnabszeß als Folge einer odontogenen Sinusitis maxillaris. Dtsch Zahnärztl Z 43:1357–1358

Wallenfang T (1984) Steroidtherapie beim entzündlichen Hirnödem. In: Grumme T (Hrsg) Das Hirnödem. De Gruyter, Berlin New York

Wunderer S, Watzke I (1984) Cerebrale Spätfolgen nach retromaxillären Eiterungsprozessen im Säuglingsalter. In: Pfeifer G, Schwenzer N (Hrsg) Fortschritte der Kiefer- und Gesichtschirurgie, Bd 9. Thieme, Stuttgart, S 144–145

Lesions of the Skull Base and Paranasal Sinuses Presenting with Unilateral Exophthalmos

K. Abbassioun, A. Amirjamshidi, and J. Amini, Tehran/Iran

In a series of 168 consecutive cases of unilateral exophthalmos, we have encountered 59 cases with various lesions involving the skull base and paranasal sinuses. They are categorized here into the three groups of most common, the interesting, and rare lesions.

Among the most common group, mucoceles of the paranasal sinuses were the most frequent type of lesion (22 cases). These involved predominantly the frontal, and the frontal and ethmoidal sinuses. A single and interesting case of mucocele affecting the sphenoid sinus and leading to unilateral exophthalmos was also seen (Fig. 1). All but the mucocele to the sphenoid sinus were treated by evacuation through a small medial eyelid incision and perforation of the lamina papyracea. Curettage of the mucocele cavity and finally recanalization of the lesion into the nasal cavity were also performed. The sphenoid sinus mucocele was evacuated by a transsphenoidal approach. There has been no recurrence in any of the cases treated in this manner over the past 14 years. The next most common lesions were malignancies involving the paranasal sinuses. There were seven primary adenocarcinomas (differentiated, four; mucoid, one; transitional, one; clear cell, one) and six malignant lymphomas, of which three also involved other systems at the time of presentation. These cases were treated either with biopsy or radical excision followed by radiation and chemotherapy.

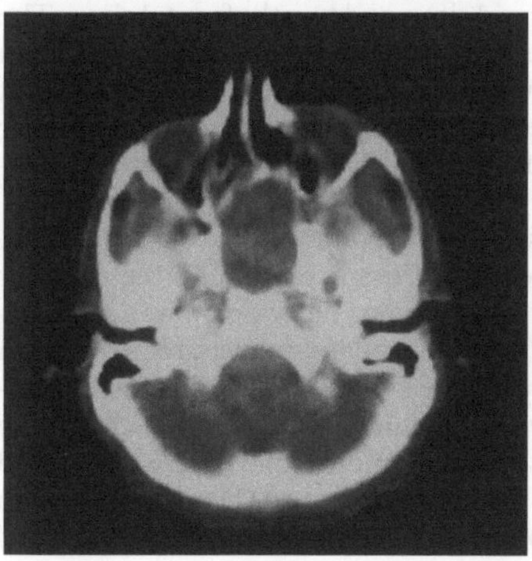

Fig. 1. A single and interesting case of mucocele affecting the sphenoid sinus and leading to unilateral exophthalmos

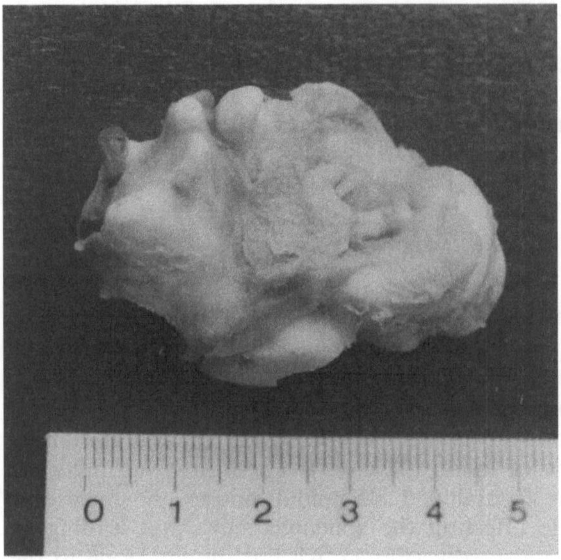

Fig. 2. The en bloc resection of an osteoma which led to permanent recovery of the exophthalmos

Among those with interesting pathology there were five cases of fibrous dysplasia of the skull base, (four women and one man), all in the 2nd decade of life and presenting with unilateral exophthalmos without notable visual disturbances. Surgical excision and decompression of the orbit followed by acrylic cranioplasty were performed in all cases. So far only one case has required a second surgical intervention for cosmetic purposes 3 years later. Histiocytosis X and eosinophilic granuloma occurred in two young girls, expanding the superior orbital plate and causing unilateral exophthalmos which reduced after craniotomy and resection of the lesion. There were three cases of osteoma, one affecting the ethmoid sinus and two the frontal sinus. The en bloc resection of the lesion led to permanent recovery of the exophthalmos (Fig. 2). Two cases of mucormycosis were diabetic elderly men with painful unilateral exophthalmos who were treated with appropriate antibiotics after drainage, biopsy, and confirmation of infection. One case of esthesioneuroblastoma was seen.

Among the rare lesions, a single case of Brown tumor of the orbit was seen. The proliferative osteoblastic and osteoclastic tissue expanding the superior orbital plate was totally removed after craniotomy. The woman was suffering from hyperparathyroidism, and a parathyroid adenoma was removed in a second operation. A 7-year-old boy presented with right unilateral exophthalmos and harbored a transitional type of meningioma (5 × 4 × 4 cm) filling up the right nasal cavity and based over the ethmoid bone (Fig. 3); a second case of paranasal sinus meningioma was also treated. There has been no recurrence 5 years after total resection of the tumor through a frontonasal

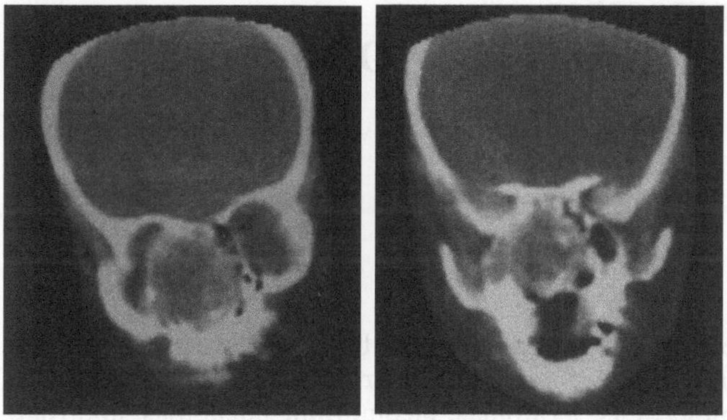

a b

Fig. 3 a, b. Two coronal views showing a transitional type of meningioma (5 × 4 × 4 cm) in a 7-year-old boy with right unilateral exophthalmos filling up the right nasal cavity and based over the ethmoid bone

approach. A pencil was traumatically driven into the medial and superior compartment of the right orbit and ethmoid sinus and caused unilateral exophthalmos in a 2-year-old boy. An exophthalmos was progressively getting worse and painful 2 months after trauma and was recovered after removing the split pencil through a frontobasal approach. In a 50-year-old man a giant epidermoid tumor involved the frontal, temporal, parietal, and most of the roof and lateral wall of the orbit (Fig. 4). It was removed totally, and cranioplasty was performed after reconstruction of the roof of the orbit using acrylic methacrylate. We also encountered single cases of plasmocytoma, invasive pituitary adenoma, traumatic arachnoid cyst, and rhinoscleroma.

It is worth mentioning that for handling these cases computed tomography played the most valuable role in preoperative diagnosis and selection of proper surgical management.

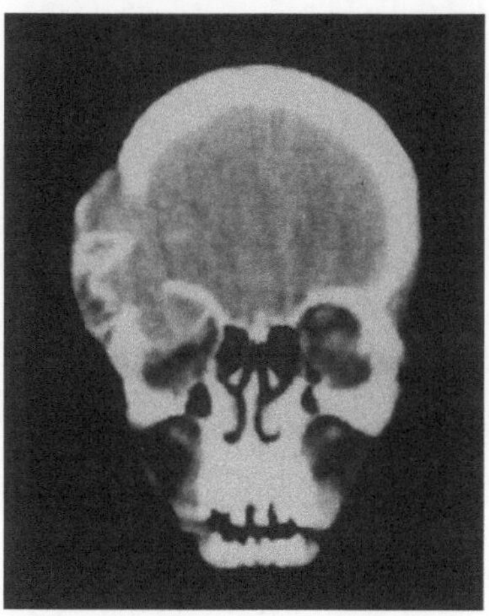

Fig. 4. A giant epidermoid tumor in a 50-year-old man involving the frontal, temporal, parietal, and most of the roof and lateral wall of the orbit

Periorbital Approaches for Resection of Tumors of the Orbita

N. SCHWENZER, Tübingen/FRG

Tumors of the orbit are fairly rare and can be classified as primary and secondary. The first group consists of neurofibromas, lymphangiomas, hemangiomas, dermoid cysts, teratomas, pleomorphic adenomas, and rhabdomyosarcomas; together with neuroblastomas these are the most common malignancies of the orbit in childhood. Primary carcinomas of the orbit originate from the lacrimal gland. Squamous cell carcinomas and basal cell epitheliomas are the most common secondary tumors of the orbit and usually invade it by direct extension. Carcinomas of other organs can also metastasize to the orbit. Sarcomas and melanomas can reach the orbit either directly or indirectly (Henderson 1980; Reese 1976; Müller 1973).

It is the goal of tumor surgery to remove the neoplasm completely without leaving disfiguring, visible scars and without injury to the eye (Walser 1956). While primary, benign tumors of the orbita can often be removed without damage to the orbital contents, enucleation is the treatment of choice for

Fig. 1. Typical skin incisions for periorbital approaches to the orbita

malignancies. Various surgical approaches for tumor removal are available and depend on the location, type, and size of the tumor. The transfrontal approach, most often used by neurosurgeons, is not discussed in this paper. Rather, the less well-known, less traumatic, periorbital approaches are described (Fig. 1).

Access to the orbita should provide good exposure, prevent injury to the eye, and not leave disfiguring scars. Orbitotomies can be classified as those with and those without bone resection. Simple orbitotomies are especially well suited for tumors to the anterior section of the orbita; nasocranial, nasocaudal, temporocranial, temporocaudal, and caudal approaches can be distinguished. The incision is performed in the eyebrow, or subciliarly in a skinfold. These incisions usually do not leave a visible scar and allow the surgeon to obtain adequate exposure without the fear of later disfigurations. For this purpose, the lower margin of the eyebrow is especially well suited. The temporocranial incisions provide deep exposures and are usually best suited for excision of most dermatoid cysts located in this region, which can cause cystic changes of the lateral orbital wall and can invade the cranial cavity or the temporal fossa (Fig. 2). Using the nasocranial approach, tumors of the mediocranial orbital region can be reached. This access is as suitable for the transethmoidal approach. Great care should be taken to preserve the canthal ligaments. A transconjunctival infraorbital incision may be used to obtain good exposure to tumors of the orbital floor (Figs. 3, 4).

Orbitotomy with temporary resection of bone, first described by Krönlein, is appropriate for tumors of the posterior orbita. For these cases, we usually prefer an S-shaped incision, similar to that described by Kocher. This incision is usually made in the eyebrow and extended laterally to the margin of the zygomatic arch (Figs. 1, 5–7). This incision, which follows the skin tension lines, provides excellent exposure without injury to the zygomatic arch or the

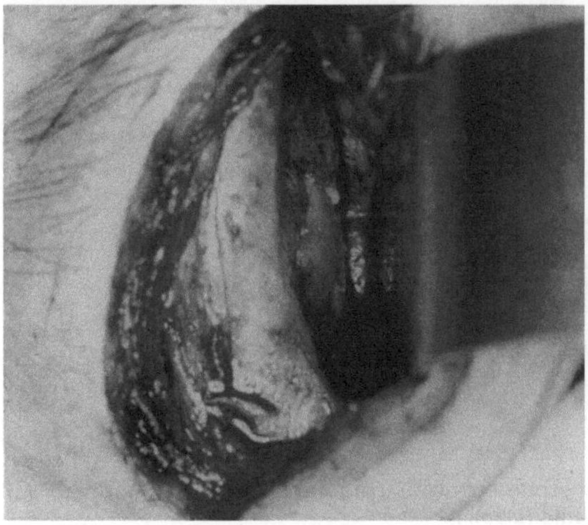

Fig. 2. Temporocranial access used for removal of a dermoid cyst of the lateral orbit

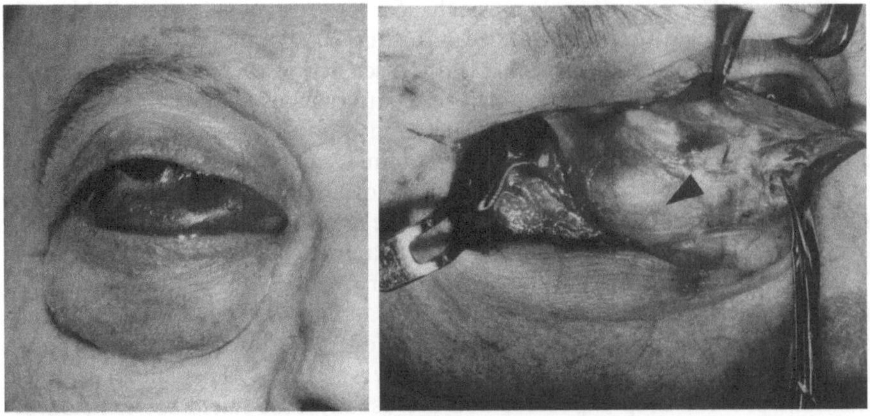

Fig. 3. A transconjunctival infraorbital incision is used to excise this spindle cell lipoma (*arrow*)

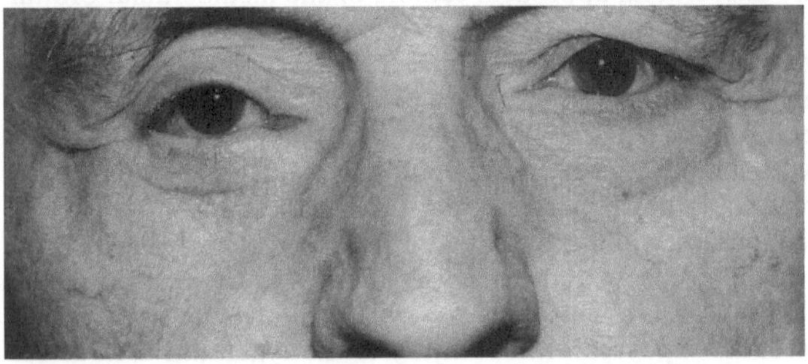

Fig. 4. The same patient as in Fig. 3 six months postoperatively

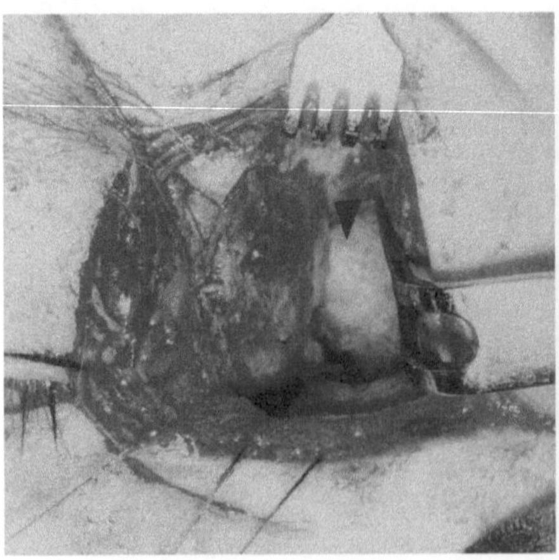

5

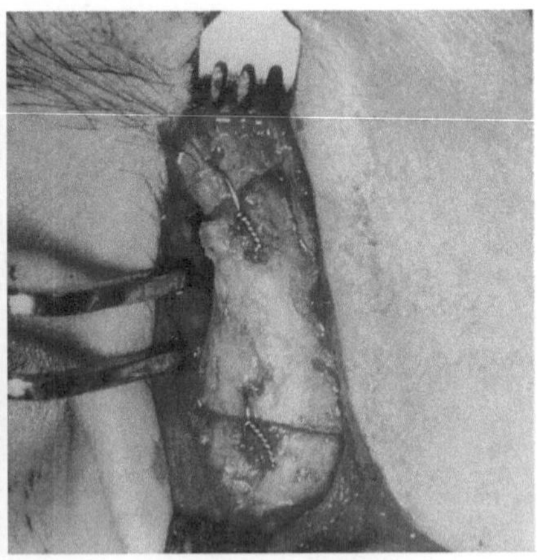

6

Fig. 5. S-shaped incision and lateral orbitotomy with temporary bone resection for excision of a retrobulbar pleomorphic adenoma (*arrow*)

Fig. 6. The resected bone fragment has been replanted and secured with two wire sutures

Fig. 7. The excised specimen (retrobulbar pleomorphic adenoma)

facial nerve. Oscillating saws are especially suitable for resections of the lateral orbital wall since their cut has only a minimal width. The transected bone fragment may remain attached to the temporal muscle or can be removed completely until it is replanted and secured with wire sutures or miniplates (Fig. 6). Temporary resections of the zygomatic arch are necessary only in the case of extensive tumors.

Except for quite rare scar revisions, no operative interventions follow orbitotomies. However, the reconstruction of an orbit, which is capable of holding a prosthesis following radical resections of malignancies, is a surgical challenge (Schwenzer 1977, 1978). Various possibilities exist for such orbitoplasties, ranging from free skin grafting to pedicle-flap plasties, tube grafts, and microvascular free grafts, which are usually necessary only for extensive maxillo-orbital defects.

In summary, orbital tumor surgery using the periorbital approach can be performed with or without temporary bone resection. It is a fairly atraumatic procedure, is indicated for benign neoplasms of the orbit, and usually does not impair the eye function or leave significant scars. If enucleation becomes necessary, the main problem is the reconstruction and care of the empty orbita.

References

Henderson JW (1980) Orbital tumors. Thieme, Stuttgart New York

Müller HK (1973) Die Erkrankungen der Orbita. In: Axenfeld T, Pau H (Hrsg) Lehrbuch und Atlas der Augenheilkunde. Fischer, Stuttgart

Reese AB (1976) Tumors of the eye. Harper & Row, New York 3rd edn

Schwenzer N (1977) Ein Beitrag zur Augenhöhlenrekonstruktion. In: Schmid E, Widmaier W, Reichert H (Hrsg) Wiederherstellung von Form und Funktion organischer Einheiten der verschiedenen Körperregionen. Thieme, Stuttgart

Schwenzer N (1978) Indikation zur chirurgischen und epithetischen Deckung von Gesichtsdefekten. Fortschr Kiefer Gesichtschir 23:21

Walser E (1956) Eingriffe im Bereich der Augenhöhle. In: Guleke N, Zenker R (Hrsg) Allgemeine und spezielle Operationslehre, Bd 4. Springer, Berlin Göttingen Heidelberg

Management of Benign Orbital Tumors Via Medial and Osteoplastic Lateral Orbitotomy

W. STOLL, H. BUSSE, and P. KROLL, Münster/FRG

Medial Orbitotomy

Indications. The main indications for medial orbitotomy via the ethmoid region are severe orbital complications, especially visual disturbance, following purulent ethmoiditis or rhinitis. Further indications involve decompression of the orbit in cases of malignant exophthalmos, resection of mucoceles and medially localized orbital tumors (Naffziger 1948; Richter et al. 1984).

Management. Adequate treatment of the presented processes requires a complete ethmosphenodectomy and resection of the frontal sinus floor. Killian's incision (Fig. 1) opens the way to these paranasal sinuses. In cases of severe visual impairment, we also remove the lamina papyracea and decompress the optic nerve, drilling down the medial wall of the optic canal under microscopic control. In this maneuver, only superficial manipulations are feasible to avoid destruction of important structures such as the carotid artery which are located deep in the lateral wall of the sphenoid cavity. Finally the frontal sinus and the frontal ethmoid cells must be drained by a flap technique

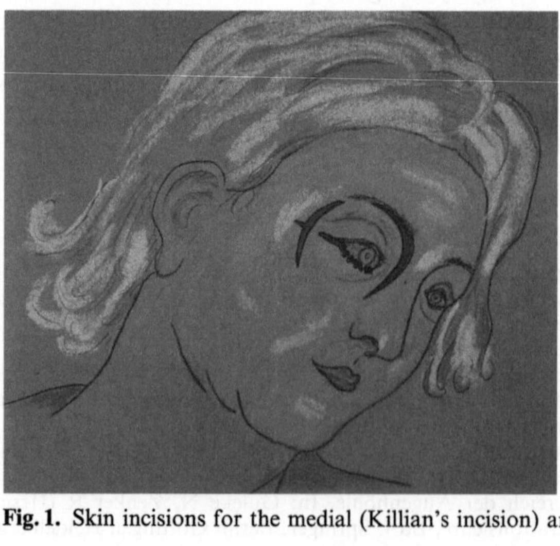

Fig. 1. Skin incisions for the medial (Killian's incision) and lateral orbitotomies

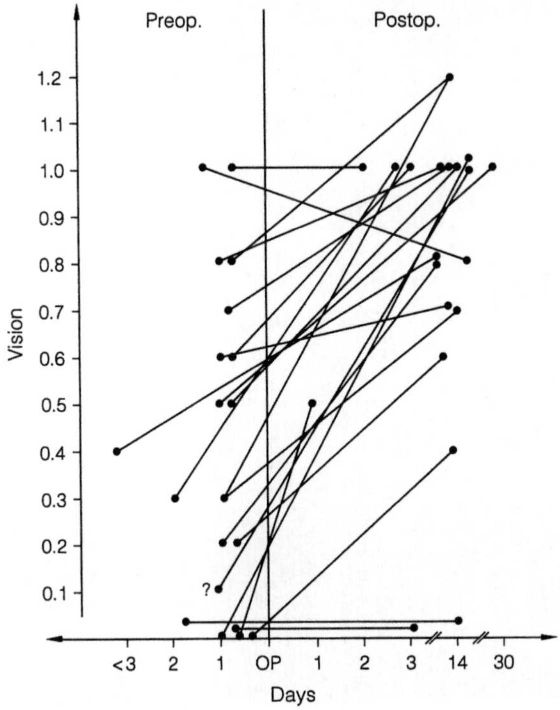

Fig. 2. Pre- and postoperative visual function after decompression of the orbit caused by infectious processes

splitting the mucocutaneous wall of the nose, as described by Uffenorde and Uffenorde in 1952.

Results. Figure 2 demonstrates the preoperative visual loss of 20 patients on account of infectious processes and the obvious visual improvement (80%) following medial orbitotomies. The results depended on the preoperative interval between the manifestation of visual disturbance and the time of surgery. Excellent improvement was achieved when surgery was started as soon as possible. The results became worse when the patients had suffered a delay of 2–3 days before surgical treatment.

Lateral Osteoplastic Orbitomy

Indication. The lateral osteoplastic orbitotomy is a suitable approach for enucleation of encapsulated benign tumors. The technique was first described by Krönlein in 1889. Among other authors, a modified Krönlein operation was presented by Birch-Hirschfeld in 1922, Berke in 1954, Kennerdell and Maroon in 1976, Buschmann and Linnert in 1978, Glanz in 1980, Stoll et al. in 1984.

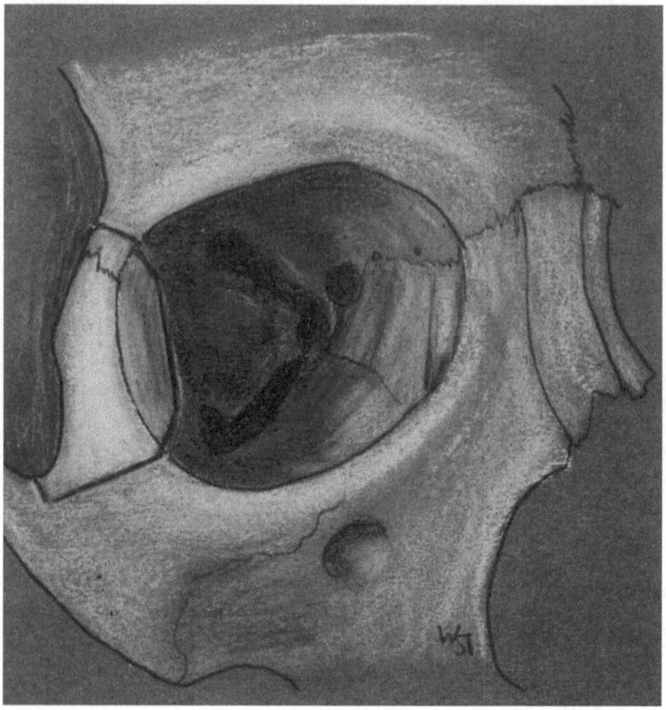

Fig. 3. The bony door of the lateral orbital wall which must be removed temporarily

Table 1. Histological findings in 16 encapsulated benign tumors (lateral orbitotomies)

Hemangiomas	4
Lymphangiomas	3
Pseudotumors	2
Dermoid	1
Cystic osteoma	1
Pleomorphous adenoma	1
Desmoid	1
Mucocele	1
Fibroma	1
Foreign body granuloma	1

Table 2. Postoperative clinical findings 1–4 weeks after lateral osteoplastic orbitotomies

Visual function preserved or improved	16
Diplopia, disturbed motility – in primary position	2
Diplopia, disturbed motility – in extreme ocular excursions	14
Enophthalmos (3 mm)	1
Mydriasis	1
Dehiscent incision	1

Management. The indication for this procedure is based on computed tomography and/or nuclear magnetic resonance imaging. Using these diagnostic possibilities, the location of the tumor in relation to the optic nerve can be determined exactly. Lateral, superior, or inferior location of the tumor in relation to the optic nerve indicates the lateral osteoplastic access. The procedure itself starts with a skin incision over the lateral orbital rim (Fig. 1). Compared to the classic Krönlein operation, we do not completely detach the lateral wall up to the inferior orbital fissure; we only cut a bony door out of the wall (Fig. 3).

For upper and lower osteotomy, we use the oscillating saw described by Feldmann in 1977. With a view to the reposition of the wall, the lower osteotomy should be wedge shaped. For the blunt dissection of the encapsulated tumor, we use our ear instrument set under microscopic control. After the enucleation, the reposed wall can be fixed with steel wires or miniplates.

Results. The histological findings in our 16 cases showed a high variability (Table 1). Fortunately, severe complications were rare. Visual function in all cases could be preserved or improved. Postoperatively, disturbed motility in extreme ocular excursion seemed to be nearly unavoidable (Table 2). However, in general this complication exists only temporarily. Nevertheless, we must refine our technique with special regard to this point.

References

1. Berke RN (1954) A modified Krönlein operation. Arch Ophthalmd (Chicago) 51:609–632
2. Birch-Hirschfeld A (1922) Operative Eingriffe im Bereich der Augenhöhle. In: Elsching A (Hrsg) Augenärztliche Operationslehre. Springer, Berlin S 1894–1959
3. Buschmann W, Linnert D (1978) Wandel der Indikationen und der Technik in der Chirurgie retrobulbärer Orbitatumoren. Klin Monatsbl Augenheilkd 172:1–12
4. Feldmann H (1977) Osteoplastische Meato-Attiko-Antrotomie. Laryngol Rhinol 56:781–786
5. Glanz H (1980) Osteoplastische Zugänge zur Stirnhöhle, inneren Nase und Orbita. Arch Otorhinolaryngol 227:641–643
6. Kennerdell JS, Maroon JC (1976) Microsurgical approach to intraorbital tumors. Arch Ophthalmol 94:1333–1336
7. Krönlein RU (1889) Zur Pathologie und operativen Behandlung der Dermoidcysten der Orbita. Beitr Klin Chir 4:149–163
8. Naffziger HC (1948) Exophthalmus: some surgical principles of surgical management from neurosurgical aspect. Ann J Surg 75:25–41
9. Richter WCh, Kley W, Buschmann W (1984) Ethmoidektomie und Orbitadekompression bei endokriner Ophthalmopathie. Laryngol Rhinol Otol 63:356–360
10. Stoll W, Busse H, Kroll P (1984) Einsatz der Feldmann-Säge in der Orbitalchirurgie. Sitzungsbericht XXXVII der 145. Vers Rheinisch Westfälischer Augenärzte in Hagen, S 179–181
11. Stoll W, Busse H, Kroll P (1987) Visusgewinn nach Orbita und Optikusdekompressionen. Laryngol Rhinol Otol 66:577–582
12. Uffenorde W, Uffenorde H (1952) Anzeige und Ausführung der Eingriffe an Ohr, Nase und Hals. Thieme, Stuttgart

Surgical Technique and Results of Orbital Decompression in Graves' Disease

G. Geyer, F. X. Brunner, W. Buschmann, and J. Müller, Würzburg/FRG

Introduction

Surgical decompression in malignant endocrine exophthalmos (Graves' disease) is indicated when the patient's health does not improve through regulation of thyroid metabolism and/or systemic corticoid therapy. A compression optic neuropathy, motility disorder, lagophthalmos, and cosmetic impairment found in these cases support the indication for a surgical approach. The rhinosurgical solution requires preparing the paranasal system either partially [3, 8, 9] or completely [1, 6, 7] as needed for the uptake of the orbital fatty tissue. Besides technical recommendations for the orbital decompression procedure, developed by Richter and Buschmann [1, 2, 6], the results of 83 decompressions are presented; these were performed from 1984 to 1988 in close cooperation with the Ophthalmic Department of the University of Würzburg.

Surgical Technique

The principle of the procedure is an extended ethmoidectomy by a fronto-orbital approach (Fig. 1). By a skin incision (Killian) the floor of the frontal sinus and parts of the nasal bone are exposed and removed. The trochlear loop and the medial lid tendon are preserved and marked for refixation at the end of the operation. The anterior and posterior ethmoidal artery are coagulated, and the lamina papyracea is removed. Additionally the medial part of the orbital floor adjacent to the infraorbital nerve is removed. The mucosa of the medial ethmoid and the maxillary sinus remain intact. The cavity is lined with lyophilized dura and sealed with fibrin glue. The ventilation of the maxillary sinus is guaranteed by a newly formed window in the medial sinus wall inferior to the lower turbinate. If a unilateral decompression is performed, the viability of the frontal sinus is maintained by a contralateral drainage (resection of the interfrontal bony septum [4]). In a bilateral procedure an additional median drainage (Mayer [5]) is necessary. Maintaining the olfactory region, the superior nasal spine and parts of the cranial nasal septum are removed. The rhinosurgical preparation of the cavity is followed by the ophthalmosurgical incisions of the periorbit. Gentle pressure on the bulb fills the cavity with fatty orbital tissue.

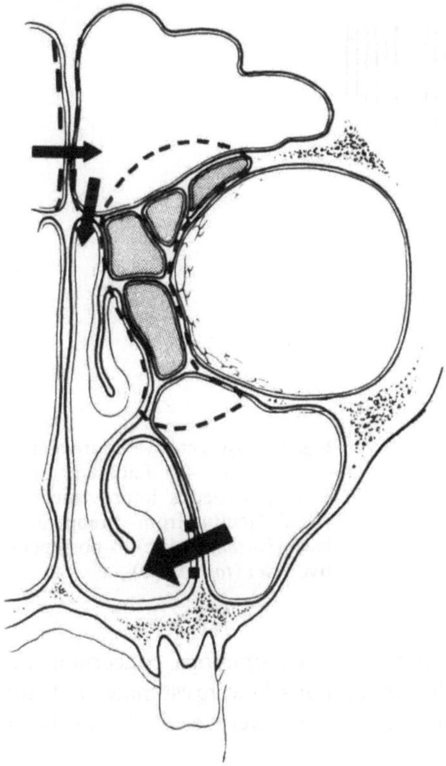

Fig. 1. The newly formed cavity for part of the orbital contents (*shaded areas*) and the drainage canals of the paranasal sinuses into the nasal cavity (*arrows*)

Results

The 47 patients who have undergone an operation were between the ages of 20 and 80 years, most of them being 50–60 years of age. There were twice as many women as men in the study.

The immediate postoperative complaints may be explained by the approach to the operative field. Within the first 3 weeks paresthesia, headache, and periorbital soft tissue swelling were common. These findings are usually to be expected after extensive paranasal surgery. After 6 months these physical complaints were reduced and were estimated by the patients to be less disturbing. In the reexamination, special attention was paid to the paranasal system. Pathological changes of the sinuses had to be taken into consideration because of the newly formed ostia. However, no complaints about their sinuses were mentioned by the patients. The viability of the paranasal windows and ducts to the nasal cavity was confirmed by endoscopic investigation. An X-ray control was therefore not necessary in most of the cases.

In six eyes exophthalmos reoccurred. In five eyes an improvement could be achieved by a subsequent medication and radiotherapy. One eye remained resistant to therapy.

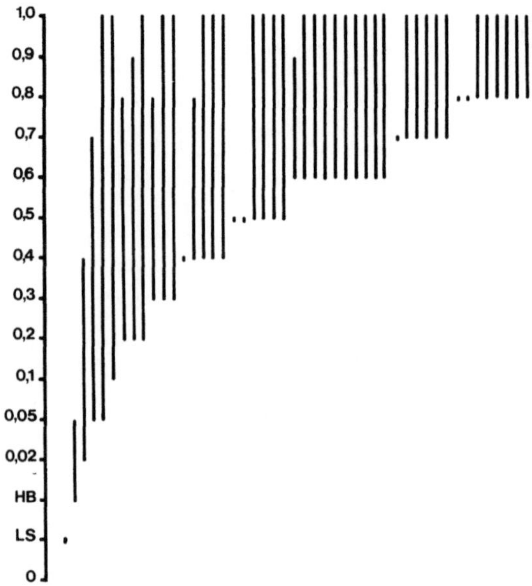

Fig. 2. Postoperative improvement of vision ($n = 47$). Each line represents a patient's improvement in visual acuity from preoperative level (*bottom of line*) to postoperative level (*top of line*)

It was important to ascertain the patients' own opinions concerning the postoperative results. Out of 83 orbital decompressions 78 were estimated as being satisfactory by the patients, and 5 decompressions were criticized as being unsatisfactory.

Conclusions

Skin sensibility alterations, paresthesia, and transitory soft tissue swelling of the periorbital region are noted as moderate postoperative impairments. Compared with the favorable operative results, these complaints appear acceptable. In general the protrusion of the bulb disappears by gentle shift of the orbital content after incision of the periorbit. An almost normal appearance is thus attained. The eye motility improves, and the dislocation is corrected by squint surgery in the course of the operation.

In our experience the described surgical method is the best approach to the problem and yields an immediate improvement of vision in most cases (Fig. 2).

References

1. Buschmann W, Richter W (1984) Ophthalmo-rhinochirurgische Entlastungsoperation bei malignem endokrinen Exophthalmus. Klin Monatsbl Augenheilkd 185:1–8
2. Buschmann W, Kruse H, Richter W (1986) Operative Therapie bei hochgradigem endokrinen Exophthalmus. Fortschr Ophthalmol 83:248–250

3. Hirsch O, Urbanek K (1930) Behandlung eines exzessiven Exophthalmus (Basedow) durch Entfernung von Orbitalfett von der Kieferhöhle aus. Monatsschr Ohrenheilkd 64:212–213
4. Kressner A (1950) Die Indikation zur Median- und Kontralateraldrainage der Stirnhöhle und deren Durchführung. Arch Otorhinolaryngol 157:28–40
5. Mayer O (1940) Über die Herstellung einer breiten Verbindung mit der Nase bei der wegen chronischer Entzündung vorgenommenen radikalen Stirnhöhlenoperation. Arch Otorhinolaryngol 148:282–290
6. Richter W, Kley W, Buschmann W (1984) Ethmoidektomie und Orbita-Dekompression bei endokriner Ophthalmopathie. Laryngol Rhinol Otol 63:356–360
7. Schmidt W, Lehnhardt E (1979) Berührungspunkte zwischen Hals-Nasen-Ohrenheilkunde und Augenheilkunde. HNO 27:86–90
8. Sewall EC (1936) Operative control of progressive exophthalmos. Arch Otolaryngol 24:621–624
9. Walsh TE, Ogura JH (1957) Transantral orbital decompression of malignant exophthalmos. Laryngoscope 67:544–568

Experiences with an Evacuable Anatomic Maxillary Sinus Implant for the Management of Orbital and Maxillary Injuries

C. Krenkel, J. Hachleitner, H. Thaller-Antlanger, and G. Lixl, Salzburg/Austria

Introduction

Fractures involving the walls of the maxillary sinus, particularly major fractures of the orbital floor, require optimal primary reduction and fixation. Conventional packing of the maxillary sinus or balloon-tipped catheters to support the orbital floor have several disadvantages: reduction may be inadequate, bleeding may occur when the packing is removed, and pressure necrosis may cause infections of the mucosal lining. To prevent these events, a special implant was developed at the Department of Maxillofacial Surgery in Salzburg (manufactured by G. Lixl, Department of Maxillofacial Surgery, Salzburg). In a pilot study, impression casts were obtained from 24 cadaveric maxillary sinuses. On the basis of the typical average shape (Schaeffer 1910), three thin-walled hollow implants of medical silicon were made for each side. To account for the variations in size, these were designed to accommodate volumes of 10, 15, and 20 cm^3 (Holzner et al. 1986). The implant fit in the maxillary sinus is critically dependent on the correct attachment of the nasal drain, which is introduced through a window cut into the medial sinus wall (Fig. 1).

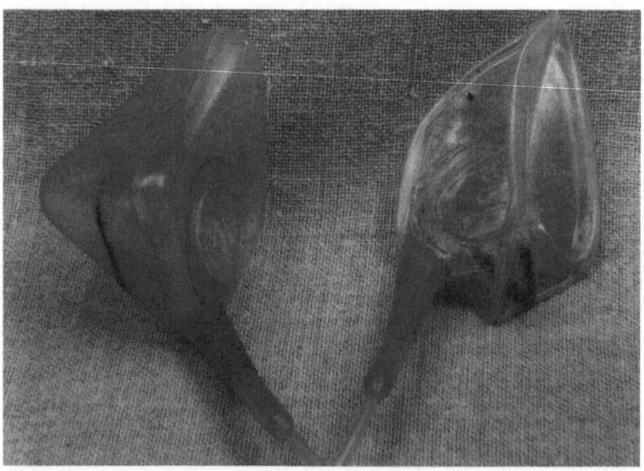

Fig. 1. A maxillary sinus implant (medium size with 15 cm^3) for the right side and an evacuated one for the left side

Technique

Severe orbital floor fractures are explored from infraorbital. If indicated, the lesions are bridged with lyophilized dura or preferably with strips of lyophilized cartilage (Sailer 1983). To support the orbital floor from caudal, the sinus implant is then introduced through a window in the facial sinus wall and filled with contrast agent through the nasal drain (Fig. 2). Using an infusion set, care is taken to ensure that the filling pressure does not exceed the diastolic blood pressure so that epithelial lesions are prevented (Figs. 3, 4). Implants are left in place for 4 weeks.

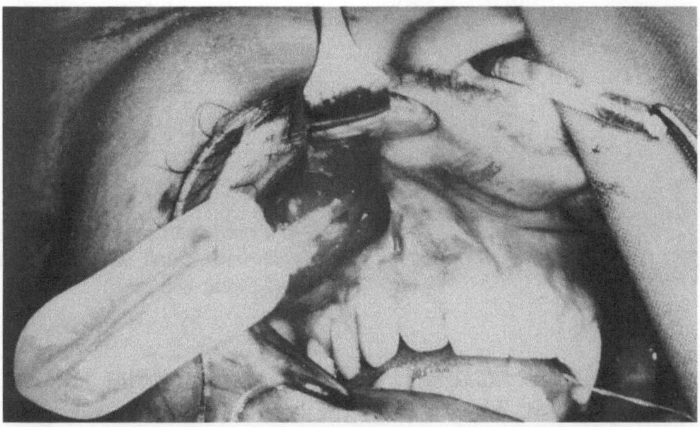

2

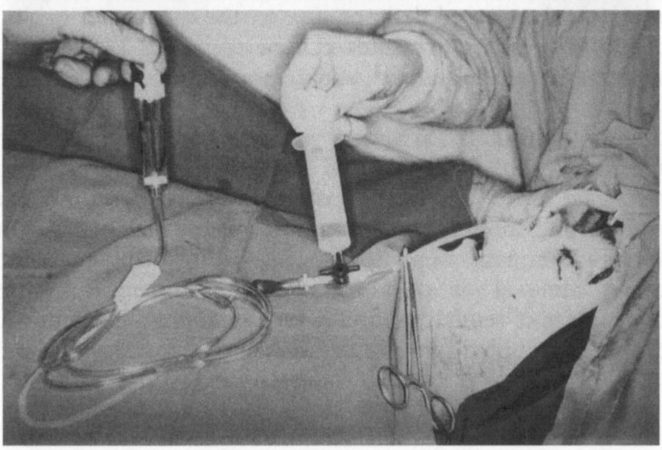

3

Fig. 2. The evacuated sinus implant before introduction through the window in the facial sinus wall

Fig. 3. For controlling the filling pressure an infusion set with a three-way valve is used. The system works on the principle of communicating vessels. It is set for a gradient of about 100 cm. This is equivalent to about 70–80 mmHg and reflects the level of the diastolic blood pressure

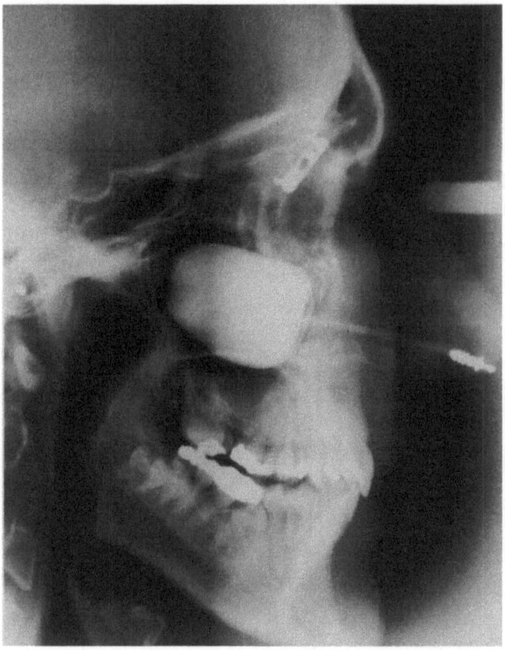

Fig. 4. Postoperative control of the sinus implant filled with contrast agent (21-year-old man)

Concomitant zygomatic or maxillary fractures require prior internal fixation by stable osteosynthesis. Elevation of the orbital floor to its original level often results in transient exophthalmos due to the associated orbital hematoma. Postoperatively, filling pressures can be corrected as required. After 4 weeks the implants are evacuated and removed through the nasal window.

At the Department of Maxillofacial Surgery in Salzburg, 97 patients with fractures of the maxillary sinus walls received 101 sinus implants between January 1985 and August 1987. Of these, 51 patients were followed up. The mean follow-up time was 17.5 months. Sinus implants had been left in place for an average of 25 days. Five implants were withdrawn early; in two patients with comminuted fractures removal on days 15 and 17 postoperatively was necessitated by nondrainage of secretions, and in the remaining three removal after about 2 weeks was prompted by ophthalmological factors.

Results

Tables 1 and 2 show the ophthalmological results. At the last follow-up enophthalmos of more than 2 mm was found to be present in five patients. One patient complained of double vision in the central visual field, and six had double vision on peripheral field testing. Radiologically, 34 of the sinuses were normal at the last follow-up (Figs. 5, 6). Five showed minor opacities along

Table 1. Results of orthoptic examination ($n = 51$)

Exophthalmometer readings (Hertel)

	> +3	+3	+2	+1	0	−1	−2	−3	< −3
Preoperative ($n = 29$)	0	2 7%	5 17%	2 7%	8 28%	1 3%	2 7%	3 10%	6 21%
Postoperative ($n = 51$)	1 2%	4 8%	6 12%	8 16%	19 37%	6 12%	5 10%	0	2 4%
Last follow-up ($n = 51$)	0	0	1 2%	5 10%	16 31%	15 29%	9 18%	4 8%	1 2%

Table 2. Results of orthoptic examination ($n = 51$)

Double vision	Preoperative ($n = 46$)	Postoperative ($n = 51$)	Last follow-up ($n = 51$)
Primary position	4 9%	2 4%	0
Central field	25 54%	15 29%	1 2%
Peripheral field	2 4%	12 24%	6 12%
No double vision	15 33%	22 43%	44 86%

5
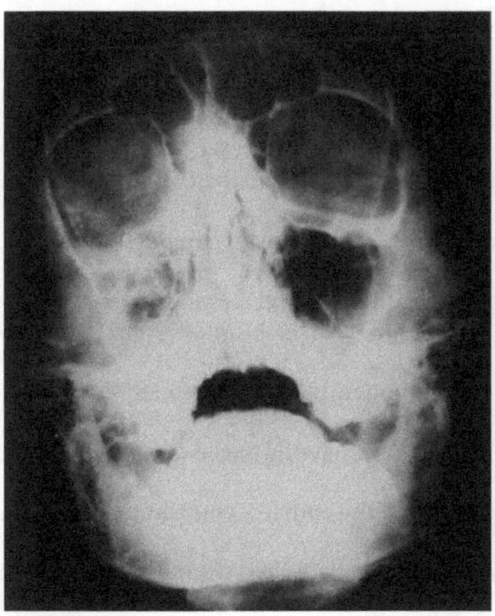

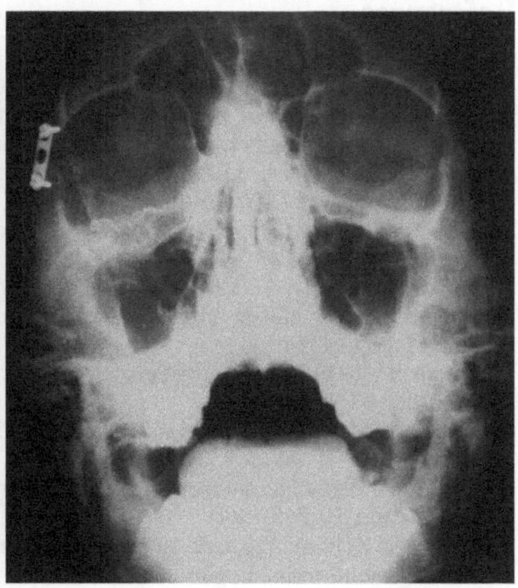
6

Fig. 5. A severe zygomatic and orbital floor fracture on the right side (same patient as in Fig. 4)

Fig. 6. X-ray control 8 months postoperatively (same patient as in Figs. 4, 5)

the sinus walls; in 12 cases the orbital floor contour was thickened; and complete opacification of the sinus was found in one patient due to mucocele formation. Four sinuses were examined endoscopically, and biopsies were taken; normal ciliated epithelium was present throughout. In one case the stroma contained inflammatory cells.

Discussion

In patients with extensive orbital floor fractures secondary subsidence of the orbital floor can often not be prevented by lyophilized dura inspite of implant support for 4 weeks. Lyophilized cartilage cut to strips of 0.7-mm thickness by a dermatome is better suited for closing defects (Sailer 1983). The procedure described (Holzner et al. 1986) is superior to the widely used bladder catheters, balloon catheters, or soft packs (Jackson et al. 1956; Jarabak and Lejeune 1959; Kamp and Bremerich 1987; Machtens 1970; Tschopp 1975; Wörle et al. 1975) because orbital floor fractures can be adequately reduced and stabilized with it up to the orbital apex. The possibility of adjusting the filling volumes postoperatively to ensure normal Hertel exophthalmometer readings is another advantage. Even after weeks, volume loss by diffusion is absent. The technique described helps avoid secondary corrective procedures. The excellent radiological results and the normal pneumatization of the maxillary sinuses found in 98% of cases amply document the successful repair of the orbitozygomatic and maxillary regions.

References

Holzner K, Krenkel C, Matras H (1986) Retention des frakturierten und reponierten Orbitabodens mit neu konzipiertem Kieferhöhlen-„Ballon“. In: Watzek G, Matejka M (Hrsg) Erkrankungen der Kieferhöhle. Springer, Wien New York, S 121–127

Jackson VR, Abbey JA, Glanz S (1956) Balloon technic for treatment of fractures of the zygomatic bone. J Oral Surg 14:14–19

Jarabak JP, Lejeune C (1959) Use of the Foley catheter in supporting zygomatic fractures. J Oral Surg 17:39–43

Kamp W, Bremerich A (1987) Die Behandlung von isolierten und kombinierten Orbitabodenfrakturen durch transantralen Ballonkatheter. Dtsch Z Mund Kiefer Gesichtschir 11:333–335

Machtens E (1970) Die Sofortversorgung der Orbitabodenfrakturen. Münch Med Wochenschr 37:1643–1649

Sailer HF (1983) Transplantation of lyophilized cartilage in maxillo-facial surgery. Karger, Basel

Schaeffer JP (1910) The sinus maxillaris and its relations in the embryo, child, and adult man. Am J Anat 10:313–368

Tschopp HN (1975) Plastisch-chirurgische Prinzipien der Wiederherstellung bei schweren Mittelgesichtsverletzungen. In: Schuchardt K (Hrsg) Fortschritte der Kiefer- und Gesichtschirurgie, Bd 19. Thieme, Stuttgart, S 137–141

Wörle M, Müller E, Tenner A (1975) Zur Reposition und Retention der lateralen Mittelgesichtsfrakturen mittels Ballonkatheter. In: Schuchardt K (Hrsg) Fortschritte der Kiefer- und Gesichtschirurgie, Bd 19. Thieme, Stuttgart, S 178–180

Chapter 2: Sellar Region

A. General Aspects

Clinical Anatomy of the Ophthalmic Artery and Cavernous Sinus

J. LANG and I. KAGEYAMA, Würzburg/FRG

Ophthalmic Artery: Area of Origin (to the Dura)

In our material the area of origin was found in 47% of cases in the subdural space, in onethird between the exit zone of the internal carotid artery and the subdural space, and in 18.3% in the dura mater of the cavernous sinus. In very rare conditions the origin is found in the sagittal course of the internal carotid artery of the sinus (Engel 1975). According to Linć et al. (1955) the origin is in 60% of cases on the rostromedial, in 26.7% on the upper medial, and in 13.3% on the lateral circumference of the artery. In our specimens we found these percentages to be 55%, 25%, and 20% (Fig. 1).

The normal width of the ophthalmic artery was found to be 1.5 mm (1.1–1.9 mm). In rare cases we found a vessel which was outside and inside the orbit 2.8 mm wide. We demonstrate one case which shows an infundibular widening of the area of origin of the ophthalmic artery. The posterior dural border of the optic canal has produced a transverse groove on the upper circumference of the optic nerve. Its lower surface was grooved by the widened ophthalmic artery. In 80% the ophthalmic artery was then found running in the subdural space, in 10% between the two dura leaves, and in 10% perforating the meningeal dural leaf in the last third anterior to the intracranial opening of the optic canal (59

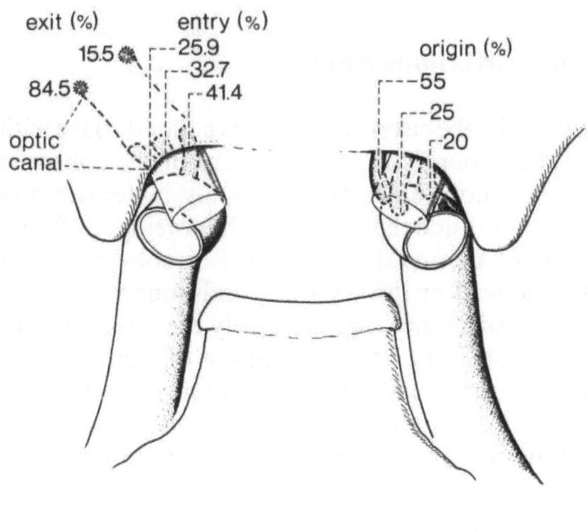

Fig. 1. Area of origin of the ophthalmic artery in the region of the anterior knee of the internal carotid artery (Kageyama 1988) and situation of the artery at the intracranial and intraorbital pore of the optic canal (Engel 1975)

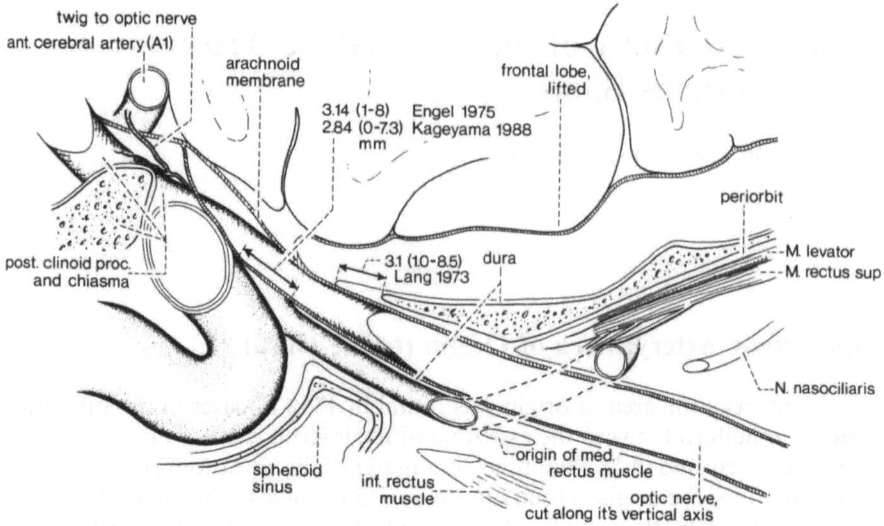

Fig. 2. Subdural course. (From Lang 1983)

head halves of our material). In earlier research on our material it was estimated that the length of the ophthalmic artery in the dural leaves is 3.14 mm (1–8 mm). We have now estimated the length of this intradural course to the optic canal (bottom area) to be 2.84 mm (0–7.3 mm; Fig. 2). In about 90% the artery perforates the inner leaf of the dura anterior to the optic canal bottom area. In 75% of cases the ophthalmic artery was found in the area of the intracranial pore of the optic canal medial to the optic nerve and in the apex of the orbit laterally (see Fig. 1). In about 85% the artery leaves the optic canal on its lateral border of the optic nerve in the apex of the orbit, in 15% on its medial border. Some variations (also of the ophthalmic canal) are described by Lang (1979).

Anterior Blood Space of the Cavernous Sinus

As described by Lang (1979), in adults the cavernous sinus is a blood space with triangular borders: capsule of the hypophysis, endocranium of the sphenoid bone, and dura mater of the cavernous sinus (Fig. 3). Only in fetuses and children did we sometimes find veins inside the sinus. In adults veins are found only in the area of the effluents of the middle superficial cerebral vein, the superior ophthalmic vein, the side wall or in very rare conditions veins from the area of the anterior perforated substance, or very rarely the upper wall of the cavernous sinus. After postnatal resorption of the walls of the veins we find trabeculae in the sinus which are surrounded by endothelium. Fatty tissue is found inside the cavernous sinus in different areas (see Fig. 26a, 27b in Lang 1979). Now we tried to determine the size of the blood space anterior to the

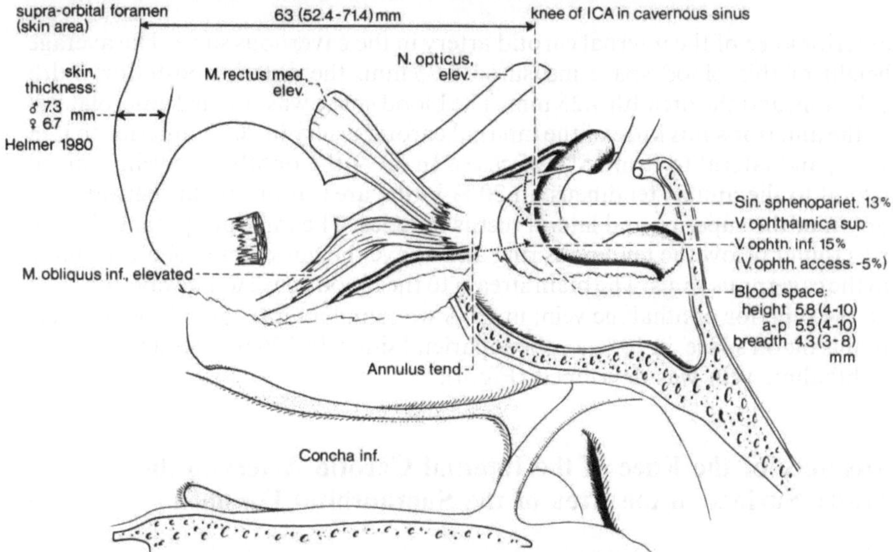

Fig. 3. Anterior blood space of the cavernous sinus, dimensions, affluents and distance of the anterior knee of the internal carotid artery to the skin area of the supraorbital foramen

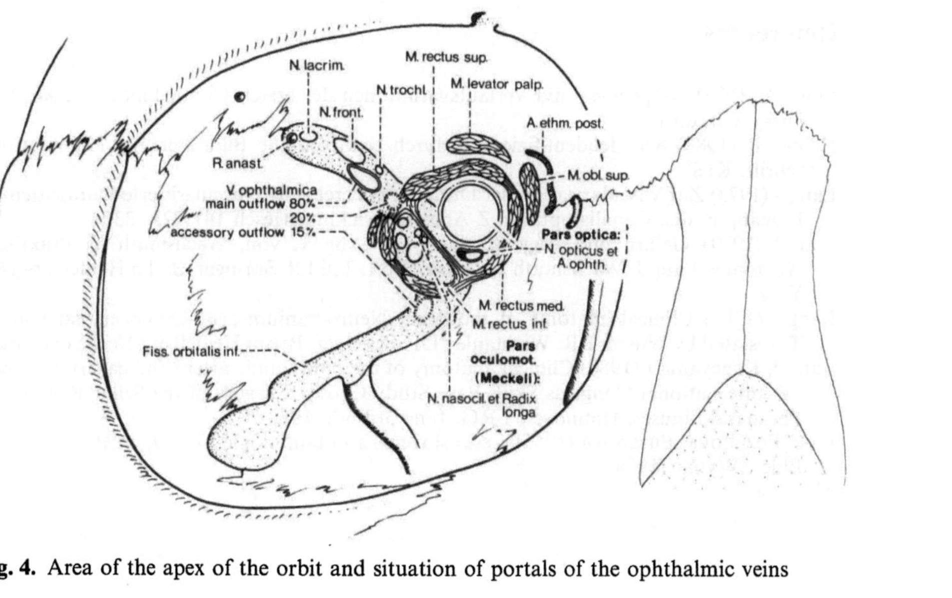

Fig. 4. Area of the apex of the orbit and situation of portals of the ophthalmic veins

anterior knee of the internal carotid artery in the cavernous sinus. The average height of this blood space measured 5.75 mm, the anterior-posterior width 5.45 mm, and the breadth 4.25 mm. The blood space was situated anterolateral to the anterior sinus knee of the internal carotid artery in 70%, anterior to it in 15%, and lateral to it in 15% of cases. In 80% the ophthalmic vein drained lateral to the anulus tendineus, in 20% in the area of the anulus between the origin of the superior and lateral rectus muscles. The inferior ophthalmic vein was found below the tendinous ring and above the musculus orbitalis running in the cavernous sinus. The main stream to the blood space was always found to be the superior ophthalmic vein; in 15% we found an inferior ophthalmic vein to this blood space, in 13% a sphenoparietal sinus. In 5% of cases an accessory ophthalmic vein was developed (Fig. 4).

Distance of the Knee of the Internal Carotid Artery to the Outer Surface in the Area of the Supraorbital Foramen

It has recently been possible to measure the blood stream of different arteries ultrasonically in the human body and head. The anterior knee of the internal carotid artery (syphon) has a distance to the supraorbital area of 63 mm (52.4–71.4 mm; see Fig. 3).

References

Engel A (1975) Ursprungs- und Verlaufsvariationen der ersten Ophthalmica-Strecke. Med Diss, Würzburg

Helmer R (1980) Schädelidentifizierung durch elektronische Bildmischung. Habilitationsschrift, Kiel

Lang J (1973) Zur Vascularisation der Dura mater cerebri. II. Vascularisierte Durazotten am Eingang in den Canalis opticus. Z Anat Entwicklungsgesch 141:223–236

Lang J (1979) Gehirn und Augenschädel. In: Lang T. von, Wachsmuth W, Praktische Anatomie. Lang J, Wachsmuth W (Hrsg), Bd 1, Teil 1 B. Springer, Berlin Heidelberg New York

Lang J (1983) Clinical anatomy of the head. Neurocranium-orbit-craniocervical regions. Translated by Wilson RR, Winstanley DP. Springer, Berlin Heidelberg New York Tokyo

Lang J, Kageyama I (1988) Clinical anatomy of the ophthalmic artery and cavernous sinus. IV. International Congress Skull Base Study Group, Surgery of the Sellar Region and Paranasal Sinuses. Hannover, FRG. June 3rd-6th, 1988

Linć, Puzanova, Puzanova (1955) Ceskoslovenska Oftalmologie 11:219, cited from Duke-Elder, Wybar (1976)

Microsurgical Anatomy of the Intracavernous Carotid Artery and Its Branches

M. Dujovny, J. I. Ausman, S. C. Ohaegbulam, F. G. Diaz, F. Holzer, L. J. Zamorano, and B. Sadasivan, Detroit/USA and Enugu/Nigeria

Introduction

After Robert Bartholow in 1872 described the clinical presentation of internal carotid artery aneurysms (probably referring to those of the cavernous carotid artery), interest in the anatomy of the cavernous carotid artery (CCA) was stimulated. The development of carotid angiography provided further impetus, paving the way in unraveling numerous pathological conditions which affect this portion of the carotid artery.

Materials and Methods

Forty cavernous sinuses in 20 tissue blocks of the skull base were studied. The tissue blocks containing the sphenoid and parts of the petrous temporal bone were removed from the cranial base of human adults at autopsy, using an electrical vibrating saw. The cranial nerves, arteries, and the pituitary stalk were severed close to the brain to preserve their relationships to the cranial base. The petrous carotid artery was exposed at its entrance into the petrous canal and was cannulated with polyethylene tubing. The arteries were then flushed with Ringer's solution, and their subarachnoid branches were ligated or coagulated with a bipolar coagulator. Polyester resin injection material (Russ Simpson, Warren, MI) was prepared as described by Umansky et al. (1984). The preparation was injected manually into the carotid artery, and the blocks were left undisturbed to harden for 3–4 h. A Zeiss OPMI surgical microscope (Carl Zeiss, New York, NY) with a Pentax K-1000 camera (Honeywell, Photographic Products Division, Littleton; for photographic documentation) was used for the microdissections.

Results

In our dissection, we found that the shape of the intracavernous carotid is not a simple S. Only the left side approximates an S shape, while the right side is a reversed S. We found that the intracavernous carotid could be subdivided into three parts – the ascending, the horizontal, and the presubarachnoid portions. The artery is directed forward, medially, and slightly upward in the petrous

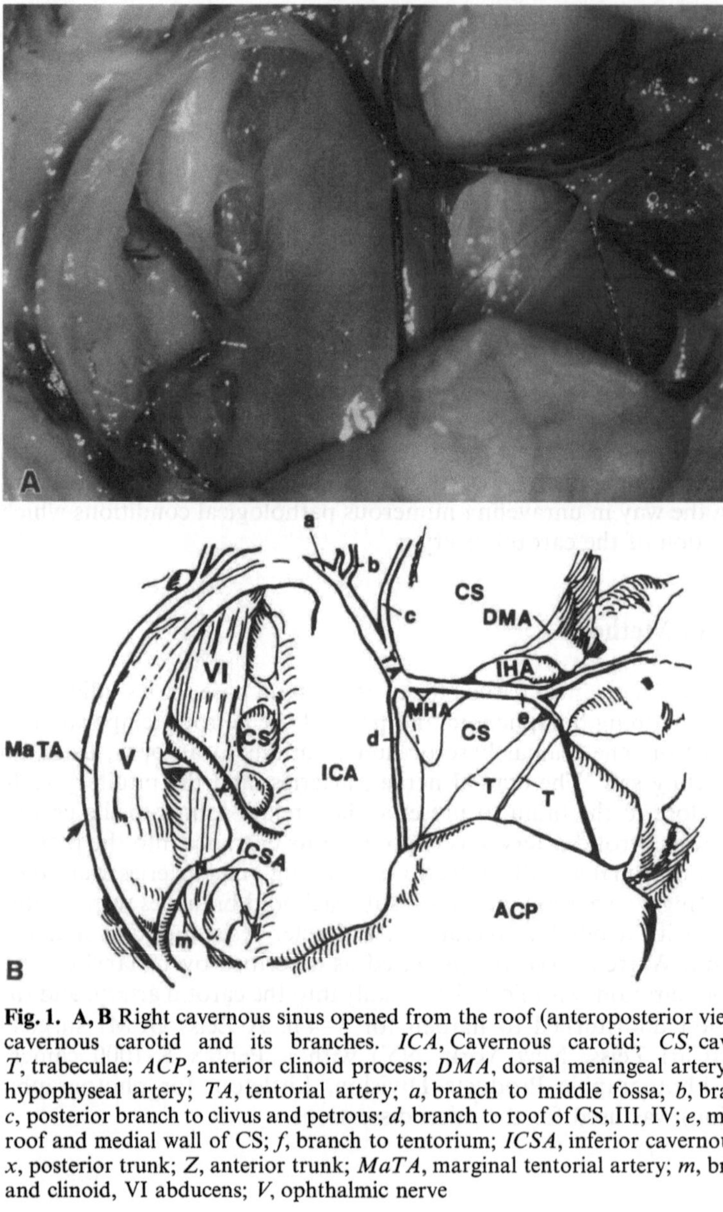

Fig. 1. A, B Right cavernous sinus opened from the roof (anteroposterior view) to show the cavernous carotid and its branches. *ICA*, Cavernous carotid; *CS*, cavernous sinus; *T*, trabeculae; *ACP*, anterior clinoid process; *DMA*, dorsal meningeal artery; *IHA*, inferior hypophyseal artery; *TA*, tentorial artery; *a*, branch to middle fossa; *b*, branch to V, VI; *c*, posterior branch to clivus and petrous; *d*, branch to roof of CS, III, IV; *e*, medial branch to roof and medial wall of CS; *f*, branch to tentorium; *ICSA*, inferior cavernous sinus artery; *x*, posterior trunk; *Z*, anterior trunk; *MaTA*, marginal tentorial artery; *m*, branch to II, IV, and clinoid, VI abducens; *V*, ophthalmic nerve

canal. It bends sharply upward with a forward convexity at the foramen lacerum, when encountering the body of the sphenoid bone as it exits from the apex of the petrous bone. It enters the cavernous sinus at this point to become the ascending section. Since the cavernous sinus extends into the petrous canal in 69% of our cases, we noted that a more reliable and consistent landmark for the boundary between the petrous and ascending portion of the CCA is the point where the artery changes direction. This can also be visualized in angiograms. The ascending portion of the CCA, which is directed upward, medially, and slightly backward, terminates under the posterior clinoid process where it bends forward forming an apex with its convexity backward to become the horizontal portion. The artery courses forward and medially from this point, then downward forward and laterally until it reaches the undersurface of the anterior clinoid process. Here it is termed the presubarachnoid section of the CCA, which turns medially and upward to lie in the carotid sulcus under the optic nerve. It passes through a fibrous ring or a bony tunnel, called the caroticoclinoid foramen in 5 of the 40 cases (13%). The vessel then loops upward and laterally under the optic nerve piercing the dura for a second time as it appears on the lateral side of the optic nerve. The anterior convexity of the artery is at the level of the commencement of the optic canal (Fig. 1). The length of the entire intracavernous carotid artery was 33.5 ± 5.1 mm (range, 26–46 mm) while the outer diameter was 5.19 ± 0.63 mm (range, 4–7 mm).

Branches of the CCA

The branches of the intracavernous carotid artery have been studied in 40 injected sinuses. The injection technique and the unfixed nature of the specimens allowed visualization of very small branches.

The meningohypophyseal artery was present in all tissue blocks studied while the inferior cavernous sinus artery was present in 95%. The other branches – McConnell's capsular artery (20%), clival or accessory dorsal meningeal artery (15%), and the intracavernous ophthalmic artery (15%) – were less frequently seen. The meningohypophyseal artery was the largest and most frequently seen of the branches. This vessel had a length of 1.14 ±0.41 mm and an outer diameter of 0.64 ± 0.27 mm. Its point of origin was consistently at the apex formed when the ascending section of the CCA curves forward and downward to become the horizontal section. The meningohypophyseal artery arises from either the posterior surface, lateral surface or the superior surface of this turning point in the cavernous carotid artery course. In none of our specimens did it arise from the medial or anteroinferior surface of this turning point. The artery was directed posteriorly and was more easily visualized from the roof of the CCA than from the side due to the exaggerated downward, forward, and lateral course of the ICA.

The meningohypophyseal artery had a well-defined trunk in 36 of the 40 sinuses studied, but no trunk was present in the remaining four sinuses. Seven typical types of branching patterns were identified. The most common type was

II, in which the main trunk split into two divisions – medial and lateral. The medial division gave rise to the dorsal meningeal artery, continuing medially to the posterior part of the sella as the inferior hypophyseal artery. The inferior hypophyseal and the dorsal meningeal arteries were present in 98 % of the cases studied. The dorsal meningeal artery supplied the abducens, the clivus, and the area around the foramen lacerum. The inferior hypophyseal artery gave off a branch to the cavernous sinus and the dura of the posterior clinoid area before entering the pituitary fossa.

The lateral division of the meningohypophyseal artery becomes the tentorial artery. The tentorial artery was present in all of the sinuses examined. The tentorial branch of this artery passes backward, entering the tentorium near the apex of the petrous temporal bone, and then running backward in the attached margin of the tentorium to ramify in the adjoining zone of the dural sheet. The remaining section of the tentorial artery gives off numerous branches. A medial branch crosses the carotid artery to be distributed to the roof and adjoining parts of the medial wall of the cavernous sinus. One or two branches run under the fifth cranial nerve into the middle fossa, where they anastomose with similar branches from the inferior cavernous sinus artery and middle meningeal artery. Small branches from the tentorial artery supply the third and fourth cranial nerves. A posterolateral branch supplied the petrous bone region, trigeminal nerve, and the roof of the sinus. The tentorial artery branching pattern varied from case to case.

The inferior cavernous sinus artery was the second largest branch of the intracavernous carotid artery, arising from the horizontal part of the carotid artery 9.61 $\pm$ 3.2 mm (range, 2.5–20.0 mm) distal to the meningohypophyseal artery origin. It was present in 38 of the 40 sinuses (95 %) and was hypoplastic in one case. The length of the vessel from its origin to its bifurcation was 0.77 $\pm$ 0.62 mm (range, 0.1–3.0 mm), and the outer diameter at the origin was 0.41 $\pm$ 0.16 mm (range, 0.1–1.0 mm). The artery traveled laterally, splitting into two trunks, which subdivided into numerous branches crossing the lateral side of the abducens nerve. These branches supplied the oculomotor, trochlear, ophthalmic, and maxillary nerves. Some branches supplied the lateral wall of the cavernous sinus, the dura of the anterior clinoid and lesser wing of the sphenoid, and the superior orbital fissure area. Other branches ramify in the middle cranial fossa anastomosing with branches of the middle meningeal artery. A prominent marginal tentorial branch ran by the side of the trochlear nerve and then posteriorly, to ramify over the tentorium near its free margin.

McConnell's capsular artery was present in 8 of the 40 sinuses (20 %). It is a small artery arising distal to the inferior cavernous sinus branch, running across the floor of the cavernous sinus toward the sella.

The accessory dorsal meningeal artery (clival artery) is present in 15 % of cases (6/40). It is the most proximal branch of the internal carotid artery, given off as soon as this artery enters the cavernous sinus. The artery originates from the dorsal surface of the ascending portion of the carotid artery just above the point of the sixth cranial nerve crossover. It splits into two branches, which then subdivide, encircling the sixth cranial nerve and ramifying over the petrous tip, foramen lacerum, the adjacent portion of the clivus, and the dura around the inferior petrosal sinus. When the dorsal meningeal branch of the meningohypophyseal artery was absent, its area of supply was taken over by this artery, as seen in one of our cases, where its external diameter reached 0.5 mm. The ophthalmic artery arose from the third portion of the intracavernous carotid artery in 15% of our cases (6/40). In four cases, the artery arose from the anteromedial surface. In the remaining two specimens, it arose from the superomedial surface of the carotid artery. It pierced the dura to enter the optic canal, lying under the optic nerve, which it accompanies to the orbit.

A sympathetic plexus was seen around the internal carotid artery in the cavernous sinus. The ganglia of the cavernous sinus were situated between the internal carotid artery and the sixth cranial nerve, between the ophthalmic and maxillary divisions and near the lower surface of the ophthalmic nerve.

Discussion

The peculiarities of shape, size, and position of the cavernous sinus often conceal structural details. Conclusions based on dissections, histological sections, or microscopic examination alone should be interpreted with caution. In this study, an approach using a combination of the above techniques and injected decalcified sections cut in coronal, sagittal, and axial lanes was used to build a complete three-dimensional picture.

Reference

Umansky F, Montoya Juarez S, Dujovny M, Ausman JI, Diaz FG, Gomes F, Mirchandani HG, Ray W (1984) Microsurgical anatomy of the proximal segments of the middle cerebral artery. J Neurosurg 61:458–467

Microsurgical Anatomy of the Upper Cranial Nerves in the Sellar Region

S. C. Ohaegbulam, Enugu/Nigeria, J. I. Ausman, M. Dujovny, F. G. Diaz, and H. G. Mirchandani, Detroit/USA

Introduction

Cranial nerves II–VI have intimate relations to the sella and cavernous sinus (CS). They are, however, variable in number and course (Nathan et al. 1973; Umansky and Nathan 1982; Lang 1983). These nerves also provide excellent anatomical landmarks for the surgical exposures of this region (Diaz et al. 1988). We used gross- and microdissection to study the surgical anatomy of the region.

Materials and Methods

Microsurgical anatomy of cranial nerves II–VI in the sella region was studied in 45 skull base blocks, and simulated surgical exposure was performed in 10 cadavers. The internal carotid artery (ICA) was cannulated and injected at its petrous portion with tinted polyester resin in 20 skull base blocks before dissection. The remaining 25 skull base blocks were prepared and studied in axial, coronal, and sagittal sections. Each skull base block contained the sphenoid bone, parts of petrous temporal bone, the pituitary gland, the CS, and related structures.

Results

The *optic nerve* (Fig. 1) emerges from the optic canal above the ICA. The ophthalmic artery is usually on its lateral side (70%) although it may be central (20%) or medial (10%). The nerve has a mean diameter of 6.0 ± 0.79 (range, 4–7.5 mm) and a mean length of 9.1 ± 1.5 mm (range, 5–12 mm).

The *oculomotor nerve* (Fig. 1) has a diameter of 3.1 ± 0.5 mm (range, 2.2–4 mm). It pierces the dura (entry pore) between the clinoid processes 5.0 ± 2.4 mm behind the anterior clinoid process (ACP) and 5.4 ± 2.4 mm in front of the posterior clinoid process (PCP). It has a variable relationship to the plane of the line joining the two PCPs, being below it in 90%, at same level in 7.5%, and above it in only 2.5%. It is anterior to this plane in 55%, posterior to it in 17.5%, and at the same level in 27.5%. Its relationship to the plane of the line joining the ACP to the PCP is also variable. It is below it in 90%, level in

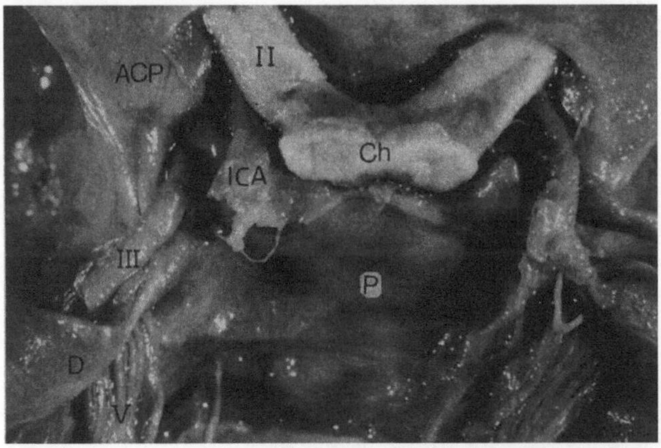

Fig. 1. Sella region viewed from above. *ACP*, anterior clinoid process; *Ch*, chiasma; *II*, optic nerve; *III*, oculomotor nerve; *V*, trigeminal nerve; *ICA*, subarachnoid part of internal carotid artery; *D*, dura of tentorium; *P*, diaphragma sella

7.5%, and above it in 2.5%. It is lateral to it in 97.5%, on the same line in 2.5%, but medial in none.

The *trochlear nerve* enters the lateral wall of the CS more posteriorly at the lower surface of the tentorial notch (78%) or the apical region of the basin of the CS which is the point of divergence of the dura folds (22%). This entry pore is 14.2 ± 2.8 mm (range, 11–20 mm) from the PCP and 11.0 ± 3.0 mm (range, 7–18 mm) from the entry pore of nerve III. Its distance to the entry pore of nerve V is 6.8 ± 2.9 mm (range, 3–10 mm), and from nerve VI is 13.4 ± 2.8 mm (range, 10–18 mm). The entry pore is a longitudinal slit, and the nerve is closely related to the tentorial artery. As it approaches nerve III, it turns downward to lie below and parallel to it. The nerve was duplicated in one instance. It is very thin, being 0.95 ± 0.32 mm in outer diameter, which may account for its easy vulnerability.

The *trigeminal nerve* crosses the apex of the petrous temporal bone to reach the middle cranial fossa at the posterolateral aspect of the CS. The ganglion overlies the ICA as it passes from its petrous portion to its ascending cavernous portion. The mandibular nerve is related to the ICA in its petrous canal from which it is separated by dura (85%) and by a thin plate of bone (15%).

The maxillary nerve forms the inferolateral wall of the CS in its posterior half, usually behind the nerve III entry pore. Anterior to this landmark, the ophthalmic division is in the lateral wall of the CS where it forms the inferior wall of Parkinson's triangle.

The *abducens nerve* perforates the dura of the posterolateral side of the clivus 9.3 ± 1.3 mm (range, 7–12 mm) below and medial to the nerve V entry pore and 21.4 ± 2.8 mm below the PCP. It proceeds forward and medially through the basilar sinus (inferior petrosal sinus) into Dorello's canal and below the superior sphenopetrosal ligament. About 6 mm from its entry pore,

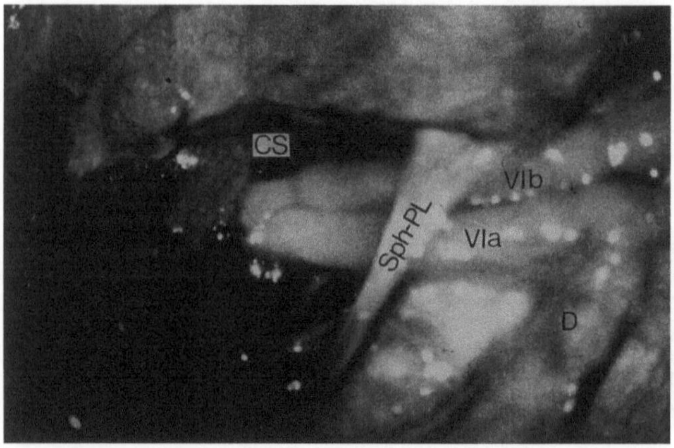

Fig. 2. Right cavernous sinus viewed through Dorello's canal. Note the duplication of abducens nerve (*VIa*, *VIb*) which rejoin in the cavernous sinus (CS). *D*, Dura; *Sph-PL*, sphenopetrosal ligament

it meets the ICA 3 mm above its (ICA) entry point into the cavernous sinus. As the ICA ascends toward the undersurface of the PCP, nerve VI skirts around its lateral surface, passing foward and downward medial to the ophthalmic nerve to lie under the horizontal part of the ICA. It then passes over the lateral side of the ICA under the ACP just as the ICA bends medially to pierce the fibrous ring. The inferior cavernous sinus artery or its branches cross the nerve. The nerve then parts company with the ICA and proceeds with the other cranial nerves toward the superior orbital fissure.

The abducens is of variable size, with a mean diameter of 1.57 ± 0.20 mm (range 1–2 mm). It is duplicated in 15% and is broken up into 3, 4, or 5 bundles in another 6.7%. In the former group, the two nerves enter through the same dura opening or through separate dura openings with the caudal root below the superior sphenopetrosal ligament and the rostral one above it. The two remain separate or reunite before leaving the cavernous sinus (Fig. 2). The abducens nerve is intracavernous as it enters the CS, but as it crosses the lateral side of the ICA, it is pushed to its lateral wall.

The *triangular space* first described by Parkinson was visualized in all our dissections. This is appreciated only after the dura of the lateral wall of the CS has been removed or reflected. The triangle is formed by nerves III and IV above, nerves V_I and VI below, and by the slope of the dorsum sellae and clivus posteriorly (Fig. 3). In two cadavers, nerve IV pursued an aberrant course by crossing the space instead of forming the superior wall. There was one instance of duplication of nerve IV.

On the lateral wall, using a subtemporal exposure, an incision 10 mm below nerve III is very safe as it avoids injury to nerve IV which may pursue an aberrant course. The lateral approach is suitable for the exposure of nerves III–VI. For exposure of the cavernous sinus, the cavernous ICA, and its branches a

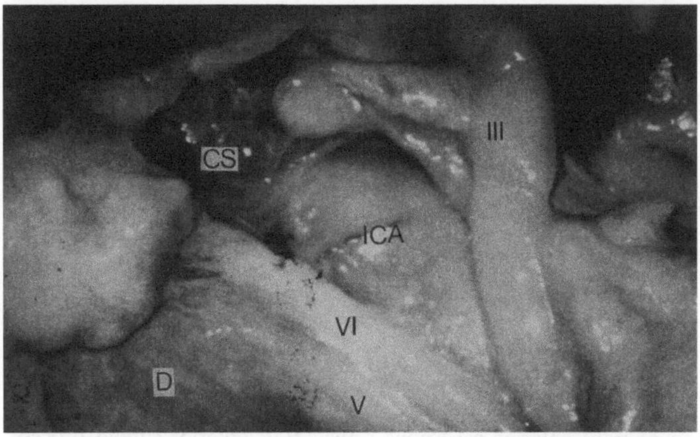

Fig. 3. Right cavernous sinus with dura of lateral wall reflected to show Parkinson's triangle. *III*, Oculomotor nerve; *V*, trigeminal nerve; *VI*, abducens; *CS*, cavity of cavernous sinus; *ICA*, intracavernous carotid artery; *D*, dura of lateral wall of CS

roof approach offers a better view. The roof of the CS is well exposed using a pterional, transsylvian approach. A generous excision of the ACP and PCP facilitates the exposure. Nerve III provides a good landmark, and an incision 4 mm medial to it and parallel to the sagittal plane is suggested.

The optic nerve is supplied by the small delicate branches of the superior hypophyseal arteries and the ophthalmic artery. These branches also supply the chiasma. In the parasellar region, these cranial nerves are supplied from the neighboring branches of the ICA. The inferior cavernous sinus artery and the tentorial branch of the meningohypophyseal artery supply nerves III–VI, but nerve VI receives additional supply from the clival artery.

Discussion

Our study used a combination of gross- and microdissection to examine the parasella and cavernous sinus region in coronal, sagittal and axial planes to facilitate the reconstruction of a three-dimensional picture (Dujovny et al. 1986).

The ACP is frequently removed to facilitate the exposure of aneurysms and the exploration of the cavernous sinus. Dolenc (1983) combined this procedure with the opening of the optic canal to gain access to the cavernous sinus. It is important to bear in mind the anatomical relationship of the cranial nerves and the carotid artery to the anterior clinoid process (Fig. 4). The lateral side of the ACP should be avoided to preserve these structures. Furthermore, the delicate arterial supplies to the optic nerve and chiasma must be safeguarded. Inadvertent and often unrecognized injury to these arteries might be the cause of the occasional unexplained visual loss that complicates surgery of this region.

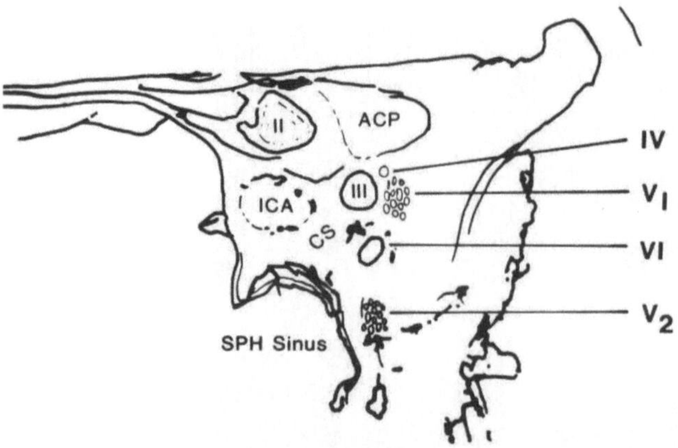

Fig. 4. Coronal section of right anterior cavernous sinus. *ACP*, anterior clinoid process; *ICA*, cavernous carotid; *CS*, cavernous sinus; *SPH Sinus*, sphenoid sinus; *II*, optic nerve; *III*, oculomotor nerve, *IV*, trochlear nerve; V_1, V_2, trigeminal divisions

The cranial nerves provide very valuable surgical landmarks. Parkinson (1973) used a subtemporal approach to enter the cavernous sinus through the triangular space. In our study we found this approach rather hazardous and tedious. Parkinson (1973) recommended an incision 4 mm below the point at which nerve III disappears into the dura for exposure of the cavernous carotid in order to avoid injury to nerve IV. We found the course of nerve IV at this point to be unpredictable (Fig. 5), hence there is a real risk of injury to the nerve if this approach is followed. Our study showed that injury to the nerve may be avoided if the incision is made 10 mm below the oculomotor nerve entry point.

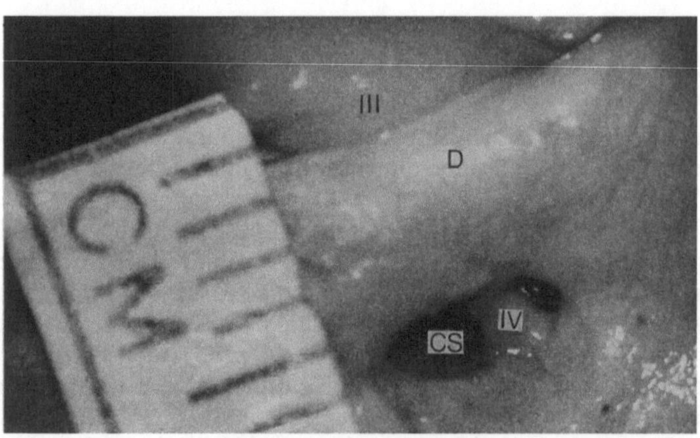

Fig. 5. Parkinson's lateral approach to right cavernous sinus (*CS*) through an incision 4 mm below oculomotor nerve (*III*). Note the vulnerability of trochlear nerve (*IV*). *D*, Dura of lateral wall of CS

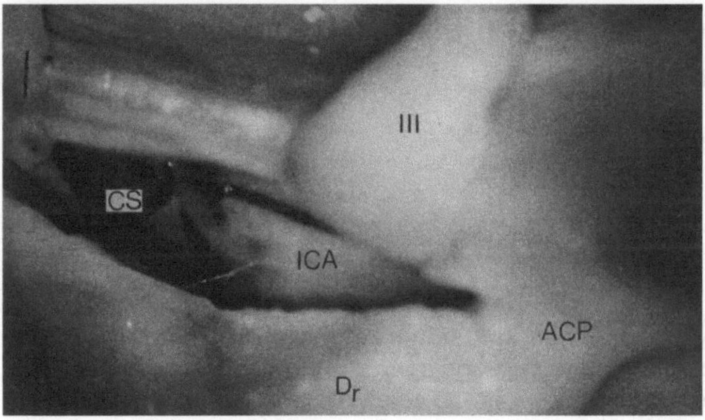

Fig. 6. Left cavernous sinus viewed from above to show the roof approach incision made medial to oculomotor nerve (*III*). *ACP*, Anterior clinoid process; *Dr*, dura of roof of CS; *CS*, cavernous sinus; *ICA*, cavernous carotid and branches

Our dissection further showed that a more convenient approach to the cavernous sinus is afforded by opening it from its roof where the dura is very thin, and cranial nerves III–VI are out of danger in the lateral wall of the CS. The roof of the CS is exposed via a pterional, transsylvian route. An incision is made on the roof just medial to the entry point of the oculomotor nerve. This affords an excellent exposure of the cavernous carotid and its branches (Fig. 6).

From our study, we believe that the lateral approach is more suitable for the exploration of the cranial nerves whereas the roof approach is better for the exploration of the cavernous carotid and its branches.

References

Diaz FG, Ohaegbulam SC, Dujovny M, Ausman JI (1988) Surgical management of aneurysms in the cavernous sinus. Acta Neurochir 911:25–28

Dolenc V. (1983) Direct microsurgical repair of intracavernous vascular lesions. J Neurosurg 58:824–831

Dujovny M, Ohaegbulam SC, Ausman JI, Diaz FG, Berman K, Holzer F, Mirchandani HG (1986) Microsurgical anatomy of the anterior clinoid process region. Surgical Forum 37:513–515

Lang J (1983) Clinical anatomy of the head, neurocranium, orbit, craniocervical regions. Springer, Berlin Heidelberg New York Tokyo

Nathan H, Ouaknine G, Kosary IZ (1973) The abducens nerve. Anatomical observations in human brains. Acta Anat 84:590–596

Parkinson D (1973) Carotid cavernous fistula: direct repair with preservation of the carotid artery. Technical note. J Neurosurg 38:99–106

Umansky F, Nathan H (1982) The lateral wall of the cavernous sinus with special reference to the nerves related to it. J Neurosurg 56:228–234

Ophthalmological Symptoms in Tumors of the Sellar Region

A. Nover, Mainz/FRG

The chiasmatic syndrome is typical of tumors of the sellar region. Visual field defects, impairment of visual acuity, and optic atrophy are the consequences of a conduction defect in the chiasm or the adjacent parts of the optic nerves or tracts. This is why most of these patients are first seen by the ophthalmologist, who determines on the basis of perimetry, examination of the visual acuity, and ophthalmoscopy whether this process is growing mostly in the sella or above it or laterally in the parasellar region.

In a large series of patients with histologically confirmed hypophyseal adenomas, 90% had scotomas before the operation (77% bilateral, 13% unilateral), 73% had decreased visual acuity (half of them bilaterally), and 54% had fundus pathologies such as partial or total optic atrophy (Table 1). Visual field defects were mostly symmetrical and bitemporal (52%), but atypical, asymmetric, or only minimal bitemporal scotomas were also seen. On the other hand, bitemporal scotomas can be found in nontumorous diseases of the chiasmal region caused by inflammatory, toxic, vascular, traumatic, or hereditary atrophic processes of the optic nerves.

Most tumors in the sellar region are hypophyseal adenomas. The most common sign of these tumors is the symmetric bitemporal hemianopia. Of course, this is seen only in advanced stages because in the very beginning the scotomas start in the temporal-superior visual field. These minimal defects were seen in 11% of our cases and had not been recognized by the patients themselves; 8% presented with less typical bilateral defects such as homonym, concentric, or binasal defects; 13% had only unilateral defects of the visual field (Fig. 1).

Comparing the optic nerve head and visual acuity, it can be seen that optic atrophy is not regularly accompanied by a cross-visual impairment; on the

Table 1. Visual fields, visual acuity, and optic fundi in cases of pituitary adenoma before and after operation

	Abnormal findings Before	After
Visual fields	90%	73%
Visual acuity	73%	59%
Optic fundi	54%	63%

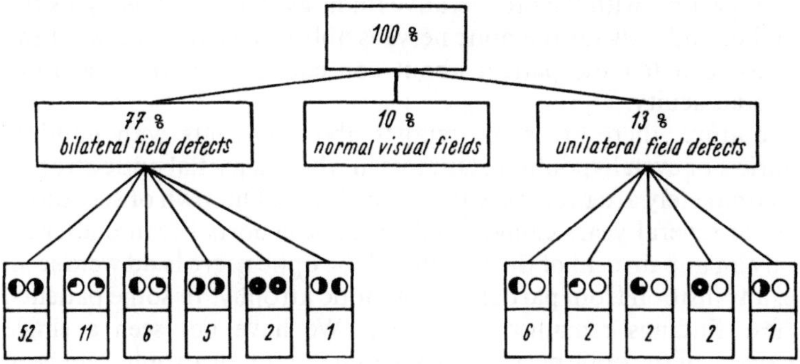

Fig. 1. Visual fields in cases of pituitary adenoma before operation

other hand, impairment of visual acuity and visual field defects could be detected without any pathologies of the optic nerve head.

Another interesting finding concerned the behavior of visual field defects, visual acuity, and findings of the optic nerve head after the operation (Fig. 2). This also highlights the complex interrelationships between these three parameters and suggests the capacity for recovery of a more or less damaged optic nerve. After the operation 73 % of any kind of visual field defects were seen as compared to 90 % before the operation; in 60 % these were bilateral and in 13 % unilateral (Fig. 2). The visual acuity was normal in 41 % as compared to 27% before the operation. Ophthalmoscopy showed nevertheless that a normal optic nerve head was present in only 37% after the operation as compared to 46 % before it. A normal optic nerve head was nearly almost seen in patients with normal visual acuity. Nevertheless, normal visual acuity was found in 43 % with a pathologic optic nerve head, mostly with partial optic atrophy.

In summary, after a successful operation pathologic findings of the optic nerve head increase, whereas visual impairment and visual field defects

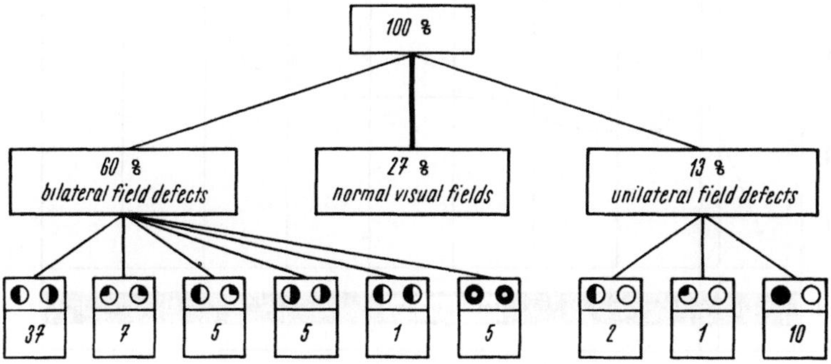

Fig. 2. Visual fields in cases of pituitary adenoma after operation

decrease in comparison with the preoperative findings. This fact highlights the question whether and how far the optic nerve is able to recover, and what the ophthalmologist can tell the patient about consequences of the operation concerning visual acuity.

In the literature, there are some reports about patients with multiple sclerosis, injuries, or even hypophyseal tumors in whom a partial optic atrophy disappeared completely after recovery from the disease. Our own observations over a period of several years cannot confirm these reports because after an operation only an impairment of the findings of the optic nerve head were seen, for example, a transition from partial to total optic atrophy. In some patients, however, these findings remained stationary. We have not seen a single complete recovery of the optic nerve head once a pallor or an atrophy had been established. It may be astonishing that an increase in optic atrophies is faced by an improvement of visual acuity and visual field defects, pointing out that the findings of the optic nerve head are not always strictly correlated with the function of the optic nerve. Impressive improvements in visual functions after an operation and a failure of progression of the optic damage over a long period of observation point out that in these cases there was only functional, not anatomical and thus irreversible damage of the optic nerve fibers. Rapid improvements of the visual acuity within a few days after an operation can be explained by sudden regression of an edema of the optic nerve. Nevertheless, in many cases it is uncertain whether over a long period of time a visible damage of the optic nerve head will be followed by an impairment of function. Two examples are given below showing that impairment of visual acuity and visual

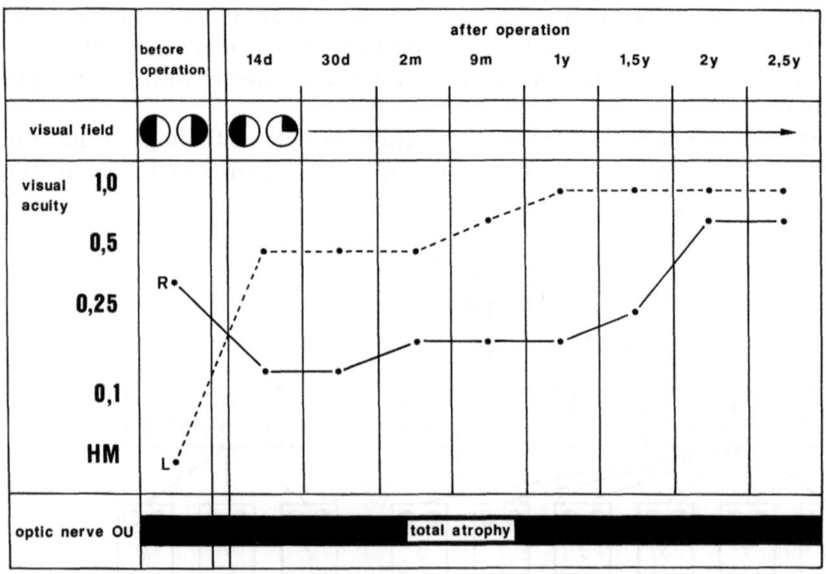

Fig. 3. Recovery of visual acuity despite total optic atrophy

field defects is possible in the first weeks and months after an operation; these examples also demonstrate that improvement in visual acuity is possible even after years, even if complete optic atrophy had been established (Fig. 3).

The following conclusions can be drawn from the above:

1. Improvement in visual functions is seen mostly in the first weeks and months after an operation, although this may occur even after years.
2. Visual functions and findings of the optic nerve head are often but not always correlated with each other.
3. Improvements in visual acuity and visual field defects are impossible if the optic nerve head does not change or even shows signs of atrophy.
4. The findings of the optic nerve head do not permit prognostic conclusions concerning improvement or impairment of its function.
5. Pathologic findings of the optic nerve head can persist at each stage after an operation eliminating its cause.
6. Complete recovery from optic atrophy cannot be expected.
7. Missing optic atrophy is a favorable precondition for the recovery of visual functions; optic atrophy impairs the prognosis.

Another important question is what the ophthalmologist can contribute to the early recognition of tumor recurrence. It has been shown that visual field defects are the most common and diagnostically important symptoms of a hypophyseal adenoma; the same holds true for the development of tumor recurrence. The most striking subjective complaints of these patients are, again, visual disturbances such as impairment of visual acuity and visual field defects

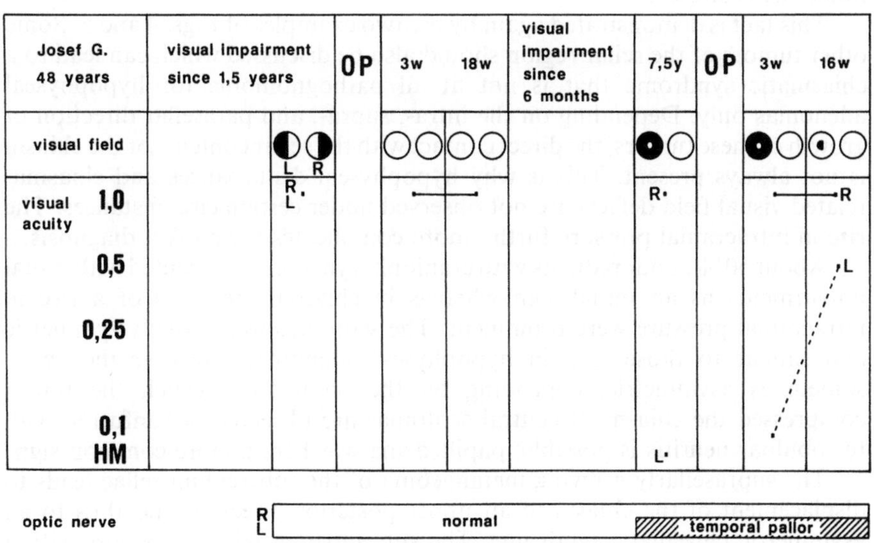

Fig. 4. Visual fields and acuity in normal optic nerve

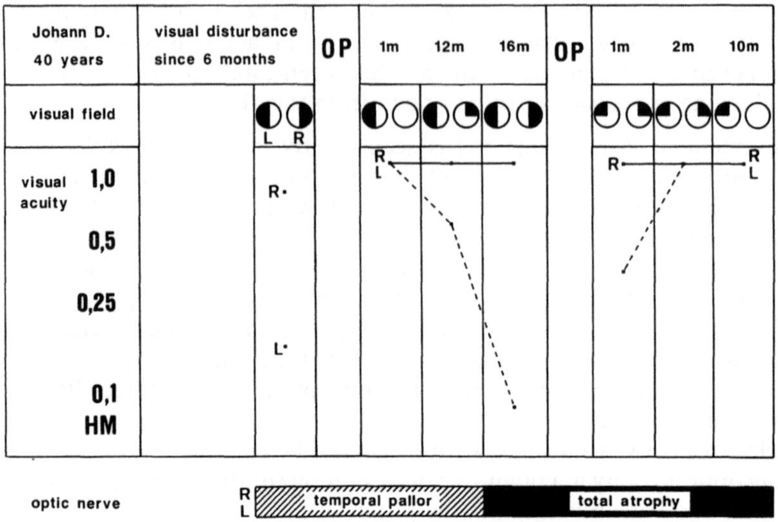

Fig. 5. Visual fields and acuity in partial and total optic atrophy

that can be easily objectified. Because the frequency of pathologic findings of the optic nerve head in tumor recurrence is higher and visual impairment and visual field defects are worse as compared to the first manifestation of such a tumor, the importance of the ophthalmologic examination and especially that of perimetry for early diagnosis and thus for salvation of the remaining visual functions is stressed.

This fact is demonstrated again by the two examples of Figs. 4 and 5. Some other tumors of the sellar region should also be discussed which can lead to a chiasmatic syndrome that is not at all pathognomonic for hypophyseal adenomas only. Depending on the intra-, supra-, and parasellar direction of growth of these tumors, the direct contact with the sellar contents or the chiasm is not always present. This is why hypophyseal disturbances and chiasma-related visual field defects are not observed under certain circumstances. The rise in intracranial pressure furthermore complicates the correct diagnosis.

About 50% of our patients with craniopharyngeomas presented with visual impairments as an initial sign whereas in children the signs of a rise in intracranial pressure were prominent. The scotomas were usually symmetric and similar to those seen in hypophyseal adenomas, however they were sometimes asymmetric, depending on the point from which the tumor compressed the chiasm. If central scotomas are observed, a confusion with retrobulbar neuritis is possible; papilledema was here a more common sign.

The suprasellarly growing meningeoma of the tuberculum sellae leads to displacement of the chiasm in an upper posterior direction and thus to an asymmetric chiasmatic syndrome. The subjective visual disturbances with a unilateral impairment of the visual acuity up to complete blindness often take a very long time and are similar to the unilateral conduction defect of retrobulbar neuritis. The optic nerve head always showed atrophy and never papilledema.

Finally, posterior olfactory meningeomas can compress the chiasm and the optic nerves, but in these tumors the ophthalmic signs have no topographic significance because the diagnosis is based mainly on uni- or bilateral anosmia. Furthermore, various tumors are located far from the chiasm but affect the chiasm by the mechanism of hydrocephaly of the third ventricle and the gliomas of the chiasm as primary tumors of the optic nerve or as a symptom of the generalized neurofibromatosis of Recklinghausen's disease, which in clinical signs are similar to the craniopharyngeomas.

Anatomy and Imaging of the Normal Sella Turcica and Pituitary Gland

F. HERAN and J. VIGNAUD, Paris/France

The investigation of the pituitary gland and the sellar region has undergone significant changes during the past two decades. Computed tomography (CT) and magnetic resonance imaging (MRI) now replace pluridirectional tomography, pneumoencephalography, and angiography.

Anatomy of the Pituitary Gland and the Sellar Region

The sella turcica is a central, saddle-shaped depression in the body of the sphenoid bone. The air-containing sphenoid sinus is directly below and anterior to the sella. The sellar floor is continuous with the tuberculum sellae anteriorly and the dorsum sellae posteriorly. The anatomy of the region of the sella and pituitary gland is known to vary considerably. The sellar floor can be asymmetrical (19%), flat (61%), concave (36%), or convex (3%) – all types related to the pneumatization of the sphenoidal sinus (sinus presellar or sellar). There is no correlation between the degree of asymmetry of the sellar floor and the volume of the sella and the pituitary gland. Minor cortical changes of the sella turcica can be noticed (local thinning of the cortical layer, defect, bulge) [1].

The normal sellar content consists of two distinct parts – the anterior lobe and the posterior lobe. *The anterior lobe* (adenohypophysis) is formed by proliferation of the anterior wall of the Rathke pouch, an ectodermal upward growth of the stomodeum or primitive oral cavity [2]. Rathke cleft cysts, a common finding at autopsy, can now be imaged as round midline structures often located in the anterosuperior portion of the sella [2]. The anterior lobe is composed of cells containing membrane-bound secretory granules specific to each type of cell. Fibrous tissue either in the form of local scars or diffusely spread can be seen in more than half of the cases occupying 1%–15% of the section area in the anterior lobe [1]. The blood supply comes from the hypophyseal portal system except for small capsular branches arising from the internal carotid artery [3]. The anterior lobe is partly wrapped around the posterior lobe and separated from this lobe by the pars intermedia, which comes from the posterior wall of the Rathke pouch. It can contain small epithelial cysts believed to be derived from a remnant of the Rathke pouch [4].

The posterior lobe (neurohypophysis) comes from a downward extension of the diencephalon and consists of intrinsic pituicytes and a large bundle of unmyelinated nerve fibers, the bodies of which lie in the supraoptic and paraventricular nuclei of the hypothalamus. The axons and axon terminals include membrane-bound neurosecretory granules that contain specific hormones (vasopressin or ADH and oxytocin) synthetized by the hypothalamus and stored and released by the neurohypophysis [3, 5, 6]. The posterior lobe is often embedded in a convexly cupped dorsum sellae or posteroinferior sella floor. It is vascularized by branches of the internal carotid artery.

The mean height of the pituitary gland is 5.5 mm (3–9 mm) in the sagittal plan [7]. Mark [8] and Bonneville [9] observed no relationship between the size of the gland and the patient's sex or age; other authors found the length, width, and height of the gland to be significantly larger in women than in men. Decrease in height could correspond to regression of the glandular tissue from decreasing function, persistent CSF pressure over time, or ischemic changes in the anterior lobe [10]. The upper border of the gland is straight (60 %), concave (28 %), or convex (10.9 %). Convexity of the upper limit of the gland should not be considered abnormal (Fig. 1).

The upper border of the gland is formed by the diaphragma sellae, a dural membrane that extends from the dorsum sellae to the tuberculum sellae and

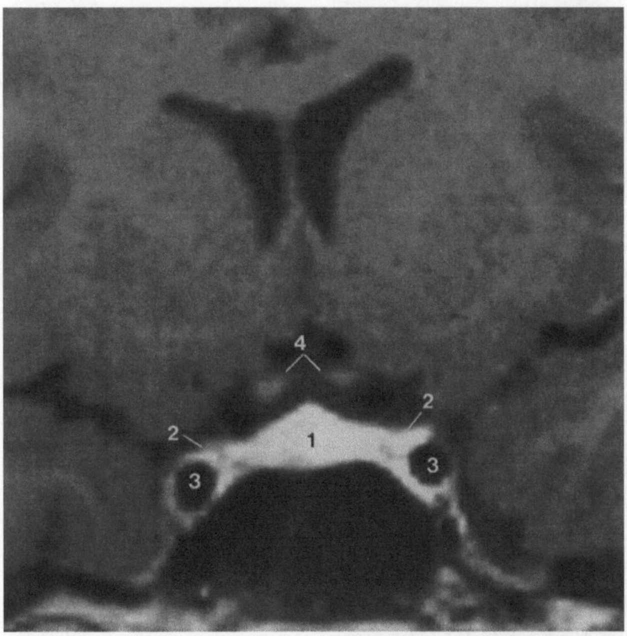

Fig. 1. T1-weighted image (TR = 600 ms, TE = 15 ms) with injection of gadolinium DTPA. Coronal view of the sellar region. The signal of the sellar content (*1*) and the cavernous sinuses (*2*) is very bright, enhanced by the paramagnetic agent. The carotid arteries (*3*) are dark because of the flow-void phenomenon. The optic pathways (*4*) remain isointense with respect to cerebral white matter. (Courtesy of Prof. A. Gaston, Department of Neuroradiology, Hôpital H. Mondor, Creteil, France)

between the medial dural walls of the cavernous sinuses. The infundibulum or stalk extends from the hypothalamus to the posterior lobe of the pituitary gland through an opening of variable size of the diaphragma.

Above the diaphragma, the CSF of the suprasellar cistern contains the superior portion of the infundibulum, the supraclinoidal carotid arteries and the circle of Willis, and the optic chiasm. The optic chiasm is formed anterior to the stalk by the optic nerves that emerge from the optic foramina and extend posteriorly, close to the inferior surface of the frontal lobe. The optic tracts run posterolaterally toward the lateral geniculate bodies. The optic chiasm has an average length of 6–10.6 mm. Its orientation in relation to the bony structures of the skull as well as to the brain is highly variable. The variability of the chiasmatic angulation combined with the variability of the distance from the hypophysis to the chiasm explain the variability of time elapsing between extension of a suprasellar tumor and mechanical compression of the chiasm initiating visual disturbances. The location of the chiasm compared to the sella is variable. In 9% of cases it is prefixed (anterior border situated at the level of the tuberculum sellae), and in 6%–11% of cases it is postfixed (anterior border

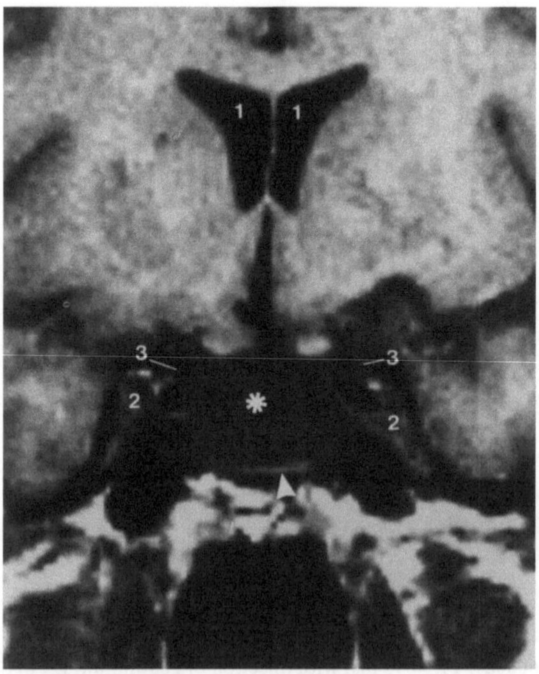

Fig. 2. T1-weighted image (TR = 600 ms, TE = 15 ms). Coronal view. The sellar content has a low signal intensity (empty sella, *) and bulges laterally toward the cavernous sinuses (*2*) and carotid arteries (*3*). Only a thin layer of pituitary glandular tissue can be seen (*arrowhead*) within the sella. (Courtesy of Prof. A. Gaston, Department of Neuroradiology, Hôpital H. Mondor, Creteil, France)

visible in front of the dorsum sellae) [9]. A posteriorly located chiasm could favor intrasellar expansion of the subarachnoid space and thus anterior cisternal herniation [11]. The sella is bordered laterally by the cavernous sinuses which contain the carotid arteries and cranial nerves III, IV, V1, V2, and VI.

The venous channels connecting the right and left cavernous sinuses (intracavernous sinuses) may be found at any site along the gland. Anterior intracavernous sinuses are reported to be twice as frequent as posterior ones (76% versus 32%).

The empty sella is defined as partly empty when subarachnoid space extend to the tuberculum sellae and occupies less than 80% of the total height of the sellae (Fig. 2). It is considered empty when the subarachnoid space occupies more than 50% of the sella, and the pituitary gland is reduced in size and shifted posteriorly and inferiorly. It can be associated with herniation of visual pathways and third ventricle, with or without visual impairment [12].

Computed Tomography

Classical Examination

After lateral scout view with the patient's head in hyperextension (supine) or hyperflexion (propine), contiguous cuts focused on the sella are performed in the coronal plane after a 100-ml bolus followed by a drip infusion of contrast material or during a bolus (angioscanning) [13]. Thin CT slices (1.5-mm coronal sections) have a good spatial resolution (0.8 mm) [20] and entail less masking effect from partial volume averaging, thus permitting a more complete evaluation. Maximal hyperextension of the head is necessary in coronal scanning of the pituitary gland to avoid artifacts from teeth. The prone position is recommended in older patients and those with a relatively short neck. Complete immobilization is necessary, for head movements can be a source of artifacts and may remove part of the gland from the scanning planes. Strict midline cuts with absence of even the slightest head rotation can be achieved most reliably with the patient in the supine position. In most cases, axial scans are not necessary for the interpretation, however if teeth artifacts cannot be avoided on direct coronal scans, coronal reconstruction of contiguous 1.5-mm axial sections is strongly recommended [14].

Dynamic Scan

It is performed in coronal plane with sections at the same level after careful determination of the ideal plane (generally the middle of the gland) and during the rapid injection of a bolus of warm contrast medium. Scan time and interslice time are chosen so as to obtain the best opacification of the carotid arteries, the capillary bed, and the venous spaces of the cavernous sinuses. This method of imaging allows the demonstration of anterior cerebral and carotid arteries, cavernous sinuses, capillary bed, and their displacements or absence in case of pituitary impairment [15] (Fig. 3).

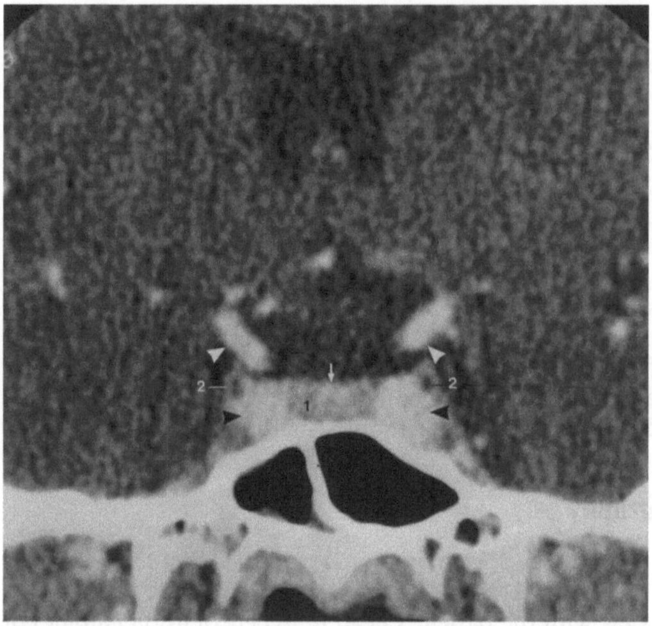

Fig. 3. Dynamic CT scan. Coronal view of the sellar region. Strong enhancement of the internal carotid arteries and their branches (*arrowhead*) and of the capillary bed of the pituitary gland (*arrow*) is well demonstrated. The pituitary gland (*1*) and the cavernous sinuses are enhanced. Notice the low intensity of cranial nerve III (*2*)

Plain Scan

Unenhanced scan may be helpful in evaluating acidophilic adenomas, which appear hyperdense relative to adjacent tissues [15].

On plain CT, the anterior lobe is hypo- or isodense with respect to brain parenchyma and enhances after contrast injection [16]. It may be homogeneous or exhibit a heterogeneous appearance, consisting of intermingled CT lucent and dense areas. The lucent areas are round (49%), oval (28%), or elongated (22%) and have a maximum height of 3.2 mm and a width of 4 mm. These are due to (a) a low or moderate degree of tissue compactness, (b) a low degree of cellular granularity, or (c) loosely packed fibrous tissue with a few cells embedded in a vascular stroma in the midpart of the anterior lobe, along the superior border and at the base of the infundibulum [17]. They disappear after injection because of their high degree of vascularity and can be responsible for the CT dense border after injection. This can be also due to an enhancing diaphragma sella or the presence of intracavernous sinuses. They may also be due to (d) extraglandular areas of adipose or connective tissue (between anterior lobe and cavernous sinus). The CT density difference between dense and lucent areas increases with contrast enhancement. Abnormality must be evocated if lucencies are associated with asymmetry.

The pars intermedia contains small colloid cysts defined as round, midline, lucent areas. It diverges on each side of the dense, centrally located, often heterogeneous posterior lobe. The most posterior diverging part, when lucent, has a typical elongated shape with parallel sides and tends to slant toward the midline, forming a tentlike configuration.

The posterior lobe is a well-defined round or oval area located in the posterior part of the sella just anterior to the dorsum sellae, often in a small depression of the bone. It can appear as an enhancing lesion particularly if the infundibulum inserts into its superior surface, resulting in a midline superior convexity [10]. Multiple round lucencies intermingled with dense areas correspond to less compact and less vascular neurohypophysal tissue. The most anterior part corresponds to compact and vascular neurohypophysal tissue, receiving vessels directly from superior and inferior vascular networks on immediately adjacent glandular surface and is the most dense area of the entire gland.

The normal stalk is best seen on axial planes as a round contrast-enhancing structure. Its diameter is inferior or equivalent to that of the basilar artery which stands behind it [18].

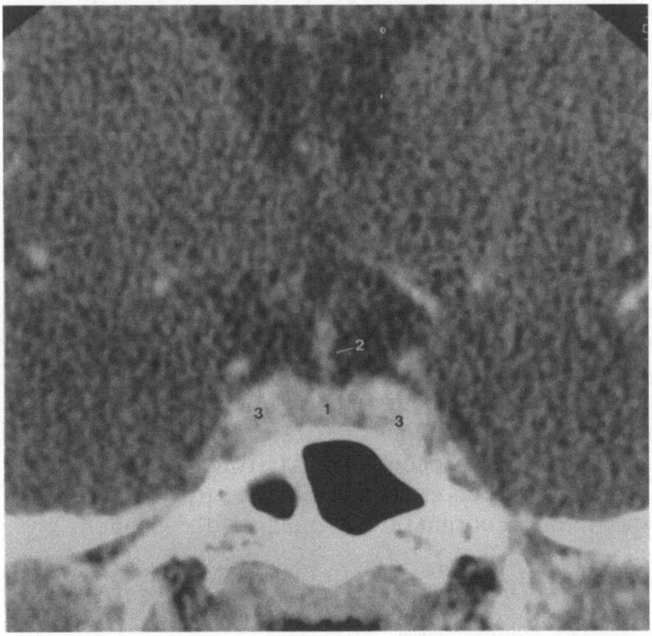

Fig. 4. Enhanced CT scan. Coronal view of the sellar region. The components of this region are clearly delineated: pituitary gland (*1*), stalk (*2*), bordered laterally by the cavernous sinuses (*3*)

The cavernous sinuses are low density areas strongly enhanced after injection. Even with a bolus injection, differentiation between venous structures and carotid arteries is difficult (Fig. 4). On coronal cuts, cranial nerve III can always be identified (Fig. 3). The cavum meckelii is hypodense and does not enhance after injection [19].

The optic chiasm is seen on coronal and axial planes as a dense structure.

Magnetic Resonance Imaging

MRI has rapidly gained acceptance for the investigation of diseases of the sellar region. MRI provides excellent delineation of the contour of the gland due to different signals of CSF, soft tissues, air, vessels, and bone marrow. Recently, thinner sections with high-field MRI [20] with good spatial resolution (1.2 mm) [7] and adequate signal-to-noise ratio has been achieved, which has allowed the use of MRI for the investigation of small lesions, especially those at the base of the skull which require both high spatial and high contrast resolution [21, 22].

The patient is positioned in a comfortable supine position, and a circumferential head coil is used. Surface coils offer no advantage because of the central location of the sella turcica. The imaging plane is selected at the option of the console operator. The pulse sequence is chosen so that the pixel values from the pituitary and brain tissue are clearly delineated from the signal of the surrounding CSF, air, and cortical bone [23].

Classical Sequences

T1-weighted sequences (TR = 600 ms, TE = 25 ms) offer the best sensitivity in displaying morphology of the infundibulum, the gland, and the adjacent structures and well demonstrate pathological internal intensity changes. As TR and TE increase (spin-echo T2-weighted sequences with TR = 1800–2000 ms, TE = 30–90 ms), contrast between the content of the pituitary fossa and the CSF decreases, making identification of the pituitary gland and structures within the suprasellar cistern more difficult [24]. Coronal sections are preferable for demonstrating internal architecture and surface topography and for discriminating intrasellar content from bordering structures [21, 25]. This plane minimizes partial volume averaging of parasellar vascular and CSF structures, which may simulate microadenomas on axial or sagittal cuts, as well as the chemical shift artifact [24]. It is easier to evaluate, partly because of the radiologists' experience in direct coronal CT scanning of the gland.

3D Sequences

These new sequences use short TR and TE (35 ms, 6 ms) and a small flip angle (40 °) and allow thin cuts (1 mm) of the volume under study. The number of cuts varies from 64 to 128, and the whole sequence lasts about 9–13 min.

Thanks to a special software, images can be rapidly reformated in any plane of the space. The signal-to-noise ratio is excellent, and definition of the pituitary gland and sella region is often of very good quality.

Use of Contrast Medium

Gadolinium DTPA is a paramagnetic compound which decreases both longitudinal (T1) and transverse (T2) relaxation times and thus enhances signal intensity on T1-weighted images. Its effect can be compared to that of iodinated contrast media used for CT, as it does not cross the brain-blood barrier (BBB) and has no effect on CSF or white matter.

The advantages of MRI are soft tissue contrast, absence of bone artifact, multiplanar imaging, no known biological side effects, no ionizing radiation, which is ideally suited for patients who require repeated follow-up studies [26, 27]. It allows direct sagittal cuts, as opposed to CT with which sagittal views are available only through reformated images with lower spatial resolution than the standard CT axial or coronal views. Moreover, reformated images require a high number of thin, strictly adjacent cuts, thereby exposing the patient to a high radiation dose [28]. The data collection time for a single CT slice is about 2–3 s, compared with 2–3.5 min. for MR. But if one considers the patient positioning time, the simultaneous (MRI) versus sequential (CT) multislices acquisition, the requirement of the contrast injection, the total examination time for one sequence of MR is similar to CT [7].

The anterior lobe of the pituitary gland occupies more than one-half of the sellar fossa and at 1.5 T is of homogeneous signal intensity and isointense with respect to cerebral white matter on both T1- and T2-weighted sequences [5, 29, 30] (Fig. 5).

The posterior part has been recently identified. It is smaller and often nested in the dorsum sellae. It frequently (69%) shows high intensity signal in T1-weighted images, indistinguishable from the signal of the fatty bone marrow [3] (Fig. 6). There is a marked decrease on intensity relative to the anterior lobe on T2-weighted images. The high signal has been first related to fat pad in the gland [3, 31] even if no chemical shift artifact could be shown (as could be expected for fat) [5], high concentration in lipoproteins [30], or slow flowing of blood into intrasellar venous channels [9]. Recent data from literature favor the hypothesis that this corresponds to the posterior pituitary gland. The high signal could be related to the presence of specific hormones, for it disappears during diabetes insipidus [3]. Moreover, a bright signal is also seen in the hypothalamic median eminence in patients with tumoral or traumatic impairment of the stalk or after hypophysectomy, and may be attributed to the development of a miniature neurohypophysis yielding a storage of ADH and oxytocin [32, 33].

The diaphragma sellae may sometimes be identified above the sella turcica as a thin band of negligible signal. The central hiatus in the band through which

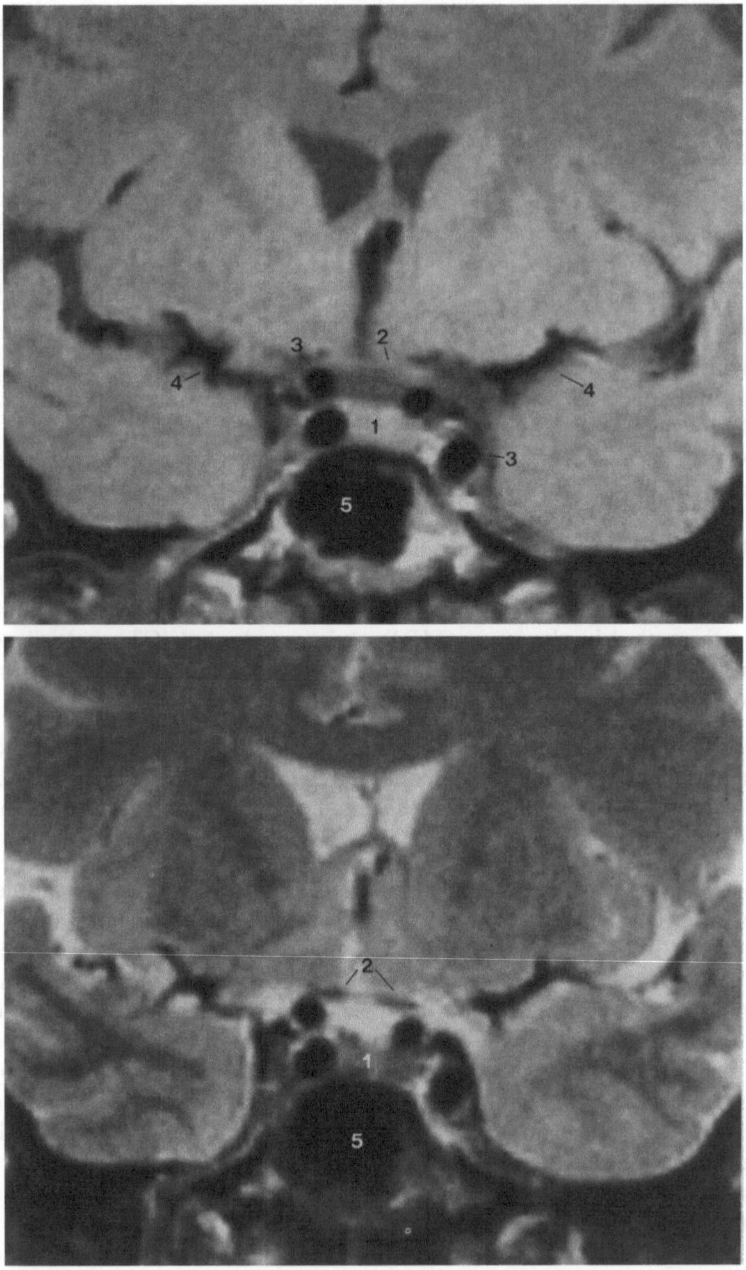

a

b

Fig. 5a, b. T2-weighted images. **a** TR = 2200 ms, TE = 30 ms. **b** TR = 2200 ms, TE = 90 ms. The signals of the pituitary gland (*1*) and the optic chiasm (*2*) are isointense with respect to the cerebral white matter. Notice the flow-void phenomenon within the carotid arteries (*3*) and their branches (*4*). (Courtesy of Prof. A. Gaston, Department of Neuroradiology, Hôpital H. Mondor, Creteil, France)

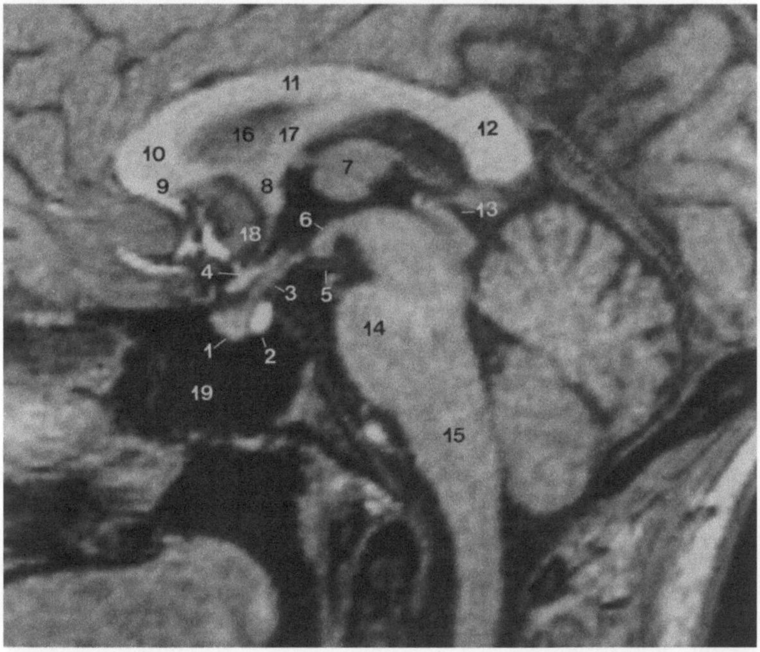

Fig. 6. Flash sequence (TR = 35 ms, TE = 6 ms, flip angle = 40 °). This sagittal view displays a very good anatomical study of the region. It clearly delineates the two components of the pituitary gland, anterior lobe (*1*) and posterior lobe (*2*), with their high signal intensities and the other component of this area, the stalk (*3*), optic chiasm (*4*), nervus oculomotorius (*5*), mammillary bodies (*6*), thalamus (*7*), commissura anterior (*8*), corpus callosum with rostrum (*9*), genu (*10*), truncus (*11*) and splenium (*12*), lamina quadrigemina (*13*), pons (*14*), medulla (*15*), septum pellucidum (*16*), fornix (*17*), lamina terminalis (*18*), the sphenoidal sinus (*19*). (Courtesy of Prof. A. Gaston, Department of Neuroradiology, Hôpital H. Mondor, Creteil, France)

the infundibulum attaches to the pituitary gland can be seen in coronal section. It is best demonstrated in long TR and short TE (proton-density) sequences. Identification can be used to help defining the upper limit of intrasellar pathology and lower limit of suprasellar pathology [31].

The optic pathways (Fig. 7) are easily evaluated on T1-weighted images because of the excellent contrast between chiasm or optic nerves and suprasellar CSF. On T2-weighted images, this contrast is lowered because of relative increased signal intensity of CSF and, thus the interfaces between pituitary gland, cistern, and optic pathways are difficult to appreciate.

The cavernous sinuses have a thin medial wall which cannot usually be resolved as a separate structure. MRI demonstrates cranial nerves III, VI, and V1 in the cavernous sinus especially in T1-weighted sequences and coronal

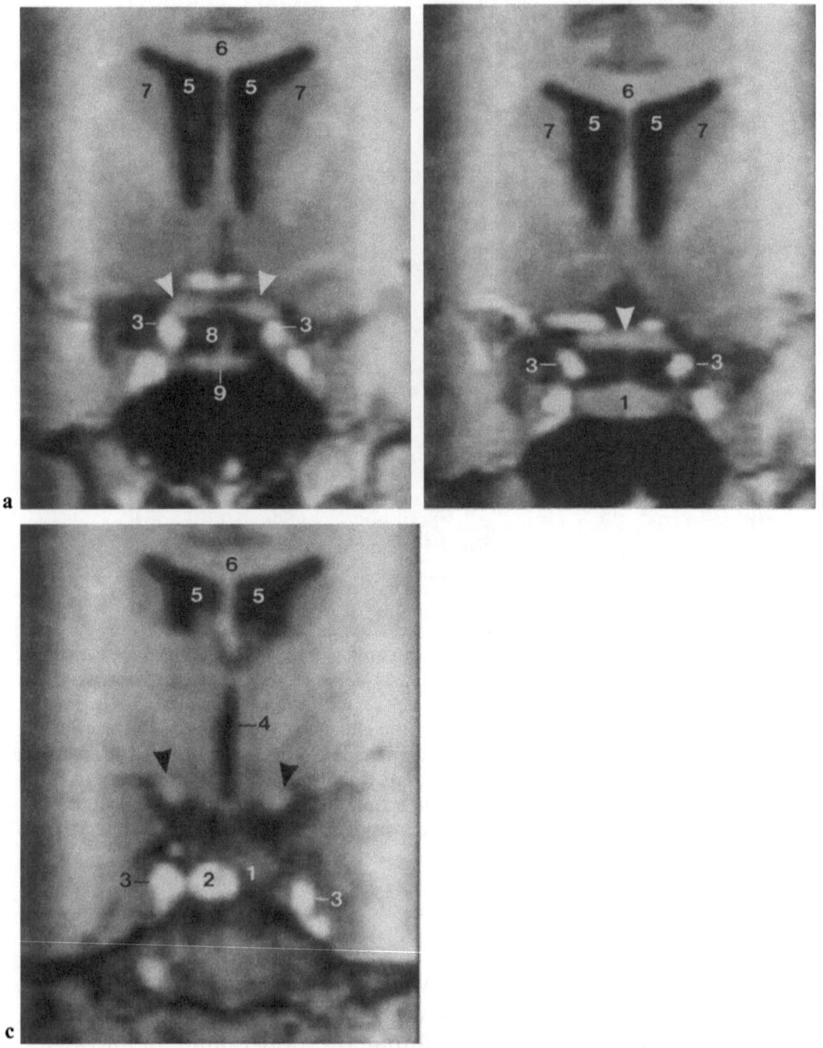

Fig. 7a–c. Flash sequence (TR = 35 ms, TE = 6 ms, flip angle = 40 °). Coronal view demonstrating optic pathways. **a** Optic nerves (*arrowhead*) are below the stalk (*8*) and pituitary gland (*9*) and under cerebral hemisphere with the lateral ventricles (*5*), corpus callosum (*6*), and caudate nuclei (*7*). On this sequence the signal of the internal carotid arteries (*3*) is high. **b** Optic chiasm (*arrowhead*) is separated from the pituitary gland (*1*) by CSF. **c**, optic tracts (*arrowhead*) are applied to the inferior surface of the frontal lobe. The third ventricle and vessels (*4*) are seen as well as the lateral ventricles (*5*) and corpus callosum (*6*). In the sella, the two components of the pituitary gland, anterior lobe (*1*) and posterior lobe (*2*), are well delineated and bordered by carotid arteries (*3*). (Courtesy of Prof. A. Gaston, Department of Neuroradiology, Hôpital H. Mondor, Creteil, France)

plane because their signal, approximatively equal to that of the corpus callosum, is surrounded by a low-intensity signal from flowing blood. The venous space between V1 and V2 has a negligible signal. The cavum meckelii appears as an oval-shaped region with slightly greater signal intensity than CSF on T1-weighted images and intense on T2-weighted images. On axial images, cranial nerves and vascular channels can always be resolved while CT often fails to delineate vascular structures in the cavernous sinus, unless dynamic scanning techniques are used [34]. One of the limitations of the CT scanner is its inability to delineate the carotid arteries from sellar and juxtasellar masses in most of the cases because both often enhance in the same way. In MRI a signal void is present in the carotid artery, which allows separation of this vessel from tumors and may be sufficient to determine modifications of the artery, making arteriography unnecessary [26]. The lack of signal from cortical bone of the sphenoid is identical to the lack of signal from aerated sinus, and detection of bone discontinuity is impossible.

After injection of gadolinium DTPA, gray matter enhances weakly (5%) and, due to absence of a BBB, normal pituitary gland and infundibulum show marked (60%–80%) [35] and relatively homogeneous enhancement. Some inhomogeneity of the enhancement pattern within the normal gland can be seen due to tissue compactness, granularity, and vascularity and may account for occasional false-positive MRI findings. This enhancement persists for about 1 h and remains strong for 30 min after injection [36, 37], but it is typically maximum at 3 min. [38, 35] (Fig. 1). This difference in dynamics mandates a concern for postinjection timing of imaging. To capitalize on selective enhancement of the normal gland, imaging should be initiated shortly after injection, and if the images are nondiagnostic, delayed images should be obtained [36]. There is no enhancement visible in the carotid artery (due to rapid blood flow). The signal intensity of the venous spaces of the cavernous sinuses increases from negligible (typical of flowing blood prior to gadolinium DTPA administration) to approximately the same as that of the pituitary gland. Flowing blood has a signal dependent on the proton density, T1 and T2, and flow velocity. A weak signal means that the flow moves resonant protons out of the image prior to data acquisition. The long T1 relaxation time allows minimal recovery of magnetization during the short TR pulse sequence. After gadolinium, signal intensity is detected in slowly flowing blood because T1 relaxation time is shortened [35]. Appreciable enhancement of the walls of the cavernous sinus, the intracavernous segments of cranial nerves III and VI, the gasserian ganglion, and the second and third divisions of cranial nerve V in the base of the skull is seen, but enhancement neither of the cisternal portion of cranial nerve V nor of the chiasm or optic nerves may be noticed.

References

1. Muhr C, Bergstrom K, Grimelius L, Larson SG (1981) A parallel study of the roentgen anatomy of the sella turcica and the histopathology of the pituitary gland in 205 autopsy specimens. Neuroradiology 21:55–65
2. Kucharczyk W, Peck WW, Kelly WM, Norman D, Newton TH (1987) Rathke cleft cysts: CT, MR imaging and pathological features. Radiology 165:491–495
3. Colombo N, Berry I, Kucharczyk J, Kucharczyk W, De Groot J, Larson T, Norman D, Newton TH (1987) Posterior pituitary gland: appearance on MR images in normal and pathological states. Radiology 165:481–485
4. Okamoto S, Handa H, Yamashita J, Ishikawa M, Nagasawa S (1985) Computed tomography in intra- and suprasellar epithelial cyst (symptomatic Rathke cleft cysts). AJNR 6:515–519
5. Nishimura K, Fujisawa K, Itoh K, Nanano Y, Itoh H, Torizuka K (1986) Posterior lobe of the pituitary: identification by lack of chemical shift artifact in MR imaging. J Comp Assist Tomogr 10:899–902
6. Gudinchet F, Brunelle F et al. (1980) MR imaging of the posterior hypophysis in children. AJNR 10:511–514
7. Wiener SN, Rzeszotarski MS, Droege RT, Pearlstein AE, Shafron MM (1985) Measurement of pituitary gland height with MR imaging. AJNR 717–722
8. Mark L, Pesch P, Daniels D, Charles C, Wiliams A, Haughton V (1984) The pituitary fossa: a correlative anatomic and MR study. Radiology (1984) 153:435–457
9. Bonneville JF, Dietmann JL (1981) Radiology of the sella turcica. Springer, Berlin Heidelberg New York
10. Roppolo HMN, Latchaw RE, Meyer JD, Curtin HD (1983) Normal pituitary gland. 1. Macroscopic anatomy – CT correlation. AJNR 4:927–935
11. Zatz LM, Janon EA, Newton TH (1969) The enlarged sella and intrasellar cistern. Radiology 93:1085–1091
12. Kaufman B, Tomsak RL et al. (1989) Herniation of the suprasellar visual system and third ventricle into empty sellae: morphologic and clinical considerations. AJNR 10:65–76
13. Davis PC, Hoffman JC, Tindall GT, Braun IF (1984) Prolactin-secreting pituitary microadenomas: inaccuracy of high-resolution CT imaging. AJNR 5:721–726
14. Sadoka K, Mukada K, Yonezawa S, Matsumura S, Yoshimoto H, Mori S, Uozumi T (1981) CT scan of pituitary adenomas. Neuroradiology 20:249–253
15. Bonneville JF, Cattin F, Dietmann JL (1986) Dynamic CT of the pituitary gland in computed tomography of the pituitary gland. Springer, Berlin Heidelberg New York Tokyo, pp 33–46
16. Boulin A (1987) La region sphenoidale. In: Tomodensitometrie cranioencephalique. Vigot, Paris, pp 420–501
17. Roppolo HMN, Latchaw RE (1983) Normal pituitary gland. 2. Microscopic anatomy – CT correlation. AJNR 4:937–944
18. Seidel FG, Towbin R, Kaufman RA (1985) Normal pituitary stalk size in children: CT study. AJNR 6:733–738
19. Vignaud J, Boulin A, Elmaleh M (1987) La loge caverneuse in Tomodensitometrie cranio-encephalique. Vigot, Paris, pp 458–470
20. Dwyzer AJ, Frank et al. (1987) Pituitary adenomas in patients with cushing disease: initial experience with Gd-DTPA-Enhanced MR imaging. Radiology 163:421–426
21. Kurchaczyk W, Davis DO, Kelly WM, Sze G, Norman D, Newton TH (1986) Pituitary adenomas: high-resolution MR imaging at 1.5 T. Radiology 161:761–765
22. Hyman RA, Gorey MT (1988) Imaging strategies for MR of the brain. Radiol Clin North Am 494
23. Kucharczyk W (1987) The pituitary gland and the sella turcica. In: Brant-Zadwadzki M, Norman D (eds) Magnetic resonance imaging of the cental nervous system. Raven, New York, pp 187–195

24. Pojunas KW, Daniels DL, Williams AL, Haughton VM (1986) MR imaging of prolactin-secreting microadenomas. AJNR 7:209–213
25. Davis PC, Hoffmen JC, Tindall GT, Braun IF (1987) MR imaging of pituitary adenoma: CT, clinical and surgical correlation. AJNR 8:107–112
26. Lee BCP, Deck MDF (1985) Sellar and juxtasellar lesion detection with MR. Radiology 157:143–147
27. Kaufman B (1984) Magnetic resonance imaging of the pituitary gland. Radiol Clin North Am 22, 4:795–803
28. Lemort M, Haesendonck P, Louryan S, Rodesch G, Christophe C, Baleriaux D (1988) Magnetic resonance imaging of the sellar region and suprasellar cisterns: normal morphology on sagittal sections. In: Gouaze A, Salamon G (eds) Brain anatomy and magnetic resonance imaging. Springer, Berlin Heidelberg New York, pp 158–163
29. Bilaniuk LT, Zimmerman RA et al. (1984) Magnetic resonance imaging of pituitary lesions using 1.0 to 1.5 T field strength. Radiology 153:415–418
30. Dietmann JL (1990) Imagerie par résonance magnétique des adénomes de l'hypophyse. Rev Im Med 2:3–11
31. Daniels DL, Pojunas KW et al. (1986) MR of the diaphragma sellae. AJNR 7:765–769
32. Brooks BS, El Gammal T, Allison JD, Hoffman WH (1989) Frequency and variation of the posterior pituitary bright signal on MR images. AJNR 10:943–948
33. Fujisawa I, Asato R et al. (1989) Hyperintense signal of the posterior pituitary on T1-weighted images. An experimental study. J Comput Assist Tomogr 13:371–377
34. Daniels DL, Pech P, Mark L, Pojunas K, Williams AL, Haughton VM (1985) Magnetic resonance imaging of the cavernous sinus. AJNR 6:187–192
35. Kilgore DP, Breger RK, Daniels DL, Pojunas KW, Williams AL, Haughton VM (1986) Cranial tissues: normal MR appearance after intravenous injection of Gd-DTPA. Radiology 160:757–761
36. Kucharczyk W, Davis OD, Kelly WM, Sze G, Norman D, Newton TH (1986) Pituitary adenomas: high resolution MR imaging at 1.5 T. Radiology 161:761–765
37. Doppman JL, Frank JA et al. (1988) Gadolinium DTPA enhanced MR imaging of ACTH-secreting microadenomas of the pituitary gland. J Comput Assist Tomogr 12:728–735
38. Newton DR, Dillon WP, Norman D, Newton TH, Wilson CB (1989) Gd-DTPA-enhanced MR imaging of pituitary adenomas. AJNR 10:949–954

New Possibilites for Computed Tomography in the Diagnosis of Pituitary Microadenomas

J. F. BONNEVILLE and F. CATTIN, Besançon/France

Introduction

Magnetic resonance imaging represents today the method of choice for imaging the hypothalamic-pituitary region, especially for suprasellar tumors. Nevertheless, it is our opinion that computed tomography (CT) remains the optimal technique for imaging intrasellar pituitary tumors on the condition that high-resolution dynamic CT is used; moreover, dynamic CT allows perfect delineation of cavernous sinus components. Direct coronal sections are virtually always superior in quality to reformated images from axial sections and should constitute the first line of approach. Axial complementary sections are necessary only when plain films suggest the presence of a very anterior or very posterior adenoma, or when coronal sections are not diagnostic. Intrasellar adenomas are nearly always less enhanced by the iodinated contrast medium than the normal pituitary. Compression of pituitary cells at the periphery of the adenoma should explain this lesser enhancement. Dynamic CT in coronal sections is essential for demonstrating small adenomas [2]. This technique allows visualization of (a) the delay in enhancement of the adenoma by comparison with the healthy pituitary tissue, (b) the contralateral displacement of the secondary pituitary capillary bed, (c), sometimes a very early and transient enhancement of the microadenoma itself corresponding to its direct arterial blood supply, and (d) the intracavernous cranial nerves and veins.

Dynamic CT of the Normal Pituitary Gland

Dynamic CT is of major benefit for diagnosis of small pituitary lesions. Visualization of the pituitary capillary bed with this technique is today as fundamental for diagnosis of intrasellar lesions as was demonstration of the calcified pineal gland or internal cerebral vein for identification of the midline of the brain in the past.

Anatomy. A brief review of pituitary blood supply is necessary to understand the utility of a dynamic scan. The various structures comprising the pituitary are directly or indirectly vascularized from two principal sources, the superior and inferior hypophyseal arteries. The inferior hypophyseal artery arises from each side of the intracavernous segment of the internal carotid artery; the superior hypophyseal artery, also paired, arises from the internal carotid immediately after its exit from the cavernous sinus following passage through the dura mater. There are transversal anastomoses between both superior hypophyseal arteries. Furthermore, the trabecular artery forms anastomoses with the superior and inferior pituitary network.

The anterior pituitary does not receive any direct arterial blood supply with exception of the trabecular artery or artery of the fibrous core which can supply pituitary adenomas [4]. At the superior part of the pituitary stalk the hypophyseal arteries give rise to highly convoluted spiral capillary networks which constitute the first capillary bed of the pituitary portal system [7]. The first capillary bed is drained by the long portal vessels which descend the stalk and terminate in the second capillary bed, constituted by the anterior lobe sinusoids. These sinusoids, larger than the capillaries, are fed solely by portal vessels carrying blood which already passed through the first capillary network in the pituitary stalk. Popa and Fielding [5], who first discovered the existence of the pituitary portal system, believed that this system functioned in an ascending fashion, i.e., from the pituitary to the hypothalamus. Proof that blood circulation actually occurs in the opposite direction, i.e., from the hypothalamus to the pituitary, was shown by subsequent morphological studies (Wislocki) [6] and by in vivo observation in animals. Dynamic CT provides confirmation of the Wislocki hypothesis in living subjects, demonstrating that enhancement is progressive and clearly separated in time: initially the carotid siphons, followed by the top of the pituitary stalk, the secondary capillary bed, and finally centrifugal enhancement of the anterior lobe.

Unlike the anterior lobe, the posterior lobe has a conventional arterial-capillary-venous circulatory system, supplied by the right and left inferior hypophyseal arteries, with sinusoids far smaller than those of the anterior lobe. In the absence of a portal system, sinusoids of the posterior lobe appear to be enhanced far earlier than those of the anterior lobe. Finally, sinusoids of both lobes terminate in small venules in the periphery of the gland, draining into the venous sinus which surrounds the pituitary.

Technique. Dynamic scanning requires rapid injection of a bolus of 80 cm^3 contrast medium at 15 cm^3/s, pushed by 30 cm^3 saline solution. Six to eight 1.5-mm sections are carried out beginning with the start of the injection. With the General Electric 9800 CT/T system, scan time is 2 s, and the interscan delay time is 2.5 s.

Results. Dynamic CT (Fig. 1) allows visualization of the intracavernous and supraclinoid internal carotid arteries, central part of the secondary pituitary capillary bed (the pituitary tuft), centrifugal enhancement of the anterior pituitary, posterior pituitary in axial section, and the venous spaces of the cavernous sinus. The intracavernous internal carotid arteries and the branches of their intracranial division are perfectly opacified 10 s after the end of injection. The high quality of information thus available concerning topography of the carotid siphons today renders carotid angiography superfluous before transsphenoidal surgery.

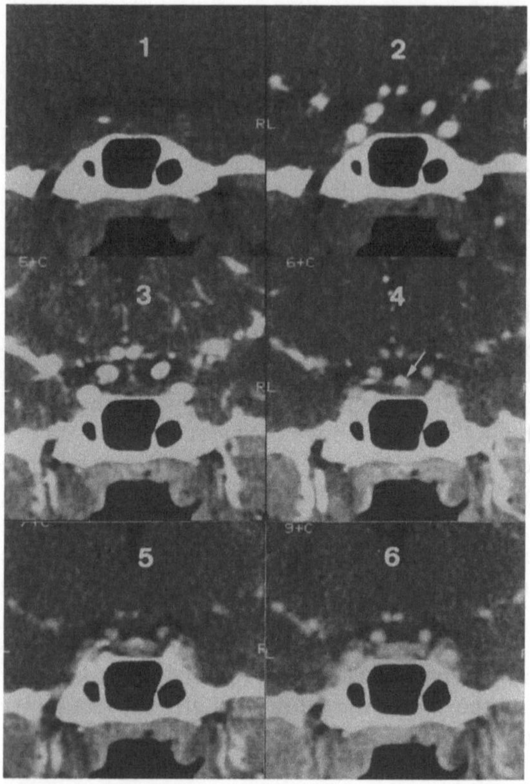

Fig. 1. Normal dynamic coronal CT scan of the pituitary gland. *1*, Before bolus injection; *2*, onset of enhancement of internal carotid arteries; *3*, optimal visualization of internal carotid arteries; *4*, opacification of the pituitary tuft (*arrow*); *5*, centrifugal enhancement of the pituitary gland; *6*, pituitary enhancement is homogeneous

Approximately 10 s after optimal enhancement of internal carotid arteries, the secondary pituitary capillary bed is visualized as an unpaired median vascular structure situated below the sellar diaphragm, in a prolongation of the direction of the pituitary stalk. This vascular tuft is virtually always regular and symmetric; however, a moderate degree of asymmetry is possible in the absence of pathological changes. When the pituitary height is small, the pituitary capillary bed may be composed of a vascular band running immediately below the superior border of the gland. Between 20 and 40 s after injection, there is a gradual, regular, and symmetrical centrifugal enhancement of the anterior pituitary from the capillary bed; 60 s after bolus injection, pituitary enhancement is homogeneous, and the pituitary capillary bed is no longer visible.

In dynamic axial sections posterior pituitary capillary enhancement occurs very soon after enhancement of the intracavernous carotid arteries due to the direct arterial blood supply. At some time after the bolus injection the posterior

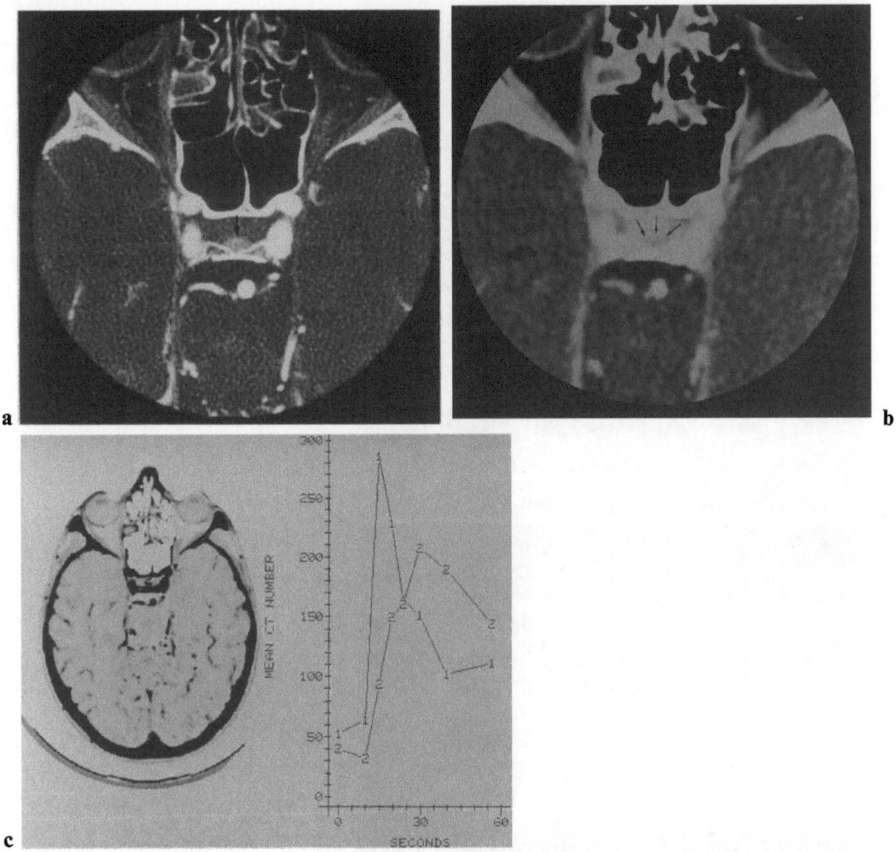

Fig. 2a–c. Axial dynamic CT scan of the pituitary gland. **a** Arterial phase. Enhancement of the posterior lobe (*arrow*) contemporary of optimal opacification of the internal carotid arteries. **b** At the end of dynamic CT scan, the posterior pituitary (*arrows*) appears less dense than the anterior pituitary. **c** Representative time/density curves for the anterior and posterior pituitary

pituitary appears less enhanced than the anterior pituitary. Indeed, when contrast media equilibrates between the intravascular and extravascular interstitial spaces, a lesser enhancement of the posterior lobe is seen, correlating with its lesser volume of the interstitial spaces (Fig. 2).

Dynamic CT of Pituitary Microadenomas

The capillary bed is regularly displaced or deformed (tuft sign) in cases of lateralized intrasellar microadenoma (Fig. 3). This sign is fundamental for the diagnosis of pituitary adenomas too small to deform the sellar floor or the

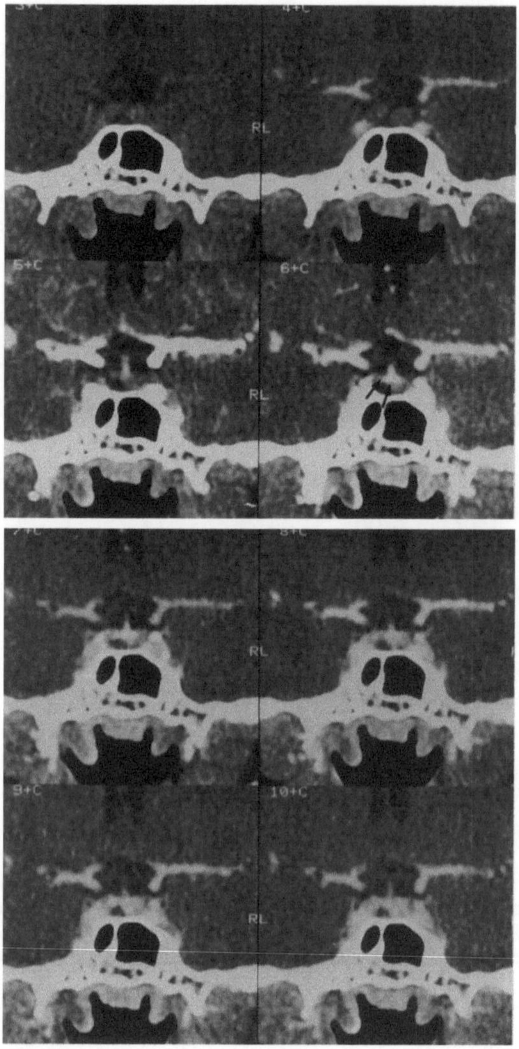

a

b

Fig. 3 a, b. Prolactinoma, 4 mm. Dynamic coronal CT scan, localized compression of the pituitary bed (*arrows*) by the hypodense microadenoma

superior pole of the pituitary. It is also essential for the diagnosis of isodense adenomas. When microadenomas are midline, the pituitary tuft is not displaced, but localized compression of its inferior part may occur. When the pituitary capillary bed is bandlike, a localized absence of opacification has the same diagnostic significance as displacement of the tuft (Fig. 4). Finally, after opacification of the pituitary bed, asymmetrical enhancement of the pituitary

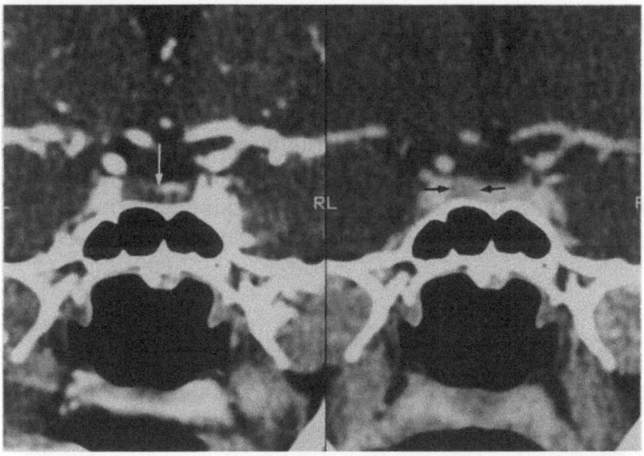

Fig. 4. Prolactinoma. Dynamic coronal CT scan. *Left*, localized defect in filling of the capillary bed (*arrow*); *right*, at the end of dynamic scanning, a hypodense 3-mm microadenoma is demonstrated (*arrows*)

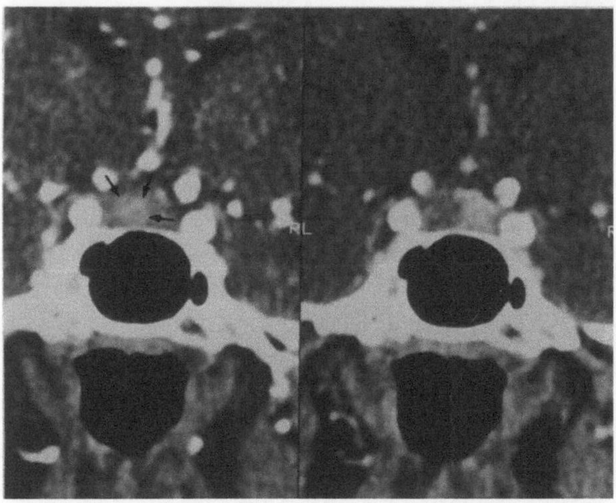

Fig. 5. Prolactinoma, 7 mm. Coronal dynamic CT scan. *Left*, early and transient enhancement of the adenoma (*arrows*); *right*, at the end of the dynamic scan the adenoma appears hypodense

constitutes a further sign for the presence of an intrasellar lesion. In general, intrasellar microadenomas are enhanced more slowly and to a lesser extent than the healthy pituitary tissue. At last, we have recently shown [1] that in two out of three cases pituitary microadenomas demonstrate very early and transient enhancement due to their direct arterial supply through the artery of the fibrous core (Fig. 5).

Dynamic CT of the Cavernous Sinus

The cavernous sinuses are generally of symmetrical topography and volume. Under normal conditions the lateral dural wall is flat or slightly concave outwards in axial sections. The dural sheath at the lateral limit of the sinus frequently shows marked contrast enhancement. The internal carotid artery is well visualized within the cavernous sinus. Due to its sigmoidal form in both coronal and sagittal planes, it is virtually never completely visualized in a single

6

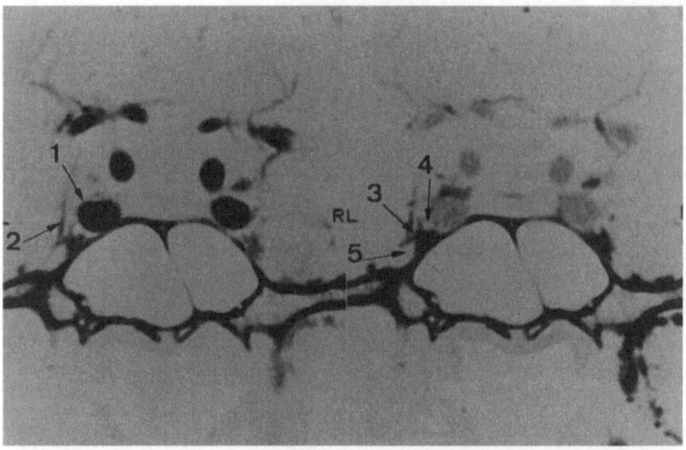

7

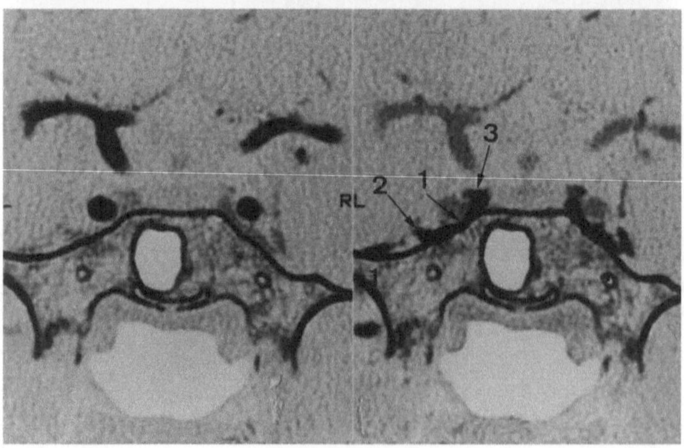

Fig. 6. Normal coronal dynamic CT scan of the cavernous sinus. *1*, Intracavernous internal carotid artery; *2*, vein of the lateral wall; *3*, inferolateral venous group; *4*, abducens nerve (VI); *5*, ophthalmic nerve (V_1)

Fig. 7. Normal coronal dynamic CT scan of the cavernous sinus. Large bilateral veins of the carotid sulcus (*1*) communicating freely with the inferolateral group (*2*) and the medial vein (*3*)

plane. The artery occupies a medial position within the cavernous sinus, and is only 1–2 mm from the lateral pituitary border. In arteriosclerosis or in the elderly or in the case of elongation of the carotid, the artery comes into direct contact with the gland and can change its shape. The laterosellar venous spaces can be perfectly defined with dynamic CT [3]. They include a constant large space inferolateral to the horizontal portion of the cavernous carotid artery (Fig. 6), a space often asymmetrical and inconsistently seen medial to the artery and a space between the artery and the sphenoid bone (the vein of the carotid sulcus) (Fig. 7). Early invasion of the cavernous sinus by a pituitary adenoma can be suspected in the cases where laterosellar veins, particularly the vein of the carotid sulcus are not opacified with dynamic CT.

References

1. Bonneville JF, Cattin F (1990) In vivo demonstration of direct arterial blood supply of pituitary microadenomas
2. Bonneville JF, Cattin F, Moussa Bacha K, Portha C (1983) Dynamic computed tomography of the pituitary gland: the "tuft sign". Radiology 149:145–148
3. Bonneville JF, Cattin F, Racle A, Bouchareb M, Boulard D, Potelon P, Tang YS (1989) Dynamic CT of the laterosellar extradural venous spaces. AJNR 10:535–542
4. Gorczyca W, Hardy J (1988) Microadenomas of the human pituitary and their vascularization. Neurosurgery 22:1–6
5. Popa G, Fielding U (1930) A portal circulation from the pituitary to the hypothalamic region. J Anat 65:88–91
6. Wislocki GB (1938) The vascular supply of the hypophysis cerebri of the rhesus monkey and man. Res Publ Assoc Nerv Ment Dis 17:48–68
7. Xuereb GB, Prichard MML, Daniel PM (1954) The hypophysial portal system of vessels in man. QJ Exp Physiol 39:219–230

Computed Tomographic Cisternography of the Sellar and Parasellar Region: A Radiological Approach to Cisternal Anatomy

M. Dickob, Ulm/FRG, M. Galanski, Hannover/FRG, and
W. Wittkowski, Münster/FRG

With the introduction of microsurgical techniques in neurosurgery new requirements have been established for the radiological management of the sellar region. For planning an operation the surgeon needs exact information about the extension of lesions and their encroachment upon surrounding structures (Yasargil 1984). Concerning the parasellar cisterns computed tomographic (CT) cisternography with metrizamide is a suitable method to visualize both nerval and vascular contents and its spatial relationships (Dickob 1987; Galanski et al. 1986). It is superior to magnetic resonance imaging (MRI) in image acuity and should be performed if more details were needed than intravenous contrast-enhanced CT or MRI can provide.

Patients and Methods

The image acuity of CT cisternography was studied with the help of postmortem air cisternograms and specially designed models. CT cisternographies with metrizamide were performed on 13 patients with various clinical symptoms. All examinations were carried out on a Tomoscan 350 scanner (Philips) using 1.5-mm or 3-mm slice thickness and continuous scanning. For exact and reproducible projections the anthropological baseline (orbitomeatal line, OML) was fixed on a lateral "scanogram" by the inferior orbital rim and the bony auditory canal. Besides the OML a plane in 15° extension to the OML and coronal projections (90° extension to the OML) were used. For cisternography 6–10 ml Solutrast 250 M (Byk Gulden) were administered by lumbar punction. Then the patient was placed on the CT unit as described by Lotz (1982).

Results

The patients suffered from only mild sideeffects such as transient headache or vertigo. No severe complications were observed. Examination duration was about 30 min for the complete sellar region.

Using the OML optic tracts and mamillary bodies are visualized in the roof of the suprasellar cisterns. Cerebral veins such as anterior cerebral vein, middle cerebral vein, interpeduncular and basal vein can be delineated. The posterior cerebral and the superior cerebellar artery are shown within the ambient cistern (see Fig. 1). Several millimeters downward the origins of the anterior and middle cerebral arteries from the internal carotid artery can be seen. Even the anterior communicating artery is visualized. Behind, the posterior cerebral artery turns around the midbrain. Laterally the middle cerebral artery is excellently outlined in the Sylvian fissure. The proximity of the pituitary stalk and optic chiasm is demonstrated in the center of the suprasellar cisterns (Fig. 2).

As optic and oculomotor nerves run in 15° extension to the OML, using this projection can improve imaging of these structures. Optic nerves are shown from the chiasm through the optic canal to the bulbus oculi on one scan (see Fig. 3). Posteriorly the infundibular recess of the third ventricle bulges downward. On both sides of the chiasm the internal carotid arteries curve into the cisterns.

Approximately 3–5 mm downward the spatial relationships between oculomotor nerve and posterior communicating artery are well outlined. In the interpeduncular cistern the oculomotor nerve crosses under the posterior cerebral artery and over the superior cerebellar artery, both originating from the basilar artery. Internal carotid artery, diaphragma sellae, and the end of the pituitary stalk are visualized. For more details, such as cavernous sinus and superior orbital fissure, see Fig. 4. The level of the pituitary gland is reached one scan below that on which the oculomotor nerve can usually be distinguished disappearing within the cavernous sinus.

Using coronal projections in 90° extension to the OML, anterior cerebral arteries with the linking anterior communicating artery can be demonstrated. The proximity of the internal carotid artery and the optic nerve is shown. Further cisternal and bony detail is visualized (Fig. 5).

Several millimeters behind, the anterior cerebral arteries and veins can be seen above the optic chiasm (Fig. 6). Below the chiasm the pituitary stalk runs directly to the pituitary gland, which normally has a concave or straight upper surface. A convex surface of the gland and inclination of the stalk on coronal scans can indicate a pituitary lesion (Nakagawa et al. 1984). To demonstrate the origin of the anterior and middle cerebral arteries from the internal carotid artery a scan with 3-mm slice thickness gives more details.

Conclusions

MRI today allows the visualization of the carotids, the pituitary stalk, and the optic nerves within the parasellar cisterns avoiding the application of any contrast media (Bilaniuk 1984). As shown here, CT cisternography provides further details. Until now it is the only method to demonstrate both the vascular and neural contents of the parasellar cisterns and the spatial relationships between even small structures. Projections parallel to the OML, in 15° extension to the OML and coronal scans (90° extension to the OML), are particularly instructive.

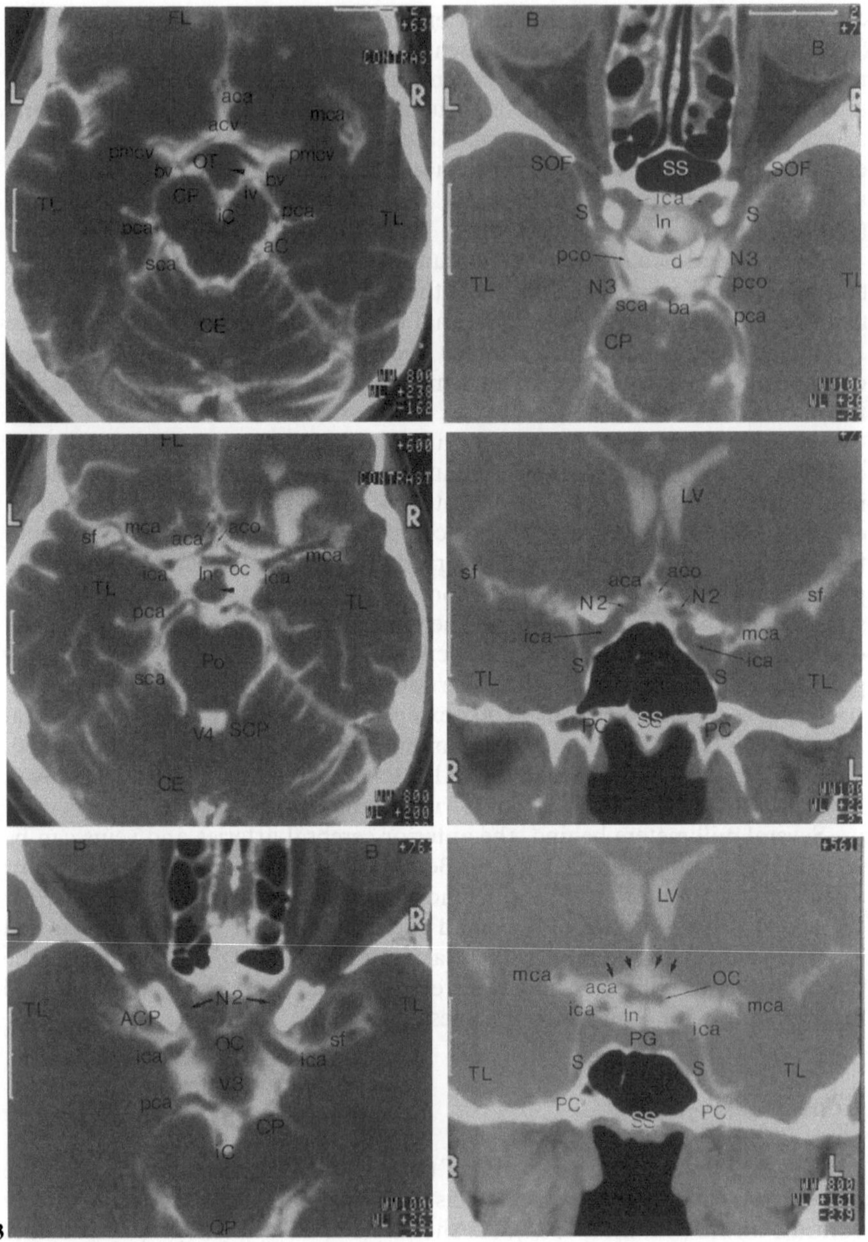
1
FL
CONTRAST
L
R
aca
acv
mca
pmcv
pmcv
bv
OT
bv
CP
lv
TL
iC
pca
TL
pca
aC
sca
CE
WW 800
WL +238
-162
2
FL
+600
CONTRAST
L
R
sf
mca
aca
aco
mca
ln
oc
ica
ica
TL
TL
pca
Po
sca
V4
SCP
CE
WW 800
WL +200
3
B
B
L
R
N2
TL
TL
ACP
OC
sf
ica
ica
V3
pca
CP
iC
OP
4
B
B
L
R
SOF
SS
SOF
ica
S
S
ln
pco
d
N3
TL
N3
pco
TL
sca
ba
pca
CP
5
LV
sf
aca
aco
sf
N2
N2
ica
mca
ica
S
S
TL
TL
PC
SS
PC
R
L
6
LV
mca
aca
OC
ica
ln
mca
ica
PG
S
S
TL
TL
PC
SS
PC
R
L
WW 800
WL +161
-239

Figs. 1–6

Although invasive, the CT cisternography should be performed if intravenous contrast-enhanced CT or MRI does not confirm an expected lesion. Indications are, for example, to show the suprasellar extension of pituitary lesions and their encroachment upon cisternal structures (Goshhajra 1981; Chui and Patel 1982). Furthermore, cases of vascular compression of cranial nerves can be demonstrated (Deeb et al. 1979; Jannetta 1980). Also, other reasons for unsolved nerve palsy can be detected by this method (Hiscott and Symon 1982; Rush and Younge 1981).

References

Bilaniuk LT (1984) Magnetic resonance imaging of pituitary lesions using 1.0 to 1.5 T field strength. Radiology 153:415–418

Chui M, Patel HM (1982) Computed tomography of sellar and parasellar lesions: indications for metrizamide cisternography. J Can Assoc Radiol 33:84–88

Deeb ZL, Jannetta PJ, Rosenbaum AE, Kerber CW, Drayer BP (1979) Tortuous vertebrobasilar arteries causing cranial nerve syndromes: screening by computed tomography. J Comput Assist Tomogr 3:774–778

Fig. 1. Scan through the roof of the suprasellar cisterns, OML. *Arrowhead*, Lesion between the optic tracts and the mamillary bodies; *aC*, ambient cistern; *aca*, anterior cerebral artery; *acv*, anterior cerebral vein; *bv*, basal vein; *CE*, cerebellum; *CP*, cerebral peduncle; *iC*, interpeduncular cistern; *iv*, interpeduncular vein; *mca*, middle cerebral artery; *OT*, optic tract; *pca*, posterior cerebral artery; *pmcv*, middle cerebral vein; *sca*, superior cerebellar artery

Fig. 2. Scan 6 mm downward from that in Fig. 1, OML. *Arrowhead*, Lesion (see Fig. 1); *aca*, anterior cerebral artery; *aco*, anterior communicating artery; *FL*, frontal lobe; *ica*, internal carotid artery; *In*, pituitary stalk; *mca*, middle cerebral artery; *oc*, optic chiasm; *pca*, posterior cerebral artery; *Po*, pons; *sca*, superior cerebellar artery; *sf*, Sylvian fissure; *V4*, fourth ventricle

Fig. 3. Scan through the optic nerves using a projection in 15° extension to the OML. *ACP*, Anterior clinoid process; *B*, bulbus oculi; *CP*, cerebral peduncle; *iC*, interpeduncular cistern; *ica*, internal carotid artery; *N2*, optic nerve; *OC*, octic chiasm; *pca*, posterior cerebral artery; *QP*, quadrigeminal plate; *sf*, Sylvian fissure; *V3*, third ventricle; *TL*, temporal lobe

Fig. 4. Scan 4.5 mm downward and parallel to that in Fig. 3. *B*, Bulbus oculi; *ba*, basilar artery; *d*, dorsum sellae; *ica*, internal carotid artery; *In*, pituitary stalk; *N3*, oculomotor nerve; *pca*, posterior cerebral artery; *pco*, posterior communicating artery; *S*, cavernous sinus; *SOF*, superior orbital fissure

Fig. 5. Coronal scan in 90° extension to the OML. *aca*, Anterior cerebral artery; *aco*, anterior communicating artery; *ica*, internal carotid artery; *LV*, lateral ventricle; *mca*, middle cerebral artery; *N2*, optic nerve; *PC*, pterygoid canal; *S*, cavernous sinus; *sf*, Sylvian fissure; *SS*, sphenoid sinus

Fig. 6. Coronal scan 6 mm that in Fig. 5. *Arrows*, Anterior cerebral veins; *aca*, anterior cerebral artery; *ica*, internal carotid artery; *In*, pituitary stalk; *LV*, lateral ventricle; *mca*, middle cerebral artery; *OC*, optic chiasm; *PC*, pterygoid canal; *PG*, pituitary gland; *S*, cavernous sinus; *SS*, sphenoid sinus

Dickob M (1987) Die Topographie der basalen Zisternen des Gehirns im CT-Zisternogramm. Diss Med Fak WWU Münster
Galanski M, Dickob M, Wittkowski W (1986) CT-Zisternographie der basalen Zisternen. Fortschr Röntgenstr 145:149–157
Goshhajra K (1981) High-resolution metrizamide CT cisternography in sellar and suprasellar abnormalities. J Neurosurg 56:232–239
Hiscott P, Symon L (1982) An unusual presentation of neurofibroma of the oculomotor nerve. J Neurosurg 56:854–856
Jannetta PJ (1980) Neurovascular compression in cranial nerve and systematic disease. Ann Surg 192:518–525
Lotz PR (1982) Intracranial delivery of metrizamide from the lumbal subarachnoid space: prone versus supine position. J Comput Assist Tomogr 6:920–923
Nakagawa Y, Matsumoto K, Fukami T, Takase K (1984) Exploration of the pituitary stalk and gland by high-resolution computed tomography. Neuroradiology 26:473–478
Rush JA, Younge BR (1981) Paralysis of cranial nerves III, IV, VI: cause and prognosis in 1000 cases. Arch Ophthalmol 99:76–79
Yasargil MG (1984) Microsurgical anatomy of the basal cisterns and vessels of the brain. In: Microneurosurgery, vol 1. Thieme, Stuttgart

Magnetic Resonance Imaging Diagnosis of Microadenoma*

K. TAKAKURA, A. TERAMOTO, A. KUROIWA, and N. SHITARA, Tokyo/Japan

Introduction

Pituitary microadenomas can be properly diagnosed by magnetic resonance imaging (MRI) and endocrinological examinations. MRI can well demonstrate the three-dimensional morphological structure of microadenomas. The resolution of MRI is much better than X-ray computed tomography (CT). In this study, MRI of surgically verified microadenomas is analyzed. The localization and size of the tumor upon MRI correlated well with surgical findings. Although gadolinium diethylene-triamine-pentaacetic acid (Gd-DTPA) enhances the tumor and demonstrates well the whole figure of the tumor, the fine intratumoral structures such as hemorrhage or necrosis can better be visualized by T_1- and T_2-weighted images without Gd-DTPA enhancement. The MRI of other lesions such as Rathke's cyst or internal carotid artery aneurysm are presented for the differential diagnosis.

Materials and Methods

Eighteen cases of surgically verified pituitary microadenomas were analyzed to compare MRI and the surgical findings. There were 6 cases of Cushing's disease (all women), five of acromegaly (one man and four women), and seven of prolactinomas (all women). A few cases of Rathke's cyst and internal carotid artery aneurysms were also analyzed for the differential diagnosis. MRI (Magnetom, 1.5 Tesla) was performed under the following conditions: spin-echo method; repetition time (TR) 500, 2000, 2500, 3000 ms; echo time (TE) 14, 20, 34, 75, 115 ms; depth and width 4 and 5 mm; coronal, sagittal, and horizontal sections. Gd-DTPA (Schering) was used for the enhancement.

Results

The morphological structure of microadenoma and normal pituitary gland as well as the surrounding structures such as pituitary stalk, optic nerve, third

* This work was supported by a Research Grant to the Study Group on Midbrain and Pituitary Disorders, by a Cancer Research Grant from the Ministry of Health and Welfare, and by a Research Grant from the Ministry of Education, Science, and Culture.

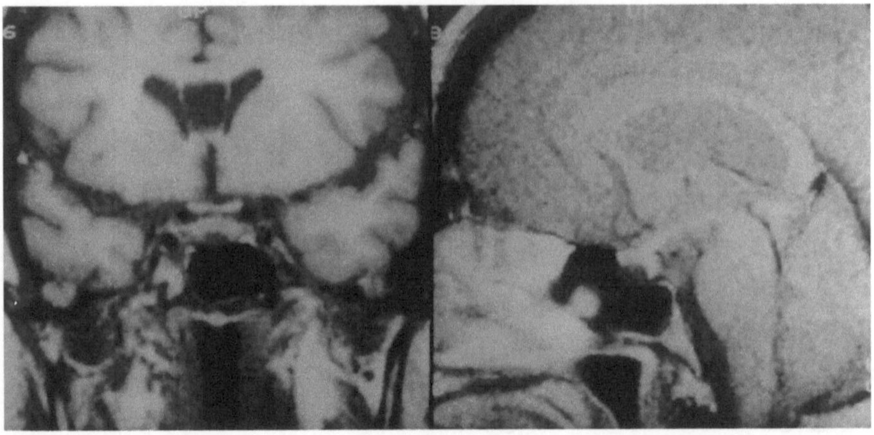

Fig. 1. Cushing disease. *Left:* T_1-weighted image. *Right:* T_2-weighted image

ventricle, and internal carotid artery are demonstrated in T_1-weighted image much better than in CT scans. MRI is, however, not suitable for detecting destruction of the sella turcica or other bony structures. Abnormal findings were detected in 12 out of 18 cases by CT scans but in all 18 cases by MRI. Tumor images were visualized in nine cases, and only enlargement of the pituitary gland was noted in three cases by CT scans, while tumor images were clearly demonstrated in 17 cases, and only enlargement of the pituitary gland was noted in one case by MRI.

Cushing's Disease. By CT scan, the low-density tumor shadow was detected in only two cases. Three cases demonstrated the enlargement of the pituitary gland without detection of the tumor. One case demonstrated no abnormality. By MRI, the low signal intensity mass was detected in all six cases by T_1-weighted image (Fig. 1 left). By T_2 images, two cases showed low signal intensity, and one case demonstrated high signal intensity.

Acromegaly. By CT scan, no abnormal finding was detected in two cases. Tumor shadow was noted in three cases (low density in two cases and high density in one case). By MRI, the tumor shadow was detected in all five cases (low signal intensity in three cases and high signal intensity in two cases) by T_1-weighted image (Fig. 2). By T_2-weighted image, all cases demonstrated high signal intensity.

Prolactinoma. By CT scan, no abnormal finding was detected in three out of seven cases. Three cases demonstrated low density tumor shadow. By MRI, six cases showed low signal intensity mass lesion in T_1-weighted image (Fig. 3). A T_2-weighted image was taken in two cases, and one case demonstrated high and the other low signal intensity mass lesion. Only one case

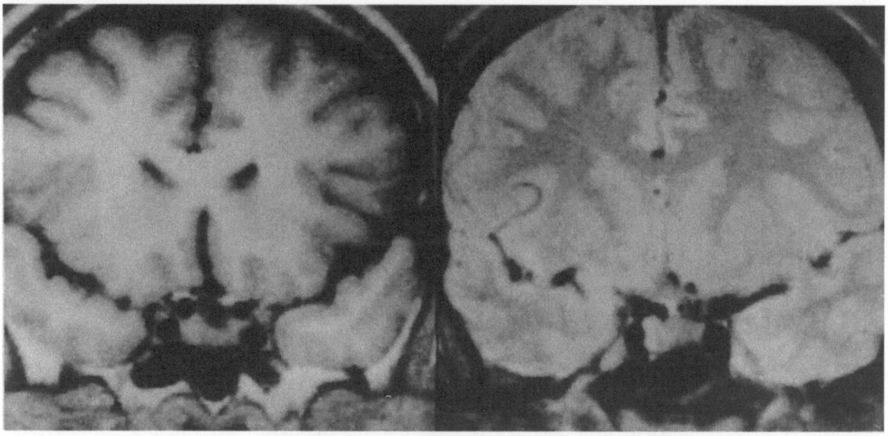

Fig. 2. Acromegaly. *Left:* T_1-weighted image. *Right:* T_2-weighted image

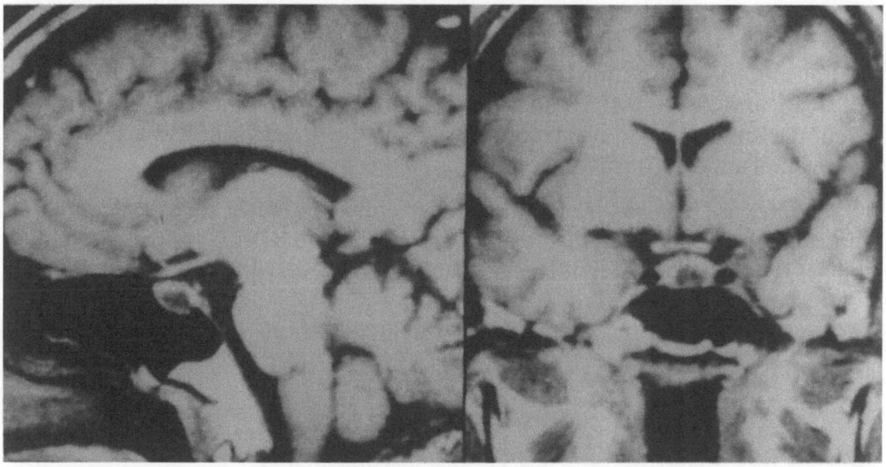

Fig. 3. Prolactinoma: T_1-weighted image

demonstrated merely enlargement of pituitary gland without tumor shadow in T_1-weighted image. The localization and size of all microadenomas in T_1-weighted images of MRI were exactly correlated with the surgical findings. All microadenomas showed enlargement of the pituitary gland, the convex-curved figure of the upper border of the gland, and often the distortion of the pituitary stalk.

Although Gd-DTPA enhanced the pituitary adenoma efficiently and demonstrated the whole figure of the adenoma quite clearly (Fig. 4), it is not suitable for detecting fine intratumoral pathological structures such as hemorrhage or necrotic tissue, since Gd-DTPA enhances so intensely and

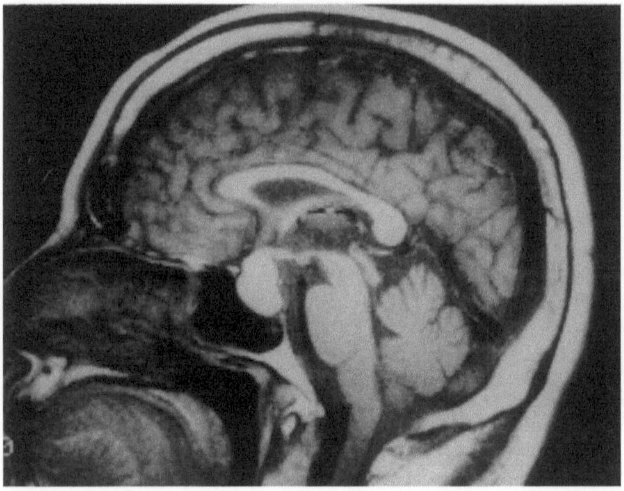

Fig. 4. Gd-DTPA enhanced MRI of prolactinoma (T_1-weighted image)

homogeneously. Although large prolactinomas can be visualized well by both CT scan and MRI, MRI can show better the three-dimensional figure of adenoma and surrounding structures such as the internal carotid artery, optic nerve, chiasm, and third ventricle.

Differential Diagnosis. Rathke's cyst showed clearly demarcated, oval-shaped, homogeneous, high-intensity mass lesions in both T_1- and T_2-weighted images (Fig. 5). The MRI of Rathke's cyst is, however, difficult to differentiate

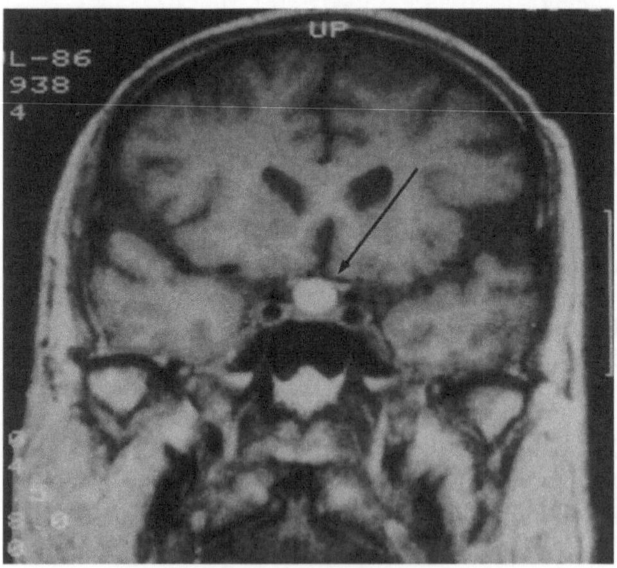

Fig. 5. Rathke's cyst (T_1-weighted image)

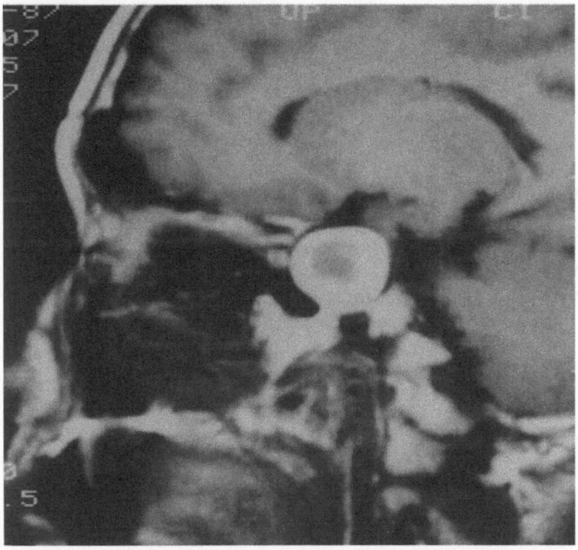

Fig. 6. Internal carotid artery aneurysm (T_1-weighted image)

from pituitary abscess by imaging diagnosis alone. Internal carotid artery aneurysm can well be visualized by MRI as a round high signal intensity mass lesion with the low signal intensity core of blood stream in T_1-weighted image (Fig. 6). Aneurysms are definitely diagnosed by serial MRI using video recording. It can show the movement of blood stream inside the aneurysm.

Discussion

It was thought until recently that pituitary microadenomas were hard to visualize by MRI due to lower resolution compared to CT scan [1–3]. The recent superconductive MRI, however, can demonstrate all endocrinologically suspected microadenomas, as shown in this paper. The signal intensity of various microadenomas (Cushing's disease, acromegaly, and prolactinoma) does not differ significantly in MRI. Therefore, it is difficult to differentiate the type of microadenoma only by imaging diagnosis. It is, however, no practical problem, since endocrinological examination can differentiate each type of adenoma definitely.

Differential diagnosis, is necessary for patients demonstrating mild hyperprolactinemia. Patients with Rathke's cyst are sometimes accompanied with hyperprolactinemia. It is, however, not difficult to differentiate from microprolactinoma by MRI as shown in the paper. No angiography is necessary for differentiating aneurysms from pituitary adenomas by the aid of MRI angiography or serial recording of blood stream by MRI.

Acknowledgement. The technical assistance of Mrs. Haruko Sano is greatly appreciated.

References

1. Hawkes RC, Holland GN, Moore WS et al. (1983) The application of NMR imaging to the evaluation of pituitary and juxtasellar tumors. AJNR 4:221–222
2. Huk WJ, Fahlbusch R (1985) Nuclear magnetic resonance imaging of the region of the sella turcica. Neurosurg Rev 8:141–150
3. Oot R, New PFJ, Buonanno FS et al. (1984) MR imaging of pituitary adenomas using a prototype resistive magnet: Preliminary assessment. AJNR 5:131–137

Magnetic Resonance Imaging of Parasellar Developed Pituitary Adenomas: New Consequences for Pituitary Surgery

R. Nistor, R. Fahlbusch, M. Buchfelder, and W. Huk, Erlangen/FRG

Introduction

Despite the classifications introduced by Hardy (1969) and Wilson (1979), which are based on the imaging capabilities of plain skull X-rays and computed tomography (CT), no exact differentiation of parasellar development of pituitary adenomas was given based on either surgical findings and imaging procedures. Previous studies (Fahlbusch et al. 1988; Kucharczyk 1987) have confirmed that magnetic resonance imaging (MRI) of the sellar and juxtasellar region is currently the best routine imaging method in the preoperative evaluation of pituitary adenomas. In the recent literature, several investigators (Ahmadi et al. 1985; Daniels et al. 1985; Kamiyama et al. 1987; Kucharczyk 1987) have described the advantages of MRI in evaluating the parasellar region focusing on the radiological standpoint. Neurosurgeons have described several types of parasellar developed pituitary adenomas (Buchfelder and Fahlbusch 1987; Hardy 1969; Wilson 1979) with special regard to the operative approach. It must be determined whether these recently available morphological features provided by MRI contribute to the classification and the surgical management of parasellar pituitary adenomas.

Patients and Method

Out of 231 patients with surgically and histologically confirmed pituitary adenomas examined by MRI we evaluated 79 patients who had a predominant parasellar tumor development toward the cavernous sinus. Direct cavernous sinus involvement was demonstrated in preoperative MRI and confirmed by surgery. In 32 adenomas we found only displacement of the cavernous sinus. In 23 patients the cavernous sinus was invaded. While these tumors were operated on via a transsphenoidal approach, the 2 intradural supraclinoidal developed tumors and the 22 giant or generally invasive adenomas underwent transcranial surgery. In three of these latter cases we used both approaches within short intervals in order to achieve complete tumor removal. Two patients did not receive surgery: one with a huge macroprolactinoma who was treated only with dopamine agonists and one with a recurrence of a large ACTH-secreting tumor (Nelson's syndrome).

Of these patients 68 were examined prior to surgery, and 11 had been operated on before. In 17 patients with macroprolactinomas who underwent pretreatment with injectable dopamine agonists prior to surgery 27 control MRIs were performed to document the efficacy of bromocriptine in shrinking this type of tumor. In eight patients early postoperative MRI controls were performed. The MRI investigations were initially performed with a Siemens Magnetom 1.0 Tesla superconducting scanner and more recently with a 1.5 Tesla field strength imager of the same type. The images were obtained using a spin-echo pulse sequence yielding T1-weighted, spin-density weighted and T2-weighted images, in coronal and sagittal planes, with a slice thickness varying between 1 and 7 mm in correlation to the tumor size.

Results

In parasellar tumor extension toward the cavernous sinus (Fig. 1) we differentiate the following types:

Type 1: tumor with localized lateral extension causing a circumscribed displacement of the medial wall of the cavernous sinus (Fig. 2).
Type 2: subcavernous extradural development (Ahmadi et al. 1985), with the adenoma extending along the basal dura into the subcavernous area (Figs. 3, 5).
Type 3: tumors causing generalized displacement and compression of the intracavernous structures (Fig. 4).
Type 4: tumors perforating the medial wall of the cavernous sinus with two varieties: localized invasion (*4a*) and with diffuse invasion (*4b*) of the intracavernous structures (Fig. 5, iC).
Type 5: tumors with supracavernous, supraclinoidal intradural extension (*5a*), in which the tumor develops between the clinoid process and the disrupted, elevated dura (Fig. 6). If there is a disruption of the continuity of the lateral wall of the cavernous sinus, as it is frequently seen in the so-called "giant adenomas" with generalized invasion tendency (*5b*), there is no doubt of the invasive growth.

In smaller tumors even the high sensitivity of MRI does not allow one to distinguish morphologically between displacement and invasion of the medial wall of the cavernous sinus, as the MRI signal of this thin wall is averaged with the signal of the other intracavernous structures (Daniels et al. 1985; Kucharczyk 1987). In our experience it is not only the morphological aspect of the parasellar developed tumors which can be described by MRI. Supplementing the morphological information, this imaging method provides information about functional aspects of the venous blood flow in the cavernous sinus. On T1-weighted images of the normal cavernous sinus the slow blood flow in the vein of the carotid sulcus or in the medial vein (Bonneville et al. 1988) shows a hyperintense signal. This is explained by the so-called phenomenon of

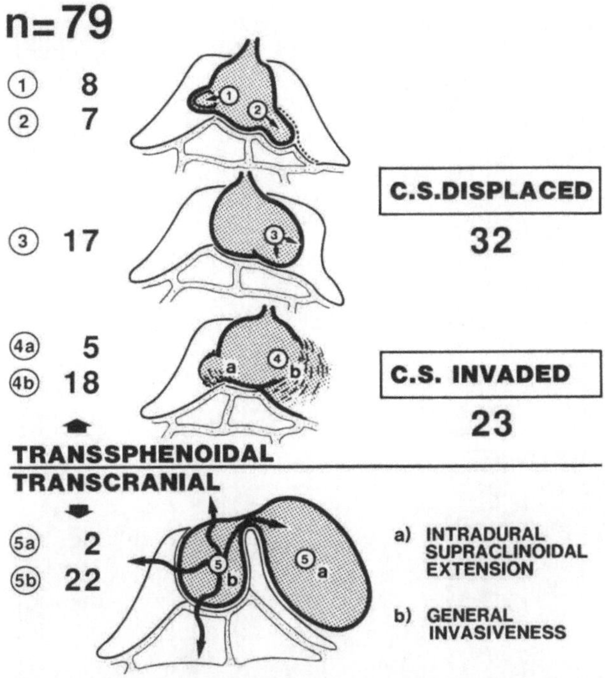

Fig. 1. MRI – surgical classification of parasellar tumor extension toward the cavernous sinus. *1*, Circumscribed displacement of the medial wall of the cavernous sinus; *2*, subcavernous extension; *3*, diffuse, massive lateral extension with displacement of the medial wall and of the intracavernous structures; *4*, tumors breaking through the medial wall with localized (*a*) and generalized (*b*) intracavernous invasion: *5 a*, supracavernous, supraclinoidal intradural extension; *5 b*, "giant adenomas" with generalized invasion tendency

paradoxical enhancement (Kamiyama et al. 1987). If changes of the venous blood flow occur due to compression or invasion of the cavernous sinus, this signal intensity pattern disappears constantly. In our series of tumors with predominant involvement of the cavernous sinus (except for the type 5 adenomas in which the involvement of the cavernous sinus was obvious) we found such changes in the signal intensity of the cavernous sinus in all the cases, regardless whether the tumor was noninvasive (32 adenomas) or invasive (23). This is in our opinion, at least in smaller adenomas, an indirect MRI sign for the involvement of the cavernous sinus (Fig. 2).

In spite of the general euphoric opinion, there still are real difficulties in appreciating directly the invasion of the cavernous sinus, although the classical angiographic signs, which now are demonstrable due to the angiographic capability of MRI (Fig. 4), as well as the clear-cut delineation of the relationship between the intra- and parasellar tumoral aspects, allow a fairly good preoperative distinction between invasion and displacement.

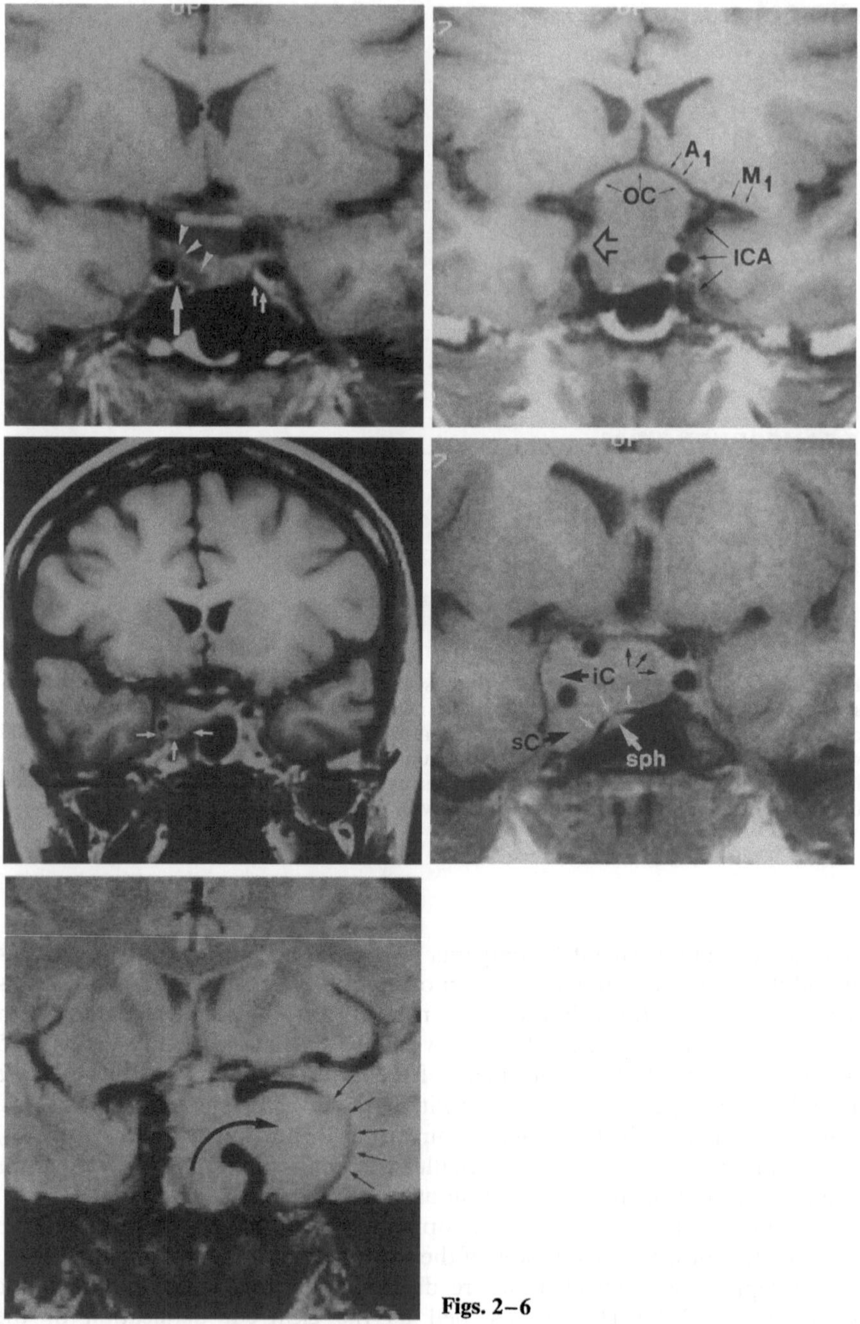

Figs. 2–6

Conclusions

Due to its multiplanar and recently its three-dimensional imaging capability, and due to the vascular imaging possibilities (Dumoulin and Hart 1986), the thin-collimation, high-field MRI is able to delineate the tumoral landmarks clearly and provides precise information about the fine anatomical structure of the cavernous sinus, accurately depicting for the first time the characteristics of tumor development. The clear-cut delineation of the relationship between the intra- and parasellar tumor aspects, and the angiographic capability of MRI lead to the following conclusions:

1. There is no further need of conventional carotid angiography in most cases.
2. On the base of these findings a new morphological differentiation of parasellar tumor extension was established. This is considered as a supplement to the existent classifications of Hardy and Wilson.
3. In analogy to Bonneville's dynamic CT scan of the sella region (Bonneville et al. 1988), demonstrating the involvement of the cavernous sinus by the obliteration (missing hyperintense signal) of the veins of the carotid sulcus or of the medial group, the MRI is able to demonstrate – in a simpler and

Fig. 2. A 5–6 mm large microprolactinoma in a 33-year-old woman, presenting with a hyperprolactinaemic amenorrhea and bromocriptine intolerance (SE; TR: 0.5 s / TE: 15 ms; SL: 3 mm). The right lateralized microadenoma (*short white arrows*) produced a circumscribed displacement of the cavernous sinus without invasion (type 1). The involvement of the intracavernous structures is clearly demonstrated by the disappearence on the right side of the hyperintense signal given by the slow blood flow in the medial venous group (*large white arrow*). Note the hyperintensity of the medial aspect of the left normal cavernous sinus (*double white arrows*)

Fig. 3. Subcavernous extradural extension (type 2) of a prolactin-secreting, noninvasive pituitary adenoma (*white arrows*) in a 37-year-old woman with a hyperprolactinemic amenorrhea (SE; TR: 0.5 s / TE: 15 ms; SL: 3 mm)

Fig. 4. T1-weighted image (SE; TR: 0.6 s / TE: 15 ms; SL: 5 mm) of an intra- and suprasellar developed nonfunctioning pituitary adenoma in a 51-year-old man with visual impairment and secondary hypogonadism. The noninvasive (type 3) tumor had also a right lateral development causing a massive displacement of the cavernous sinus (*large open arrow*). Note the angiographic capability of MRI (*ICA*, *A1*, *M1*) and the sharply delineated compressed optic chiasm (*OC*)

Fig. 5. Intra- (*iC*), subcavernous (*sC*), and sphenoidal (*sph*) development of a GH-secreting macroadenoma in a 44-year-old acromegalic woman (SE; TR: 0.5 s / TE: 15 ms; SL: 5 mm). The sella floor shows hypointense (*short white arrows*); the tumor is well delineated on the top by the slightly hyperintense anterior pituitary (*short black arrows*)

Fig. 6. Supracavernous, supraclinoidal intradural extension (type 5 a) of a prolactin-secreting adenoma in a 44-year-old man with visual impairment and secondary hypogonadism (SE; TR: 2.5 s/TE: 35 ms; SL: 6 mm). The tumor is breaking through between the clinoid process and the supraclinoidal aspect of the internal carotid artery (*curved black arrow*). Note the well circumscribed tumor capsule (*black arrows*)

more rapid manner, and without the use of contrast medium – the same phenomenon. As this pattern appears constantly in parasellar developed adenomas its presence should be evaluated as a pathognomonic MRI sign for the involvement of the cavernous sinus, although it does not allow the differentiation between compressive and invasive development of the pituitary adenomas.

4. As a result of these imaging possibilities the neurosurgeon has the advantage of more selective definite criteria for transsphenoidal surgery and transcranial surgery.

References

Ahmadi J, North C, Segall HD, Zee CS, Weiss MH (1985) Cavernous sinus invasion by pituitary adenomas. AJNR 6:893–898

Bonneville JF, Cattin F, Bouchareb M, Potelon P, Boulard D, Racle A (1988) The veins of the cavernous sinus: usefulness of their demonstration for the diagnosis of invasive pituitary adenomas. In: Landolt AM et al. (eds) Advances in pituitary adenoma research. Advances in the biosciences, vol 69. Pergamon, Oxford New York, pp 151–153

Buchfelder M, Fahlbusch R (1987) Transsphenoidal surgery of pituitary adenomas developed towards the cavernous sinus. In: Dolenc VV (ed) The cavernous sinus. A multidisciplinary approach to vascular and tumorous lesions. Springer, Wien New York, pp 404–414

Daniels DL, Pech P, Mark L et al. (1985) Magnetic resonance imaging of the cavernous sinus. AJNR 6:187–191

Dumoulin CL, Hart HR (1986) Magnetic resonance angiography. Radiology 161:717–720

Fahlbusch R, Nistor R, Buchfelder M, Huk W (1988) Magnetic resonance imaging (MRI) in the preoperative diagnosis of pituitary adenomas: a comparison with CT. In: Landolt AM et al. (eds) Advances in pituitary adenoma research. Advances in the biosciences, vol 69. Pergamon, Oxford New York, pp 117–126

Hardy J (1969) Transsphenoidal microsurgery of the normal and pathological pituitary. Clin Neurosurg 16:185–217

Kamiyama M, Yasui T, Baba M, Hakuba A, Nishimura S, Nemoto Y, Inoue Y (1987) In: Dolenc VV (ed) The cavernous sinus. A multidisciplinary approach to vascular and tumorous lesions. Springer, Wien New York, pp 152–161

Kucharczyk W (1987) The pituitary and sella turcica. In: Brant-Zawadzki M, Norman D (eds) Magnetic resonance imaging of the central nervous system. Raven, New York, pp 187–208

Wilson CB (1979) Neurosurgical management of large and invasive pituitary tumors. In: Tindall GT, Collins WF (eds) Clinical management of pituitary disorders. Raven, New York, pp 335–342

Magnetic Resonance in Modern Neuroimaging of Skull Base Neoplasms with Particular Reference to the Evaluation of Complications of Medically and Surgically Treated Pituitary Adenomas

N. Roosen, San Francisco/USA, T. Kahn, J. C. W. Kiwit, G. Fürst, and E. Lins, Düsseldorf/FRG

Introduction

After the introduction of magnetic resonance imaging (MRI) into clinical practice in the early 1980s (e.g., Crooks et al. 1983), its high efficiency in the visualization and diagnosis of skull base pathology soon became apparent (Hawkes et al. 1983; Lee and Deck 1985). Although initially imaging of osseous structures seemed to be rather difficult (Han et al. 1984), the development of more sophisticated MRI technology reduced this drawback (Davis et al. 1987; Hirsch et al. 1988). MRI has been evaluated as a primary diagnostic technique in sellar pathology by several authors, including ourselves (Davis et al. 1987; Peck et al. 1988; Roosen et al. 1988), but information on the diagnostic potential of MRI in cases of treatment complications is scarce. We therefore present our experience with MRI in the evaluation of complications related to medical and surgical therapy of pituitary adenomas.

Neuroimaging Methods

All MRI studies were performed using a supraconductive Diasonics MT/S imager operating at 0.35 T. T_1-, T_2-, and proton density weighted spin-echo (SE) pulse sequences were routinely used. In selected cases the paramagnetic contrast agent gadolinium-DTPA was administered intravenously. Slice thickness was 5 or 2.5 mm, and sections were obtained in the axial, and sagittal planes at the occasion of each examination.

The computed tomography (CT) scans were obtained with a Philips Tomoscan 300. After axial pre- and postcontrast investigations direct coronal CT studies were performed. Slice thickness for CT examinations of the sellar region was 3 mm.

Study Population

In this paper we focus on posttreatment complications related to the medical and surgical management of sellar lesions. These cases were collected from a study population of 34 patients similar to the one previously described by our group (Roosen et al. 1988). Twelve patients were examined in the immediate or early postoperative phase after transsphenoidal surgery, three of them because of significant deterioration of ophthalmological status (i.e., diminishing visual acuity and/or decreasing visual fields). We present briefly one patient with postsurgical complications and one patient with a bromocriptine-related rhinoliquorrhea.

Case 1. This 48-year-old woman had a large para- and suprasellar adenoma, which on clinical and endocrinological examination was diagnosed to be a macroprolactinoma. CT delineated the extent of the tumor at the time of initial diagnosis (Fig. 1). Despite considerable suprasellar extension, there was only a small visual field defect. Therefore, we adopted a conservative approach, and

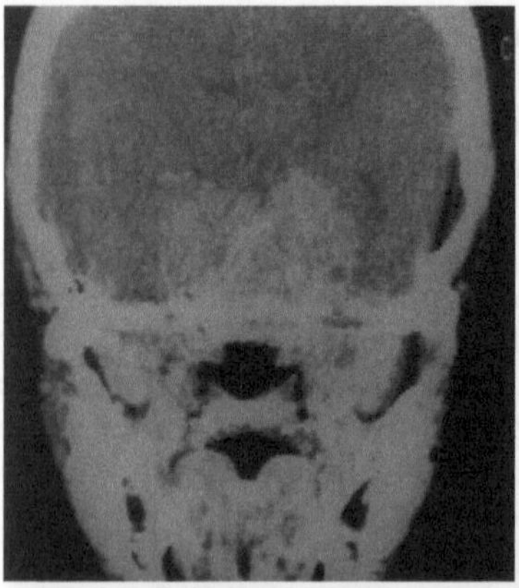

1

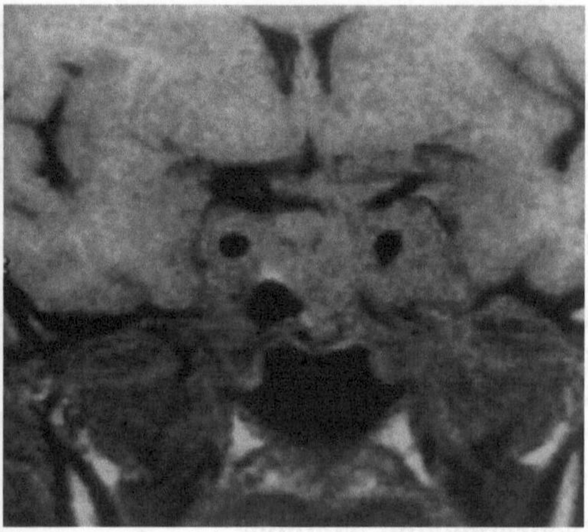

3

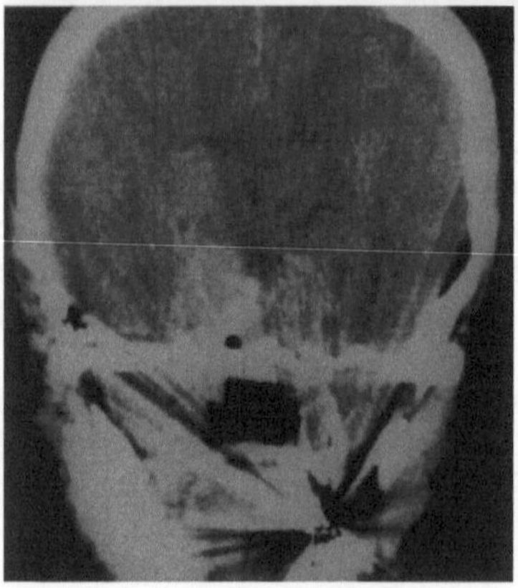

2

Fig. 1. This contrast enhanced, directly coronal CT demonstrates a huge sellar tumor with marked supra- and juxtasellar extension

Fig. 2. After bromocriptine therapy the macroprolactinoma diminished in size and a partial empty sella with a hypodense inclusion may be seen. Clinically, rhinoliquorrhea was present

Fig. 3. T_1-weighted SE MRI beautifully depicts the vascular encasement involving the parasellar internal carotid arteries. Suprasellar tumor extension is no longer present. The intrasellar tumor tissue has a somewhat heterogenous appearance

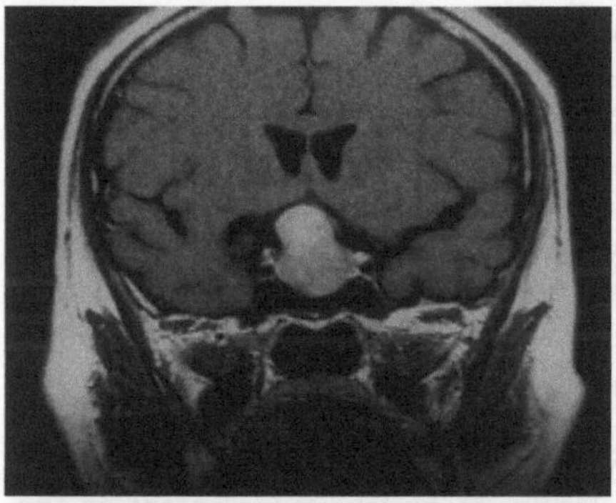

4

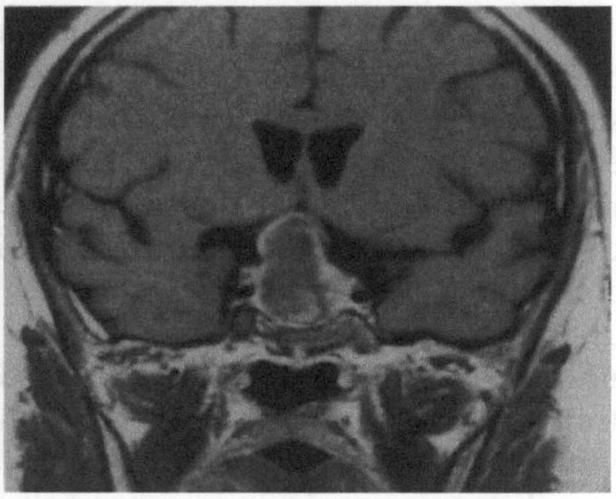

6

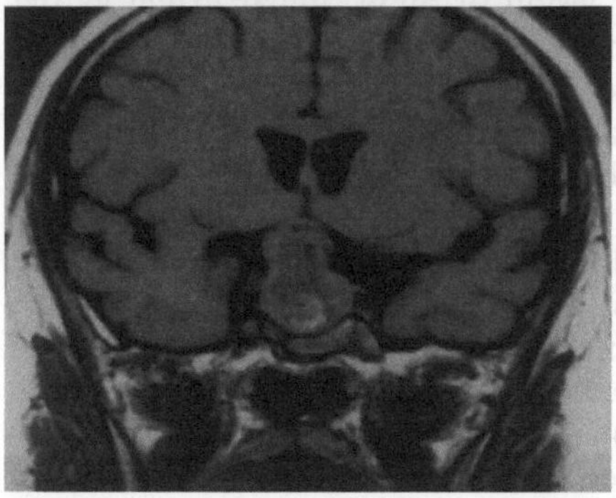

5

Fig. 4. T_1-weighted SE MRI after intravenous administration of gadolinium-DTPA reveals a large intra- and suprasellar pituitary adenoma. There is only minimal parasellar tumor growth. The adenomatous tissue shows a relatively homogeneous increase in signal intensity, which is maximal in the upper parts of the adenoma. The optic chiasm is compressed and flattened by tumor

Fig. 5. The postoperative aspect is similar to the preoperative findings with regard to gross adenoma morphology and optic chiasm deformation (unenhanced T_1-weighted SE MRI). There is a small isointense rim surrounding a more hypointense central area, which however is somewhat hyperintense in its inferior part

Fig. 6. Gadolinium-DTPA causes an increase in signal intensity of the remaining tumor capsule, whereas the central parts are unchanged (contrast enhanced T_1-weighted SE MRI). The central hypointense area suggests the presence of a fresh hematoma, impregnated muscle tissue left in situ for hemostasis

bromocriptine therapy was instituted to achieve tumor shrinkage and facilitate future surgical resection. During pharmacological therapy rhinoliquorrhea developed. CT revealed a considerable reduction in tumor mass (Fig. 2). Intrasellar adenoma tissue had a structurally loosened appearance, and suprasellar extension was minimal. These same features could be appreciated even better on MR (Fig. 3). Rhinoliquorrhea forced us to explore the sella surgically, to resect the residual adenoma tissue, and to perform a water-tight closure of the operative wound. Transsphenoidal surgery was successful in obliterating the cerebrospinal fluid fistula.

Case 2. An intra- and suprasellar nonfunctioning macroadenoma causing bitemporal hemianopsia was diagnosed in this 51-year-old man. The marked suprasellar extension was visualized on MRI (Fig. 4). This gadolinium-DTPA enhanced MRI demonstrated a relatively homogeneous increase in signal intensity of the adenoma tissue with a clearly distinguishable maximum in the dome of the tumor (Fig. 4). Intraoperatively a good decompression of the optic chiasm and diencephalic structures could be achieved by a transsphenoidal approach. Immediately after surgery the visual fields of this patient were normalized, and he was doing perfectly well. However, about 20 h after the transsphenoidal operation there was an acute onset headache and a rapid visual field narrowing. An MRI study was obtained. Nonenhanced MRI revealed a space-occupying suprasellar lesion of quite similar extent as the original adenoma (Fig. 5). Signal intensity was rather low and was heterogeneous; there was a peripheral isointense rim on these T_1-weighted MRI scans. A study following administration of gadolinium-DTPA disclosed a marked increase in signal intensity of this peripheral rim (Fig. 6). The MRI study was thought to be consistent with postoperative intrasellar hemorrhage, which had occurred despite intrasellar tamponade with hemostatic muscle tissue; the contrast enhanced peripheral rim was interpreted as the adenoma capsule remaining in situ. Fortunately, a continuous amelioration of the visual field defect was noted very soon after this acute complication, and surgery could be postponed, even turning out to be not necessary. The visual status of this patient was normalized a few days later, and MRI demonstrated a reduced space-occupying effect due to shrinkage of the intracapsular material consisting of blood impregnated muscle tissue.

Discussion

Numerous reports in the recent radiological literature attest to the efficiency of MRI in the imaging of sellar lesions (Davis et al. 1987; Peck et al. 1988; Roosen et al. 1988). Earlier studies involving macroadenomas of the pituitary gland (Hawkes et al. 1983; Bilaniuk et al. 1984; Kaufman et al. 1984; Lee and Deck 1985) were succeeded by investigations on MRI of microadenomas (Dwyer et al. 1987; Kulkarni et al. 1988; Peck et al. 1988). All these reports concern the diagnostic potential of MRI during the primary work-up of patients who present with symptoms and signs suggesting pituitary disease. Postoperative studies are not as extensively addressed in the literature as preoperative studies are. In a CT examination of the posthypophysectomy pituitary mass, Teng et al. (1988) described a slowly progressive reduction in size of this pituitary mass, and suggested that the best time for CT follow-up examination is 3–4 months after transsphenoidal surgery. Our experience with postoperative MRI in noncomplicated cases of pituitary surgery parallels their results and is described in another report.

Bromocriptine is known to cause tumor reduction, especially in cases of macroprolactinoma (Thorner et al. 1984). Neuroimaging with CT and in particular with MRI clearly visualizes this tumor shrinkage and should allow early consideration of surgical therapy to prevent the occurrence of rhinoliquorrhea. MRI is superior in this regard because of the lack of artifacts, which often decrease image quality in CT.

Early postoperative deterioration is advantageously examined by MRI. Even in the immediately postoperative stage high-quality images of the sellar region may be obtained directly in different planes of section; reconstruction is not necessary. Compression of the optic chiasm and diencephalic structures is evaluated easily. Due to the characteristic time course of hemoglobin degradation and methemoglobin formation in a fresh bleeding (Zimmerman et al. 1988), the change of hypointensity to hyperintensity within the sella or adenoma capsule or its contents allows a good estimate of the age of a postoperative hematoma.

Conclusions

In summary, MRI is the diagnostic imaging method of first choice both in the primary evaluation of patients with suspected sellar disease as well as for follow-up during medical or after surgical therapy. MRI is superior to CT in the delineation of tumor masses, the visualization of soft tissue infiltration, and the depiction of cavernous sinus invasion and vascular encasement. Only the evaluation of bony structures, especially in the postsurgical situation, may be performed more accurately with CT than with MRI.

References

Bilaniuk LT, Zimmerman RA, Wehrli FW et al. (1984) Magnetic resonance imaging of pituitary lesions using 1.0 to 1.5 T field strength. Radiology 153:413–418

Crooks LE, Ortendahl DA, Kaufman L et al. (1983) Clinical efficiency of nuclear magnetic resonance imaging. Radiology 146:123–128

Davis PC, Hoffman JC Jr, Malko JA et al. (1987) Gadolinium-DTPA and MR imaging of pituitary adenoma: a preliminary report. AJNR 8:817–823

Dwyer AJ, Frank JA, Doppman JL et al. (1987) Pituitary adenomas in patients with Cushing's disease: initial experience with Gd-DTPA-enhanced MR imaging. Radiology 163:421–426

Han JS, Huss RG, Benson JE et al. (1984) MR imaging of the skull base. J Comput Assist Tomogr 8:944–952

Hawkes RC, Holland GN, Moore WS et al. (1983) The application of NMR imaging to the evaluation of pituitary and juxtasellar tumors. AJNR 4:221–222

Hirsch WL Jr, Hryshko FG, Sekhar LN et al (1988) Comparison of MR imaging, CT, and angiography in the evaluation of the enlarged cavernous sinus. AJNR 9:907–915

Kaufman B (1984) Magnetic resonance imaging of the pituitary gland. Radiol Clin North Am 22:795–803

Kulkarni MV, Lee KF, McArdle C et al. (1988) 1.5 T MR imaging of pituitary microadenomas: technical considerations and CT correlation. AJNR 9:5–11

Lee BCP, Deck MDF (1985) Sellar and juxtasellar lesion detection with MR. Radiology 157:143:147
Peck WW, Dillon WP, Norman D et al. (1988) High-resolution MR imaging of microadenomas at 1.5 T: experience with Cushing disease. AJNR 9:1085–1091
Roosen N, Keck E, Lins E et al. (1988) Magnetic resonance (MR) imaging of sellar lesions with particular emphasis on the assessment of pituitary adenomas. Adv Biosci 69:127–130
Teng MMH, Huang CI, Chang T (1988) The pituitary mass after transsphenoidal hypophysectomy. AJNR 9:23–26
Thorner MO, Evans WS, Vance ML (1984) Medical management of prolactinomas. I. Prog Endocrinol Res Therapy 1:53–63
Zimmerman RD, Heier LA, Snow RB et al. (1988) Intracranial hemorrhage: intensity changes on sequential MR scans at 0.5 T. AJNR 9:47–57

Diagnostic and Therapeutic Angiography of the Sellar and Parasellar Region

G. Rodesch, K. Terbrugge, P. Lasjaunias, and P. Pruvost,
Kremlin Bicetre/France and Toronto/Canada

The parasellar and sellar region is one of the most important areas of the skull base. It includes the orbital apex, the sphenoid sinus, the clivus, the petrous apex, the chiasmatic region, and the anterior portion of the tentorium cerebelli. It corresponds to a geometric center that concentrates on a small volume of important nervous and vascular structures. In this area modern angiographic study using computed tomography (CT) and magnetic resonance imaging (MRI) must be as precise as possible in order to establish therapeutic possibilities.

Vascularization of the Parasellar Region

Arteries. The vascularization of the sella turcica and its surroundings is complex both from the internal and the external carotid arteries (Fig. 1). A large number of anastomoses between these two territories exist. The main vascular channel of this area is represented by the internal carotid artery (ICA) in its intracavernous (C5, C4, C3) segments. Several branches arise from the C5 portion of the carotid siphon. These include (a) the posteroinferior hypophyseal artery giving off the medial artery of the clivus, anastomosing with its contralateral homologue; (b) the lateral artery of the clivus that divides into a medial branch (anastomosing with the jugular branch of the ascending pharyngeal artery) and a lateral branch (participating to the vasculature of the tentorium and anastomosed with the middle meningeal artery); (c) the recurrent artery of the foramen lacerum accompanying the ICA into the foramen lacerum and anastomosing with the carotid branch of the ascending pharyngeal artery, the inferolateral trunk, and the cavernous branch of the middle meningeal artery; and (d) the lateral artery of the trigeminal ganglion.

From the C4 portion of the carotid siphon there arise capsular arteries, anastomosing on the midline and giving off a descending branch to supply the sphenoid sinus, and the inferolateral trunk (ILT). The latter divides into four branches (in its complete form): (a) the superior, for the roof of the cavernous sinus and the tentorium, (b) the anteromedial, passing through the orbital fissure and anastomosed with the ophthalmic system (in 8% of cases this branch represents the ophthalmic artery, the so-called OPH arising from the intracavernous segment of the ICA); (c) the anterolateral, the artery of the foramen rotundum, running through the foramen rotundum and anastomosing with the distal internal maxillary artery; and (d) the posterior, passing through the foramen oval and under the trigeminal ganglion, it anastomoses with the cavernous branch of the accessory meningeal artery. A balance exists in this area between the internal carotid branches and the arteries depending from the ECA. If there is a dominance of the ICA, the ILT gives off the four described branches. If the ECA is dominant, it is a branch of the accessory meningeal artery that supplies the whole territory, giving branches similar to those of the ILT (20% of cases).

An angiography (or anatomic) protocol for a vascular study of the sellar and parasellar region must include both the ipsilateral ICA and the distal ECA but also the ascending pharyngeal artery. The role played by the vertebral

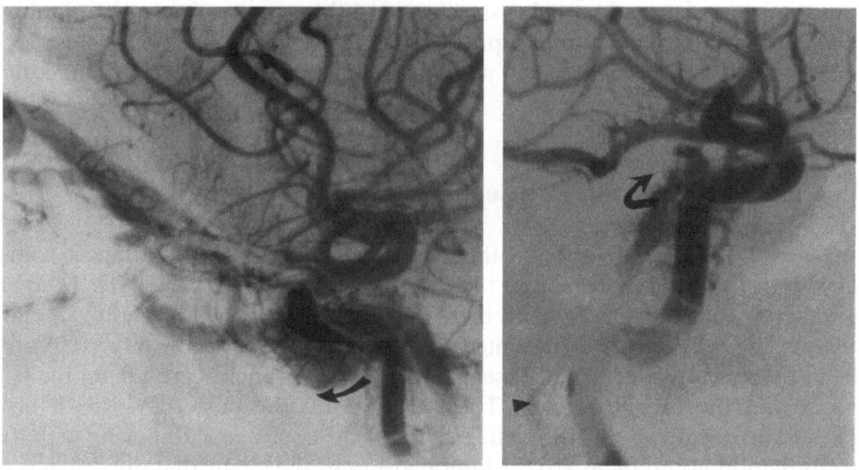

a b

Fig. 1a–k. A 50-year-old man presenting with progressive left exophthalmus, conjunctival hyperemia, diplopia and retro-ocular pain. A left cavernous sinus DAVM is demonstrated on txe left (**a**) and right (**b**) internal carotid angiograms. The drainage is anterior into the left cavernous sinus (*curvea arrow*) and ipsilateral ophthalmic vein and posterior (*broken arrow*) into the left jugular vein (*arrowhead*). **c** The middle meningeal artery injection at the foramen spinosum (*asterisk*) demonstrates the DVAM (*arrow*) and reflux into the accessory meningeal and foramen rotondum arteries (*open curved arrow*). **d, e** Contrast injections in the posterior branches (*asterisk*) of the ascending pharyngeal arteries bilaterally demonstrate the posterior component of the DVAM (*arrows*). Note the odontoid process arterial arch (*curved arrow*). Following PVA particle embolization of the left middle meningeal and ascending pharyngeal arteries the angiographic control (**f**) into the internal carotid artery the immediate clinical changes were satisfactory (*double arrow* indicates almost nonresidual shunt into the DAVM). The patient's symptoms improved dramatically but recurred 1 month later. Repeat right internal carotid (**g**) and ascending pharyngeal (**h**) angiograms demonstrated a prominent posterior component of the DVAM (*arrowhead*). Ipsilateral ascending pharyngeal and middle meningeal arteries were still occluded (not shown). IBCA embolization via the opposite (*right*) ascending pharyngeal artery was achieved following a negative provocative test. **i** Note the radiopaque glue in the posterior cavernous sinus (*arrowhead*). **jk** Final control angiogram shows no supply (*arrowhead*) from either internal carotid siphon. (From Lasjaunias and Berenstein 1987, with permission)

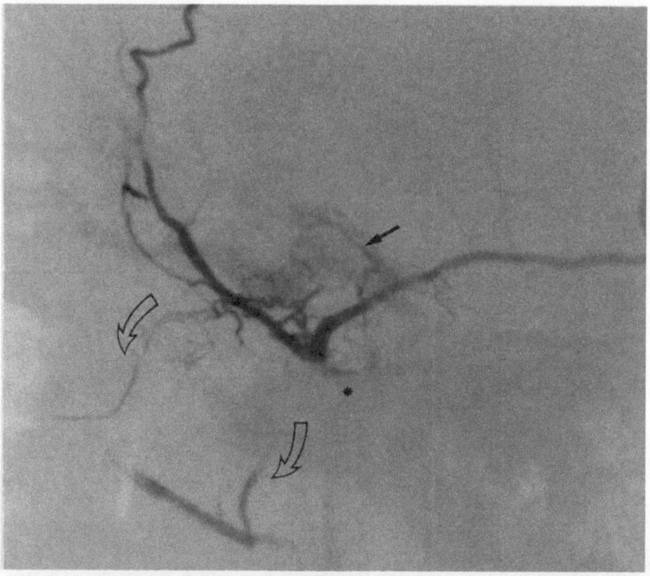

c

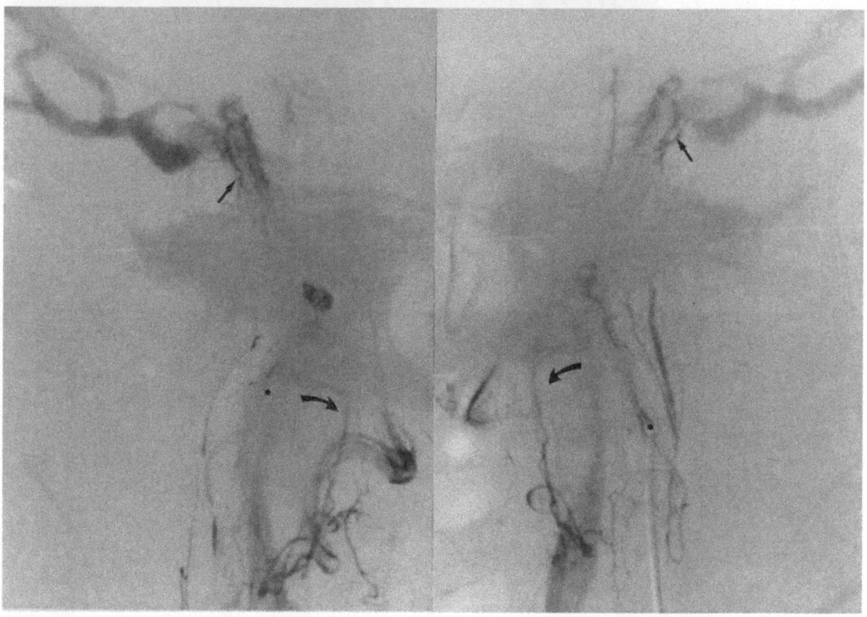

d e

Fig. 1 c–e

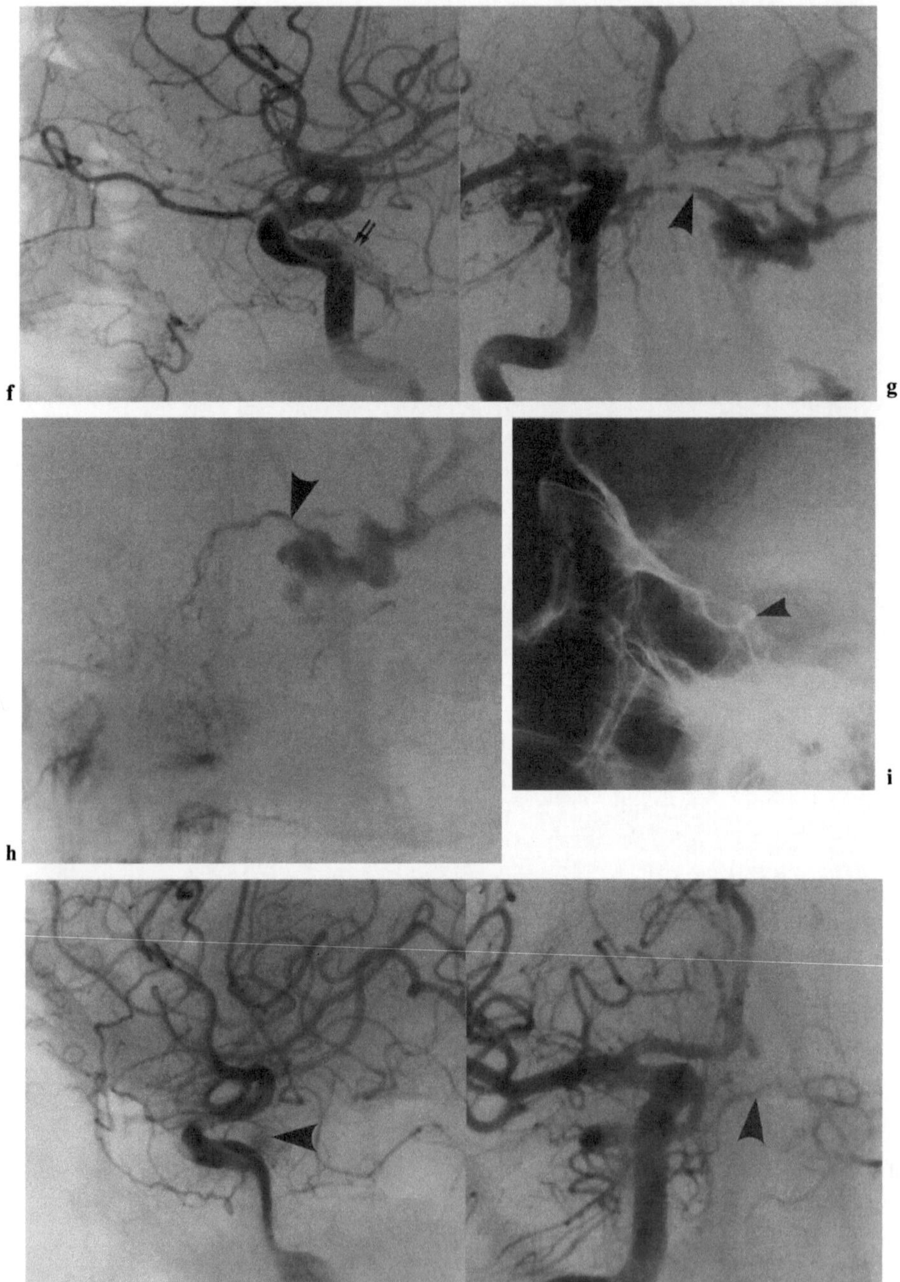

Fig. 1 f–k

artery is of less importance, except in the case of persistence of the embryonic trigeminal artery. The contralateral ICA siphon and its anastomoses through the C5 branches for the clivus must also be seen.

Veins. The parasellar region is a real venous confluence localized at the base of the skull. It communicates with five different regions through tributary veins: (a) anteriorly with the orbit, through the superior orbital fissure by the superior ophthalmic vein, (b) anteroinferiorly, through the foramen rotundum, with the pterygopalatine fossa, (c) posteriorly, through the inferior and superior petrosal sinuses, with the ipsilateral jugular vein and sigmoid sinus, (d) receiving blood from cerebral hemispheres by the deep and superficial Sylvian veins, and (e) being anastomosed with the contralateral cavernous sinus by the transverse occipital sinus (dorsal to the clivus) and the coronary sinuses, within the sellar. These two cavernous sinuses represent nonexpandable extradural venous plexus, formed by venous channels of various calibers. They represent venous outlining structures that constitute the intrinsic anatomy of the parasellar space. In some places these venous channels are confluent and form larger venous cavities. The venous phase of the ICA always provides a satisfactory visualization of the afferents to the region. Some anatomic dispositions are unfavorable for cavernous sinus opacification and should be recognized.

Pathology: Indications for Angiography

Angiography remains the examination of choice for the study of the vascular pathology at the skull base.

Aneurysms. The pretherapeutic examination includes plain films, CT, and angiography, the latter always being combined with functional tests. A four-vessel study must be performed, including compression maneuvers to look for associated lesions and to assess the competency of the circle of Willis. The collateral circulation to the carotid siphon should be studied by distal ECA and ascending pharyngeal artery injections. If treatment is contemplated during the same angiography session and prior to embolization, a tolerance test to acute carotid occlusion is performed. If this transient occlusion is not tolerated, an EC-IC bypass should be done prior to embolization. If the decision is taken to treat this aneurysm by embolization, the attempt should be made to preserve the ICA, if the angiography or CT study does not show any intraluminal thrombus. If the ICA is to be sacrificed, the occlusion is made by means of a balloon placed in front of the aneurysmal neck and with a second one at the origin of the ICA to avoid a stump effect. The postembolization care requires 36 h of strict bed rest and a moderated physical activity for 1 month after the procedure. Blood pressure and cardiac output must be sufficiently elevated to avoid neurologic deficits (above the level at which the test has been performed). In 100% of cases pain and proptosis totally disappear after endovascular treatment, diplopia improves in more than 50% of patients. We have not observed delayed complications in our experience.

Carotidocavernous Fistulas. One generally finds carotidocavernous fistulas (CCFs) to be posttraumatic and to result in the laceration of the intracavernous carotid siphon or of its branches. The venous drainage in the cavernous sinus is then reversed and refluxes into the adjacent venous channels. The pretherapeutic evaluation of CCFs includes CT (visualization of associated traumatic lesions) and angiography to localize precisely the carotid tear, collateral circulation, associated vascular lesions, venous drainage of the fistula, and circulation of the normal brain. We believe that surgery has no indication in the treatment of CCF unless embolization is unavailable. Endovascular approach with detachable balloons is actually the method of choice for the treatment of pathology. This is made in emergency only in case of rapid visual loss, epistaxis, sphenoid sinus false aneurysms, and comatose patient. By femoral approach and under heparinization, a latex balloon is placed on the venous side of the fistula and detached, preserving the ICA flow; this can be achieved now in 80% of cases. If the drainage permits, the CCF can also be occluded by a venous approach (inferior petrosal sinus or retrograde ophthalmic vein catheterization). IBCA or NBCA has little use in the management of CCF.

Dural Arteriovenous Malformations. It is generally considered that dural arteriovenous malformations (DAVM) are acquired lesions that developed secondary to venous obstruction. A DAVM may be localized anywhere on the region, but its clinical manifestations do not depend on the site of the lesion. Its venous drainage (into a sinus or a meningeal vein, into a sinus with reflux in venous lakes) and its flow characteristics determine the symptoms: cranial nerve palsies, proptosis, conjunctival hyperemia, hypertensive glaucoma, decreased visual acuity, seizure, deficit, and subarachnoid hemorrhage. More general additional symptoms may also be detected: bruit, hydrocephalus, and congestive heart failure in the pediatric population. The natural history of DAVM is unknown; some may regress while other progress spontaneously. DAVMs of the floor of the anterior cranial fossa and of the tentorial area have the highest incidence of hemorrhage (over 70%). They must therefore be treated quickly and efficiently. Angiography should be performed according to the previously described protocol. Attention should be paid to the dangerous anastomoses, venous drainage, arterial supply, and venous drainage of the adjacent territories. Embolization is the primary treatment of DAVMs today. When performed by well-trained operators, it carries a high rate of cure with very low risk of complication. In the parasellar region, one should work for a perfect embolization of the nidus to avoid collateral circulation from the ICA branches. In our experience, stable and complete occlusion is obtained only with liquid agents (NBCA). The nidus must be reached to obtain the thrombosis of the entire malformation. In our experience, when this is obtained, no embolization is required on the ICA branches. Single ligatures of feeding arteries should be avoided because of the rapid collateral circulation.

Parasellar Tumors. Meningioma is the most frequent tumor of this region. Angiography provides information about the arterial supply, tumoral vascularity, and site of the tumor attachment. Endovascular treatment can be considered as a therapy per se when one is dealing with inoperable lesions; it may stop the growth of the lesion and provide relief of cranial nerve dysfunction. Performed presurgically, it reduces the preoperative hemorrhage and allows an easier excision. The embolization is performed with the smallest particles possible to reach the tumoral bed; thus, necrosis of the tumor can be obtained. Surgery is performed 3–10 days after the embolization procedure. Sacrifice of the ICA can be proposed in certain circumstances, after the tolerance test has been satisfactory.

Juvenile Angiofibromas. Usually developing in boys at the time of puberty, juvenile angiofibromas are benign vascular tumors of the nasopharynx. They extend submucosally and may grow into the sphenoid sinus to reach the sellar and parasellar region. This extension is well seen on CT and MRI, but angiography allows precise distinction between true and false intracranial extension. This is an important point because it affects surgical resectability and should not lead to unnecessary aggressive embolization in the ICA. True intracranial extension corresponds to a blush projecting above the base of the skull on anteroposterior and/or lateral views; this blush is seen during internal maxillary artery and ICA injections. The late phase of the internal carotid angiogram verifies the patency of the cavernous sinus. False intracranial extension is seen when the tumoral blush projects outside the cranial cavity, although it is obtained from extradural internal carotid artery branches (intrapetrous or intracavernous). These branches leave the cranial cavity and supply the tumor, which is adjacent to the skull base but is extradural and usually extracranial. The cavernous sinus plexus is normal in this case.

Sphenoid sinus extension may simulate an intracavernous extension on lateral views but never on frontal views. The therapeutic goal with JAF is the eradication of the tumor. Surgery is the best available technique; it should be preceded by embolization to avoid intraoperative bleeding and allow complete resection. The technical considerations of the endovascular procedure are the same as those described with meningiomas (see above). There was a 0.6% rate of complications in our series of 350 patients representing more than 500 procedures in Bicêtre in the external carotid system; two-thirds of these were cranial nerve palsies. There was no significant permanent neurologic deficit related to embolization and no mortality.

Reference

Lasjaunias P, Berenstein A (1987) Surgical neuroangiography, vol 1 (Functional anatomy of the craniofacial arteries) and vol 2 (Endovascular treatment of craniofacial lesions). Springer, Berlin Heidelberg New York Tokyo

Value of Visual Evoked Potentials in Indicating an Operation in Sellar Space-Occupying Processes

H.-E. NAU (†), M. FOERSTER, F. RAUHUT, W. ENGEL, and F. RIMBACH,
Essen/FRG and Berlin/FRG

Introduction

The value of visual evoked potentials (VEP) in the control and therapy of processes afflicting the visual pathways is not clear. The response can be altered by a number of factors which hinder an accurate interpretation (Barnett et al. 1980; Blom et al. 1980; Dustman and Beck 1969; Harter and White 1968; Hawkes and Stow 1981; Sokol et al. 1981). Such factors include check size, luminescence, visual field, state of refraction, and pupillary diameter. Their influence on the various VEPs – pattern reversal (PREP) and flash-evoked potentials (FEP), half- or full-field stimulation – is well known but difficult to distinguish from other influences, e.g., from pathological influences on the visual response. Differing results have therefore been published concerning especially the validity of side differences in localizing the pathological process. Some authors recommend VEP as the most sensitive procedure in the control of tumor patients, even in preserving visual acuity during operation (Feinsod et al. 1976).

Patients and Methods

A total of 40 patients with tumors in the sellar region were included (16 men and 24 women; mean age 54.85 years). Their histological diagnoses were: 24 cases of adenoma, 5 of meningeoma, 4 of craniopharyngeoma, and 7 of miscellaneous other tumors. All patients underwent surgery after clinical evaluation, which included neurological, radiological, and ophthalmological examinations. Electrophysiological studies were carried out in accordance with the guidelines of the American Electroencephalographic Society (1984).

Results

Among the 40 patients, 30 showed disturbances in the examination, either of the visual field or of the visual acuity. Out of the 24 patients with pituitary adenomas 19 had ophthalmological disturbances, the others endocrinological disorders. A reduction of visual acuity was seen in 24 patients, in 17 in one eye, in 7 in both eyes. Visual field defects were found in 27, in 5 in one eye, in 22 in both eyes. A combination of visual field defect and reduction of visual acuity was seen in 21 cases. Visual acuity was reduced in three as a single symptom, visual field defect without other neuro-ophthalmological disorders in six cases. Other neuro-ophthalmological symptoms were found in two cases (one paresis of oculomotor nerve and one of abducens nerve). Nine patients had no neuro-ophthalmological disturbances. They were admitted because of endocrinological symptoms (mainly acromegaly).

PREP studies could not be carried out in five patients due to a great reduction in visual acuity. In ten patients pathological latencies were found in PREP, in seven in one eye only, in three in both eyes. In FEP studies, pathological latencies were found in seven patients, in five in one eye only, in two in both eyes; in four cases we had no reproducible FEP.

Interpeak latencies were pathological (i.e., > 2.5 SD) for different peak-to-peak latencies without any prevalence. Amplitude differences (those greater than 50%) were seen in two patients. The most valid pathological sign was an increased latency of P1.

In 13 patients, visual disturbances and pathological evoked responses were seen simultaneously. In 17 patients, only a pathological neuro-ophthalmological disorder was found without evoked potential abnormalities; in one patient, only a pathological evoked response was seen. In 17 patients, normal visual evoked responses were found although a reduction of visual acuity (in 13 cases) or a visual field defect (in 15 cases) was noted (11 patients showed abnormalities in visual acuity and field).

Discussion

The data on 40 patients with tumors in the sellar region show that PREPs are most influenced by a decrease in visual acuity and FEPs by field defects. There is no significant relationship between visual acuity and field defects, on the one hand, and evoked potential abnormalities, on the other; small changes may not influence the electrical response. The most frequent pathological sign in evoked potentials was an increase in P1 latency. This is in accordance with findings of Halliday and coworkers (1976). Significant amplitude differences are rare. The exact neuro-ophthalmological examination was superior to electrophysiological responses. This corresponds to findings of nerve conduction in peripheral nerves. The sensitivity of the electrophysiological method depends on the topography and extension of the lesion. Growth of a spongioblastoma in the same fascicle in the anterior-posterior direction cannot influence the electrical answer. For these reasons, VEPs are not suitable for influencing the indication for operation.

Summary

Forty patients with tumors in the sellar region were analysed. All patients underwent surgery after clinical evaluation, which included neurological, radiological, and ophthalmological examinations. Pattern reversal and flash evoked potentials were done. Pattern reversal evoked responses are mostly influenced by a decrease in visual acuity, and flash evoked responses by field defects. There is no valid connection between visual acuity and field defects on the one hand, evoked potential abnormalities on the other. The most frequent pathological sign in evoked potentials was an increase in P1 latency. The exact neuro-ophthalmological examination was superior to electrophysiological responses. Visual evoked responses are not suitable to influence the indication for operation.

References

American Electroencephalographic Society (1984) American electroencephalographic guidelines for clinical evoked potential studies. J Clin Neurophysiol 1:3–53

Barnett AB, Friedman SL, Weiss IP, Ohlrich ES, Shanks B, Lodge A (1980) VEP development in infancy and early childhood. A longitudinal study. Electroencephalogr Clin Neurophysiol 49:467–489

Blom JL, Barth PG, Visser SL (1980) The visual evoked potential in the first six years of life. Electroencephalogr Clin Neurophysiol 48:395–405

Dustman RE, Beck EC (1969) The effect of the maturation and aging on the wave form of visually evoked potentials. Electroencephalogr Clin Neurophysiol 26:2–11

Feinsod M, Selhorst JB, Hoyt WF, Wilson CB (1976) Monitoring optic nerve function during craniotomy. J Neurosurg 44:29–31

Halliday AM, Halliday E, Kriss A, McDonald WI, Mushin J (1976) The pattern-evoked potential in compression of the anterior visual pathways. Brain 99:357–374

Harter MR, White CT (1968) Effects of contour sharpness and check-size on visually evoked potentials. Vis Res 8:701–711

Hawkes CH, Stow B (1981) Pupil size and the pattern evoked visual response. J Neurol 44:90–91

Sokol S, Moskowitz A, Towle VL (1981) Age-related changes in the latency of the visual evoked potential: influence of check size. Electroencephalogr Clin Neurophysiol 51:559–562

Preoperative Grading of Visual Function by Pattern Evoked Electroretinogram and Visual Evoked Cortical Potentials in Patients with Sellar and Parasellar Tumors

R. G. LORENZ, W. STEUDEL, W. HEIDER, and R. LORENZ,
Frankfurt/M./FRG

Introduction

One of the major effects on the health of patients with sellar or parasellar tumors is impairment of the visual system by direct or indirect damage to the optic nerve, chiasma, or optic tract. Despite successful surgical intervention many of these patients do not regain normal visual function. A persistent visual impairment is, as a rule, caused by an irreversible degeneration of nerve fibers with subsequent degeneration of retinal ganglion cells. Preoperative ophthalmological examinations in these patients are usually limited to determining visual acuity and visual fields and to the ophthalmoscopic inspection of the optic disc.

Such data, however, often do not enable us to distinguish between a visual deficit caused by transitory pressure to the optic pathway or a visual deficit caused by a definitive degeneration of nerve fibers. As the recovery possibilities of the visual system in cases of tumor-related optic pathway damage depend largely on the degree of loss of nerve fibers, i.e., retinal ganglion cells, evaluation of this loss seems to be particularly important for establishing a prognosis for the postoperative development of visual function.

Besides morphological methods such as ophthalmoscopic inspection of the optic disc or red-light-free fundus photography, there is now a new method available to check the functional integrity of human retinal ganglion cells: the pattern evoked electroretinogram (PERG). In contrast to the conventional luminance (flash) ERG, which is an electrical potential of the cones and/or rods of the retina elicited by exposing the eye to a short flash of light (Armington 1974), and to which retinal ganglion cells do not contribute (Granit 1962, 1963), the electroretinogram to a moving pattern stimulus with constant mean luminance (Riggs et al. 1964) originates in the inner retinal layers, including retinal ganglion cells (Maffei and Fiorentini 1981; Zrenner et al. 1986). Several clinical studies have shown that the PERG is a valuable tool to check retinal ganglion cell function in diseases such as glaucoma (Wanger and Persson 1983; Papst et al. 1984) or descending optic atrophy of various origin (Groneberg and Teping 1980; Dawson et al. 1982; Seiple et al. 1983; Lorenz et al. 1988, 1989).

The aim of this study was preoperatively to assess the state of the visual system in patients with sellar and parasellar tumors additional to clinical examinations at two levels of information processing, namely at the level of retinal ganglion cells by recording the PERG and at the cortical level by recording the pattern visual evoked cortical potential (PVECP). Furthermore, we report on the postoperative development of the visual function in these patients and its relationship to preoperative electro-ophthalmological tests.

Patients and Methods

We used these tests in a clinically controlled prospective study on 35 patients with sellar and parasellar tumors (16 meningiomas, 14 pituitary adenomas, and 5 other pituitary tumors). Patients were asked to fixate monocularly the center of a reversing black and white checkerboard pattern presented on a television screen subtending a visual field of 14° × 18° (check size 50′ of arc, reversal rate 4/s, mean luminance 30 cd/m^2, modulation depth 0.97). Retinal responses were recorded by means of a gold foil electrode (Arden 1979) placed into the lower conjunctival fornix of the tested eye and referenced to the ipsilateral outer canthus (Berninger and Schuurmans 1985). Cortical potentials were recorded by means of a gold cup electrode placed 2 cm above the inion. Retinal and cortical potentials were fed into a Medelec ER94a averager, and packages of 64 single averaged sweeps were stored in a computer for further off-line averaging to a total of 256 sweeps of each response.

Resultant PERG and PVECP were analyzed digitally with regard to the peak times and amplitude of the major components. Compared to an age- and sex-matched control group, PERG were regarded as pathological if the amplitude dropped out of the 98.8% confidence interval, or if there was an interocular amplitude difference greater than 30%. PVECP were regarded as pathological if the peak time dropped out of the upper limit of the confidence interval, or if the peak time differed interocularly for more than 10%, or if the amplitude difference between the two eyes was greater than 30%.

Excluded from this study were all patients with a history of previous ophthalmological or neuro-ophthalmological disorder. All patients were tested preoperatively, immediately after surgery and after 1, 3, and 6 months in the postoperative recovery period. Ophthalmological examinations included that of visual acuity, computerized static perimetry or kinetic perimetry, and funduscopy.

Results

Based on the results of preoperative electro-ophthalmological tests, the eyes of the patients were classified into three categories: category 0, regular PERG and regular PVECP; category 1, regular PERG and irregular PVECP; category 2, irregular PERG and irregular PVECP. We assigned 24 eyes to category 0, 18 eyes to category 1, and 20 eyes to category 2. Neither meningiomas nor pituitary adenomas or other pituitary tumors were found to be associated with classification into one of the three categories to a greater extent than the others.

Figure 1 shows examples for the preoperative electro-ophthalmological recordings in three patients of category 0, 1, and 2. For the patient of category 0

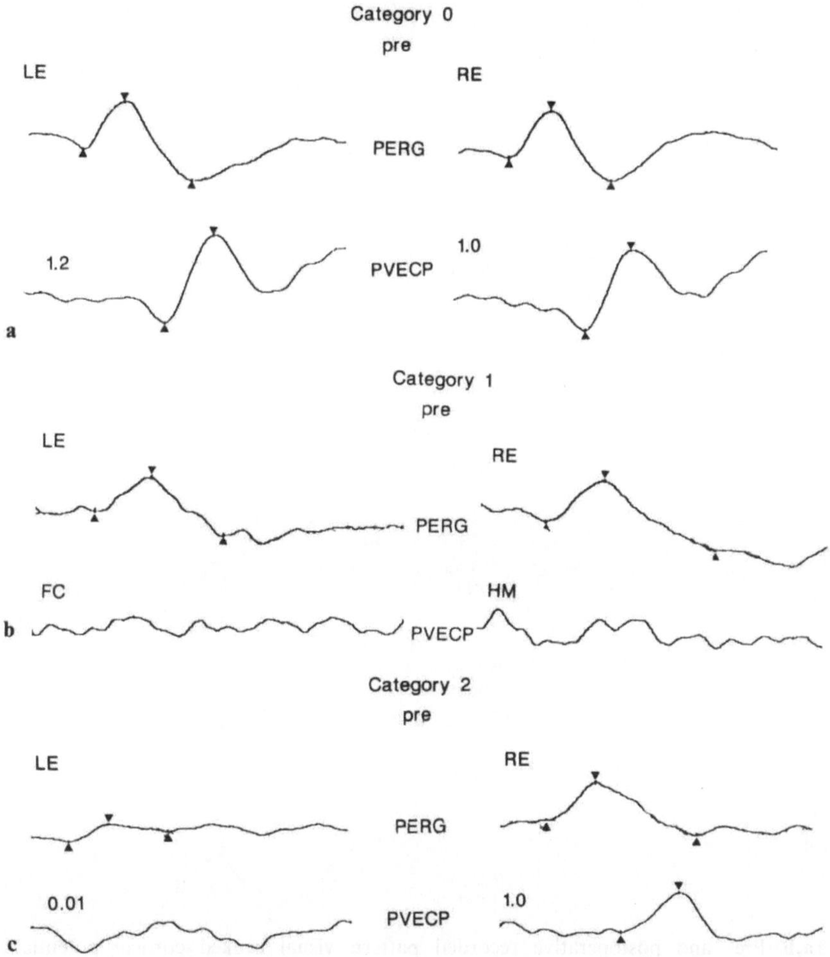

Fig. 1 a–c. Simultaneous recorded pattern electroretinograms (PERG) and pattern visual evoked cortical potentials (PVECP) and the corresponding visual acuity in three patients (six eyes) with tumors of the sella region before surgery. *Triangles*, the major PERG and PVECP components. **a** Regular PERG and regular PVECP in both eyes (category 0). **b** Regular PERG and irregular PVECP in both eyes (category 1). **c**, Irregular PERG and irregular PVECP in the left eye (category 2)

(Fig. 1 a) the normal PVECP and PERG points to a regular transmission of visual informations in the optic pathway undisturbed by the pituitary tumor. Also, clinical tests of visual acuity and visual field did not exhibit any sign of visual function impairment. The patient of category 1 (Fig. 1 b) is a 46-year-old woman with a craniopharyngioma whose visual acuity preoperatively was reduced to the perception of hand movements in the right and finger counting in the left eye. Preoperative electro-ophthalmological testing revealed an

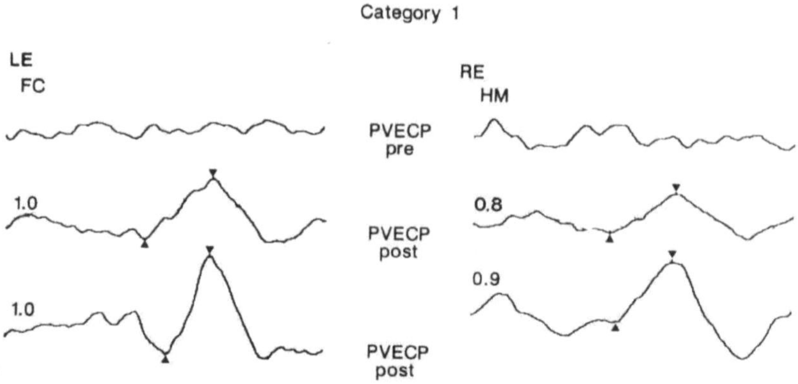

a

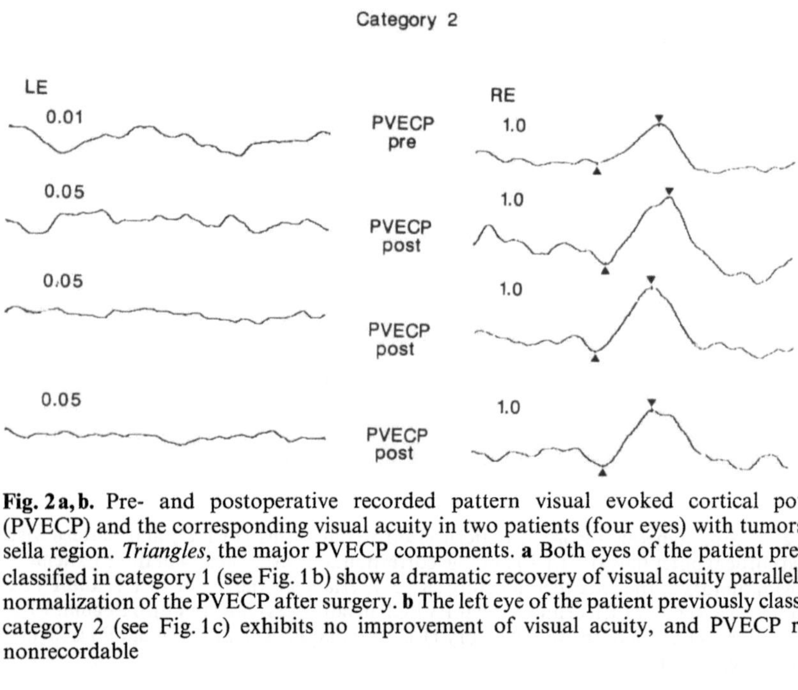

b

Fig. 2a, b. Pre- and postoperative recorded pattern visual evoked cortical potentials (PVECP) and the corresponding visual acuity in two patients (four eyes) with tumors of the sella region. *Triangles*, the major PVECP components. **a** Both eyes of the patient previously classified in category 1 (see Fig. 1b) show a dramatic recovery of visual acuity paralleled by a normalization of the PVECP after surgery. **b** The left eye of the patient previously classified in category 2 (see Fig. 1c) exhibits no improvement of visual acuity, and PVECP remains nonrecordable

almost extinguished cortical potential to pattern stimuli indicating a total block of retinocortical transmission in the optic pathway. Retinal responses, however, were still normal, giving no sign of descending retinal ganglion cell degeneration. Consequently the patient's eyes were classified as category 1. The patient of category 2 (Fig. 1c) is a 53-year-old woman with a pituitary adenoma. Preoperative visual acuity was normal in the right and severely reduced in the left eye. Electro-ophthalmological responses were normal on the right side on cortical and retinal level. On the left side not only the PVECP was nonrecordable but also the PERG was distinctly reduced. This combination indicates a tumor-related block of signal transmission in the optic pathway with a consecutive descending degeneration of retinal ganglion cells. The patient's left eye was classified as category 2.

The postoperative development of visual function, i.e., visual acuity and the PVECP in the two patients (eyes) of categories 1 and 2, is shown in Fig. 2. Postoperative follow-up of the patient classified with both eyes in category 1 (regular PERG, irregular PVECP) showed a dramatic recovery of visual acuity to 0.9 and 1.0, respectively, 3 months after surgery paralleled by a total normalization of the PVECP. In contrast, the postoperative controls of the patient, whose left eye had been classified in category 2 (irregular PERG and irregular PVECP), even 10 months after successful resection of the tumor did not exhibit any improvement of visual acuity, and PVECP remained nonrecordable.

Discussion

Delayed or reduced VECP to pattern stimuli or flashes of light are a well-known hallmark of diseases affecting the postretinal optic pathway. These potentials are widely used to support or enable clinical diagnoses of demyelinating diseases (Halliday et al. 1972). Reports on tumor-related optic pathway lesions and their effect on the VECP are less frequently (Holder 1978). While a delayed VECP generally supports the diagnosis of a demyelinating disorder such as multiple sclerosis (Halliday 1973), the VECP alterations in compressive optic pathway diseases are less uniform, as either a delay or a reduction of the potential or a combination of both can be found (van Lith 1980). Moreover, there is no common pattern of VECP alteration, which allows distinction between the effect of a transitory optic nerve compression or a persistent damage of nerve fibers and ganglion cells.

Our data show that patients even with severe VECP alterations such as a nonrecordable response can regain normal visual function after a successful treatment of their tumor disease. Thus, the role of VECP in the diagnosis and management of tumor-related visual disturbances is mainly to check whether or not a tumor of the sellar region disturbs the retinocortical signal transmission. By using patterned stimuli, the VECP determined primarily by the macular retinal area (Groneberg 1980) is also correlated to the visual acuity (Teping 1981) and thus is useful for following a possible postoperative rise in visual acuity, as shown by our own results.

While the PERG was initially developed to reduce the effect of stray light which contaminated the ERG elicited by unstructured flashes of light (Riggs et al. 1964), later clinical (Groneberg and Teping 1980) and experimental observations (Maffei and Fiorentini 1981) gave first hints that the sources of the luminance ERG and of the pattern ERG are different. Numerous studies have now shown that the PERG is selectively affected by diseases of the retinal ganglion cells where the conventional LERG is unaltered (Dodt 1987). Our results show that patients in whom a compressive optic nerve disorder has led to a descending ganglion cell degeneration exhibit reduced PERG and thus support this view. Unlike the PVECP, the PERG showed an alteration only if a descending ganglion cell degeneration had already taken place, thus pointing to the irreversibility of the lesion. Although the postoperative observation period is not yet complete for all of the patients (eyes) included in our study, the preliminary results show that a reduced PERG in these cases is a bad prognostic sign for the recovery possibilities of the visual system, while a preserved PERG even with severe PVECP alterations allow us to expect a normalization of central visual function. However, it is still disadvantageous that a descending ganglion cell degeneration in man takes months to be complete (Kupfer 1963). We may thus record an absolutely normal PERG preoperatively although a descending ganglion cell degeneration has already started but is not yet detectable by PERG. Consequently, in further studies it will be necessary to check ganglion cell function not only at one time but over a certain span of time during the course of the disease to establish a definitive prognosis for the recovery possibilities of the visual system.

Summary

In 35 patients with sellar or parasellar tumors pre- and postoperatively the pattern electroretinogram (PERG) and the pattern visual evoked cortical potential (PVECP) were investigated in addition to clinical tests of visual acuity and visual field to check the degree of visual function impairment caused by the tumor and the postoperative development of the visual system. Based on the preoperative results of the electro-ophthalmological tests a new grading for the degree of visual system damage was established. While patients even with severe PVECP alterations but preserved PERG regained good central visual function, those with additional PERG changes indicating irreversible degeneration of retinal ganglion cells did not show any recovery of central visual function. The preliminary results show that this electro-ophthalmological grading of visual function seems to be a valuable tool to evaluate the recovery possibilities of the visual system in patients with tumor-related optic pathway damage.

References

Arden GB, Carter RM, Hogg C, Siegel JM, Margolis S (1979) A gold-foil electrode extending the horizons for clinical electroretinography. Invest Ophthalmol Vis Sci 18:421–426

Armington JC (1974) The electroretinogram. Academic Press, New York

Berninger T and Schuurmans RP (1985) Spatial tuning of the pattern ERG across temporal frequency. Doc Ophthalmol Proc Ser 61:17–25

Dawson W, Maida T, Rubin R (1982) Human pattern evoked retinal responses are altered by optic atrophy. Invest Ophthalmol Vis Sci 22:796–803

Dodt E (1987) The electrical responses of the human eye to patterned stimuli: clinical observations. Doc Ophthalmol Proc Ser 65:271–286
Granit R (1962) The visual pathway. In: Davson M (ed) The eye. Academic Press, New York
Granit R (1963) Sensory mechanisms of the retina. Hafner, New York
Groneberg A (1980) Simultaneously recorded retinal and cortical potentials elicited by checkerboard stimuli. Doc Ophthalmol Proc Ser 23:255–262
Groneberg A, Teping C (1980) Topodiagnostik von Sehstörungen durch Ableitung retinaler und kortikaler Antworten auf Umkehr-Kontrastmuster. Ber Dtsch Ophthalmol Ges 77:409–415
Halliday AM, McDonald WI, Mushin J (1972) Delayed visual evoked responses in optic neuritis. Lancet I:982–985
Halliday AM, McDonald WI, Mushin J (1973) Visual evoked response in diagnosis of multiple sclerosis. Br Med J IV:661–664
Holder GE (1978) The effects of chiasmal compression on the pattern visual evoked potential. Electroencephalogr Clin Neurophysiol 45:278–280
Kupfer C (1963) Retinal ganglion cell degeneration following chiasmal lesions in man. Arch Ophthalmol 70:256–260
Lorenz RG, Dodt E, Heider W (1989) Pattern electroretinogram peak time as a clinical means of discriminating retinal from optic nerve disease. Doc Ophthalmol 71:307–320
Lorenz RG, Niepel G, Heider W (1988) Amplituden- und Latenzänderungen im Muster-Elektroretinogramm (M-ERG) bei Netzhaut- und Sehnervenerkrankungen. Fortschr Ophthalmol 85,3:296–300
Maffei L, Fiorentini A (1981) Electroretinographic responses to alternating gratings before and after section of the optic nerve. Science 211:953–955
Papst N, Bopp M, Schnaudigel OE (1984) The pattern evoked electroretinogram associated with elevated intraocular pressure. Graefes Arch Clin Exp Ophthalmol 222:34–37
Riggs LA, Johnson EP, Schick AML (1964) Electrical responses of the human eye to moving stimulus pattern. Science 144:567
Seiple W, Price IM, Kupersmith M, Siegel IM, Carr RE (1983) The pattern electroretinogram in optic nerve disease. Ophthalmology 90:1127–1132
Teping C (1981) Visusbestimmung mit Hilfe des visuell evozierten kortikalen Potentials. Klin Monatsbl Augenheilkd 179:169
van Lith GHM (1980) Application of electroophthalmology. In: Michaelson IC (ed) Textbook of the fundus of the eye. Churchill Livingstone, New York, p. 667
Wanger P, Persson HE (1983) Pattern-reversal electroretinograms in unilateral glaucoma. Invest Ophthalmol Vis Sci 24:749–753
Zrenner E, Baker CL, Hess RF, Olsen BT (1986) Current source density analysis of linear and non-linear components of the primate electroretinogram. Invest Ophthalmol Vis Sci [Suppl] 242:82

B. Pituitary Adenomas

Surgery of Sellar Lesions: Experience from 38 Years

B. RAMAMURTHI, Madras/India

Neurosurgeons in developing countries must often deal with large and massive pituitary tumors that impair vision. As surgery in these cases carries a fair amount of risk, special care and attention are required to deal successfully with these lesions. In the early days of neurosurgery in India four decades ago, most of the patients who were referred to the neurosurgeon were blind, and mortality was high [5, 6]. Today, with the increase in the number of neurosurgeons in the country and greater awareness among physicians and the public, many cases are referred early, and it is possible to deal with many of them by transsphenoidal microsurgical techniques [7].

Experience gained in the treatment of 484 pituitary tumors over the past 38 years is presented here. These include 290 nonsecreting adenomas, 93 acromegalies, 88 prolactinomas, 6 basophil tumors, and 7 other tumors.

Nonsecreting Adenomas

The majority of patients in the first two decades presented with well-marked visual signs, and many were blind in the first decade. A comparison of visual field defects over the four decades shows clearly that there has been a notable improvement in early referrals. In the first decade (1950–1959) 11/45 patients were blind, as opposed to 2/98 in the most recent decade (Table 1). The corresponding figures for bitemporal hemianopia were 22/45 as opposed to 29/98. A further 29 of these patients in the most recent decade (but none in the first) showed no visual symptoms at all. Also regarding fundus appearances we have observed such a trend: in the years 1950–1970 35/107 patients had bilateral optic atrophy, but in the years 1971–1988 only 7/183 (Table 2). It

Table 1. Field loss in nonsecreting pituitary adenomas ($n = 273$)

	n	No loss	Uni-temporal	Bi-temporal	Concentric contraction	Blind in one eye	Blind in both eyes
1950–1959	45	0	6	22	3	3	11
1960–1969	36	3	3	17	3	2	8
1970–1979	94	15	21	39	11	4	4
1980–1988	98	29	20	29	12	6	2

Table 2. Fundus appearance in nonsecreting pituitary adenomas ($n = 290$)

	n	Normal	Pallor unilateral	Pallor bilateral	Optic atrophy unilateral	Optic atrophy bilateral	Optic atrophy with pap	Pap bilateral
1950–1970	107	6	7	29	20	35	5	5
1971–1988	183	44	41	32	54	7	2	3

therefore seems that in India today there is a greater awareness among physicians and the public alike, and that more cases come early for treatment. Still, in our country we often see large pituitary tumors presenting with large extrasellar extensions. Education of general physicians and ophthalmologists by the neurosurgeons should continue as vigorously as ever.

Out of the 290 cases of nonsecreting adenomas, 28 were completely intrasellar, 110 were both intra- and suprasellar (Fig. 1), 79 had parasellar extensions, and 73 had large extensions in different directions (Figs. 2–4; Table 3). As early as 1940, Jefferson [2] discussed the problem of extrasellar extensions of pituitary adenomas and pointed out the difficulty in surgical treatment of these tumors and the high rate of associated mortality. Modern advances in anesthesia and neurosurgery have only slightly improved the prognosis in these massive tumors.

Although these tumors are termed nonsecreting tumors, the majority of patients show some signs of endocrine deficiency. In recent years, as facilities for endocrine testing have increased, some of these patients have shown

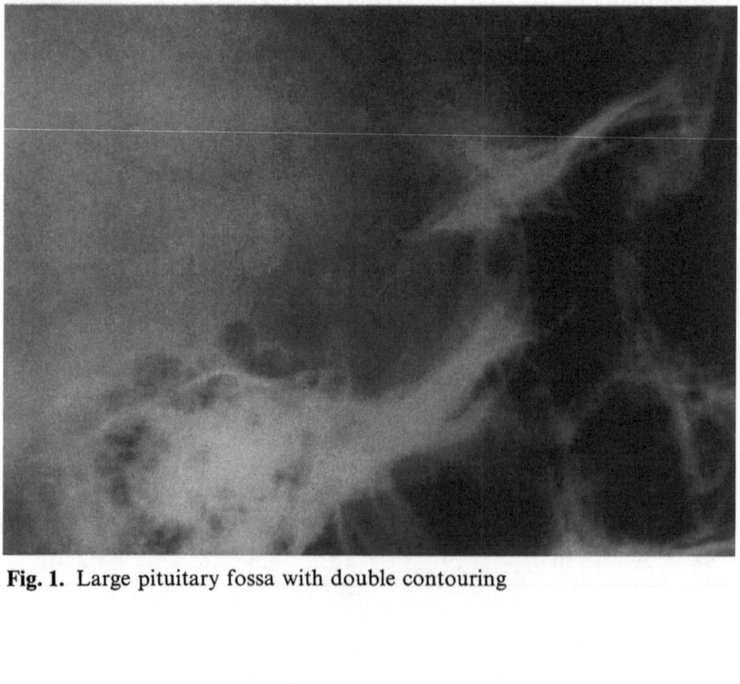

Fig. 1. Large pituitary fossa with double contouring

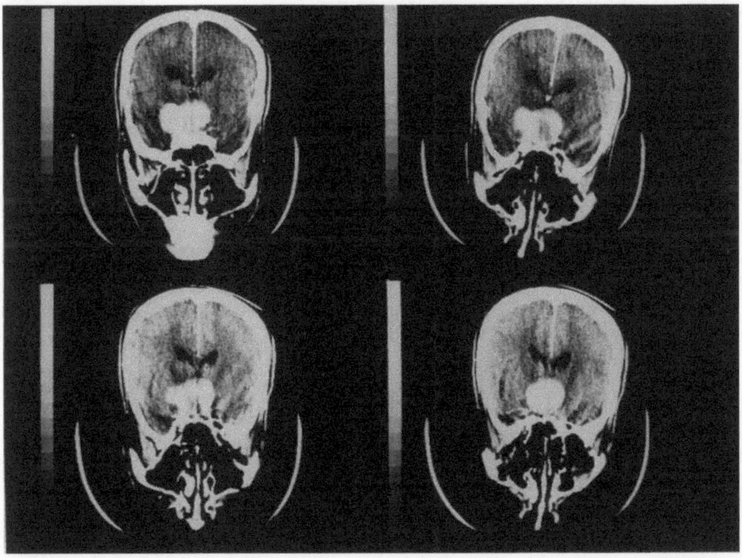

Fig. 2. Coronal section of CT scan showing the upward extension of a large pituitary tumor

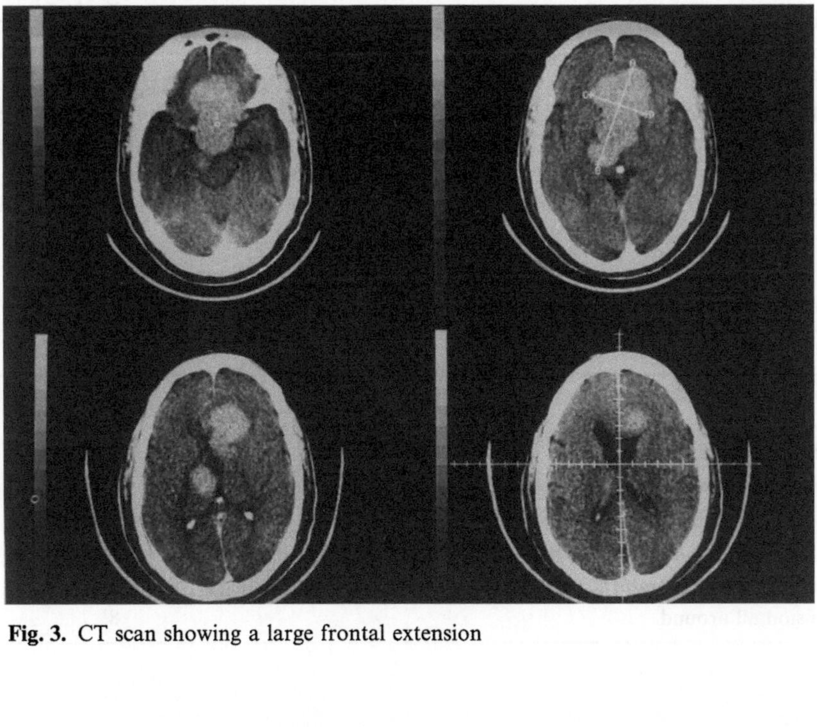

Fig. 3. CT scan showing a large frontal extension

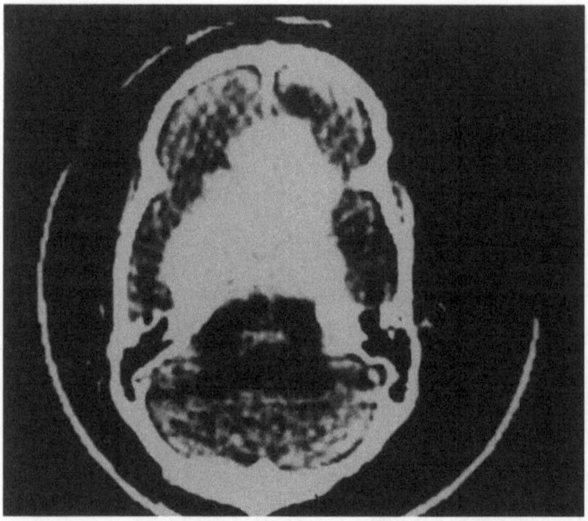

Fig. 4. CT scan showing extensive spread of a pituitary tumor along the base of the brain

elevated prolactin levels. Histochemical investigations of the postoperative specimens of these tumors show different varieties of secreting cells.

Until 1980 contrast investigations such as pneumoencephalograms and carotid angiograms were used routinely in patients suspected of extrasellar extension of the pituitary mass. Since the advent of computed tomography, air studies and angiograms are used only occasionally to define the extent of these lesions and their blood supply.

Should all massive tumors be operated on, or is it sometimes possible to use another method of therapy? Jefferson mentioned that "operation on massive tumors is a course of doubtful wisdom." When there is some doubt about the possible prognosis, it may be wise to administer cobalt therapy preoperatively. This was tried in nine of these massive cases: three patients' vision deteriorated

Table 3. Extrasellar extensions in nonsecreting pituitary adenomas ($n = 290$)

	n
Upward in front of the chiasma	31
Upward and forward under the frontal lobes and into the inter-hemispheric fissure	9
Upward and posteriorly filling the third ventricle	9
Laterally into the cavernous sinus	8
Laterally into the middle cranial fossa	6
Posteriorly into the ambient cistern and in front of the midbrain	2
Extension all around	8

rapidly and they had to be operated on; the others responded well to radiotherapy, and in two of them it was possible to operate after 6 weeks. Treatment with large doses of bromocriptine has been advocated as a preoperative measure in large pituitary tumors, but we have used this in only two cases of large prolactin-secreting adenomas. When such measures fail, the neurosurgeon has no choice other than surgical excision [1, 4].

Preoperative administration of steroids is necessary. Similarly, any anemia, if present, needs correction as do other endocrine defects. For a few days preoperatively, it may be necessary to give the patient thyroid extract and testosterone and estrogens as indicated.

Excellent anaesthesia is important, with appropriate ventilation techniques to keep the intracranial tension low. Intravenous mannitol may be used whenever necessary. In the early days, to reduce intracranial tension further, we punctured the anterior horn on the exposed side, which is usually the right side. Continuous lumbar drainage may also be instituted. In patients in whom the intracranial pressure is found to be high, the drainage of fluid is started after the dura has been opened to prevent tentorial herniation.

Pituitary tumors with large extrasellar extensions are best approached through a transfrontal route. In cases with purely temporal extension, a subtemporal approach may be indicated. The side of approach is determined by the direction of the extrasellar extension. The frontal lobe must be retracted gently and slowly to prevent postoperative seizures or mental confusion. The operating microscope at this stage reveals an entirely new world of anatomical clarity compared to what was seen 25 years ago. The capsule is opened and the contents gradually removed, till the tumor gradually reduces in size. Sucker tips curved in different directions are useful at this stage. Brisk bleeding is often encountered within the capsule. This must be controlled at every stage with bipolar coagulation, if possible, or with packing, hydrogen peroxide, or surgicel. The capsule is then separated gently from the chiasma and optic nerves, with great care being taken to preserve the small blood vessels supplying these structures. It is here that the great advantage of the operating microscope becomes clear in preserving and improving vision. Other areas that require care are the area of the anterior communicating artery that may be stretched over the anterior aspect of the tumor and of the basilar artery and its branches, if the tumor extends posteriorly. Leaving large parts of the tumor unremoved and merely decompressing the optic nerves and the chiasma may lead to severe postoperative swelling of the tumor with resulting mortality. There is no place for fiddling with massive pituitary adenomas [3]. Symon et al. [8] have suggested limited biopsy followed by shunt and radiotherapy, if the tumor is tough and fibrous and not easily removable.

In an operation that has proceeded well, and where careful and gentle techniques have been used, the postoperative period is usually smooth. Steroids must be continued and diabetes insipidus treated if it occurs. Postoperative mental confusion may be present for a few days; epilepsy is uncommon.

The mortality rate for all pituitary surgery in the first decade was 30%, and in the subsequent decade for massive tumors with large extensions it was 21%.

Since 1971 mortality has been 6% in massive tumors and under 1% in others; this reflects the great advances in early diagnosis and in technical aspects.

Improvement in vision very often occurs in the eye that was partially affected. Perception of light may rarely return to the blind eye, if the eye had been blind only for a short time.

Until recently postoperative radiotherapy was advised for all patients who had a pituitary tumor removed. Today, however, when one is quite sure of having removed the tumor completely, either transsphenoidally or transcranially, postoperative radiotherapy is not advised by some surgeons. In my opinion, on the other hand, postoperative radiotherapy is a must in all adenomas except microadenomas.

Following transcranial surgery, recurrence is seen in 10%–27% of cases in different series. Our recurrence rate has been 15%, and the patients have returned between 2 and 6 years after the first surgery. Reoperation helps these patients. Postoperative radiotherapy reduces the risk of recurrence.

There are many centers in India and other developing countries where microsurgery is routine for moderately sized nonsecreting tumors and for all secreting tumors. These neurosurgeons operate with no additional aids such as television monitoring or image intensifier; intraoperative plain X-rays are used whenever necessary to guide the surgical approach. One learns to adopt many gimmicks in the developing world, where a good job must be done with whatever facilities are available. Neurosurgeons in developing countries must still contend with a large number of massive tumors with whatever modern facilities are available.

Acromegaly

Before transsphenoidal microsurgical techniques were adopted for the treatment of acromegaly, the results were unsatisfactory. Prior to 1976, two types of therapy were available, transfrontal excision of the tumor and radiotherapy. In the majority of cases where the lesion was confined to the sella, transfrontal excision of the tumor could not be satisfactorily performed because of technical difficulties. Results were better when the tumor had become extrasellar. Of the 66 cases of acromegaly prior to 1976, 37 patients had evidence of extrasellar extension with pressure on the optic chiasma. All these were operated on, and the optic nerves were decompressed. Although vision improved, improvement in endocrine status was only marginal. One benefit from such surgery was relief from intractable headache, which many of these patients had. They were subjected to radiotherapy and were given a dose of 60–70 Gy. Endocrine improvement was still unsatisfactory. In the 29 patients without visual impairment surgery was not adopted, but a full course of radiotherapy was given. Here the endocrine improvement was also marginal, and relief from headache was achieved in only 18 patients. Thus it is clear that until about 10 years ago the treatment of acromegaly did not provide the patient any appreciable benefit.

The picture changed remarkably with the introduction of transsphenoidal microsurgical techniques. In the past decade in our department, 27 cases of acromegaly have been treated; in 19 of them (70%) we have achieved a reduction in growth hormone levels and a change in the physical features. Of these 27 patients, 9 had high blood sugar levels, not controlled with insulin, and after surgery, the diabetes came under control with small doses of antidiabetic drugs. Headache disappeared in all cases. In eight cases, it was not possible to achieve satisfactory reduction in growth hormone levels, and four of them were given bromocriptine. Somatostatin analogues have not been tried due to the high cost of the drug.

Prolactinomas

Although is has been known for many years that many patients with "chromophobe adenomas" show a loss of libido and some galactorrhea, the role of prolactin hypersecretion has been fully recognized only in the past two decades. Since the methods of estimating serum prolactin levels became standardized and easily available, the diagnosis of prolactin-secreting adenomas has been made more often and the appropriate treatment given. Hardy popularized the transsphenoidal microsurgical approach to small tumors of the pituitary, and now excision of small microadenomas has become possible.

In the period 1975–1988, 88 prolactinomas were encountered; the majority of referrals were from gynaecologists. These included 64 microadenomas; 44 were treated successfully with bromocriptine, and 20 were subjected to surgical excision. Normalization of prolactin levels was achieved in 13 of these patients, and in 7 it was necessary to continue bromocriptine. In women, disappearance of galactorrhea and achievement of pregnancy were the goals. Deficiency of thyroid function is often present in these patients and needs evaluation and treatment, along with bromocriptine administration. Surgery for microadenomas was advised when the patient could not afford the cost of bromocriptine (9 patients), or when prolactin levels did not fall after adequate bromocriptine therapy (11 patients).

Twenty-four patients had mesoadenomas or macroadenomas, with radiological evidence of a sellar lesion, such as double flooring of the sella being seen on skull X-ray. Eight of these patients desired a trial of medical therapy and were given bromocriptine and kept under observation for endocrine and visual disturbances. Three of them required surgery a few months later. When patients with macroadenomas were put on trial treatment with bromocriptine, they were advised to have their visual fields checked periodically, especially if they became pregnant. Nineteen patients were subjected to transsphenoidal microsurgery, and all attempts were made to remove the maximum possible amount of tumor. We have not used the technique of lumbar injection of air into the chiasmatic cistern to push down remnants of tumor. Toward the end of the procedure, we request the anaesthetist to lighten the anaesthesia, so that the patient regains partial consciousness. At this stage some leftover remnants of the tumor often make themselves visible for removal.

All patients in whom there is doubt about the complete removal of the tumor and those who show presence of the tumor on the postoperative scan are given radiotherapy. It must be emphasized that many patients cannot afford postoperative scan as a routine. My own personal conviction is that all these patients must be given postoperative radiotherapy, although my junior colleagues do not agree with this opinion. Six of the 19 patients required continuing bromocriptine therapy.

It is apparent that the prolactin-secreting tumors of the pituitary have many facets, and the type of treatment suited to each patient must be carefully worked out, with a good follow-up.

Conclusion

One can conclude this narration of 38 years of experience of pituitary surgery by expressing satisfaction on the tremendous advances that have taken place in our investigative and operative techniques, and in endocrinology which have resulted in much better prospects for life and happiness in our patients.

References

1. Guiot D, Derome P (1976) Surgical problems of pituitary adenomas. In: Krayenbühl H (ed) Advances and technical standards in neurosurgery, vol 3. Springer, Vienne New York, pp 3–33
2. Jefferson G (1940) Extrasellar extensions of pituitary adenomas. President's address. Proc Roy Soc Med 33:433–438
3. Jeffrys RV (1984) Surgical treatment of massive tumors. In: Belchetz PE (ed) Management of pituitary disease. Chapman & Hall, London, pp 424–435
4. Kunicki A, Grochowski W, Czerniki ZB (1975) Szerzeina sie Gruzolakow Pyzsad ki mozgowoj poza siedeko tureckie. Neurologia, Neurochirurgia. Pdska 9:626–629
5. Ramamurthi B (1960) Tumors of the pituitary region. Personal experiences and experiences of others. Neurol India 8:43–75
6. Ramamurthi B (1970) Chromophobe adenomas. Neurol India 18 [Suppl]: 65–74
7. Ramamurthi B (1986) Experience with large pituitary adenomas in India. Neurol India 34:195–201
8. Symon L, Jakubowski J, Kendall B (1979) Surgical treatment of giant pituitary adenomas. J Neurol Neurosurg Psychiatry 42:973–982

Pituitary Tumors in the Aged

K. D. Post, C. Loftus, and J. Lustgarten, New York and Iowa City/USA

Introduction

In this study we retrospectively looked at a series of pituitary tumors surgically treated between 1976 and 1987 at Tufts New England Medical Center and the Neurological Institute at Columbia-Presbyterian Medical Center. An arbitrary age cut-off of 60 years was selected. A total of 64 cases were reviewed, representing approximately 10% of the surgical series from which it was selected. We carried out this study in an effort to determine whether pituitary tumors present in a different fashion in this age population, and whether their treatment offers different options than does that of the general younger population.

Clinical Features

The age range was 60–93 years; most patients, however, were aged 65–75 (60–64, 19; 65–69, 21; 70–74, 15; 75–79, 3; over 80, 6). As opposed to the younger population, in whom endocrinological abnormalities are the most common features, in this age group visual loss is clearly the predominant presenting symptom. 41% presented with visual loss, most commonly a bitemporal hemianopsia. Five patients were seen because of apoplectic events with almost complete blindness, and another patient presented with only an extraocular movement disturbance with visual acuity and fields intact. Eight in the group presented with hypersecretory syndromes, of which six were acromegalic and two were cushingoid. Five patients had hyperprolactinemia, but none was symptomatic because of this. Three patients presented with headache and one with a seizure, and four were incidental findings in a metastatic tumor work-up. The 12 remaining patients presented with various complaints including meningitis, disturbed mental status, and dysequilibrium. The average duration of symptoms was long, with 36% extending over 3 years prior to treatment.

Preoperative visual status showed normal function in 23 patients, complete blindness in 2, bitemporal field defect in 39, extraocular movement disturbance in 7, and acuity loss in 16.

The endocrinological work-up included assays for thyroid function, prolactin, cortisol, and gonadotropins and testosterone. These were not performed in all patients routinely as they would have been in a younger population. Several patients presented with acute abnormalities of vision and were promptly evaluated radiologically and treated surgically before adequate endocrinological data could be gathered. Eleven patients presented with symptoms secondary to hypopituitarism including fainting spells, and acute deterioration during other nonassociated illnesses. Computed tomography (CT) scans were almost uniformly abnormal and demonstrated clearly whether the process was solely intrasellar, or whether there was infrasellar, suprasellar, or parasellar extension. Currently, magnetic resonance imaging is preferred.

Preoperative associated risks included cardiac disease in 41, hypertension in 13, diabetes mellitus in 5, and pulmonary disease in 7. These required consideration before surgical therapy but in the majority did not yield any problem.

Five patients presented with apoplectic events, one of whom had been blind for 48 h before surgical intervention. Another patient with apoplexy was found to have metastatic lung carcinoma within the pituitary adenoma, and it was not clear whether the apoplexy was secondary to the metastatic portion or the pituitary portion. One 93-year-old woman was operated upon transsphenoidally despite her age because of severe progressive visual loss. The diagnosis was metastatic breast carcinoma.

Seven patients had been previously treated with transcranial surgery and eight with radiation therapy. Five had received prior bromocriptine therapy, and 20 had received replacement hormones. Two patient had undergone cataract removal in an attempt to improve vision.

Surgical Treatment

The surgical approach was transsphenoidal in 63 patients and transcranial in one. In 26 a gross total removal was achieved as judged by the operating surgeon, and in 38 only a subtotal removal was carried out.

One patient underwent a transcranial operation because the CT appearance of the tumor was of a small sella and a "waisted" appearance in the coronal view. The preoperative diagnosis was meningioma, and in view of a small sella it was felt that transspheniodal intervention would be contraindicated even if it were an adenoma. Pathology showed a pituitary adenoma, and intraoperatively there were no obvious complications, however postoperatively the patient remained in a coma and expired with a large posterior third ventricular and diencephalic hemorrhage. This was the only death in the series.

Several patients presented with seizures and CT showed marked eccentric extension into the middle fossa (Fig. 1). It was clear that this could not be totally removed transsphenoidally, however because of the patient's age, it was felt that transsphenoidal decompression and tissue diagnosis would be adequate, and that transcranial surgery would not result in a complete removal either. The transsphenoidal route was selected because of the far lower morbidity and mortality rate.

a

b

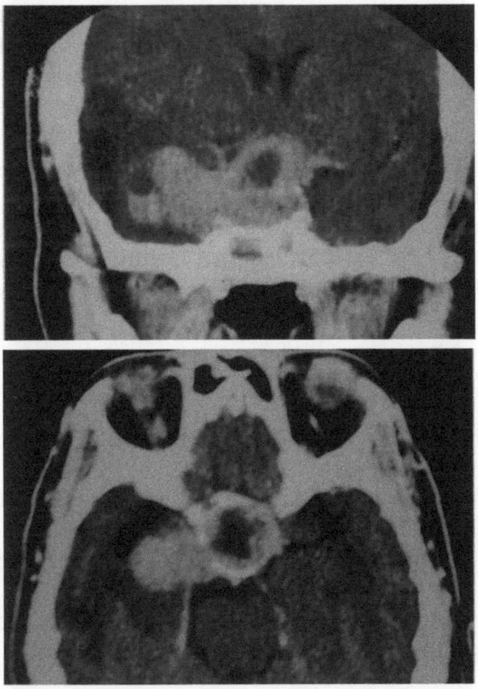

Fig. 1a, b. Pituitary adenoma with a large right middle fossa extension. **a** Coronal enhanced CT scan. **b** Axial enhanced CT scan

The tumor consistency was generally soft, as anticipated (69%), and was hemorrhagic in five, cystic in seven, firm in seven, and clearly invasive in 26.

Results

There was marked improvement in visual field and acuity in 33 (79%), no changes in seven, and deterioration in two. Extraocular movement dysfunction improved in six of seven.

The hypersecretory state was normalized in all six growth hormone secretors, neither of the two ACTH secretors, and in one of five prolactin secretors.

The only death in the series was that of the craniotomy patient, who had an intraventricular and diencephalic hemorrhage. Three patients died in the follow-up period. One patient with Cushing's disease aged 62 at treatment died 3 years later of other causes. One prolactinoma patient aged 69 died 21 months postoperatively from an acute myocardial infarction. The 77-year-old patient with apoplexy and carcinoma metastatic to the pituitary adenoma died 3 months postoperatively from his pulmonary carcinoma.

Complications included four cases of transient diabetes insipidus. Two patients had increased visual loss. One patient developed complete loss of vision in the left eye following surgery and was reexplored transsphenoidally anticipating a hematoma, but none was found. The loss of vision in that eye was permanent. A second patient had a very mild increased bitemporal defect. There were no cranial nerve palsies and no cerebrospinal fluid leaks. There was one pulmonary embolus but no myocardial infarctions or pneumonias.

Adjunctive postoperative therapy included radiation therapy in 28 patients (44%) and bromocriptine therapy in 3 (5%). Replacement hormone therapy was used in 40% preoperatively and was necessary in 58% postoperatively.

Conclusions

1. Advanced age and concurrent risks are not contraindications to the successful treatment of pituitary adenomas.
2. Visual improvement was obtained in 85% of patients, with a mortality of 1.5% and morbidity of 4.7%.
3. There were no complications referable to patient's age or preoperative medical risks.
4. Surgical approach via the transsphenoidal route is a sound choice in the management of pituitary masses in the elderly.

Pituitary Adenomas in Childhood and Adolescence

G. MAIRA and C. ANILE, Rome/Italy

Introduction

Pituitary adenomas are rare in children. In large series they account for only 1%–10% of adenomas of all ages, and some of these series include not only children but also patients in whom it was documented that the onset of symptoms was before the age of 20 years. Although many authors maintain that pituitary adenomas in children tend to be extrasellar and invasive, and that a subfrontal approach is mandatory in the majority of them, the natural history of the disease in this age group is not well known.

Clinical Material

The patients of this series are 356 operated on for adenomas between July 1976 and December 1986 (Fig. 1). The distribution of adenomas of our series as related to the hormone secretion shows that, as in the majority of clinical series (Hardy 1975; Wilson 1984), the prolactin (PRL) secreting adenomas are predominant (45%), followed by the growth hormone (GH) secreting adenomas (26%), nonsecreting adenomas (24%), and adrenocorticotropic hormone (ACTH) secreting adenomas (5%). The exact diagnosis, other than as based on preoperative hormonal and radiological data, was histologically verified using immunohistochemical hormone staining. The anatomic-radiological alterations of the sella turcica were classified according to Hardy and Vezina (1976).

In all patients pre- and postsurgical hormonal studies were performed. In particular, we have analyzed all the pituitary functions in basal conditions and with dynamic tests. The modalities of sample collection and the radioimmunoassay method used for hormone determination were reported in our previous papers (Barbarino et al. 1978; De Marinis et al. 1984).

To obtain information on the peculiar aspects of adenomas in this age group, we divided our patients into three groups (Fig. 2): group 1, 52 patients (15%; 18 boys and 34 girls, aged 7–20 years) operated on before the age of 21 years; group 2, 52 patients (15%; 2 men and 50 women, aged 21–40 years) treated after the age of 20 but in whom the clinical symptoms appeared before the age of 21 years; group 3, 252 patients (70%; 80 men and 172 women) with surgery and symptoms after the age of 20 years.

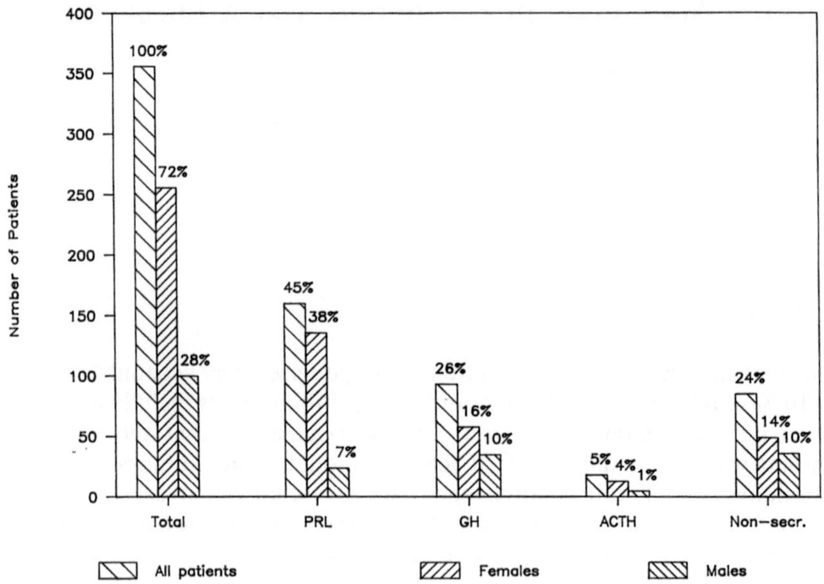

Fig. 1. Pituitary adenomas by type and sex

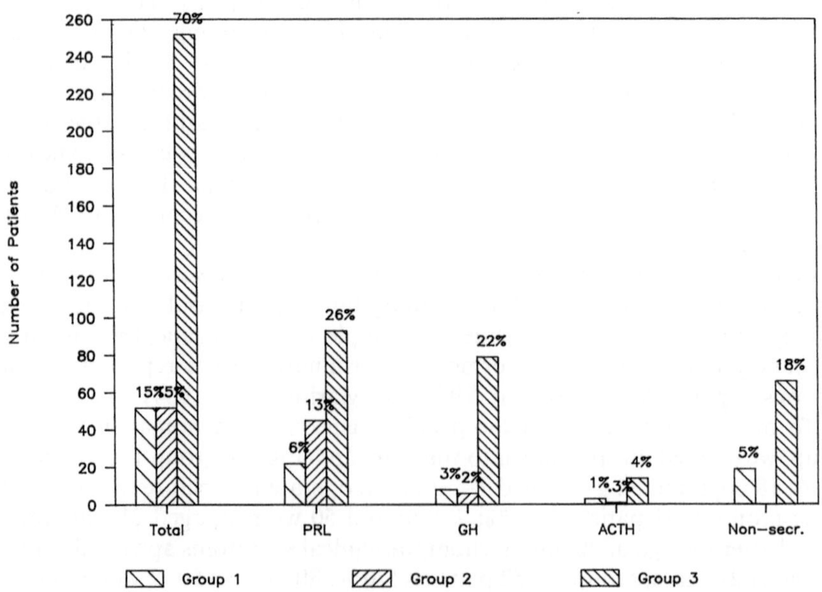

Fig. 2. Pituitary adenomas by patient group

The frequency and type of hypersecretion are similar to the distribution found in adults, with a preponderance of PRL-secreting adenomas.

We analyzed in detail each of these groups, considering their anatomic-surgical features and the possibility of cure by means of surgery, to obtain the normalization of the clinicohormonal picture.

Results

PRL-Secreting Adenomas. Twenty-two (18 females and 4 males) of the 160 patients (14%) operated on for PRL-secreting adenomas were aged under 21 years; 45 (all females, 28% of all PRL adenomas) were operated on after the age of 20 years, but in them the clinical symptoms started before the age of 21 years; in 93 patients (73 females and 20 males, 58% of all PRL adenomas) the clinical history started after the age of 20 years. The incidence of suprasellar expansion, visual defect, and invasive character in these patients are reported in Fig. 3. All the patients with PRL-secreting adenomas were operated on using a transsphenoidal approach; none required operation via a frontal approach. The possibility of successful surgery, i.e., a normalization of the PRL basal values, without signs of tumor recurrence during a follow-up of 1–9 years, was analyzed in 139 patients, comparing groups similar for sellar grade; the results are reported in Fig. 4.

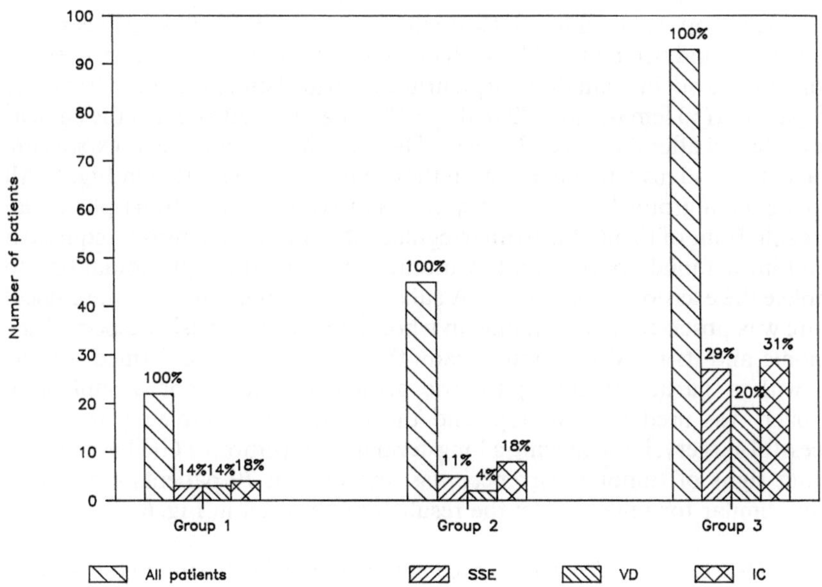

Fig. 3. Incidence of suprasellar expansion (*SSE*), visual defect (*VD*), and invasive character (*IC*) in patients with PRL-secreting adenoma ($n = 160$)

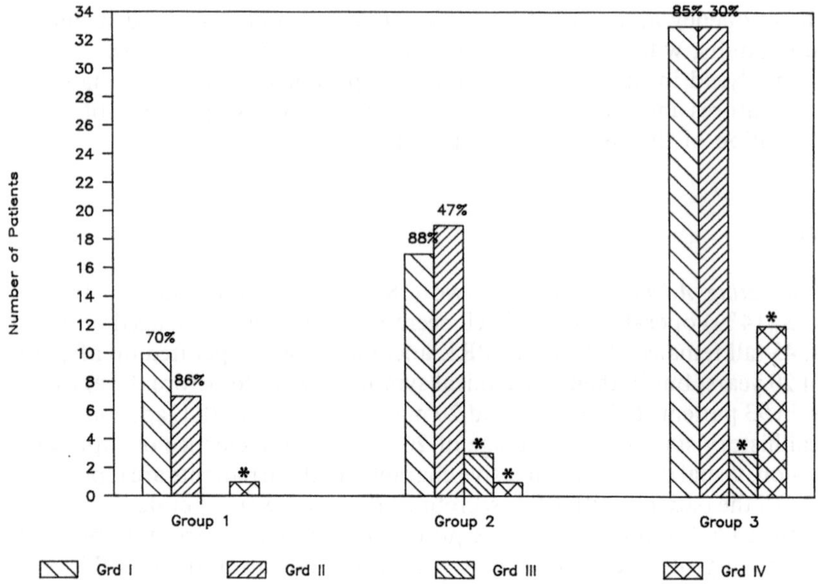

Fig. 4. Cure rate in patients with PRL-secreting adenoma in relation to sellar grade (*Grd*). *, No patient cured ($n = 139$)

GH-Secreting Adenomas. Eight (four females and four males) of the 93 patients (9%) operated on for GH secreting adenomas were under 21 years; six (four females and two males, 7% of all GH patients) were operated on after the age of 20 years but in them the clinical history started before the age of 21 years; in 79 patients (50 females and 29 males, 84% of all GH adenomas) the clinical history started after the age of 20 years. The incidence of suprasellar expansion, visual defect and invasive character in these patients are reported in Fig. 5. All the patients in group 1 except one were operated on using a transsphenoidal approach. Four of them, due to an irregular intracranial expansion, required a second intracranial approach a few months after the transsphenoidal one to complete the excision of the tumor. A subsequent normalization of the clinical picture was observed in all but one, in whom, due to the invasive aspect of the adenoma and its rapid recurrence, radiotherapy was also performed. In one case with a particular suprapituitaric growth of the tumor a subfrontal approach was used as first step, and the tumor was completely removed. Successful surgery, i.e. achieving a long-term normalization of GH basal values without signs of tumor recurrence, was analyzed in 79 patients comparing groups similar for sellar grade; the results are reported in Fig. 6.

ACTH-Secreting Adenomas. Three patients (one female and two males) of the 18 patients (17%) operated on for ACTH-secreting adenomas were aged under 21 years; one (a female, 5% of the ACTH patients) was operated on after

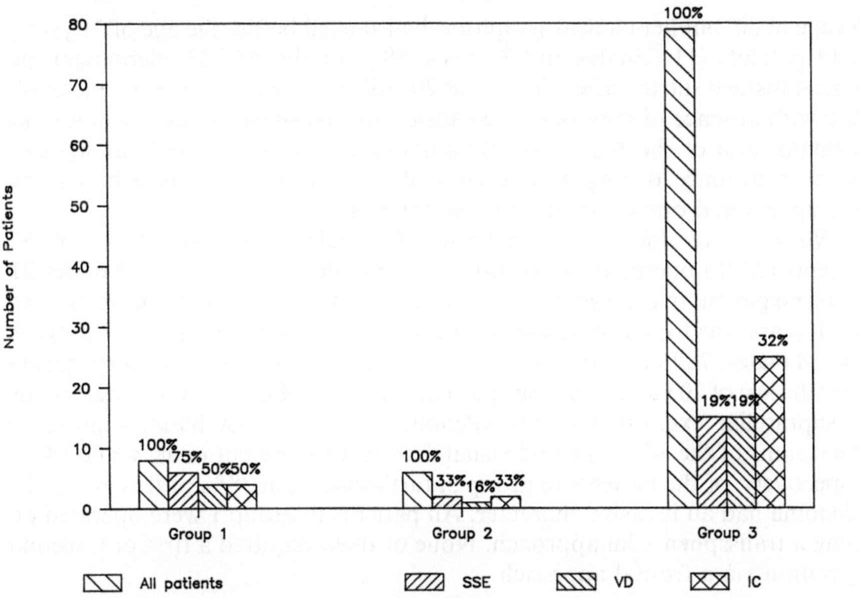

Fig. 5. Incidence of suprasellar expansion (*SSE*), visual defect (*VD*), and invasive character (*IC*) in patients with GH-secreting adenoma ($n = 93$)

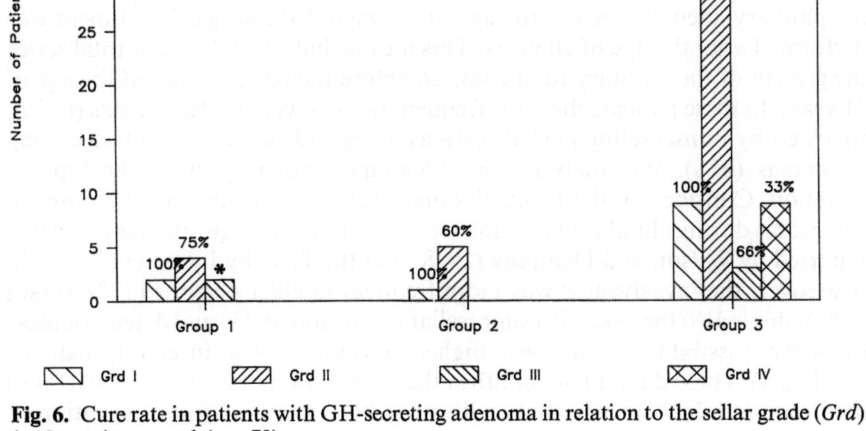

Fig. 6. Cure rate in patients with GH-secreting adenoma in relation to the sellar grade (*Grd*). *, No patient cured ($n = 79$)

the age of 20, but her clinical symptoms had started before the age of 21 years; in 14 patients (11 females and 3 males, 78% of the ACTH adenomas) the clinical history started after the age of 20. All these adenomas were enclosed, thus with absence of suprasellar expansion or invasive character. Long-term normalization of the ACTH basal values and cortisolemia without signs of tumor recurrence during a follow-up of at least 1 year was achieved by transsphenoidal operation in all these patients.

Nonsecreting Adenomas. Nineteen (11 females and 8 males) of the 85 patients (22%) operated on for nonsecreting adenomas were aged under 21 years; no patient over 20 years of age but with onset of symptoms before the age of 21 years was found in this series; in the remaining 66 patients (38 females and 28 males, 78% of the nonsecreting adenomas) the clinical history started after the age of 20 years. Only one patient in group 1 showed a visual defect due to suprasellar expansion of the adenoma. On the other hand, suprasellar expansions of the adenoma and visual defects were present in 68% and 64%, respectively, of the patients in group 3; furthermore, in 38% of this group the adenoma had an invasive character. All patients in group 1 were operated on using a transsphenoidal approach. None of them required a first or a second operation via a frontal approach.

Discussion

Pituitary adenomas are generally considered as uncommon tumors in children and adolescents, with a high tendency to involve contiguous and suprasellar structures. A subfrontal approach, rather than the safer transsphenoidal one, is considered mandatory in the majority of these cases (Gaini et al. 1977; Ortiz-Suarez and Erickson 1975). The analysis of our series, however, shows that these adenomas are less rare in childhood and adolescence than has hitherto been suspected. In our series of 356 cases, 52 patients (15%) were operated on under the age of 21 years, and another 52 (15%) presented clinical symptoms of the pituitary adenoma before the age of 21, even if the surgical treatment was performed after the age of 20 years. This means that in 30% of our total series the growth of the pituitary tumor started before the patient reached the age of 21 years. In these patients the most frequent tumors were prolactinomas (65%), followed by nonsecreting (18%), GH-secreting (13%), and ACTH-secreting adenomas (4%). We analyzed the adenomas with respect to the type of secretion. Considering the prolactinomas, 42% of patients had the onset of symptoms during childhood or adolescence – a much higher rate than the 10% reported by Wilson and Dempsey (1978) and the 11% by Lucas et al. (1980). Moreover, the invasiveness was more frequent in elder persons (31% versus 18%); this is also the case with suprasellar expansion and visual defect. In these cases the possibility of cure was higher in younger than in elderly patients (see Fig. 4). These data do not confirm the observation of Ludecke et al. (1986) who reported 75% invasiveness with only 13% hormonal normalization in young patients with PRL-secreting adenomas.

GH-secreting tumors are the only adenomas in our series in which a tendency to suprasellar expansion and invasive character was predominant in children, in comparison with patients of other ages. This peculiarity is confirmed by other authors. Ludecke et al. (1986) reported 86% suprasellar expansion and 57% invasiveness; Laws et al. (1986) reported 67% macroadenomas and 33% invasiveness. Only in this group of patients were intracranial approaches required. Only one patient in our series with GH invasive adenoma required a radiotherapy; similar data are reported in the literature (Ludecke et al. 1986). A postoperative normalization of hormone secretion was high in all groups of patients, without differences related to age. Regarding our patients with Cushing's disease, all of them, due to the possibility of an early endocrinological diagnosis, had small intrapituitaric adenomas. Therefore, signs of invasiveness or suprasellar expansion were absent. All these cases were cured, as frequently happens in adenomas operated on via a transsphenoidal approach when they are still completely enclosed in the pituitary gland.

Among our patients with nonsecreting adenoma we operated on 19 younger than 21 years, with a range between 7 and 18 years. Many of these cases belong to a series already described (Maira et al. 1980) in whom growth retardation due to an isolated deficit of GH was the predominant clinical sign that permitted early diagnosis. The recognition of this very early endocrinological sign explains the absence of suprasellar expansion and invasiveness in these younger patients compared to older ones, in the majority of whom the diagnosis was possible only after the appearance of neurological symptoms. This seems to indicate, at least in nonsecreting adenomas, that in the majority of cases the expansion and invasiveness of the tumor does not depend on a biological peculiarity. The discrepancies between these results and data reported in the literature on a high incidence of large and invasive lesions in childhood and adolescence (Lucas et al. 1980; Ludecke et al. 1986) may be due to a time lag between the beginning of symptoms of tumor growth and the treatment. Moreover, in agreement with other authors, we observed no histological differences between invasive and enclosed adenomas; the term "invasive" represents merely an involvement of contiguous structures and does not imply malignancy (Gaini et al. 1977; Martins et al. 1965; Ortiz-Suarez and Erickson 1975).

In conclusion, as for the pituitary adenomas in adults, an early diagnosis is essential to reduce the incidence of suprasellar extension and invasiveness and to increase the probability of surgical success.

References

Barbarino A, De Marinis L, Maira G, et al. (1978) Serum prolactin response to thyrotropin-releasing hormone and metoclopramide in patients with prolactin-secreting tumors before and after transsphenoidal surgery. J Clin Endocrinol Metab 47:1148–1151

De Marinis L, Mancini A, Maira G, et al. (1984) Postoperative evaluation of dopaminergic tone in prolactinoma patients. II. Plasma thyrotropin response to metoclopramide. J Clin Endocrinol Metab 58:405–409

Gaini SM, Giovanelli M, Forni C et al. (1977) Pituitary adenomas in infancy and childhood. In: Villani R, Giovanelli M, Gaini SM (eds) Modern problems in paediatrics. Karger, Basel, pp 220–225

Hardy J (1975) Transsphenoidal microsurgical removal of pituitary microadenoma. In: Krayenbuhl H, Maspes PE, Sweet WH (eds) Progress in neurological surgery. Karger, Basel, pp 200–216

Hardy J, Vezina JL (1976) Transsphenoidal neurosurgery of intracranial neoplasm. In: Thompson RA, Greene JR (eds) Advances in neurology. Raven, New York, pp 261–264

Laws ER, Scheithauer BW, Groover RV (1986) Pituitary adenomas in childhood and adolescence. Prog Exp Tumor Res 30:359–361

Lucas C, Guibout M, Jaquet P, et al. (1980) Aspects diagnostiques et évolutifs de l'adénome à prolactine chez l'enfant. Arch Fr Pediatr 37:79–86

Ludecke DK, Herrmann HD, Schulte FJ (1986) Special problems with neurosurgical treatment of hormone-secreting pituitary adenomas in children. Prog Exp Tumor Res 30:362–370

Maira G, Di Rocco C, Borrelli P, et al. (1980) Pituitary microadenomas in children: a cause of growth failure. J Microsurg 1:276–287

Martins A, Hayes GJ, Kempe LG (1965) Invasive pituitary adenomas. J Neurosurg 22:268–276

Ortiz-Suarez H, Erickson DL (1975) Pituitary adenomas of adolescents. J Neurosurg 43:437–439

Wilson CB, Dempsey LC (1978) Transsphenoidal microsurgical removal of 250 pituitary adenomas. J Neurosurg 48:13–22

Wilson CB (1984) A decade of pituitary microsurgery. The Herbert Olivecrona lecture. J Neurosurg 61:814–833

Recurrent Pituitary Adenomas

R. FAHLBUSCH and M. BUCHFELDER, Erlangen/FRG

Introduction

Since the term "recurrence" is used to describe extremely different conditions, the authors would like to define this term by using the model of a secreting pituitary adenoma (Fig. 1). For the evaluation of surgical results in such hormonally active pituitary adenomas it is essential to distinguish between an immediately postoperative normalization of elevated hormone levels and a persistent hormone excess. Even when the hormone excess is improved but not normalized, this should still be regarded as a persistent active disease. Spontaneous remissions, sometimes also referred to as spontaneous cure [2] may very rarely occur as a result of tumor infarction or hemorrhage.

As opposed to a persistent normalization of the hormone excess for years or even decades, real recurrences are defined as a relapse of the hormone excess following initial normalization. If, however, the tumors lack hormonal markers, it is less easy to distinguish between the regrowth of persistent tumor tissue and real recurrences. In earlier series only clinical signs of visual function were used to differentiate both entities. At present, repeat neuroradiological examinations by computed tomography (CT) or magnetic resonance imaging (MRI) with high resolution allows more accurate differentiation between tumor progression (Fig. 2) and recurrence in these cases.

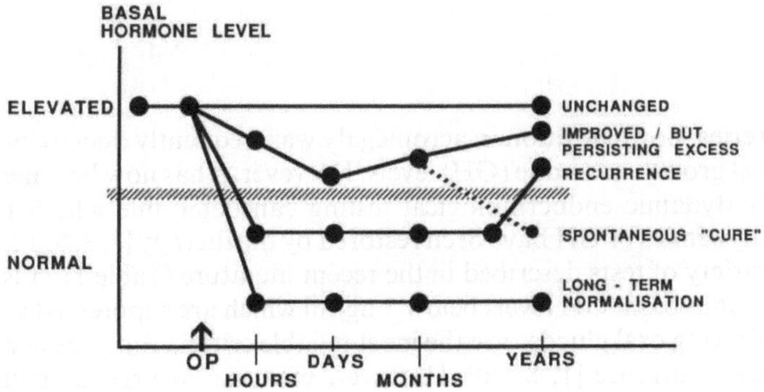

Fig. 1. Basic principles and definitions of operative results in hormonally active pituitary adenomas

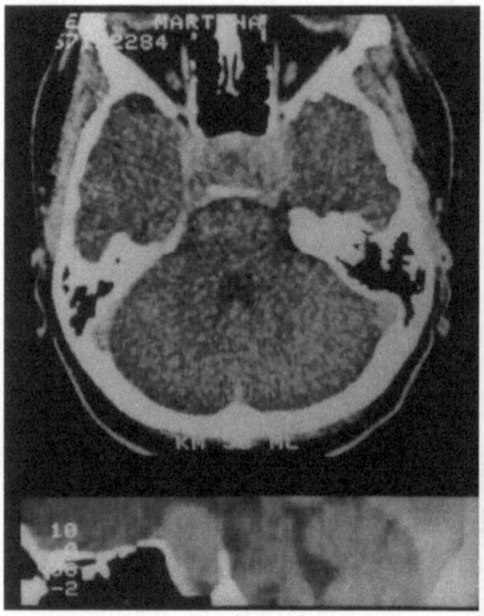
a

b

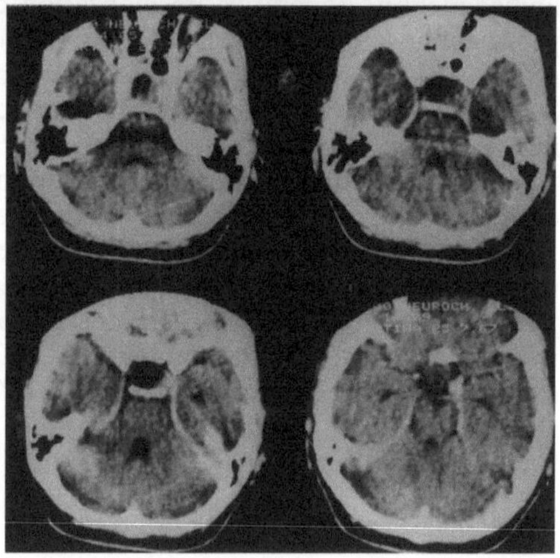
c

Fig. 2a–c. Pre- and postoperative cranial CT scans of a 16-year-old girl with a hormonally inactive intra-, supra-, and parasellar pituitary adenoma. **a** Preoperative CT scan. **b** CT scan obtained 3 months after transsphenoidal surgery, suggestive of some residual tumor at the tentorial edge. **c** CT scan obtained 6 months thereafter, documenting rapid progression of this residual invasive tumor necessitating transcranial surgery and postoperative irradiation

Acromegaly

In most surgical series, normalization in acromegaly was frequently documented by normal basal growth hormone (GH) levels. However, it has now become obvious that only dynamic endocrinological testing can determine whether normal secretion dynamics of GH have been restored by the therapy [1, 3, 8, 12, 25]. Among the variety of tests described in the recent literature (Table 1), it is generally accepted that basal GH levels below 5 ng/ml which are suppressed to below 2 ng/ml following oral glucose are the most reliable criteria for assessing normalization and recurrence [1, 8, 20]. However, most neurosurgeons still quote their results using normalization of basal GH levels to below 5 ng/ml as the sole criterion.

Table 1. Recurrences after transsphenoidal surgery for acromegaly (a survey of the current literature)

Authors	Number of patients followed up	Follow-up period mean (range) in years	Criteria of remission	Recurrence rate absolute
Serri et al. 1985 [39]	21	8.9 (5–11)	OGTT < 2.5 ng/ml	3/21 (14%)
Roelfsema et al. 1985 [33]	43	3.3 (0.5–7)	TRH + GnRH	0/43 (0%)
Grisoli et al. 1985 [20]	60	? (2–6)	Basal GH < 5 ng/ml TRH + GnRH	6/60 (10%)
Arafah et al. 1987 [1]	30 21	7.6 (?) 8.1 (?)	Basal GH < 5 ng/ml OGTT + SmC	5/30 (16%) 0/21 (0%)
Artia et al. 1988 [4]	41 20	? (0.25–6.2) ? (0.25–6.2)	Basal GH < 5 ng/ml TRH + GnRH	2/41 (4.8%) 0/20 (0%)
Buchfelder et al. 1988 [8]	60 43	5.1 (1–14) 5.1 (1–14)	Basal GH < 5 ng/ml OGTT < 2 ng/ml	4/60 (6.6%) 1/43 (2.3%)

OGTT, Oral glucose tolerance test; TRH, thyrotropin-releasing hormone; GH, growth hormone; GnRH, gonadotropin-releasing hormone.

Although the GH excess in microadenomas could be normalized in 81% of the 234 patients whom we operated on between 1979 and 1987 [17], the overall normalization rate was only 64%. The extent of surgical resection technically permissible in order to eliminate the cause of the hormone excess is often debated. Undoubtedly, radical resection of all tumor tissue detected inside, outside, and within the tumor capsule itself (Table 2) helps to improve the

Table 2. Causes of persistent postoperative hormonal hypersecretion and recurrences

Tumor residuals

- Inside the tumor "capsule"
- Outside the tumor "capsule"
  - Cavernous sinus
  - Sphenoid sinus
  - Bony skull base
- In the tumor "capsule" itself
  (Shaffi and Wrightson 1975 [40], Selman et al. 1986 [37])

Hypothalamic factors

- Hypothalamic dysregulation (Quabbe 1978 [30])
- Ectopic releasing hormone secretion
  (Thorner et al. 1982 [43], Von Werder et al. 1984 [46])

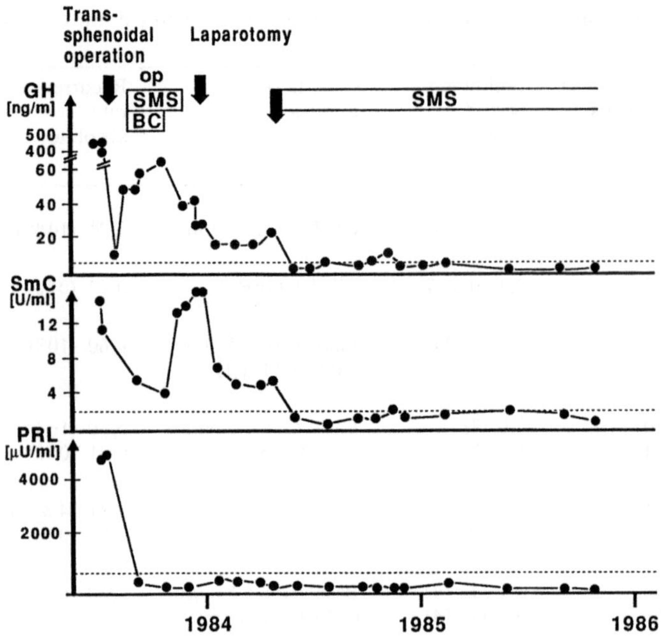

Fig. 3. Course of growth hormone (*GH*), somatomedin C (*SmC*), and prolactin (*PRL*) levels in a 14-year-old girl with acromegaly resulting from ectopic GH secretion. *BC*, Bromocriptine; *SMS*, somatostatin analogue SMS 201-995. (Modified from [46])

hormonal results achieved by surgery. We showed that if in cases of laterally invasive adenomas the tumor resection was extended into the cavernous sinus [16], the growth hormone excess could be normalized in 3 out of 13 macroadenomas. The hypothalamus is only exceptionally responsible for GH hypersecretion [30]. However, in the recent literature, extrahypothalamic GRH secretion has been reported to be a rare cause of acromegaly [43, 44].

In 1983 we removed a large intra- and suprasellar tumor in a 14-year-old girl acromegalic. After surgery, there was no residual tumor detected in the plain and contrast-enhanced CT scan. Initially we observed a rapid decrease in GH levels. A total dose of 50 Gy, in daily fractions of 2 Gy, was used. Similar period, GH and somatomedin C levels again increased (Fig. 3). Retrospectively this could be explained by high GRH levels. The cause was an ectopic GRH secreting pancreatic tumor with metastases in the liver, which was later controlled by the administration of the somatostatin analogue SMS 201-995 (Sandoz). As expected, the histological diagnosis of the pituitary tumor was that of somatotrophic hyperplasia [46]. However, our other surgical failures could not be explained by elevated GRH levels or ectopic GRH secretion.

Is postoperative endocrinological testing of any prognostic value? We have observed that the lowest GH levels following oral glucose tolerance testing obtained as early as 1 week postoperatively are comparable to the values obtained 3 months after surgery. In the majority of our patients GH levels were suppressed not only to below 2 ng/ml but even to below 1 ng/ml. In these cases, somatomedin C levels were also found to be within the normal range, which is not always the case if GH is not suppressed to below 1 ng/ml. Applying these strict criteria, we have followed up 43 patients among a series of 60 normalized patients with GH levels suppressed to below 2 ng/ml 3 months after surgery. Only a few patients were lost to follow-up in the course of several years [8]. Two doubtful "recurrences" occurred in the 2nd year. Both patients had normal basal GH levels hours, days, and 1 week following surgery. After 3 months, however, suppressed GH levels were borderline. In the first case the basal GH value of 3.6 ng/ml was suppressed to 2 ng/ml following an oral glucose load. In the second patient the values were 1.8 ng/ml and 2.0 ng/ml, respectively. However, after 2 years, failure of suppression to below 2 ng/ml was found in only one case. This patient also developed clearly abnormal somatomedin C levels.

Intra-operative monitoring of GH levels using a rapid assay as described by Lüdecke [26] may help in improving the surgical outcome in individual cases. This investigation, however, is certainly not of predictive value since dynamic testing of GH secretion is not possible during surgery.

At this stage, we would like to stress the point that sophisticated endocrinological testing alone is not the sole criterion which guides the clinician in the management of an acromegalic patient. Initial tumor size and extent of the GH disturbances after therapy also play a significant role. In general, additional treatment is indicated in all acromegalic patients with persistent or recurrent GH hypersecretion, as persistent pathological metabolism reduces life expectancy and quality.

Transsphenoidal reoperation should always be considered first, particularly in smaller tumors. We have performed second operations within a short interval in 21 patients (Fig. 4), in whom GH levels did not decrease to within the normal range after primary surgery. Normalization occurred in 11 of 21 patients after the second operation, and in one case after the third intervention. In all the other patients a significant GH decrease was also obtained, thus providing a favourable basis for further treatment. We recommend transsphenoidal reoperation especially in tumors without extrasellar extension. In three out of four microadenomas, normal GH dynamics, as demonstrated by oral glucose testing, was achieved. In the majority of these cases the indication for surgery was based only on functional endocrinological testing, independently of CT or MRI results which did not demonstrate tumor tissue.

Apart from the use of the well-known dopamine-agonists such as bromocriptine and lisuride, a new approach in medical treatment is now available in the form of the somatostatin analogue SMS 201-995 (Sandoz). Unfortunately, only parenteral administration is possible. Since GH levels increase again within a few hours following subcutaneous injection, repeated injections, three or four times daily, are required [45]. Continuous administration by means of an implanted pump offers some advantage. The importance of somatostatin analogues in the supplementary treatment of acromegaly is

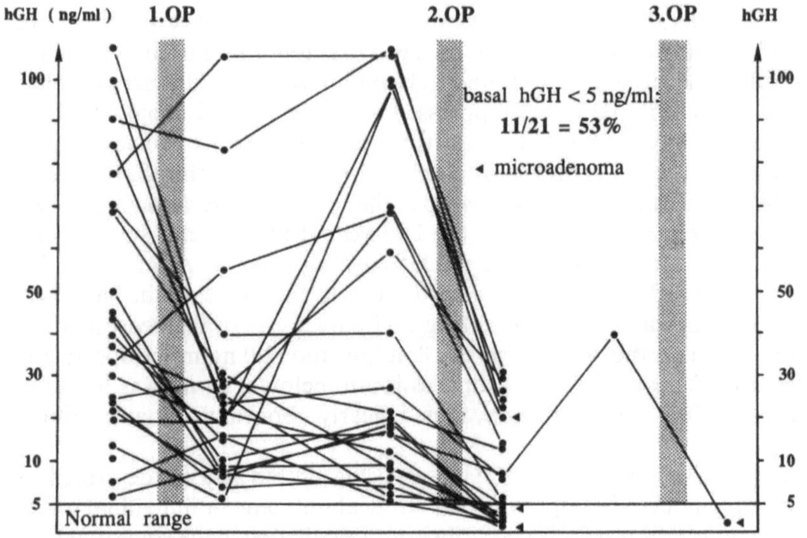

Fig. 4. Results after transsphenoidal reoperations for persistent acromegaly ($n = 21$)

still debatable. At present, there is general agreement that SMS should be used only if all other modes of additional treatment have failed. External megavoltage radiotherapy was performed in all our patients with tumors invading the cavernous sinus or the skull base, as well as in patients who still had elevated GH levels. A total dose of 50 Gy, in daily fractions of 2 Gy, was used. Similar effects can be obtained by the interstitial implantation of radioactive seeds, which at the same time allows for reexploration of the sella. GH levels were seen to decrease progressively over years, but normalization within 6 years was achieved only in those who initially had GH levels below 30 ng/ml. Giovanelli et al. [19] obtained similar results.

The main disadvantage of radiotherapy is that of pituitary insufficiency.

Cushing's Disease

Like acromegaly, Cushing's disease carries a high risk of morbidity and considerably reduces life expectancy. Therefore, permanent elimination of hormonal oversecretion is mandatory. We define normalization in Cushing's disease as the restitution of a normal secretory pattern of ACTH and cortisol, including a normal diurnal rhythm. Cortisol levels should be suppressed to below 2 µg/ml following the overnight administration of 2 mg dexamethasone [15]. This endocrinological remission is, in all cases, associated with clinical remission. In our series of 142 patients with Cushing's disease in whom selective adenomectomy was performed, endocrinological remission occurred in 102 out of 139 patients (73%). Other authors report higher remission rates (Table 3).

Table 3. Recurrences after transsphenoidal surgery for Cushing's disease (a survey of the current literature)

Authors	Number of patients followed up	Follow-up period mean in years	Recurrence rate absolute	Operative procedure
Hardy 1982 [22]	63	1.75	2/63 (3%)	partial or total hypophysectomy
Boggan et al. 1983 [6]	85?	4.6	4/85 (4.7%)	Selective adenomectomy
Kageyama et al. 1984 [24]	82	?	5/82 (6%)	Selective adenomectomy
Fahlbusch et al. 1986 [15]	71	3.2	5/71 (7%)	Selective adenomectomy
Fahlbusch et al. 1988 [17]	71	5.1	9/71 (12.6%)	Selective adenomectomy
Derome et al. 1988 [11]	124	3.8	19/124 (15%)	Selective adenomectomy
Mampalam et al. 1988 [28]	141	3.9	8/141 (6%)	Selective adenomectomy
	16	3.9	1/16 (6.5%)	Total hypophysectomy

The reported rates of intraoperatively discovered adenomas vary considerably, from 90% in our series [15] to 75% in Hardy's series [22]. Furthermore, not all authors have followed up their patients closely and therefore did not perform sequential 2-mg dexamethasone suppression tests for documenting remission. Other factors such as age and sex of the patient as well as the extent of surgical manipulation within the sella must be taken into consideration.

In Cushing's disease, hypothalamic factors seem to play a more important role than in acromegaly. This accounts for the less favorable surgical outcome in Cushing's disease. Considering the high frequency of microadenomas (and/or nodular hyperplasias), one would anticipate a higher normalization rate in Cushing's disease as compared to acromegaly. Furthermore, in contrast to acromegaly, true recurrences are observed more frequently. This raises the question of whether Cushing's disease is dependent on an autonomous ACTH-secreting pituitary adenoma or on a deficient hypothalamic control. In our series we observed 9 recurrences among 71 patients (12.6%), followed up between 3 and 16 years, with an average follow-up period of 5.1 years (Figure 5). Nakane et al. [29] report on 8 recurrences among 86 patients (9.3%). This can be regarded as a similar result, since their mean observation time was only 3.2 years. We have reoperated seven out of nine patients with recurrences, and in five patients the tumor was found in the same location as in

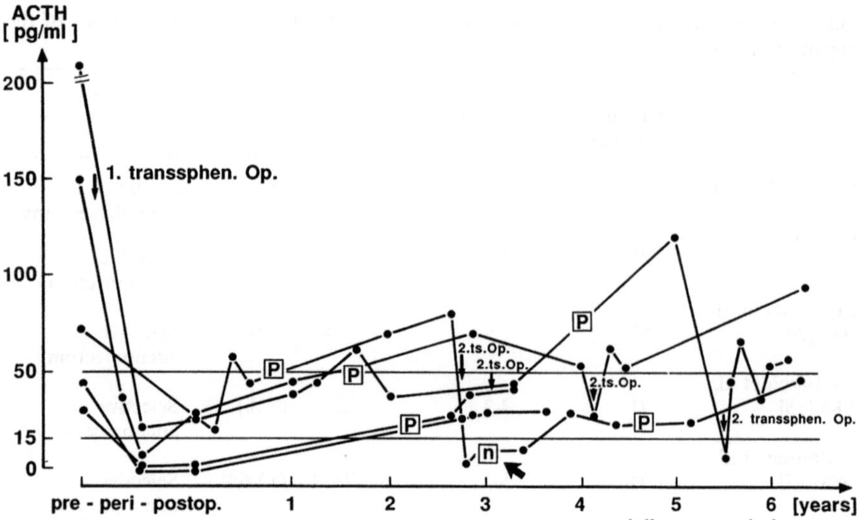

a

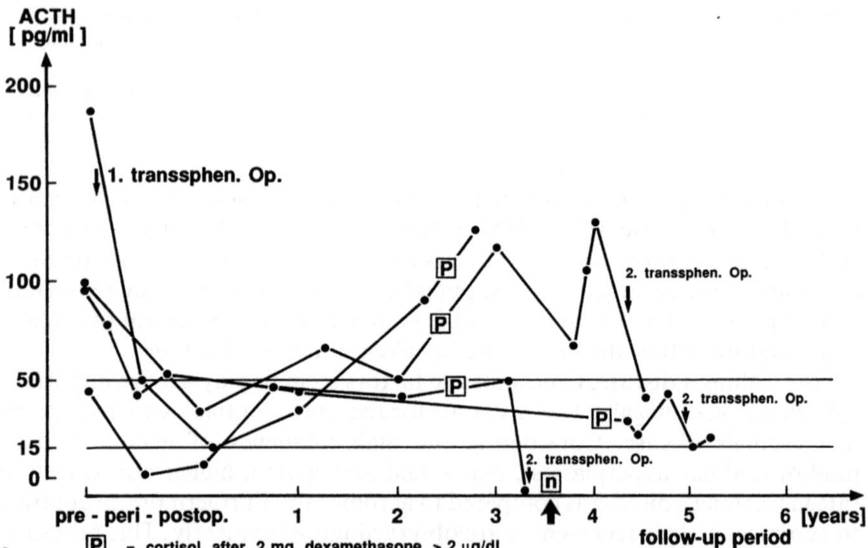

b

Fig. 5a, b. Real recurrences after successful adenomectomy of pituitary adenomas in Cushing's disease (**a** $n = 5$; **b** $n = 4$)

the first intervention. Only two out of these seven patients had another endocrinological remission. Small tumor residuals may thus account for some of the "recurrences" occurring in Cushing's disease. However, not all recurrences can be thus explained. The pathophysiology of this disease still remains an enigma.

In attempting to elicit possible predictive factors for the long-term outcome after surgery of patients with Cushing's disease we have studied pre- and postoperative serum CRH (corticotropin releasing hormone)-levels [41], the CRF induced ACTH increase [36], and the prevalence of autoantibodies against ACTH secreting cells [35]. The results obtained from these studies, however, have not been found to be indicative of the surgical outcome and are not helpful in predicting the development of recurrences.

Prolactinomas

A hypothalamic influence is less frequently discussed in prolactinomas. Prolactin is generally regarded as a tumor marker. However, nontumorous hyperprolactinemia may develop after surgery of hormonally inactive adenomas as a result of either cisternal herniation or scar tissue around the pituitary stalk. Nontumorous hypersecretion is explained by the disturbed passage of the hypothalamic prolactin inhibiting factor via the pituitary stalk to the anterior pituitary. However, we think that tumor residuals mostly account for the fact that we were able to normalize prolactin levels in only 34% of all our patients with prolactinomas [14]. The normalization rate is higher in microadenomas and lower in large macroadenomas [14, 42].

Normalization of hyperprolactinemia is defined as basal serum prolactin levels below 500 μE/ml. In general, normalization was associated with the

Table 4. Recurrences after transsphenoidal surgery for prolactinomas (a survey of the current literature)

Authors	Number of patients followed up	Follow-up period mean in years	Recurrence rate absolute	Tumors
Faglia et al. 1983 [13]	39	3.5	6/39 (15%)	Microadenomas
Serri et al. 1983 [38]	24	6.2	12/24 (50%)	Microadenomas
Rodman et al. 1984 [32]	29	4.2	5/29 (17%)	Microadenomas
	5	3.4	1/5 (20%)	Macroadenomas
Fahlbusch et al. 1985 [14]	50	4.1	8/50 (16%)	Microadenomas
	15	4.1	2/15 (13%)	Macroadenomas
Fahlbusch et al. 1988 [17]	62	3.0	10/62 (16%)	Microadenomas

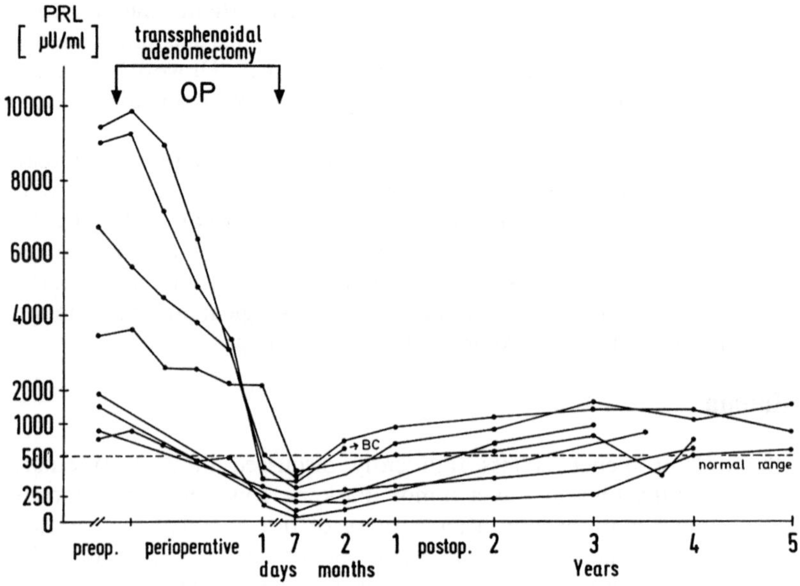

Fig. 6. Real recurrences in eight microprolactinomas, (From [14])

restoration of sexual functions. A reincrease in prolactin-levels, following initial normalization within 3 months of the surgical removal of prolactinomas, was observed in 2 out of 15 macroadenomas (1%), during a mean follow-up period of 4.1 years (Table 4). In microadenomas we [7] observed recurrences in 8 out of 50 patients (Fig. 6) and in a subsequent larger series [17] in 10 out of 62 patients (16%). In this later study, however, the follow-up period was only 3 years, thus not allowing definite conclusions to be drawn. While some other authors [13, 32] report over a similar observation period, Serri et al. [38] have followed up Hardy's patients for 6.2 years and found relapses in 12 out of 24 patients (50%).

In contrast to GH hypersecretion in acromegaly and hypercortisolism in Cushing's disease, hyperprolactinemia itself does not cause any serious morbidity. In no reported cases [7] have tumor residuals been documented by CT. In our series, only four out of eight patients with microprolactinomas causing recurrent hyperprolactinemia assessed for sexual function had recurrence of galactorrhea, and two were found to have anovulatory cycles. However, two patients became pregnant again without further treatment, and prolactin levels following delivery were found to be the same as before pregnancy (Fig. 4).

Medical treatment by dopamine agonists is advocated only in small tumors if symptomatic treatment is indicated. In macroadenomas persistent or recurrent hyperprolactinemia necessitates long-term therapy with dopamine agonists. In these cases the treatment is not only symptomatic but also has an antiproliferative effect and offers the chance of normoprolactinemia even after the withdrawal of drugs after years of treatment [14]. Irradiation is still mandatory in invasive adenomas, particularly if the dopamine agonists fail to restore normoprolactinemia.

TSH-Secreting Adenomas

The clinical and laboratory findings of 11 TSH-secreting adenomas are compiled in the report of Benoit et al. [5]. These lesions usually present with signs of hyperthyroidism in the presence of high or inappropriately suppressed TSH levels. Benoit et al. suggest that they be treated by surgery and irradiation. However, none of the three patients, in whom we selectively removed TSH-secreting adenomas and who showed endocrinological remission, exhibited recurrence of TSH-hypersecretion during a follow-up period of 9, 6, and 3 years, respectively.

Hormonally Inactive Adenomas

In contrast to the hormone-secreting adenomas, there is no measurable tumor marker in the serum to document total resection of a nonsecretory pituitary adenoma. Prior to the introduction of prolactin assay, those tumors not associated with clinical signs of acromegaly or hypercortisolism were usually grouped together with actual prolactin-, gonadotropin- or α-subunit secreting tumors and called "chromophobe" adenomas. In using data from historical series, one must be aware of the fact that as many as 30%–40% of these so-called "inactive" adenomas were in fact prolactinomas. In the historical series prior to the CT era, when the tumors were operated on without the use of microsurgical techniques [23, 27, 47], preexisting visual symptoms recurred or new ones developed. The diagnosis of such a "clinical" recurrence was not based on objective documentation of total tumor removal. Nowadays a normal postoperative CT scan and the CT documentation of tumor regrowth are prerequisites for classifying such a case as a "real recurrence" of a hormone inactive pituitary adenoma. However, CT scanning has now been available for only about 15 years, and it is well known that most of these tumors grow slowly. Therefore, the current literature still refers to data from the historical series, where total tumor removal was hardly ever accomplished and is not well documented. Today, much lower rates of tumor recurrence are to be expected following transsphenoidal surgery utilizing microsurgical techniques. This was demonstrated by Ciric et al. [10] who describe their experience with the transsphenoidal approach in 108 cases with tumors of at least 20 mm in diameter. Although they compiled hormone-secreting and nonsecreting tumors, their results may be used as a representative series of modern transsphenoidal surgery in large adenomas. The postoperative CT scan showed no evidence of residual tumor in 57 out of 97 patients. Furthermore, they report

Table 5. Recurrences after surgical and radiation therapy of "nonfunctioning" pituitary adenomas (a review of selected literature)

Authors	Number of patients	Recurrence rate		Follow-up period (years),	
		Without radiotherapy	With radiotherapy	mean	(range)
Henderson 1939 [23]	205	56%	13%	5	
Ray and Patterson 1971 [31]	146	22%	8%	n.s.	
MacCarty et al. 1973 [27]	96	31.8%	3.0%	>5	
Wirth et al. 1974 [47]	157	25.8%	11.7%	5.3	
Salmi et al. 1982 [34]	56	–	20%	5	(0.5–6)
Ciric et al. 1983 [10]	99	28%	6%	?	(0.5–14)

an overall recurrence rate of 12.7% among 102 patients followed up for 0.5–14 years. While in 32 patients who did not receive irradiation 9 recurred, only 4 out of 67 patients who were postoperatively irradiated eventually developed a recurrence. While we documented residual tumor after initial surgery (requiring further therapy) in 11% of our patients operated on for hormonally inactive pituitary adenomas between 1983 and 1988, the recurrence rate as documented by repeated CT and MRI investigations hardly exceeds 5%. However, the follow-up interval is too short to draw definitive conclusions. Postoperative radiotherapy had already played an important role in the management of Cushing's own patients, as pointed out by Henderson's follow-up study in 1939 [23]. He significantly reduced the rate of symptomatic recurrences, as depicted in Table 5. In 1978, Guiot [21], in particular, drew attention to the value of postoperative radiotherapy and suggested that "every time there are doubts regarding the complete removal of a tumor, the patient should be submitted to postoperative cobalt therapy." Clinging strictly to this policy, he claimed to have a recurrence rate of below 1% among 813 operated cases.

Summary

In this survey the incidence of recurrences in pituitary adenomas was described for hormone-secreting and for hormonally inactive tumors. The biological significance of these recurrences and possible causative factors were discussed. The therapeutic strategies vary considerably among the various subtypes of pituitary adenomas and are summarized in Table 6. These differ little from the

Table 6. Treatment of recurrences after pituitary adenoma surgery

	Endocrinological type (hypersecretion)			
	"Inactive"	GH	PRL	ACTH
Observation	–	–	X	–
Dopamine agonists	–	(X)	XX	–
Somatostatin analogues	–	X	–	–
Radiotherapy	X	X	–	–
Reoperation				
Transcranial	X	–	–	–
	or			
Transsphenoidal	X	XX	–	(X)

XX, very successful; X, successful; (X), helpful. Treatment of ACTH-secreting pituitary adenomas may also include adrenolytic therapy and bilateral adrenalectomy.

management of persistent pituitary adenomas after initial operative therapy. Surgical reinterventions and radiotherapy still play a major role. While dopamine agonists have proven to be extremely effective in the treatment of prolactinomas, they are less helpful in acromegaly. Here somatostatin analogues are expected to be of increasing importance. However, today reoperation, particularly in small adenomas and radiotherapy, is considered the treatment of choice. Since transsphenoidal reoperations for relapses in Cushing's disease do not produce comparably favorable results, bilateral adrenalectomy is in many cases performed to combat hypercortisolism [15]. Asymptomatic recurrent hyperprolactinemia in microprolactinomas does not necessarily require reoperation but observation [7].

Since the value of postoperative dynamic endocrine testing up to now is unfortunately limited to confirming the current hormonal secretory state and to assessing the present activity of the disease, but does not help to indicate recurrence of such a tumor prospectively, and since conventional and immunohistochemistry fail to contribute information in this respect, only further developments may reveal the tendency of an adenoma to recur. Investigations that assess the proliferation activity of an adenoma, such as flow cytoflourometry [18] or immunostaining for Ki-67 [9], may lead to some progress in this field.

References

1. Arafah BM, Rosenzweig JL, Fenstermaker R, Salazar R, McBride CE, Selman W (1987) Value of growth hormone dynamics and somatomedin C (insulin-like growth factor I) levels in predicting the long-term benefit after transsphenoidal surgery for acromegaly. J Lab Clin Med 109:346–354
2. Arisaka O, Hall R, Hughes IA (1983) Spontaneous endocrine cure of gigantism due to pituitary apoplexy. Br Med J 287:1007–1008
3. Arosio M, Giovanelli MA, Riva E, Nava C, Ambrosi B, Faglia G (1983) Clinical use of pre- and postsurgical evaluation of abnormal GH responses in acromegaly. J Neurosurg 59:402–408

4. Artia N, Mori S, Saitoh Y, Nagatani M, Ushio Y, Hyakawa T, Mogami H (1988) Transsphenoidal surgery for acromegaly – follow-up results. In: Landolt AM (ed) Advances in pituitary adenoma research. Pergamon Press, London, pp 265–266
5. Benoit R, Pearson-Murphy BE, Robert F (1980) Hyperthyroidism due to a pituitary TSH secreting tumour with amenorrhea and galactorrhea. Clin Endocrinol (Oxford) 12:11–19
6. Boggan JE, Tyrrell JB, Wilson CB (1983) Transsphenoidal microsurgical management of Cushing's disease. Report of 100 cases. J Neurosurg 59:195–200
7. Buchfelder M, Lierheimer A, Schrell U, Von Werder K, Fahlbusch R (1985) Long-term follow-up of surgically normalized microprolactinomas. In: Auer LM, Leb G, Tscherne G, Urdl W, Walter GF (eds) Prolactinomas – an interdisciplinary approach. de Gruyter, Berlin New York, pp 183–187
8. Buchfelder M, Brockmeier S, Schrell U, Pichl J, Fahlbusch R (1988) Long-term results of transsphenoidal selective adenomectomy in patients with acromegaly. Acta Endocrinol 117 [Suppl 287]:106
9. Burger PC, Shibata T, Kleihues P (1986) The use of the monoclonal antibody Ki-67 in the identification of proliferating cells: application to surgical neuropathology. Am J Surg Pathol 10:611–617
10. Ciric I, Mikhael M, Stafford T, Lawson L, Garces R (1983) Transsphenoidal microsurgery of pituitary macroadenomas with long-term follow-up results. J Neurosurg 59:395–401
11. Derome PJ, Delalande O, Visot A, Jedynac CP, Dupuy M (1988) Short and long term results after transsphenoidal surgery for Cushing's disease. In: Landolt AM (ed) Advances in pituitary adenoma research. Pergamon, London, pp 375–379
12. Eversmann T, Lüdecke U, Fahlbusch R, von Werder K (1986) TRH-stimulierte Wachstumshormonsekretion bei Akromegalie. Deutsch Med Wochenschr 111:1091–1096
13. Faglia G, Moriondo P, Travaglini P, Giovanelli MA (1983) Influence of previous bromocriptine therapy on surgery for microprolactinoma. Lancet I:133–134
14. Fahlbusch R, Buchfelder M (1985) Present status of neurosurgery in the treatment of prolactinomas. Neurosurg Rev 8:195–205
15. Fahlbusch R, Buchfelder M, Müller OA (1986) Transsphenoidal surgery for Cushing's disease. J Roy Soc Med 79:262–269
16. Fahlbusch R, Buchfelder M (1988) Transsphenoidal surgery of parasellar pituitary adenomas. Acta Neurochir 92:93–99
17. Fahlbusch R, Buchfelder M, Schrell U, Nistor R (1988) Recurrences in hormonally active pituitary adenomas. In: Shizume K, Shibasaki T (eds) Recent issues on pituitary adenomas. Sandoz Japan, Tokio, pp 20–28
18. Fitzgibbons PL, Appley AJ, Turner RR, Bishop PC, Parker JW, Breeze RE, Weiss MH, Apuzzo MLJ (1988) Flow cytometric analysis of pituitary tumors. Cancer 62:1556–1560
19. Giovanelli MA, Gaini SM, Tomei G, Motti EDF, Villiani R (1980) Acromegaly: surgical failures and recurrences. In: Derome PJ, Jedynac CP, Peillon F (eds) Pituitary adenomas. Biology, physiopathology and treatment. Asclepios France, Paris
20. Grisoli F, Leclerq T, Jaquet P, Guibot M, Winteler JP, Hassoun J, Vincentelli F (1985) Transsphenoidal surgery for acromegaly – long-term results in 100 patients. Surg Neurol 23:513–519
21. Guiot G (1978) Considerations on the surgical treatment of pituitary adenomas. In: Fahlbusch R, von Werder K (eds) Treatment of pituitary adenomas. Thieme, Stuttgart, pp 202–218
22. Hardy J (1982) Cushing's disease: 50 years later. Can J Neurol Sci 9:375–380
23. Henderson WR (1939) The pituitary adenomata: a follow-up study of surgical results in 338 cases (Dr. Harvey Cushing's series). Br J Surg 26:811–921
24. Kageyama N, Kuwayama A, Takahashi T, Ichihara K, Kato T, Yokoe T (1984) Microsurgical results in 90 cases of Cushing's disease. In: Lamberts SWJ, Tilders FJH, Van der Veen EA, Assies J (eds) Trends in diagnosis and treatment of pituitary adenomas. Free University Press, Amsterdam, pp 325–333

25. Karashima T, Kato K, Nawata H, Ikuyama S, Ibayashi H, Nakagaki H, Kitamura K (1986) Postoperative plasma GH levels and restoration of GH dynamics in acromegalic patients surgically treated by the transsphenoidal approach. Clin Endocrinol 25:157–163
26. Lüdecke DK (1985) Recent developments in the treatment of acromegaly. Neurosurg Rev 8:167–173
27. MacCarty CS, Hanson EJ jr, Randall RV, Scanlon PW (1973) Indications for and results of surgical treatment of pituitary tumors by the transfrontal approach. In: Kohler PO, Ross GT (eds) Diagnosis and treatment of pituitary tumors. Excerpta Medica/American Elsevier, New York, pp 139–145
28. Mampalam TJ, Tyrell JB, Wilson CB (1988) Transsphenoidal microsurgery for Cushing's disease. Ann Int Med 109:487–493
29. Nakane T, Kuwayama A, Watanabe M, Takahashi T, Kato T, Ichihara K, Kageyama N (1987) Long term results of transsphenoidal adenomectomy in patients with Cushing's disease. Neurosurgery 21:218–222
30. Quabbe HJ (1978) Endocrinology of growth hormone producing tumors. In: Fahlbusch R, Werder K von (eds) Treatment of pituitary adenomas. Thieme, Stuttgart, pp 47–60
31. Ray BS, Patterson RH (1971) Surgical experience with chromophobe adenomas of the pituitary gland. J Neurosurg 34:726–729
32. Rodman EF, Molitch ME, Post KD, Biller BJ, Reichlin S (1984) Long-term follow-up of transsphenoidal selective adenomectomy for prolactinoma. JAMA 252:921–924
33. Roelfsema F, van Dulken H, Frölich M (1985) Long-term results of transsphenoidal pituitary microsurgery in 60 acromegalic patients. Clin Endocrinol 23:555–565
34. Salmi J, Grahne B, Valtonen S, Pelkonen R (1982) Recurrence of chromophobe adenomas after operation and postoperative radiotherapy. Acta Neurol Scand 66:681–689
35. Scherbaum UA, Schrell U, Glück M, Fahlbusch R, Pfeiffer EF (1987) Autoantibodies to pituitary corticotropin cells: possible markers for unfavourable outcome after pituitary microsurgery for Cushing's disease. Lancet 1:1394–1398
36. Schrell U, Fahlbusch R, Buchfelder M, Riedl S, Stalla GK, Müller OA (1987) Corticotropin-releasing hormone stimulation test before and after transsphenoidal selective adenomectomy in 30 patients with Cushing's disease. J Clin Endocrinol Metab 64:1150–1159
37. Selman WR, Laws ER, Scheithauer BW, Carpenter SM (1986) The occurrence of dural invasion in pituitary adenomas. J Neurosurg 64:402–407
38. Serri O, Rasio E, Beauregard H, Hardy J, Somma M (1983) Recurrence of hyperprolactinemia after selective transsphenoidal adenomectomy in women with prolactinoma. N Engl J Med 309:280–283
39. Serri O, Somma M, Comtois R, Rasio E, Beauregard H, Jilwan N, Hardy J (1985) Acromegaly: biochemical assessment of cure after long term follow-up of transsphenoidal selective adenomectomy. J Clin Endocrinol Metab 61:1185–1189
40. Shaffi OM, Wrightson P (1975) Dural invasion by pituitary adenomas. N Z Med J 81:386–390
41. Stalla GK, Stalla J, Von Werder K, Müller OA, Lüdecke DK, Schrell U, Fahlbusch R (1987) Corticotropin releasing hormone in plasma of patients with Cushing's disease. Klin Wochenschr 65:529
42. Tindall GT, McLanahan CS, Christy JH (1978) Transsphenoidal microsurgery for pituitary adenomas associated with hyperprolactinemia. J Neurosurg 48:849–860
43. Thorner MO, Perryman RL, Cronin MJ, Rogol AD, Draznin M, Johanson A, Vale W, Horvath E, Kovacs K (1982) Somatotroph hyperplasia: successful treatment of acromegaly by removal of a pancreatic islet tumor secreting growth hormone releasing factor. J Clin Invest 70:965–968
44. Thorner MO, Frohman LA, Leong DA, Thominet J, Downs T, Hellmann P, Chitwood J, Vaufhan JM, Vale W, Besser GM, Lytras M, Edwards CRW, Schaaf M, Gelato M, Krieger DT, Marcovitz S, Ituarte E, Boyd AE, Malarkey WB, Blackard WG, Prioleau G, Melmed S, Charest NJ (1984) Extrahypothalamic growth-hormone releasing

factor (GRF) secretion is a rare cause of acromegaly: plasma GRF levels in 177 acromegalic patients. J Clin Endocrinol Metab 59:846–849
45. Tolis G, Yotis A, del Pozo E, Pitoulis S (1986) Therapeutic efficacy of a somatostatin analogue (SMS 201-995) in active acromegaly. J Neurosurg 65:37–40
46. Von Werder K, Losa M, Müller OA, Schweiberer L, Fahlbusch R, Del Pozo E (1984) Treatment of metastasizing GRF-producing tumor with a long acting somatostatin analogue. Lancet 2:282–283
47. Wirth FP, Schwartz HG, Schwetschenau PR (1974) Pituitary adenomas: factors in treatment. Clin Neurosurg 21:8–25

A Clinical, Endocrinological, and Morphological Study of Pituitary Tumor Recurrence

F. Rauhut, H.-E. Clar, H.-E. Nau (†), L. Gerhard, and M. Bamberg,
Essen and Ingolstadt/FRG

Introduction

Depending on the surgical approach, involving either total or subtotal extirpation of pituitary tumors, the recurrence rates have varied from 10.4% (intra- and suprasellar tumors without parasellar enlargement) to 35% (large tumors with parasellar configuration) in patients examined by us (Rauhut et al. 1986). In the literature there are many reports of a high risk of tumor recurrence in cases of a primary supra- and parasellar tumor configuration with infiltration of the neighboring structures (Fahlbusch et al. 1978; Guiot 1978; Wilson 1984).

Patients and Methods

In the present study 74 patients, 33 of whom suffered from a tumor relapse, received combined surgery and radiotherapy. Local irradiation with two parallel opposed fields using a 5.7 MeV linear accelerator was performed infiltration of the neighboring structures (Fahlbusch et al. 1978; Guiot 1978; Wilson 1984).

Results

At present the median follow-up time after combined surgery and radiotherapy (74 cases) is 62.4 months. Six patients have died, but only one has died directly due to the pituitary tumor (Fig. 1). Thirty-three patients suffered from a tumor relapse, which developed after an average of 38 months. After combined treatment the follow-up time varied from 18 to 162 months; the median follow-up time was 61.5 months. Most importantly, the interval between first presentation and tumor recurrence operation was shorter in many individual patients than the tumor-free interval of the present follow-up after reoperation with subsequent radiotherapy (Fig. 2). The most frequent tumor relapse developed in 45% of cases in the first 2 years. Subsequently, the recurrence rate decreased significantly (Fig. 3).

An endocrinological investigation of 34 patients with subtotally extirpated hypophysomas showed a pituitary deficiency in 35%–40% of cases at

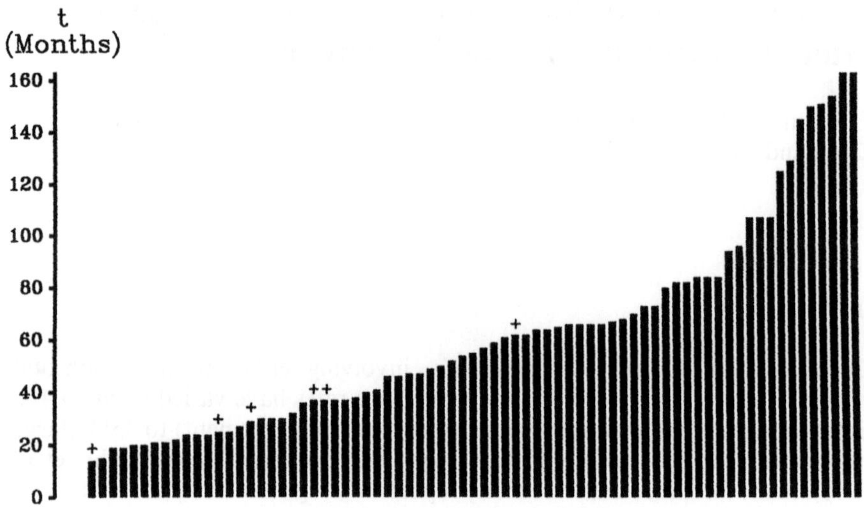

Fig. 1. Follow-up times after combined surgery and radiotherapy ($n = 74$)

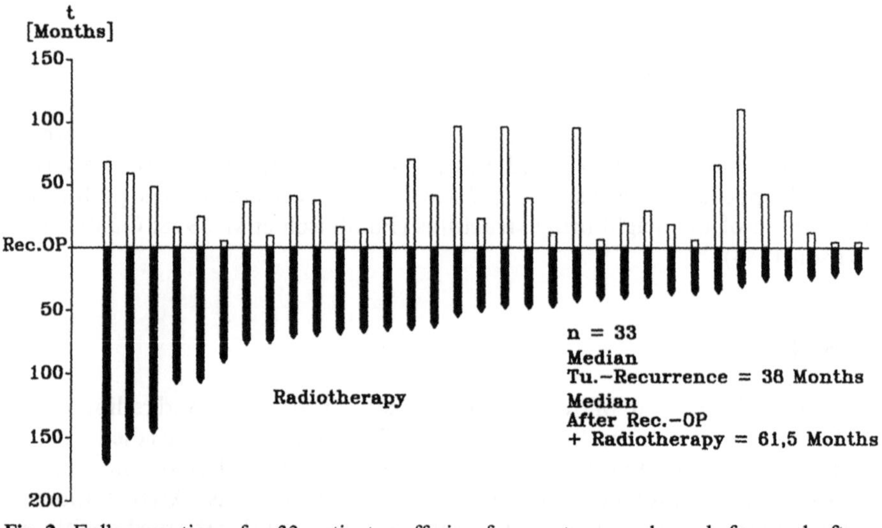

Fig. 2. Follow-up times for 33 patients suffering from a tumor relapse before and after combined surgery and radiotherapy

diagnosis. The disturbance to the gonadal axis and posterior pituitary increased postoperatively in 12% of cases. Subsequent radiotherapy produced additional gonadal deficiency in four cases to the same degree. When recurrence was diagnosed, the disturbance to adrenal, thyroid, and gonadal axes varied from 57% to 71%. However, this high rate of pituitary deficiency did not decrease significantly after combined surgery and radiotherapy. After

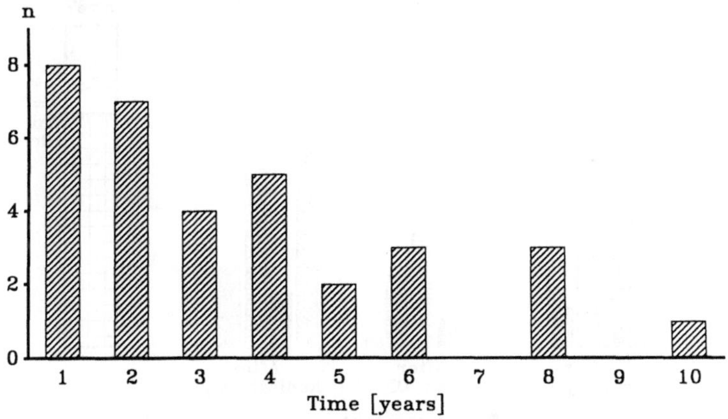

Fig. 3. Frequency of tumor recurrence over time following primary operation ($n = 33$). The main regrowth (45%) developed in the first 2 years

reoperation, only the posterior lobe showed a 9.4% increase in vasopressin insufficiency, albeit in three cases.

Prolactinomas were diagnosed in 11 cases after combined surgery and radiotherapy. Here it should be stressed that the prolactin (PRL) serum level decreased in most cases and was only seldom normal, and this level decreased over a few years. In the acromegalic group the human growth hormone (HGH) serum level was distinctly lowered to 10 ng. However, in these patients hormone secretion in some cases decreased 2–4 years after radiotherapy. Finally, no bromocriptine therapy was given. At the time of tumor recurrence progressive loss of vision was the main symptom. Clinical symptoms and computed tomography (CT) findings were generally found to correlate. Loss of vision could be demonstrated earlier than in CT scans in 20% of cases only. Vision and visual field in 67 patients were investigated. The extent of visual field defects after the single therapy steps is shown in Fig. 4. It seems that visual impairment was greater in the tumor recurrence group pre- and postoperatively than in that involving primary hypophysomas.

According to the guidelines of the German Ophthalmological Society, vision was graded before and after combined treatment. In 43 cases (66%) there was no or slight loss of vision, and 24 patients (34%) showed further medium or severe loss of vision after surgery and radiotherapy. It is of interest that after primary operation and radiotherapy, 60% of the patients had no or stable visual acuity. Visual defects improved in 31.4% and worsened in 8.6% of cases. After recurrence operation and radiotherapy, 42% of patients had no or stable visual disturbance. Vision improved in 16.7% and worsened in 30.5% of cases. The visual changes were due only to the surgical procedure; radiotherapy had no marked effect on visual changes.

The results of visual evoked potential studies in 40 patients showed no reliable changes which could allow a follow-up and give an indication as to

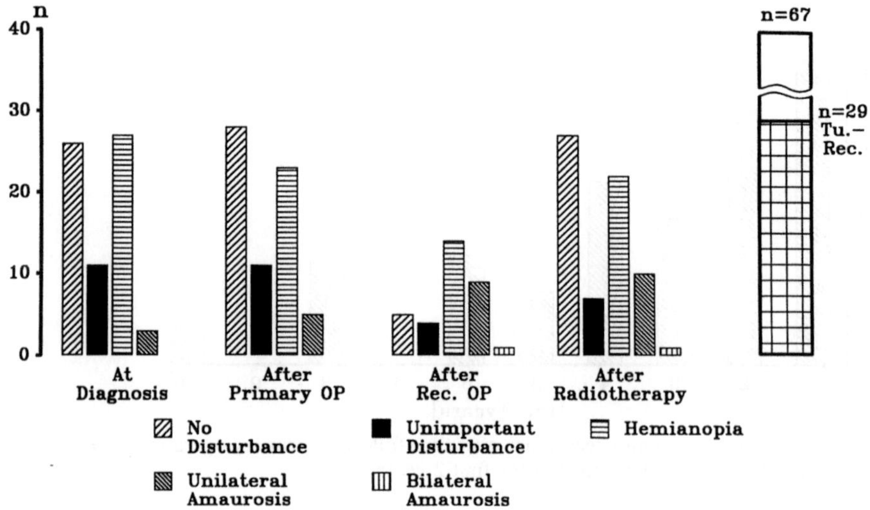

Fig. 4. Visual field defects after the single therapy steps

therapy (Nau et al., this volume). The morphological investigation of primary hypophysomas and pituitary tumor recurrence showed no difference in cytology and architecture. In particular, there were no signs of progressive malignancy in the tumor periphery of the relapsed adenomas although an invasion of granular tissue with pathological vascularization and calcification was observed. The practical value of the immunohistochemical examination is the exact differentiation of tumor and normal hypophyseal tissue, as well as a precise endocrinological gradation of adenomas with a small endocrine active cell portion.

Conclusions

The combined modality of surgery and radiotherapy can decrease the frequency of tumor relapse and reduce morbidity in the treatment of hypophysomas. Sheline and Tyrrell (1983) reported a 5-year determinate recurrence rate of 4% for resection plus radiation therapy. The endocrine pituitary deficiency increased slightly after additional radiotherapy. In the literature the incidence of radiation-induced hypopituitarism varies from 10% to 38% (Tindall and Barrow 1986). Serum hormone levels of prolactinomas and HGH-producing tumors decrease after 2–4 years. Hormone follow-up for up to 90 months after radiotherapy resulted only in a 50% normalization of the PRL-serum level, but HGH generally decreased to below 10 ng (Bamberg et al. 1988).Up to now in our patients radiotherapy has shown no negative effect on the optic nerve. When a single dose of 1.5–1.8 Gy and a total dose of 45–50 Gy were not exceeded, a radiogenic chiasma or optic nerve damage developed in only 1% of cases (Harris and Levene 1976). The effects of irradiation on pituitary tumors seem to be due to vascular obliteration and/or a biochemical and physiological alteration in the permeability of the cell membrane (Anniko 1981).

The main symptom of tumor recurrence appears to be progressive loss of vision. Studies of visual potentials are of no significance on postoperative follow-up.

In accordance with reports in the literature (Jänisch et al. 1988; Tindall and Barrow 1986), our morphological studies of pituitary tumor recurrences do not result in progressive malignancy in cytology or tumor architecture. An invasion of granular tissue with pathological vascularization in the hypophysoma periphery was remarkable, which could explain the higher vulnerability of endocrinological and visual structures at reoperation.

References

Anniko M (1981) Early morphological changes following gamma irradiation: a comparison of human pituitary tumours and human acoustic neuromas (schwannomas). Acta Pathol Microbiol Scand 89:113

Bamberg M, Langrock J, Rauhut F, Hoederath A, Sack H (1988) Indikationsstellung und Ergebnisse bei der Radiotherapie von Hyophysenadenomen. In: Bamberg M, Sack H (Hrsg) Therapie primärer Hirntumoren. Zuckschwerdt, München Bern Wien San Francisco, S 139–146

Fahlbusch R, Rjosk HK, Werder KV (1978) Operative treatment of prolactin-producing adenomas. In: Fahlbusch R, Werder KV (eds) Treatment of pituitary adenomas. Thieme, Stuttgart, pp 225–237

Guiot G (1978) Consideration on the Surgical Treatment of Pituitary Adenomas. In: Fahlbusch R, Werder KV (eds) Treatment of pituitary adenomas. Thieme, Stuttgart, pp 202–218

Harris JR, Levene MB (1976) Visual complication following irradiation for pituitary adenomas and craniopharyngeomas. Radiology 120:167–171

Jänisch W, Schreiber D, Güthert H (1988) Neuropathologie: Hirntumoren des Nervensystems. Fischer, Stuttgart, S 316–317

Rauhut F, Clar HE, Bamberg M, Benker G, Grote W (1986) Diagnostic criteria in pituitary tumour recurrence – combined modality of surgery and radiotherapy. Acta Neurochir 80:73–78

Sheline GE, Tyrrell B (1983) Pituitary Adenomas. In: Phillips TL, Pistennaa (eds) Radiation oncology annual. Raven, New York, pp 1–35

Tindall GT, Barrow DL (1986) Disorders of the pituitary. Mosby, St. Louis Toronto Princeton pp 410, 428–429

Wilson CB (1984) A decade of pituitary microsurgery. J Neurosurg 61:814–833

The Growth Rate in Pituitary Adenomas: Measurement by Proliferation Marker Ki 67

E. Knosp, K. Kitz, A. Perneczky, and A. Korn, Vienna/Austria, and Mainz/FRG

Conventional histological criteria such as cellular and nuclear polymorphism and rarely observed mitotic figures do not sufficiently explain the biological behavior of pituitary adenomas. Although pituitary adenomas are generally described as benign tumors, they show locally invasive growth which can be demonstrated surgically, radiologically, and histologically (Jefferson 1972; Martins et al. 1965; Landolt 1980; Selman et al. 1986; Scheithauer et al. 1986). Invasive growth has been related to distinct ultrastructural and secretory patterns (Kovacs 1984), to immunohistological findings (Scheithauer et al. 1986), and to tumor size (Selmann et al. 1986). Invasive growth influences the surgical outcome (Balagara et al. 1981; Wilson 1984) as well as the postoperative management of these patients in terms whether radiation therapy should be given or not (Ross and Wilson 1988).

None of these means, however, permit a definitive decision concerning growth characteristics. The mouse monoclonal antibody Ki 67 on the other hand recognizes a nuclear antigen associated with proliferating cells (Gerdes et al. 1983; 1984), and enables us to estimate the growth rate in pituitary adenomas. Conventional histological staining techniques enable us to demonstrate cells only during the mitotic phase, whereas the monoclonal antibody Ki 67 is capable of recognizing nuclei of cycling cells during the G1, S, G2, and mitotic phases. This fact increases the sensitivity for determining the proliferation rate of tumors. Unlike other methods, Ki 67 can be applied to fresh-frozen biopsy material and does not entail the ethical problems of preoperative drug administration with, for example, bromodeoxyuridine (Hoshino et al. 1986) or radioactive labeled thymidine (Johnson et al. 1960).

We measured the proliferation activity of tumor cells in 62 pituitary adenomas by means of the proliferation marker Ki 67 (Knosp et al. 1989). The immunohistological staining with Ki 67 was carried out on smear and dried cryostat sections of biopsy material (Fig. 1). Proliferation rate was determined by counting 1500–2000 adenoma cells and calculating the percentage of stained nuclei. Proliferation activity could be measured in all adenomas, ranging from 0.1% to 2.8% of cells, with an overall mean of 1.1% of proliferating cells. Table 1 demonstrates the results of proliferation rates in 62 pituitary adenomas in relation to hormonal activity. This represents an average distribution of adenomas operated on in our institution (Knosp et al. 1988). We could not demonstrate a statistically significant difference between hormonal

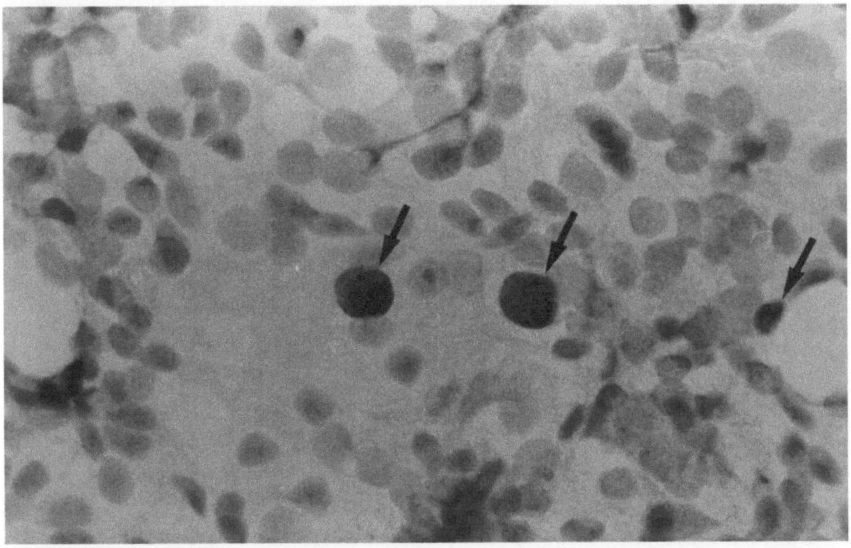

Fig. 1. Smear specimen of a pituitary adenoma. The immunohistological staining with Ki 67, demonstrating proliferation nuclei (*arrows*). (×750)

Table 1. Proliferation rate in 62 pituitary adenomas

	Proliferation	
	n	Rate
Nonfunctioning adenomas	28	0.98% (0.2%–2.8%)
PRL-secreting adenomas	17	
Men		1.39% (0.4%–2.8%)
Women		1.06% (0.3%–1.86%)
GH-secreting adenomas	12	1.11% (0.2%–2.7%)
ACTH-secreting adenomas	2	0.65% (0.1%, 1.2%)
Nelson's tumors	2	2.10% (1.6%, 2.6%)
LH-secreting adenomas	1	0.48%

PRL, Prolactin; GH, growth hormone; ACTH, adrenocorticotropic hormone; LH, luteinizing hormone

activity and proliferation rate reported by others (Landolt et al. 1987), and in our material there was no correlation between tumor size and proliferation rate.

In 27 cases of transsphenoidally operated pituitary adenomas the dura of the sella floor overlying the adenoma was investigated histologically. In two-thirds of these adenomas, invasive growth could be demonstrated, and statistically the proliferation activity was significantly higher ($p < 0.05$) as compared to the noninvasive group. The mean proliferation rate was 0.7% in the noninvasive group versus 1.4% in the invasive group (Fig. 2).

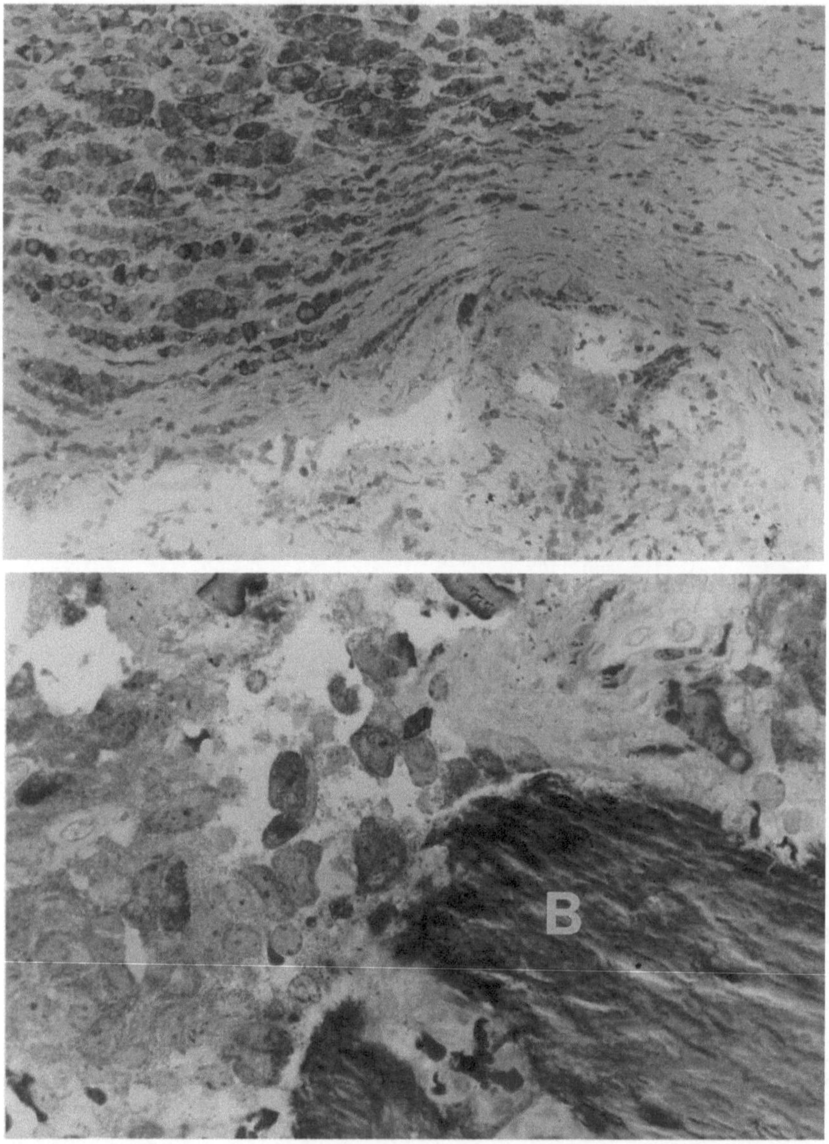

a

b

Fig. 2. Invasive growth into dura (**a**) and bone (**b**). **a** Semithin section demonstrating dura invasion by the adenoma. The dural layers are split by nests of adenoma cells (toloidine blue staining; × 300). **b** Invasiveness of a pituitary adenoma into the bone of the sphenoid: bone (*B*), adenoma cells with typical secretory granula; semithin section (toloidine blue staining; × 600)

In conclusion, we can observe the following:

- There is a wide range of growth rates among pituitary adenomas, from 0.1% to 2.8% of cells.
- There is no correlation between hormonal activity and growth rate.
- There is no correlation between the size of the adenoma and growth rate.
- There is a significantly higher growth rate in those adenomas in which dura invasiveness is demonstrated histologically.

Careful follow-up and examination of recurrent tumors will hopefully enable us to find those who would benefit from adjunctive radiotherapy. While in functioning adenomas cure rates or recurrences are documented by plasma hormonal access, in nonfunctioning adenomas only repeated computed tomography or magnetic resonance imaging investigations demonstrate tumor recurrences. Especially in this group of nonfunctioning adenomas the measurement of proliferation activity gives us a guideline in postoperative decision making.

References

Balagara S, Derome P, Guiot G (1981) Acromegaly: analysis of 132 cases treated surgically. Neurosurg 8:413–416

Cooper EH, Frank GL, Wright DH (1966) Cell proliferation in Burkitt tumors. Eur J Cancer 2:377–384

Gerdes J, Schwab U, Lemke H, Stein H (1983) Production of a mouse monoclonal antibody reactive with a human nuclear antigen associated with cell proliferation. Int J Cancer 31:13–20

Gerdes J, Lemke H, Baisch H, Wacker HH, Schwab U, Stein H (1984) Cell cycle analysis of a cell proliferation-associated human nuclear antigen defined by monoclonal antibody Ki 67. J Immunol 133:1710–1715

Hoshino T, Nagashima T, Murovic JA (1986) In situ cell kinetics studies on human neuroectodermal tumors with bromodeoxyuridine labeling. J Neurosurg 64:453–459

Jefferson G (1972) The invasive adenomas of the anterior pituitary. Thomas, Springfield, IL, pp 56–60

Johnson HA, Haymaker WE, Rubini JR (1960) A radioautographic study of a human brain and glioblastoma multiforme after in vivo uptake of tritiated thymidine. Cancer 13:636–642

Knosp E, Krisch K, Schmidbauer M, Budka H (1988) Immunologischer Hormonnachweis bei Hypophysenadenomen: Korrelation von Serumhormonbefunden mit immunzytochemischen Hormonbefund am Tumorschnitt. Wien Klin Wochenschr 100:322–325

Knosp E, Kitz K, Perneczky A (1989) Proliferation activity in pituitary adenomas: measurement by monoclonal antibody Ki-67. Neurosurg 25:927–930

Kovacs K (1984) Light and electron microscopic pathology of pituitary tumors: immunhistochemistry. In: McL Black P, Zervas NT, Ridgeway EC, Martin JB (eds). Secretory tumors of the pituitary gland. Raven, New York, pp 365–375

Landolt AM (1980) Biology of pituitary microadenomas. In: Faglia G, Giovanelli MA, Mac Leod RM (eds). Pituitary microadenomas. Academic Press London, pp 107–122

Landolt AM, Shibata T, Kleihues P (1987) Growth rate of human pituitary adenomas. J Neurosurg 67:803–806

Martins AN, Hayes GJ, Kempe LG (1965) Invasive pituitary adenomas. J Neurosurg 22:268–276

Ross DA, Wilson CB (1988) Results of transsphenoidal microsurgery for growth hormone-secreting pituitary adenomas in a series of 214 patients. J Neurosurg 68:854–867

Selman WR, Laws ER, Scheithauer BW, Carpenter SM (1986) The occurrence of dural invasion in pituitary adenomas. J Neurosurg 64:402–408

Scheithauer BW, Kovacs KT, Laws ER, Randall RV (1986) Pathology of invasive pituitary tumors with special reference to functional classification. J Neurosurg 65:733–745

Schloffer H (1907) Erfolgreiche Operation eines Hypophysentumors auf nasalem Wege. Wien Klin Wochenschr 20:621–624

Wilson CHB (1984) A decade of pituitary microsurgery. The Herbert Olivecrona lecture. J Neurosurg 61:814–834

Therapeutic Considerations in Pituitary Tumor Recurrence

H.-E. CLAR, Ingolstadt/FRG and F. RAUHUT, Essen/FRG

Transcranial and transsphenoidal approaches in pituitary tumors mean incomplete excisions in many cases. Depending upon the operative technique, the following regions are difficult to reach by transsphenoidal approach: suprasellar hypothalamic, lateral chiasmatic, retrosellar, parasellar, and cavernosal (Fig. 1). The corresponding regions with regard to the transcranial approach are: intrasellar (lateral and frontal), cavernosal, and subsellar (sphenoidal) (Fig. 2). More often than surgeons realize, tumor residuals remain after so-called total excision with either approach. In some cases only subtotal removal is possible due to local problems. This means tumor recurrence in sellar tumors in the range of 10–35 percent. The risk of regrowth is high in the

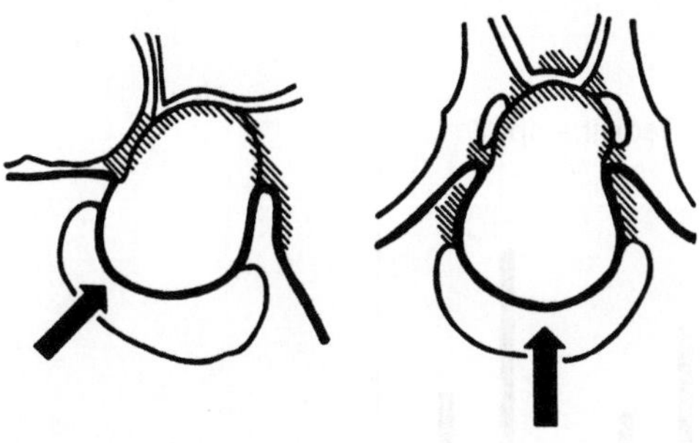

Fig. 1. The transsphenoidal approach, lateral and frontal view, with regions of possible incomplete tumor resection

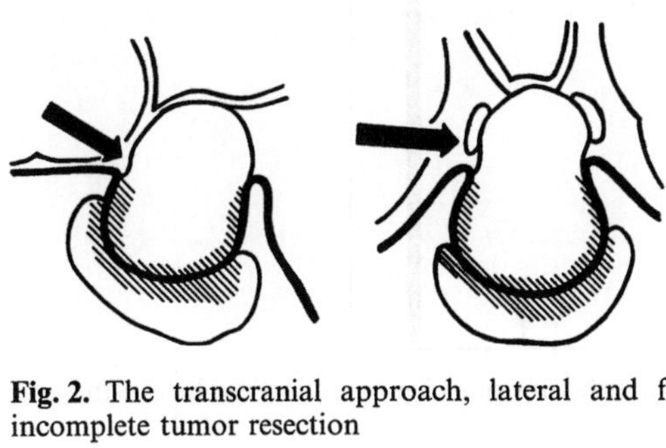

Fig. 2. The transcranial approach, lateral and frontal view, with regions of possible incomplete tumor resection

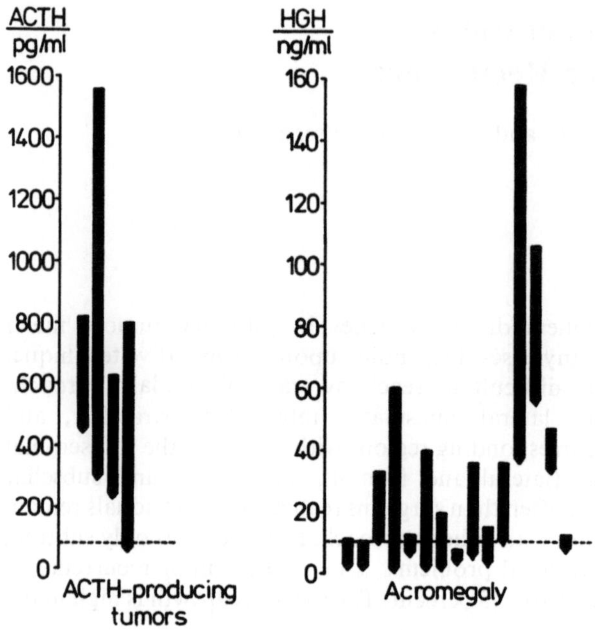

Fig. 3. Decrease of hormone levels in hormone-active tumors, before and after operation

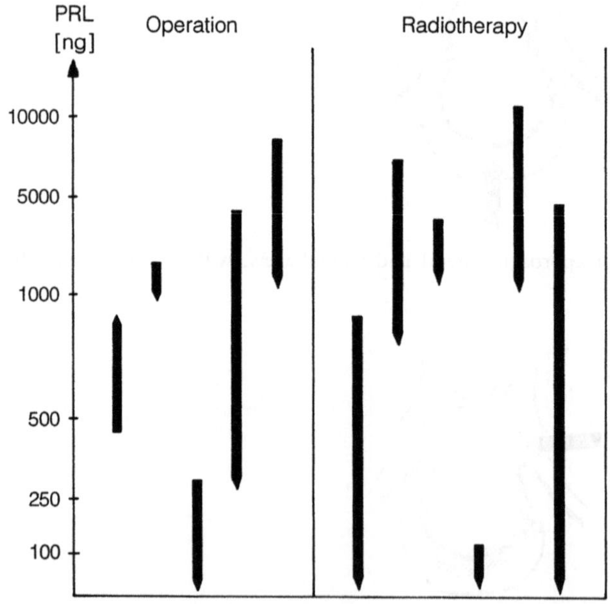

Fig. 4. Prolactin levels decreasing in patients before and after operation and before and after radiotherapy

1st year after operation and decreases in the following years. But late recurrence cases are observed even after a period of decades due to the slow growth rate of hypophyseal tumor cells (Burian 1976).

The recurrence rate of smaller tumors operated on mainly by the transsphenoidal approach was about 10% (10 out of 96) with an average time of 43 months; in larger tumors operated on transcranially, the recurrence rate was 20% (14 out of 64) after 78 months. In all cases with subtotal excision a regrowth rate of 33% was found (Rauhut et al. 1986, 1988).

Hypophyseal hormones serve as tumor markers and indicate the factor of active tumor mass. In cases with hormone secretion this can be demonstrated by the decrease in hypophyseal hormones after operation. In cases with incomplete reduction of hormones postoperatively residual tumor parts which are not removed are the only reason for high hormone levels (Hardy 1980). Recurrence or subtotal excision in hormone-active tumors is diagnosed by clinical symptoms, computed tomography, or magnetic resonance imaging. The different tumor forms require different activities in tumor recurrence. ACTH- and HGH-producing tumors require active methods since these hormones are vital factors in hypersecreting cases (Fig. 3). The value of radiation is confirmed by measuring prolactin levels before and after radiation, which shows a marked decrease in the tumor marker (Fig. 4).

Normalization of hormone secretion should be the final therapeutic result. Reoperation and radiation treatment should follow incomplete resection or increasing hormone activity even with missing mass lesion (Fig. 5). Since

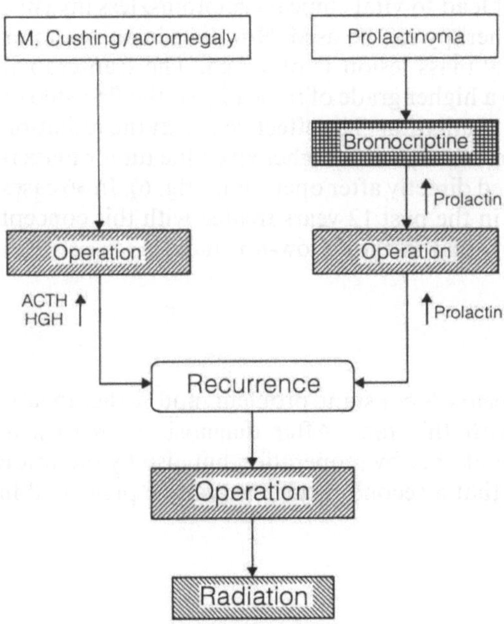

Fig. 5. Therapeutic concept in hormone-active tumors

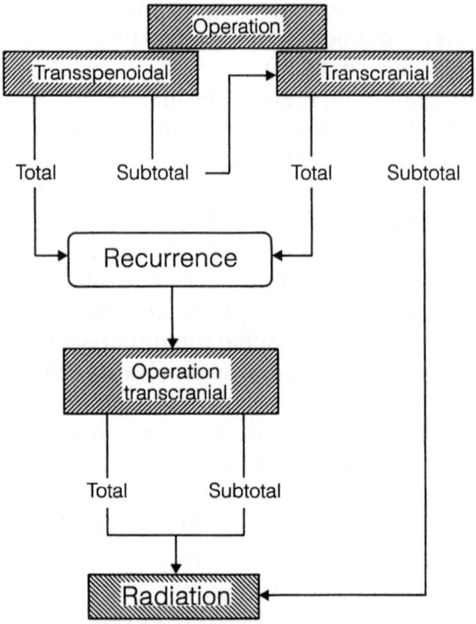

Fig. 6. Therapeutic concept in hormone-inactive tumors

prolactin hypersecretion does not lead to vital clinical symptoms, less invasive methods such as bromocriptine therapy can be used. Hormone inactive tumor recurrence is treated when a new mass lesion is observed. The transcranial approach (which seems to lead to a higher grade of radicality) is the first step of therapy. The second step is local irradiation. The effectiveness of the radiation depends upon the active tumor cell mass; this is higher when the tumor mass is low. Radiation is therefore planned directly after operation (Fig. 6). In 36 cases with tumor recurrence observed in the past 12 years treated with this concept no signs of regrowth have been found in the follow-up study.

Conclusion

Recurrence in sellar tumors appears a constant problem, and a therapeutic concept is necessary to deal with this fact. After diagnosis of regrowth, subsequent therapy is necessary not only by reoperation but also by radiation therapy. By this means it seems that a second recurrence can be prevented in most cases.

References

Burian K (1976) A two-stage neurorhinosurgical procedure in the treatment of pituitary tumors. In: Koos WT, Böck FW, Spetzler RF (eds) Clinical microneurosurgery. Thieme, Stuttgart, pp 15–22

Hardy J (1980) Discussion on the surgical results, therapeutic failures and recurrences of oversecreting pituitary tumors. In: Derome PJ, Jedynak CP, Peillon F (eds) Pituitary adenomas. Asclepios, Paris, pp 223–225

Rauhut F, Clar H-E, Bamberg M, Becker G, Grote W (1986) Diagnostic criteria in pituitary tumor recurrence – combined modality or surgery and radiotherapy. Acta Neurochir 80:73–78

Rauhut F, Clar H-E, Bamberg M, Becker G, Grote W (1988) Kombiniertes Vorgehen bei Makroadenomen der Hypophyse – Indikation und Therapieergebnisse. In: Bamberg M, Sack H (Hrsg) Therapie primärer Hirntumoren. Zuckschwerdt, München Bern Wien San Francisco, S 179–182

Aggressive Behavior and Changing Histology in a Pituitary Adenoma

J. J. A. Mooij, Groningen/The Netherlands

A 50-year-old woman presented in 1984 with several months' history of nausea, headache, vision loss of the right eye with exophthalmus and ptosis. Treatment for hypopituitarism had already been started; high levels of prolactin (>80000 mU/l) were found. Computed tomography (CT) showed a large tumor in the sella turcica, with extension into the skull base and the middle fossa (Fig. 1). Craniotomy was performed, and the tumor was removed incompletely, leaving tumor parts in the skull base. Although histology was not conclusive at first, with later staining it was confirmed to be a prolactin-secreting adenoma (antiprolactin staining positive, anti-ACTH staining negative; Fig. 2).

The patient made an uneventful recovery, and the tumor was well controlled with adjuvant radiotherapy (50 Gy) and bromocriptine medication.

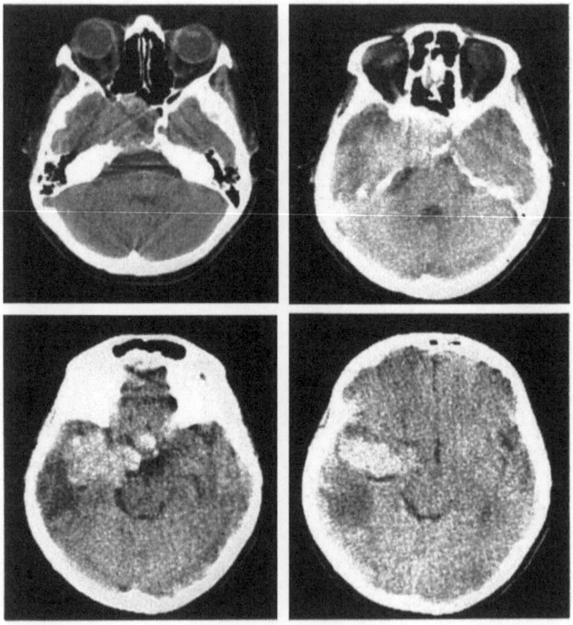

Fig. 1. CT scan, pituitary adenoma prior to treatment

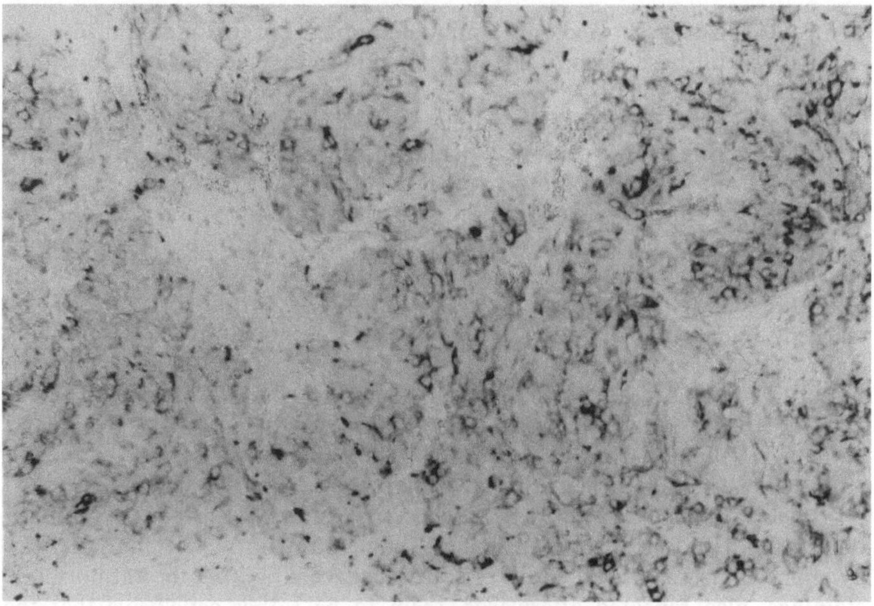

Fig. 2. Antiprolactin staining in a tumor sample, first operation (×140)

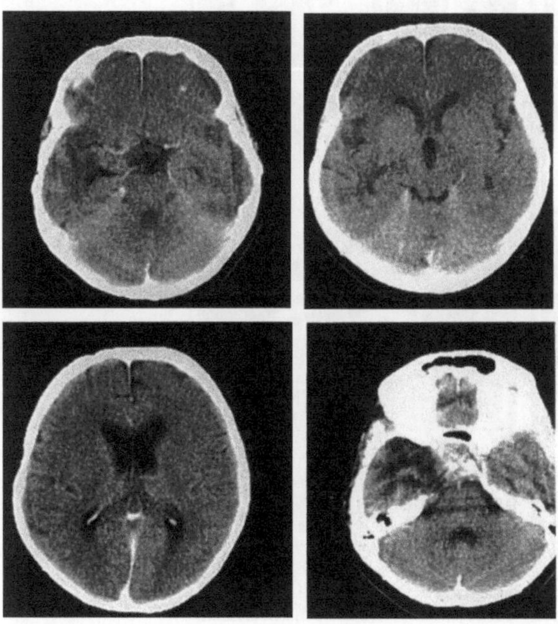

Fig. 3. CT scan, 2 years after surgery and radiotherapy, patient on bromocriptine

Control CT in 1986 showed only a minimal remnant of tumor confined to the sellar and skull base region (Fig. 3). Prolactin levels had declined to values under 2000 mU/l.

One year later, however, this patient developed new and serious signs and symptoms within a few weeks: gait disturbances, loss of vision, and a changing appearance, her face becoming swollen and puffy. CT scan showed a large recurrence of tumor with extension into the posterior fossa and possible invasion into the pons (Fig. 4). Hydrocephalus was prominent; laboratory investigations showed elevated levels of glucose and a low potassium concentration. The prolactin level was low (1221 mU/l), but the cortisol level was markedly elevated (2260 nmol/l). Diagnosis was conclusive for Cushing's disease. Treatment consisted of CSF shunting and partial tumor removal by surgery. The tumor was shown to be indeed invasive into the brain stem and to be hemorrhagic. Antiprolactin staining of tissue specimens was negative, and anti-ACTH staining was positive (Fig. 5). The patient died within a few weeks. At autopsy, the hemorrhagic invasive character of the tumor was evident.

Regarding this case a few remarks should be added. Recurrence of pituitary tumors is not seldom, but a change of histological and hormonal characteristics is very peculiar. Of course, one may doubt such a change, but the antihormone stainings, also repeated afterwards, seem to be rather conclusive that the

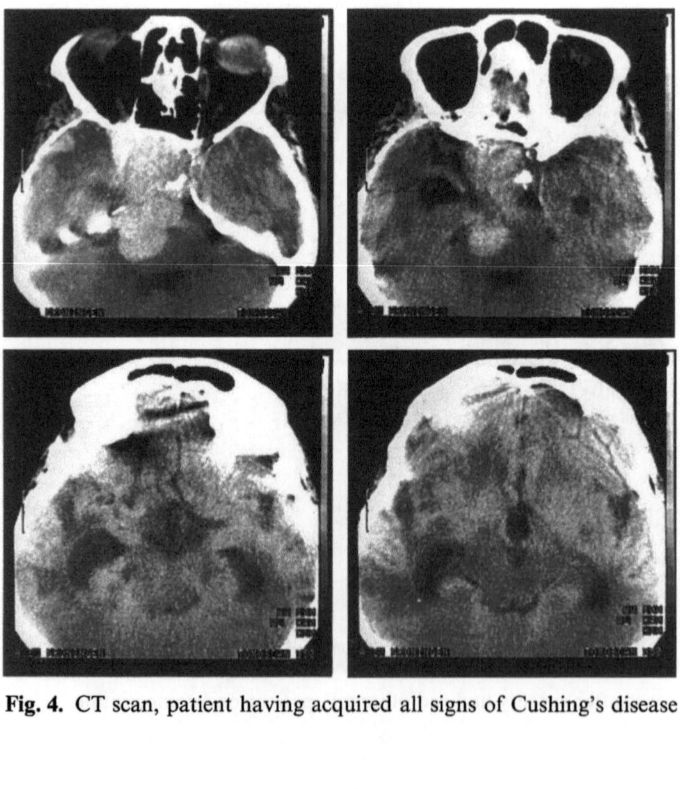

Fig. 4. CT scan, patient having acquired all signs of Cushing's disease

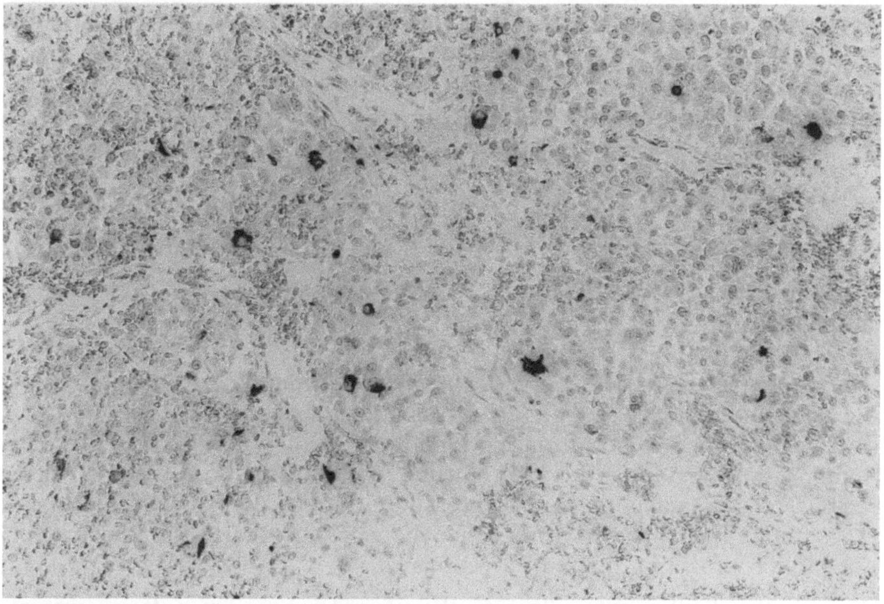

Fig. 5. Anti-ACTH staining in sample of recurrent tumor (×140)

regrowth was really of another cell type than were found at the first exploration. The aggressive behavior of the recurrent tumor, invading the brain stem, might be concomitant with malignancy rather than with a benign adenoma. However, in all specimens the tumor retained the characteristics of a benign adenoma. It might therefore be concluded that this tumor behaved differently than recurrences in general with respect to histology and aggressiveness.

Pituitary Adenoma with Cavernous Sinus Involvement

A. Basso, S. Berner, E. Caputi, and C. Yampolsky,
Buenos Aires/Argentina

Introduction

During the past few years, because of a better understanding of the anatomy of the sellar and parasellar regions and of the cavernous sinus, it has been possible to develop a series of microsurgical techniques that allow access to vascular and tumor lesions involving the anteromedial region of the skull base. Of these lesions, pituitary adenomas are the most important because of their frequency. These tumors may present with invasive characteristics that project them out of the pituitary fossa in different directions:

1. over the diaphragma sella into the suprasellar region, anterior cranial fossa, or posteriorly towards the clivus;
2. through the sellar floor into the sphenoidal sinus and nasal cavities;
3. laterally towards the cavernous sinus, either displacing or invading it and in this direction extending under the dura into temporal fossa or Meckel's cavum.

The clinical features of adenomas involving the cavernous sinus were first described by Foix (1922), who attributed severe pain in the region of the first trigeminal branch to a pituitary adenoma. The tumor was also associated with ipsilateral paralysis of the third, fourth, and sixth cranial nerves. Later, classical papers described these clinical signs and developed concepts that are still valid today in any discussion of pathological features of invading adenomas (Weinberger et al. 1940; Jefferson 1955).

Material and Methods

This paper presents our experience in the treatment of pituitary adenomas with lateral expansion and discusses the different microsurgical approaches and results obtained. Between 1983 and 1987, 226 adenomas were operated on by the same surgeon in the Neurosurgical Unit of the Hospital Santa Lucia. Twenty of these tumors (8.8%) showed at surgery invasion of the cavernous sinus. Ten were in men and ten were in women; the age range was 12–61 years.

All patients presented with fronto-orbital headache; six also complained of facial pain in the distribution of the ophthalmic nerve and 11 of occasional diplopia. In five cases there was paralysis of the oculomotor nerve; in two there

Table 1. Clinical findings at surgery

Case no.	Age (years), sex	Presenting symptoms	Cranial nerves disturbances	Hormonal levels
1	57, M	h, d, i	N	FSH, IH↓
2	60, F	h, d	↓va, ch s, III	PRL↑
3	30, F	h, a, g	↓va, ch s	PRL↑
4	56, F	h, d, tp	↓va, ch s, NV1	PRL↑
5	48, M	h, i	N	FSH, IH↑
6	44, F	h, a, g	↓va, ch s	N
7	61, F	h, d	↓va, ch s	GH↑
8	52, M	h, d, i, fp, hip	↓va, ch s	FSH, IH↓
9	36, M	h, i	↓va, ch s	PRL↑
10	30, M	h, d, fp	↓va, ch s, NV1, III	ACTH↑ Cortisol↑
11	17, M	h, d, i, g, hip	↓va, ch s, VI	PRL↑
12	41, F	h, a, fp	↓va, ch s	N
13	44, F	h, a, tp	↓va, ch s	N
14	12, F	h, d, a, tp, hip	↓va, ch s, NV1, III, VI	ACTH↑
15	32, F	h, d, a, tp	NV1, III	N
16	29, F	h, a, tp	↓va, ch s, NV1,2	ACTH↑
17	17, M	h, fp	↓va, ch s	PRL↑
18	37, M	h, d, fp	↓va, ch s	PRL↑
19	44, M	h, d, i, tp	↓va, ch s, NV1, III	FSH, IH↓
20	49, M	h, i, fp	↓va, ch s	FSH, IH↓

h, Headache (frontoorbital); d, diplopia; i, impotence; a, amenorrhea; g, galactorrhea; hip, high intracranial pressure; tp, trigeminal pain; fp, facial pain; NV1,2, numbness in first or second root trigeminal nerve; III, III nerve palsy; VI, VI nerve palsy; ↓va, visual acuity loss; ch s, chiasmal syndrome; N = normal

was paralysis of the abducens nerve. Five patients had hypoesthesia in the distribution of the ophthalmic nerve; one had hypoesthesia in the distribution of the ophthalmic and maxillary nerves (Table 1). Seventeen patients had diminished visual fields. Seven men complained of impotence. Six women suffered secondary amenorrhea. Galactorrhea was detected in three patients. Three patients showed evidence of raised intracranial pressure (Table 1).

All patients were studied by plain radiology, linear polytomography of the sella, computed tomography (CT) using axial and coronal sections with and without contrast and bilateral carotid angiography. Four cases were studied by magnetic resonance imaging (MRI). Fifteen patients belonged to grade 3 or 4 sellar type, according to Vezina and Sutton's classification. In all the cases angiography showed distortion of the cavernous carotid artery, and in three of the cases there was also a narrowing of the intracavernous portion of the carotid artery. In all cases CT scan allowed a diagnosis of sellar tumor with parasellar extension to be made (Fig. 1). Suprasellar extension was detected in 15 cases, retrosellar growth in two, and invasion of the sphenoidal sinus in five (Table 2).

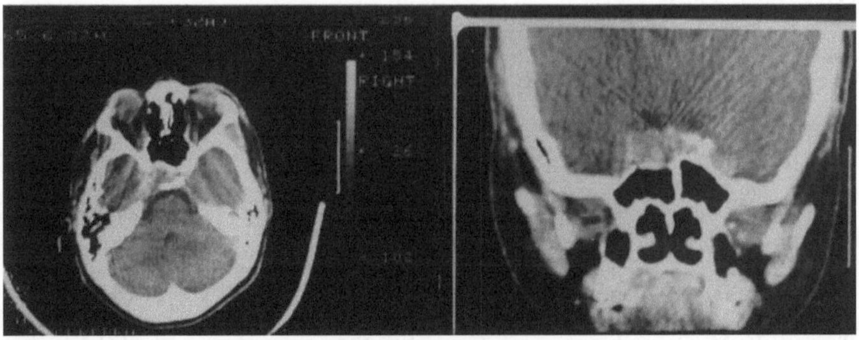

a b

Fig. 1. Preoperative CT scan (case 18) demonstrating the lateral extension of a pituitary tumor. Axial (*A*) and coronal (*B*) sections

Endocrine function was evaluated biochemically by testing for levels of PRL, GH, LH, TSH, thyroid hormones, ACTH, and cortisol levels. Seven patients had PRL values over 100 ng/ml. Three showed increased ACTH and cortisol levels. One patient had an extremely high level of GH. One patient had a marginally increased GH level. Five patients had normal hormone levels. Three patients had levels consistent with panhypopituitarism.

Fourteen patients were operated on through the transcranial approach and six through the transsphenoidal route. Six cases needed reoperations, four

Table 2. Tumor extension and surgical approach

Case no.	Tumor extension	Surgical approach
1	i, p, s	Tc
2	i, p, s	Tc
3	i, p, s	Tc
4	i, p, sph	Ts, Tc
5	i, p	Tc
6	i, p, s	Tc
7	i, p, s	Tc
8	i, p, r, s	Tc
9	i, p	Tc
10	i, p, s	Tc
11	i, p, r, s	Tc
12	i, p, s	Tc
13	i, p, s, sph	Ts, Tc
14	i, p, s	Ts, Tc
15	i, p, sph	Ts, Tc
16	i, p, s	Tc
17	i, p, s	Tc
18	i, p, s	Tc
19	i, p, s, sph	Ts, Ta + Ts
20	i, p, sph	Ts, Ta + Ts

i, Intrasellar; p, parasellar; r, retrosellar; s, suprasellar; sph, sphenoid; Tc, transcranial; Ts, transsphenoidal; Ta, transantral through a transcranial approach and two through a combined transantral transsphenoidal approach (Table 2).

Histological studies were carried out by light microscopy using conventional fixation methods. Immunohistochemistry studies were performed using the immunoperoxidase method.

Surgical Technique

From our surgical experience with over 1100 pituitary adenomas we favor a transcranial approach for the surgical treatment of tumors with lateral extension, leaving the transsphenoidal and the transantral routes for those cases in which there is clear invasion of the sphenoidal sinus and related cavities.

The selection of a surgical technique aims at obtaining maximum surgical effectiveness with a minimum of risk and sequelae for the patient. In selecting a particular technique it is necessary to understand the surgical anatomy in relation to the invading adenoma. The pituitary adenoma originates within the gland and is soon surrounded by a pseudocapsule formed by a palisade of parenchymal cells, a discrete stromal reaction, and the pituitary capsule. When the adenoma extends beyond the sella, the direction of the extension is a result of the anatomical features of the bone and dural structures around the sella. In the case of adenoma with pure lateral expansion this is usually the result of a nonyielding sellar diaphragm, as well as a sphenoidal bone spongy enough to allow lowering of the floor without perforation into the sinus. Under these circumstances the tumor first displaces the internal dural wall of the cavernous sinus, traversing and invading it through its arteriovenous orifices (Weinberger et al. 1940).

In 14 cases a frontopterional craniotomy was performed. The microsurgical resection was performed through the prechiasmatic area, through the interoptical carotid space, and if necessary dissecting the anterior clinoid to gain access to the cavernous sinus through the upper wall. Suction, ring curettes, and lateral dissectors were used for resection. In six cases where the lateral extension was much lower we used the transsphenoidal route. In two cases we combined the transsphenoidal route with the Caldwell-Luc transantral approach.

Results

There was no mortality in this series of 20 patients. The follow-up period ranged from 4 months to 4 years. After surgery 17 patients showed improvement in their headache. In three out of five cases with facial pain, the pain disappeared. In four of the six patients with trigeminal neuralgia and hyoesthesia there was improvement. Of the five patients with oculomotor palsy three showed complete recovery. Two patients developed transient third nerve palsy, and two patients developed transient sixth nerve palsy after surgery. Fifteen patients showed improvement in their visual fields after the surgery (Table 3).

The three patients with Cushing's disease and the patient with acromegaly had clear lowering of hormone levels. Five out of the seven patients with preoperatively elevated plasma PRL levels showed levels under 25 ng/ml after surgery. The other two patients continued to have high PRL levels and remained on bromocriptine treatment. Four of the patients with amenorrhea recovered their menstrual cycles. The patients with panhypopituitarism required substitute hormone therapy. Postsurgical CT scan showed no evidence of tumor in 14 cases. In six cases there were changes compatible with dural infiltration. These patients had recurrence of their tumors and were reoperated on. In four of them the transcranial route was taken and in two of them a combined transantral-transsphenoidal approach was used.

Postsurgical radiotherapy was indicated in all patients. One patient did not have radiotherapy, and this patient had a recurrence 1 year after surgery.

Light microscopy histological examination confirmed the diagnosis of pituitary adenoma. Immunocytological technique showed PRL-secreting cells in seven cases and ACTH-secreting cells in one case. Dural infiltration was detected in eight cases. Isolated mitosis and anisocariosis were seen in two cases. There was intratumoral bleeding in one case.

Table 3. Postoperative findings in 20 patients

Case no.	Neurological (III, V, VI nerves)	Ophthalmological (visual fields acuity)	Endocrinological
1	o	o	++
2	++	+	++
3	o	+	++
4	++	+	++
5	o	o	++
6	o	+	++
7	o	++	++
8	++	o	o
9	o	+	++
10	++	+	++
11	++	o	++
12	o	+	o
13	o	++	o
14	++	+	+
15	o	o	o
16	++	+	+
17	o	++	o
18	o	+	++
19	o	+	+
20	++	+	o

o, No change; +, partial recovery; ++, total recovery

Discussion

Of the 20 patients, 15 showed clinical signs associated with cavernous sinus involvement. Ten patients had occasional diplopia, six had pain in the trigeminal territory, and five had atypical pain. A common feature in all the patients was frontal headache radiating to the orbital region. Sixteen patients had visual field disturbances due to suprasellar expansion of the tumor.

In our experience it is difficult to correlate the extension of an adenoma with a clinical sign. Individual variation was wide and at times identical tomograms showed different clinical signs. This is probably due to the fact that sometimes the tumor does not invade but displaces the cavernous sinus (Ahmadi et al. 1986). Frontal headache, which was present in all cases, is possibly a result of irritation of the surrounding dura and vessels (Friedman et al. 1982). Third nerve palsy does not necessarily indicate invasion or displacement of the cavernous sinus, since the third nerve may be involved before it enters the cavernous sinus. Sixth nerve palsy and pure trigeminal pain with hypoesthesia are more likely to be associated with cavernous sinus invasion.

Seven patients (five men and two women) showed a marked increase in PRL plasma levels with the corresponding clinical manifestations. Three patients had Cushing's disease. One developed acromegaly. One had an increase in serial gonadotrophin levels.

In the literature there is controversy as to whether invading adenomas should be defined according to histological features (Scheithauer et al. 1986). The analysis of our series regarding recurrence and histological type does not allow us to attribute predominance of recurrence to any one of the five histological types (NS-, PRL-, GH-, ACTH-, or GH-secreting cells). Of the 20 adenomas with invasion that were operated on there were seven recurrences (four nonsecreting, one PRL, one ACTH, and one gonadotrophin cell). The series, however, is too small to have statistical significance. All patients with functioning adenomas were treated medically over various periods of time before surgery. The best results were obtained in prolactinomas treated with bromocriptine. The evident difficulty of radical resection of macroadenomas makes bromocriptine therapy especially important before and after surgery in prolactinomas (Basso et al. 1986; Fahlbusch et al. 1987; Hubbard et al. 1987).

Imaging diagnostic methods allow us to identify all invading adenomas of the cavernous sinus. We consider polytomography indispensable in determining the bone structure of the sella. Carotid angiography is needed to show the pathological relationship of the intra- and extracavernous carotid artery with the tumor. CT scan was extremely important in diagnosing extrasellar extension. In the four cases studied by MRI there was better definition of the intracavernous extension in relation to the carotid artery and cranial nerves (Kaufman et al. 1987).

As to the concept of invading adenoma, it is evident that the study and diagnosis of pituitary adenomas has evolved from a morphological to a functional point of view. The classification of an adenoma must nowadays involve integration of data: on one hand, the cytomorphological features allowing the classification of tumor type and, on the other, the cytoimmunological study to assign the tumor a functional type. A third aspect completing the classification of the tumor is the extent of the tumor according to neurological signs and images from conventional radiology, CT scan, and MRI. Endocrinological data from relevant laboratory tests complete the description of the tumor.

In the diagnosis of an adenoma the classical cytological and architectural differences allow the differentiation of benign and malignant processes (Guiot 1958). The endocrinological findings guide the hormonal treatment needed to correct altered functions. The radiological and clinical findings are the guide for planning surgical strategy.

Every adenoma has a limit, whether capsulated or not. The pituitary adenoma is related to: (a) the gland itself, (b) the glandular capsule, (c) the osteoaponeurotic structure surrounding it, (d) the mucosa of the sphenoidal sinus or the arachnoid, and (e) paranasal sinuses or other distant regions. These elements must be taken into account in the microscopic examinations because they have great prognostic significance and allow evaluation of the level of tumor penetration. We call these elements reference levels because of the prognosis they offer with respect to the biological behavior of the pituitary adenoma.

The significance of the word aggressive is not clear since it does not tell us about the histological character of the tumor. It is an indefinite and imprecise topographical denomination indicating only that a tumor has attacked neighboring structures. We prefer to speak of invasion or penetration. With respect to pituitary carcinomas or pituitary adenocarcinomas we do not think it is necessary to wait for a tumor to give rise to metastasis before it is classified as malignant. Using the standard histological criteria we classify tumors as benign or malignant.

We have found immunohistological techniques useful in defining the levels of tumor penetration, since the application of regular histological techniques does not allow a clear identification of the different cells. This is especially important in reoperations where the adenoma tissue is intermixed with scar tissue.

Because of the low cure rate with surgery alone in the treatment of adenomas with invasion it is mandatory to apply in all cases, additional radiotherapy or chemotherapy. With respect to the surgical approach we believe a great number of intracavernous extensions of the adenoma can be excised through the interoptical carotid area or in those extensions which are mainly basal through a transsphenoidal, transantral, or combined route. We do not believe it is justified to approach the cavernous sinus through the lateral wall (Parkinson's triangle) because of the danger of damaging the oculomotor nerves (Mac Kay and Hosobuchi 1978).

References

Ahmadi J, North CM, Segall HD, Zee CS, Weiss MH (1986) Cavernous sinus invasion by pituitary adenomas. AJR 146:257–262

Basso A, Vitale M, Yampolsky C, Chervin A, Kreutel A, Caputi E, Guitelman A (1986) Clinical and surgical management of giant pituitary adenomas. In: Samii M (ed) Surgery in and around the brain stem and the third ventricle. Springer, Berlin Heidelberg New York Tokyo, pp 375–382

Fahlbusch R, Buchfelder M, Schrell U (1987) Short term preoperative treatment of macroprolactinomas by dopamine agonists. J Neurosurg 67:807–815

Foix C (1922) Syndrome de paroi externe des sinus caverneux. Amélioration considérable par le traitement radiothérapique. Rev Neurol 38:827

Friedman AH, Wilkins RH, Kenan PD, Olanow CW, Dubois PJ (1982) Pituitary adenomas presenting as facial pain: report of 2 cases and review of the literature. Neurosurg 10:742–745

Guiot G (1958) Adenomes hypophysaires. Masson, Paris, pp 160–161

Kaufman B, Kaufman BA, Arafah BV, Roessmann V, Selman WR (1987) Large pituitary gland adenomas evaluated with magnetic resonance imaging. Neurosurg 21:540–546

Hubbard JL, Scheithauer BW, Abboud LF, Laws ER (1987) Prolactin secreting adenomas: the preoperative response to bromocriptine and surgical outcome. J Neurosurg 67:816–821

Jefferson G (1955) The invasive adenomas of the anterior pituitary. Thomas, Springfield, IL, pp 56–60

Mac Kay A, Hosobuchi Y (1978) Treatment of intracavernous extensions of pituitary adenomas. Surg Neurol 10:377–383

Scheithauer BW, Kovacs KT, Laws ER, Randall RV (1986) Pathology of invasive pituitary tumors with special reference to functional classification. J Neurosurg 65:733–744

Weinberger LM, Adler FH, Grant FC (1940) Primary pituitary adenoma and the syndrome of the cavernous sinus. Arch Ophthalmol 24:1197–1236

Cushing's Disease

K. D. Post and J. Habas, New York/USA

Introduction

More than any other secretory tumor of the anterior pituitary gland, basophilic ACTH-secreting adenomas continue to provide the clinician with diagnostic and therapeutic challenges.

The syndrome results from sustained hypercortisolemia and is usually easy to recognize but not always simple to define etiologically [25, 12, 7, 4, 1, 15]. Measurements of plasma and urinary cortisol and cortisol derivatives, basally and in response to dexamethasone suppression, as well as determination of plasma ACTH often suggest either a primary adrenal, pituitary, or ectopic neoplastic source of disease [10, 16]. If these data are equivocal, CRH [22] measurement, CRH stimulation testing [3], and petrosal venous sinus sampling [9] with or without CRH administration [18], are often employed to provide additional biochemical evidence for the diagnosis of Cushing's disease. Standard radiologic imaging, including magnetic resonance imaging (MRI) and computed tomography (CT) scans of the pituitary gland in patients diagnosed with Cushing's disease, are usually negative. When a questionable abnormality is demonstrated, its location often does not correspond to that of the tumor found intraoperatively. The use of gadolinium-enhanced MRI may prove helpful [6], but its role has not yet been thoroughly evaluated.

In patients with an excess of other hormones (most commonly prolactin and growth hormone) the hypersecretion is annoying and may ultimately be harmful, but it is not immediately life threatening. Those with Cushing's disease, however, face significant morbidity related to cortisol-induced or accelerated atherosclerosis, hypertension, diabetes mellitus, osteoporosis, infections, and perhaps peptic ulcer disease and thrombosis [5, 11].

Clinical Material

Thirty-three patients (27 women and 6 men) were included in our analysis. Their average age was 37.0 years with a range of 19–58 years. Clinical features present at the time of diagnosis are detailed in Table 1. Our observations agree with those of other published series. Not surprisingly, the cushingoid habitus, including truncal obesity and moon facies, is the most consistently noted manifestion, present in 30 of the 33 patients. Hypertension, easy bruisability,

Table 1. Clinical manifestations

Feature	Number of patients
Obesity	30
Hypertension	29
Moon facies	30
Bruising	24
Fatigue, muscle weakness	23
Striae	23
Glucose intolerance, diabetes mellitus	15
Osteoporosis	9
Supraclavicular fat pad, dorsal fat pad	22
Edema	6
Headaches	2
Mental changes	7
Hyperpimentation	2
Galactorrhea	3
Change in menses	13
Lack of virilization	1
Infertility	1

fatigue, striae, and dorsal and supraclavicular fat pads are also frequent accompaniments of the disease.

Since most patients were referred for neurosurgery from outside of the institution, the nature and extent of the laboratory evaluation on which the diagnosis of Cushing's disease was made varied considerably. The mean preoperative serum cortisol was 32.4 mg/dl, and the ACTH was 98.3 pg/ml. Urinary free cortisols ranged from 124 mg/24 h (nearly normal) to 8097 mg/24 h, with an average of 718 mg/24 h.

A classical response to dexamethasone (suppression of urinary cortisol or metabolites to $\leq 20\%$ of baseline with 2 days of high but not with 2 days of low doses of dexamethasone) was seen in 17 of 31 cases in which sufficient data were available to make such an evaluation (Table 2). Three patients had equivocal results, and 11 had atypical patterns (usually consisting of failure to suppress on the high dose of dexamethasone). Of these 17 patients with a classical response, all ultimately were shown to have Cushing's disease. Of the three patients with inconclusive patterns, two did in fact have Cushing's disease, and

Table 2. Dexamethasone suppression test results

Response	Number	Result
Classical	17	All had adenomas
Equivocal	3	2 had adenomas 1 may have had obesity and depression
Atypical	11	10 had adenomas 1 probably had ectopic secretion

one patient's final diagnosis is not yet certain. This is a patient whose frozen section, but not final pathology, showed an adenoma, and who appeared to be cured for a short period after the surgery. Her levels are now at the upper end of normal; she has unmeasurable to low ACTH in her peripheral and petrosal venous blood which does not stimulate with CRH. She has recently been started on ketoconazole, currently has a normal cortisol, and notes some (but not marked) change in weight and appearance. Of the 11 clearly atypical patterns, all but one have been proven at surgery to have Cushing's disease, and they have been cured. The one patient in whom no tumor was found, and who was not cured by total hypophysectomy probably has an ectopic source of ACTH.

Preoperative ACTH levels were measured in 31 of 33 patients. The majority were in the normal range (<100 pg/ml), but six were over 100 pg/ml and three over 200 pg/ml (Table 3). Very high levels are suggestive of ectopic secretion but clearly do not exclude the possibility of adenoma. Those ACTH levels in the normal range are inappropriately high for the hypercortisolemic state.

At the time of initial surgery tumor was identified in 26 of the 33 patients (Table 4). In 24, a microadenoma was encountered and selectively removed. None of these patients had a total hypophysectomy, and all patients were clinically and biochemically cured. One had a subsequent recurrence, nearly 7 years later. A small irregularity in the gland was removed at the time of her second surgery, and she appears to be cured. In one patient, cure seemed to have been obtained initially, but she subsequently resumed her mild hypercortisolemia, and her true diagnosis is unclear.

Two patients had large invasive macroadenomas. One initially had been operated upon at another institution, had undergone radiation therapy, and presented here with continued hypercortisolemia and a large tumor mass. He underwent a second debulking, and is on adrenolytic agents while awaiting responses to radiation therapy received less than 12 months ago. A second patient with a macroadenoma and ACTH levels in the mid 300 pg/ml range underwent transsphenoidal surgery during which her tumor was partially resected. She presently is undergoing radiation therapy to the residual mass.

No tumor was found in seven patients. In one, partial hypophysectomy was performed and deemed curative. Five patients were cured by hypophysectomy;

Table 3. Preoperative ACTH measurements in Cushing's disease ($n = 31$)

ACTH (pg/ml)	Number of patients
20–40	8
40–60	2
60–80	7
80–100	8
100–200	3
200–400	3

Table 4. Surgical results in Cushing's disease

	Number of patients	Number cured by selective surgery	Number cured by hypophysectomy	Total number cured
Microadenomas	24[a]	23	ND	23
Macroadenomas	2[b]	0	0	0
No tumor	7[c]	1	5	6
	33			29 (88%)

ND, Not done.

[a] In one patient, a small adenoma was found during surgery, but clinical remission was not achieved. Postoperative petrosal sinus sampling failed to show a gradient. One patient, in remission for 6 years following transsphenoidal adenomectomy, relapsed and appears again to be cured 6 months after a second surgery.

[b] Both patients with macroadenomas underwent surgical debulking, followed by radiation therapy and ketoconazole. Neither is cured.

[c] Two patients undergoing hypophysectomy as first surgery were cured. Five patients had subtotal hypophysectomy as the first surgery; one of these was cured. The remaining four are in remission following a second surgery with total hypophysectomy

Tabel 5. Complications in Cushing's disease ($n = 33$)

Transient diabetes insipidus	2
Local bleeding (no transfusions)	2
CSF leak	0
Meningitis	0
Neurological deficit	0
Death	0

in one case, tumor was identified during a second operation, but the patient stipulated that hypophysectomy be performed. In one patient no tumor was found at first surgery, and he underwent partial pituitary resection. This patient remained hypercortisolemic and had a second transsphenoidal exploration. At that time, the entire gland was removed, but his Cushing's syndrome persisted. Venous blood obtained by catheterization of the petrosal sinuses had an ACTH of 77 pg/ml on the right and 73 pg/ml on the left. ACTH simultaneously obtained from peripheral samples was 72 pg/ml. The search for an ectopic source of ACTH production is under way.

There were no major complications in any of the surgical procedures. Several transient episodes of diabetes insipidus, and local bleeding (not requiring transfusion) occurred in two patients (Table 5).

Discussion

The etiology of Cushing's syndrome is most commonly an ACTH-secreting pituitary adenoma (Cushing's disease). However, iatrogenic causes as well as adrenal tumor and ectopic ACTH secretion must be excluded from consideration. The standard biochemical testing, including plasma and urinary cortisol measurements, plasma ACTH, and the response to dexamethasone stimulation are often but not always adequate. CRH measurements and CRH stimulation may also be needed to confirm the diagnosis.

In some situations the diagnosis remains unclear, and retrograde petrosal sinus sampling may be of benefit. Bilateral simultaneous catheters are placed in the inferior petrosal sinuses and samples for ACTH levels are taken simultaneously from these catheters and the peripheral blood. If a gradient exists between the sinus and peripheral samples, the ACTH is presumed to come from the pituitary gland rather than an ectopic source. We do not perform this test routinely, but we do advise petrosal sinus sampling when the etiology of Cushing's syndrome is uncertain and for cases in which surgical removal of an adenoma has failed, forcing consideration of hypophysectomy.

Once the diagnosis of pituitary adenoma induced Cushing's syndrome has been made, the choice of therapy is not always straightforward. Since Hardy reintroduced transsphenoidal adenomectomy in 1969 [14], this procedure has been shown to be safe and effective. In many institutions, including our own, this has become the preferred treatment for patients with Cushing's disease. As previously noted, patients with Cushing's disease often have small tumors not visible on even the most sophisticated scans and quite frequently difficult to find during surgical exploration and microdissection of the gland. Incidental (nonfunctioning) adenomas, or small inhomogeneities in the gland or sella are common in the general population. If present in patients with pituitary Cushing's disease, they may guide the surgeon toward an area of the gland which, while abnormal, is not part of the disease process. Review of the literature suggests that the lesion responsible for excess ACTH secretions is identifiable in 70%–85% of cases [26, 13, 8, 23, 2, 20].

Total hypophysectomy, an option for patients where no tumor is clearly identified during exploration of the gland, is particularly unpalatable in the population in whom Cushing's disease is the most prevalent – young women in their child-bearing years. We do not routinely perform total hypophysectomy at the time of first surgery unless the patient specifically requests that it be done, desiring certainty of postoperative cure, even at the price of hypopituitarism.

Petrosal sinus sampling for ACTH may show lateralization of secretion [21, 24]. This has led some surgeons to perform a hemihypophysectomy based upon this lateralization in an attempt to cure the disease, rather than perform a complete hypophysectomy. Our experience has not proven this reliable. However, if the patient is not cured by the first surgical procedure, and an adenoma is not found, then total anterior hypophysectomy is recommended.

There is no drug treatment as specifically effective for Cushing's disease as bromocriptine is for prolactinomas [12, 19]. Drugs that interfere with cortisol production exist, but they are poorly tolerated, and their efficacy is variable. We use these drugs only in patients whose disease has not been controlled by surgery, while waiting for radiation therapy to take effect, or when a delay exists in definitive diagnosis, or between diagnosis and surgery.

Bilateral adrenalectomy is never considered as initial therapy in patients with Cushing's disease. We would employ such a procedure only with patients who had failed surgery, radiation therapy, and drug therapy.

Radiation therapy is effective but too slow to be used as the first therapeutic intervention. It is reserved for patients with large tumors incompletely removed by surgery and is usually used in conjunction with adrenolytic drugs. Until recently, opinion was divided as to whether radiation therapy or transsphenoidal adenomectomy should be the initial treatment of choice for children. Most reports now favor transsphenoidal microsurgery in pediatric cases [20].

Conclusions

1. Even with highly sophisticated radiologic imaging techniques, it is still difficult preoperatively to demonstrate a pituitary tumor causing Cushing's disease. Indeed, gland asymmetry or inhomogeneity is common in patients without Cushing's disease. Such findings tend to be poorly correlated with intraoperative tumor localization, and thus abnormal scans need to be interpreted cautiously.
2. Tumors causing Cushing's disease tend to be small. Thus, even at the time of exploration, it is possible to locate the tumor in only 79% of the cases.
3. Using standard biochemical parameters, errors in the diagnosis of Cushing's disease are made in 1% of the cases. In our series, CRH stimulation testing has been helpful in resolving some diagnostic ambiguity. In one patient with standard testing which was suggestive but, we felt, not diagnostic of a pituitary source of disease, CRH stimulation failed to show the robust rise in ACTH and cortisol usually seen in patients with Cushing's disease. Therefore, transsphenoidal surgery was not performed, and the patient continues under observation. In three cases with similarly equivocal laboratory data, response to CRH stimulation was characteristic of Cushing's disease, surgery was performed, and microadenomas responsible for the hypercortisolemia were found and successfully removed.
4. Petrosal venous sampling confirmed a questionable diagnosis of Cushing's disease in two patients by showing a clear step-up between peripheral and petrosal blood levels of ACTH. In two others not cured by pituitary surgery, petrosal sampling failed to demonstrate a pituitary/peripheral gradient. In fact, these patients were not felt to have had Cushing's disease. One probably has ectopic ACTH; the other may simply be depressed.
5. Transsphenoidal surgery is the overwhelming choice of therapy, unless surgery is clearly contraindicated for other reasons. Selective adenomectomy is preferred. If an adenoma is not found in an adult, then hypophysectomy is advised. In a child with Cushing's disease, if an adenoma is not found, then radiation, adrenalectomy, or medical treatment is preferred before hypophysectomy.

References

1. Burch W (1985) Cushing's disease, a review. Arch Intern Med 145:1106–1111
2. Chandler WF, Schteingart DE, Lloyd RV (1987) Surgical treatment of Cushing's disease. J Neurosurg 66:204–212
3. Chrousos GP, Schulte HM, Oldfield EH et al. (1984) The Corticotropin-Releasing Factor Stimulation Test. N Eng J Med 310:622–651
4. Crapo L (1979) Cushing's syndrome: a review of diagnostic testing. Metabolism 28:955–977
5. David DS, Grieco MH, Cushman P (1970) Adrenal glucocorticoids after twenty years. A review of their clinically relevant consequences. J Chron Dis 22:637–711
6. Dwyer AJ, Frank JA, Doppman JL et al. (1987) Pituitary adenoma in patients with Cushing disease: initial experience with Gd-DTPA-enhanced MR imaging. Neuroradiology 163:421–426
7. Eddy RL, Jones AL, Gilliland PF et al. (1973) Cushing's syndrome: a prospective study of diagnostic methods. Am J Med 55:621–630
8. Fahlbusch R, Buchfelder M, Muller OA (1966) Transsphenoidal surgery for Cushing's disease. J Roy Soc Med 79:262–269
9. Findling JW, Aron DC, Tyrell JB et al. (1981) Selective venous sampling for ACTH in Cushing's syndrome. Differentiation between Cushing's disease and the ectopic ACTH syndrome. Ann Intern Med 94:647–652
10. Findling JW, Tyrell JB (1986) Occult ectopic secretion of corticotropin. Arch Intern Med 146:929–933
11. Glenn F, Mannix H (1967) Diagnosis and prognosis of Cushing's syndrome. Surg Gynecol Obstet 4:765–776
12. Gold EM (1979) The Cushing syndromes: changing views of diagnosis and treatment. Ann Intern Med 90:829–844
13. Guilhaume B, Bertagna X, Thomsen M et al. (1988) Transsphenoidal pituitary surgery for the treatment of Cushing's disease: results in 64 patients and long term follow-up studies. J Clin Endocrinol Metab 66:1056–1064
14. Hardy JH (1982) Presidential address: XVII Canadian Congress of Neurological Sciences. Cushing's disease: 50 years later. J Can Des Sci Neurol 11:375–380
15. Howlett TA, Rees LH (1985) Is it possible to diagnose pituitary-dependent Cushing's disease? Ann Clin Biochem 22:550–558
16. Howlett TA, Drury PL, Perry L et al. (1986) Diagnosis and management of ACTH-dependent Cushing's syndrome: comparison of the features in ectopic and pituitary ACTH production. Clin Endocrinology 24:699–713
17. Jennings AS, Liddle GW, Orth DN (1977) Results of treating childhood Cushing's disease with pituitary irradiation. N Engl J Med 297:957–962
18. Landolt AM, Valavanis A, Girard J, Eberle AN (1986) Corticotrophin-Releasing Factor-Test used with bilateral, simultaneous inferior petrosal sinus bloodsampling for the diagnosis of pituitary-dependent Cushing's disease. Clin Endocrinol 25:687–696
19. Loli P, Berselli ME, Tagliaferri M (1986) Use of ketoconazole in the treatment of Cushing's syndrome. J Clin Endocrinol Metabolism 63:1365–1371
20. Mampalam TJ, Tyrrell JB, Wilson CB (1988) Transsphenoidal microsurgery for Cushing disease, a report of 216 cases. Ann Intern Med 15:487–446
21. Manni A, Latshaw RF, Page R, Santen RJ (1983) Simultaneous bilateral venous sampling for adrenocorticotropin in pituitary-dependent Cushing's disease: evidence for lateralization of pituitary venous drainage. J Clin Endocrinol Metab 57:1070–1073
22. MGH Case Records (Case 52–1987) (1987) A 20 year old woman with Cushing's disease and a pulmonary nodule (Dluky and Mark) 317:1648–1655
23. Nakane T, Kuwayama A, Watanabe M et al. (1987) Long term results of transsphenoidal adenomectomy in patients with Cushing's disease. Neurosurgery 21:218–222
24. Oldfield EH, Chrousos GP, Schulte HM et al. (1985) Preoperative lateralization of ACTH-selecting pituitary microadenomas by bilateral and simultaneous inferior petrosal venous sinus sampling. N Engl J Med 312:98–103

25. Ross EJ, Marshall-Jones P, Friedmann M (1966) Cushing's syndrome: diagnostic Criteria. Q J Med 35:149–193
26. Salassa RM, Laws ER, Carpenter PC, Northcutt RC (1978) Transsphenoidal removal of pituitary microadenoma in Cushing's disease. Mayo Clin Proc 53:24–28
27. Styne DM, Grumbach MM, Kaplan SL et al. (1977) Treatment of Cushing's disease in childhood and adolescence by transsphenoidal microadenomectomy. N Engl J Med 310:889–893

Skull Base Involvement in Pituitary Adenomas

J. Šteňo, Bratislava/Czechoslovakia

Pituitary adenomas which infiltrate their dural envelope sometimes perforate it and invade the neighboring structures: the sphenoid sinus, the cancellous bone of the clivus, the cavernous sinus, or the intracranial spaces. Radical removal of such invasive adenomas is problematic, their endocrinological cure is exceptional, and recurrence in nonirradiated patients is common (Guiot 1978). The evaluation and grading of invasiveness of pituitary adenomas are based on the radiological appearance of bony structures of the sella (Hardy 1982).

The purpose of the present study was to evaluate the results of surgical therapy of pituitary adenomas destroying the structures of the skull base. Over the past 20 years, 198 patients with pituitary adenomas were operated on in our department. The transsphenoidal approach, which allows determination of the exact extent of skull base involvement, we have used only since the beginning of 1986. Therefore only 25 patients operated on by transnasal transsphenoidal approach are analyzed in the present study.

Results

Destruction of the sellar floor was found in seven out of 21 primary operations and in three out of four patients with recurrent tumors after transcranial operations. In six of them (five primary, one recurrent adenoma), the entire floor and anterior wall of the sella was absent, and the sphenoid sinus was filled by the tumor up to its anterior wall. The latter was destroyed in one patient, in whom a firm prolactinoma involved also the ethmoidal sinus and both cavernous sinuses (Table 1). The anterior and basal surfaces of the tumor in other patients (three prolactinomas, two inactive adenomas) were covered by the dura, which was not perforated. This anatomical condition allowed radical removal of the intrasphenoidal and intrasellar portions of the tumor as well as its A and B type suprasellar extension (Figs. 1, 2). In primary operations even selective radical adenomectomy was possible (Table 1). Also, during the operation in one of two cases with C type suprasellar tumor growth we were under the impression that the tumor had been removed radically. However, postoperative hyperprolactinemia was indicative of some residual tumor tissue left in place, probably inside or in the vicinity of the cavernous sinuses.

In only one case with a giant inactive adenoma (Fig. 3) was a combined, two-stage operation necessary. Three weeks after radical removal of basal and

Table 1. Clinical, radiological, and anatomical characteristics and operative results in patients with pituitary adenomas involving the skull base

Patient, age (years), sex	Grade of invasiveness	Type of suprasellar growth	Invasion of				Hormonal activity (ng/ml)	Operation	Result of the operation	Notes
			Sphen. sinus	Clivus	Ethm. sinus	Cavernous sinus				
J.A., 29, F	IV	A	+	–	–	–	PRL 51.5	Radical selective	PRL 6.0, regular menstr. cycle	
M.R., 63, F	IV	B	+	–	–	Displaced (invaded?)	PRL 42.1	Radical selective	PRL 6.3, improvement of oculomotor disturbances	Preoperative PRL measured after tumor apoplexy
I.Š., 50, F	IV	B	+	–	–	–	Inactive	Radical	Improvement of vision	Recurrent tumor (6 years after craniotomy)
J.B., 48, M	IV	C	+	–	–	Displaced (invaded?)	PRL > 263,15	Subtotal	PRL 263,15	Bromocriptine postoperatively
M.R., 44, M	IV	C	+	+	–	Displaced (invaded?)	Inactive	Combined radical	Improvement of vision and oculomotor disturbances	Transient mental disturb. after both operations
A.L., 53, F	IV	–	+	–	+	Invaded	PRL > 263,15	Partial	PRL > 263,15, headache disappeared	Irradiation and bromocriptine postoperatively
P.D. 21, M	III	A	–	–	–	Invaded	PRL 65.0 GH 29.1	Subtotal	PRL 7.0 GH 5.9	
A.K., 51, M	III	B	–	–	–	Invaded	PRL 70.5	Radical	PRL 0	

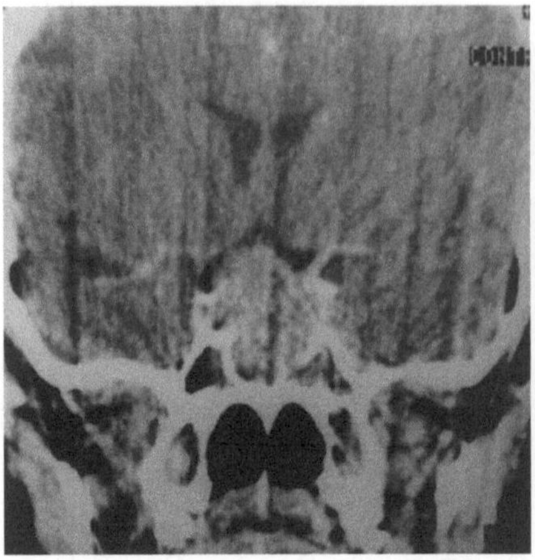

1

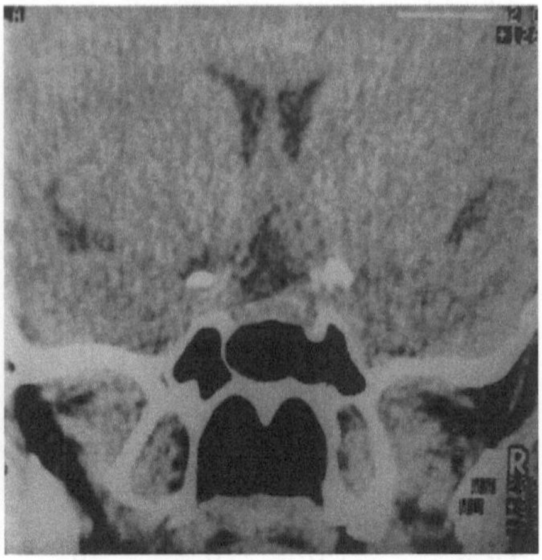

2

Fig. 1. Direct coronal CT scan of prolactinoma destroying the sellar floor, growing into the sphenoid sinus (case J. A.)

Fig. 2. The same case as in Fig. 1, 2 weeks after selective adenomectomy

posterior portions of the tumor by the transsphenoidal route (Fig. 4) a right subfrontal approach was used to remove the anterosuperior portion of the tumor, which proved to be an almost separate tumor nodule located in the cavity of the third ventricle, into which it had spread in an unusual manner through the lamina terminalis (Fig. 5).

In four out of six patients with the tumors growing into the sphenoid sinus and in two other patients without sphenoid sinus involvement a lateral extension of the tumor from the sella towards the cavernous sinus was observed on computed tomography (CT). The differentiation between the displacement and the invasion of the sinus was questionable in three cases. Dynamic CT scans (Bonneville et al. 1986) and magnetic resonance imaging (Nistor et al., this volume) probably would have been helpful in these cases. A considerable enlargement of the sinus and marked convexity of its lateral wall in a further three cases strongly indicated the invasion of the sinus cavity (Fig. 6).

Very firm consistency of the tumor in one of these cases precluded the removal of lateral tumor extensions, while in the other two patients soft tumor

Fig. 3. Midsagittal pneumoencephalotomogram and axial CT scan of a giant inactive adenoma destroying the upper part of the clivus. Benign midline cerebellar cyst not obstructing CSF pathways (case M. R., 44, M)

Fig. 4. The same case as in Fig. 3, 3 days after removal of the basal and posterior parts of the tumor by transsphenoidal approach

Fig. 5. The same case as in Figs. 3 and 4, 2 months after removal of the residual tumor from the cavity of the third ventricle through right subfrontal approach

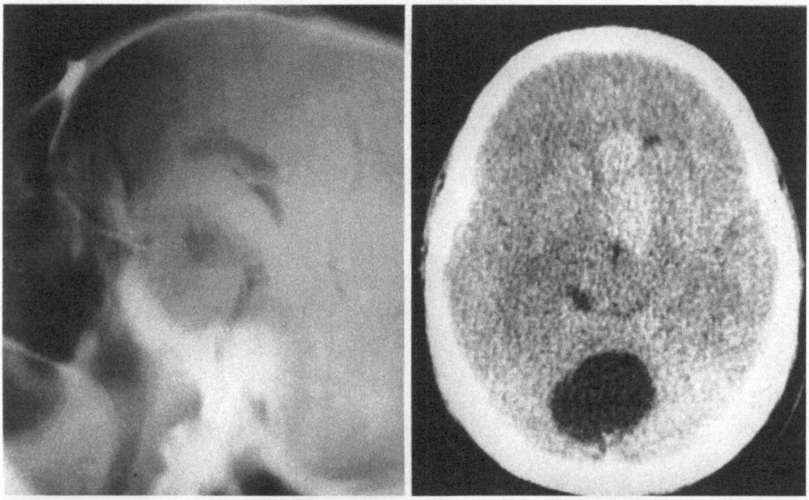
3

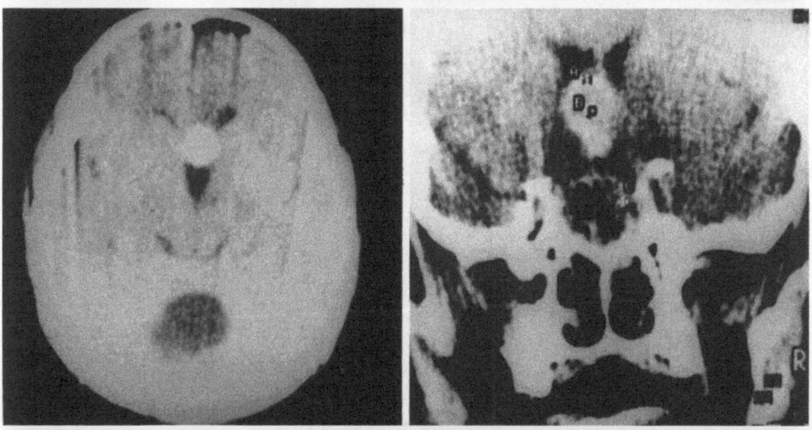

4

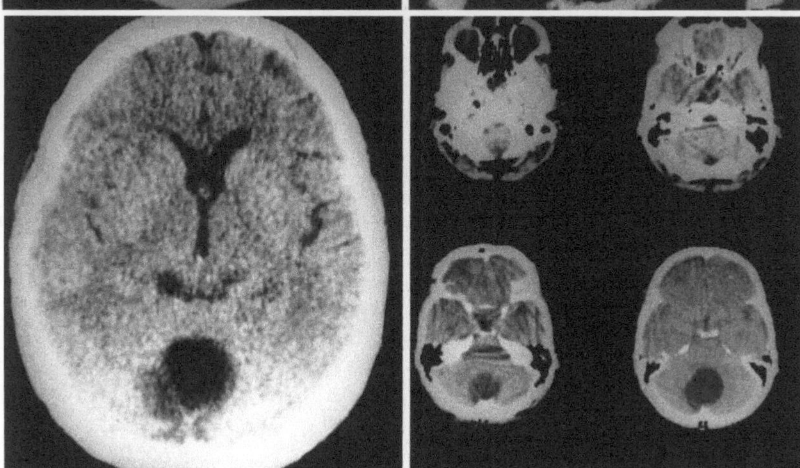
5

Figs. 3–5

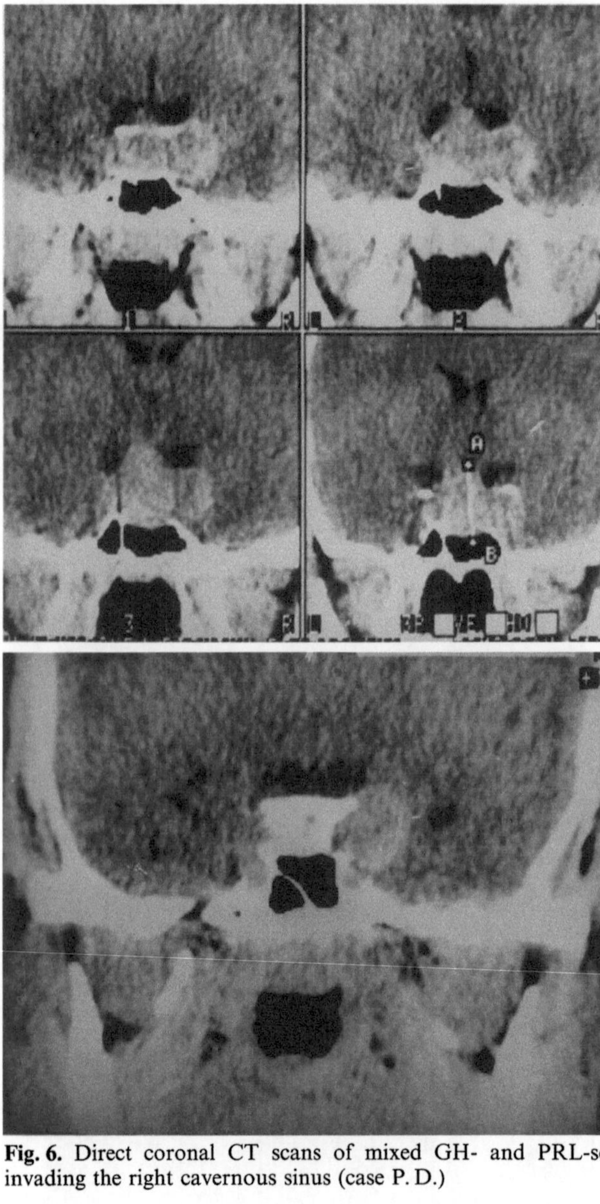

Fig. 6. Direct coronal CT scans of mixed GH- and PRL-secreting pituitary adenoma invading the right cavernous sinus (case P. D.)

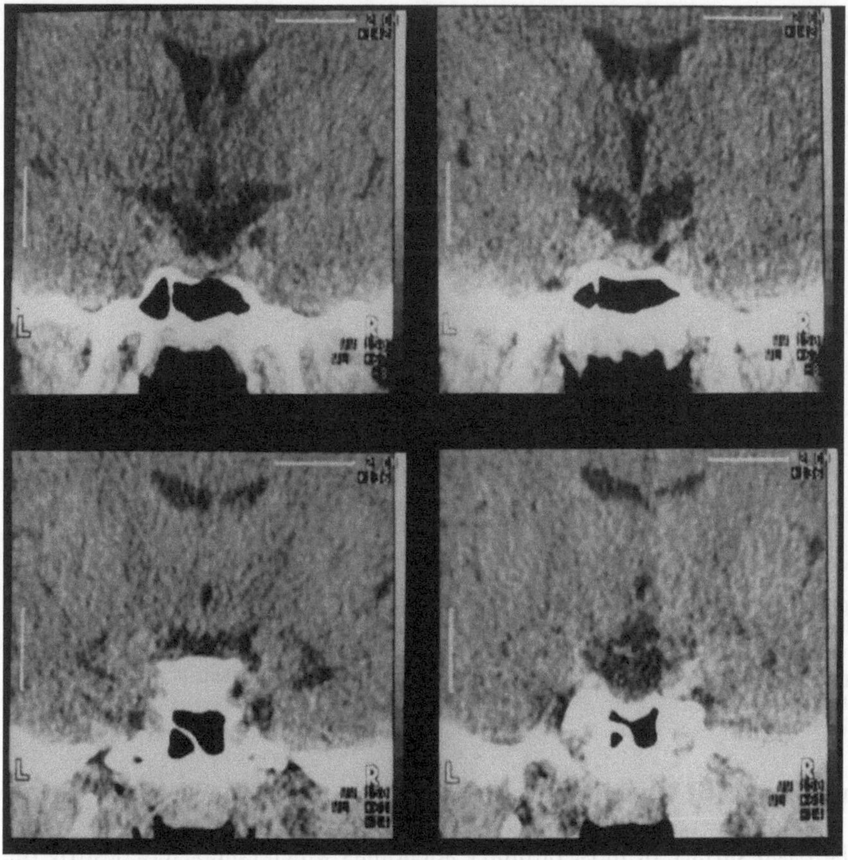

Fig. 7. The same case as in Fig. 6, 7 weeks after selective removal of the tumor

masses could also be removed from the cavernous sinus (Fig. 7). As pointed out by Buchfelder and Fahlbusch (1987), this is not accompanied by severe bleeding. Although postoperative CT scans in these patients did not reveal residual tumor, in the patient with mixed GH- and PRL-secreting adenoma only the PRL level reduced to normal. An insufficient decrease in GH may be explained by our observation from the anatomical study of another series of autopsy cases, that adenoma expanding laterally from the pituitary fossa after breaking through the medial wall of the cavernous sinus may grow also laterally from cavernous carotid artery (Fig. 8). The vessel then becomes completely enveloped by the tumor mass, the most lateral part of which is not exposed to direct visual control during midline trans-basal approaches.

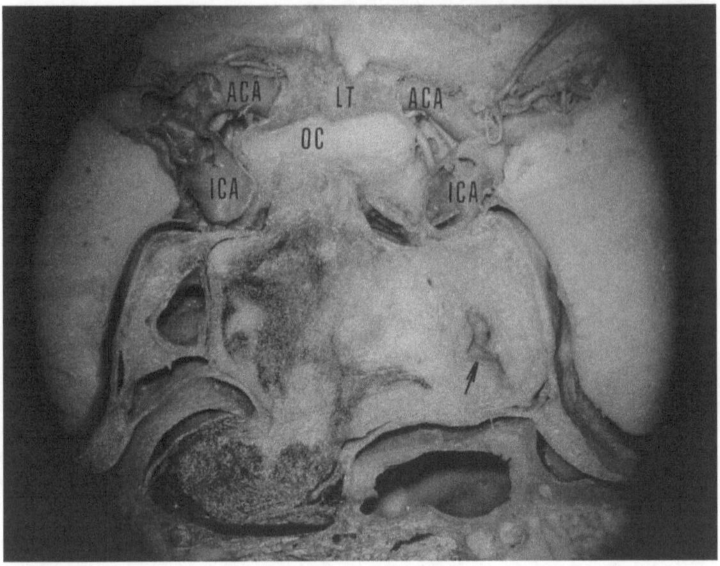

Fig. 8. Frontal section through an autopsy specimen of a GH-secreting hemorrhagic pituitary adenoma invading the right cavernous sinus and left side of the sphenoid sinus. Right cavernous carotid artery (*arrow*) completely enveloped by the tumor. *OC*, Optic chiasm; *LT*, lamina terminalis; *ICA*, supraclinoid carotid artery; *ACA*, A1 segment of anterior cerebral artery

Summary and Conclusions

The basal surface of most pituitary adenomas which completely destroy the sellar floor and grow into the sphenoid sinus is covered by unperforated dura. Radical removal of such tumors is possible even in cases with gross suprasellar extension.

The most lateral part of tumors growing into the cavity of cavernous sinus may grow laterally from cavernous carotid artery and thus be hidden from direct visual control during midline transbasal approaches.

References

Bonneville JF, Cattin F, Dietemann JL (1986) Computed tomography of the pituitary gland. Springer, Berlin Heidelberg New York Tokyo

Buchfelder M, Fahlbusch R (1987) Transsphenoidal surgery of pituitary adenomas developed towards the cavernous sinus. In: Dolenc VV (ed) The cavernous sinus. A multidisciplinary approach to vascular and tumorous lesions. Springer, Wien New York, pp 404–414

Guiot G (1978) Considerations on the surgical treatment of pituitary adenomas. In: Fahlbusch R, Werder K v. (eds) Treatment of pituitary adenomas. Thieme, Stuttgart, pp 202–218

Hardy J (1982) Transsphenoidal operations on the pituitary. Codman and Shurtleff, Inc. Randolph, p 13

Sub- and Retrochiasmatic Approach for Microsurgical Removal of Large Suprasellar Tumors

K. E. TUREL, Bombay/India, M. SAMII, Hannover/FRG, and
R. RAMINA, Curitiba Paraná/Brasil

Introduction

The suprasellar masses, by virtue of their intimate anatomic relationship with the optic apparatus, the carotid vessel complex, and the third ventricle and hypothalamus, assume considerable significance. Hence the larger the solid mass is, the greater are the surgical risks. There are additional problems when tumors penetrate the cavernous sinus, erode the skull base, or invade the sphenoid sinus. This discussion is limited to the surgical management of large suprasellar tumors, namely those involving the third ventricle. We have evaluated the last 100 sellar region tumors treated at the Neurosurgical Clinic in Hannover. Nearly half (48) of them were found to be large (Table 1).

Of all suprasellar growth processes by far the most common are the pituitary adenomas. They formed nearly 80% of the tumors in this region. Other, less frequent though no less important lesions, include craniopharyngiomas, suprasellar meningiomas, and gliomas of the optic apparatus and the hypothalamus (Table 2).

Table 1. Sellar region tumors ($n = 100$)

Pituitary adenoma	77
Tuberculum sella meningioma	7
Craniopharyngioma	6
Optic sheath meningioma	3
Optic nerve glioma	1
Optic nerve teratoma	1
Hypothalamic glioma	1
Skull base plasmacytoma	1
Intrasellar cyst	1
Carcinoma	1
Metastasis from lung	1

Table 2. Large sellar region tumors ($n = 48$)

Pituitary adenoma	39 (81%)
Craniopharyngioma	4
Tuberculum sella meningioma	3
Optic nerve glioma	1
Hypothalamic glioma	1

The most common preoperative symptom was visual deficit, present in 90% of these patients. Hormonal dysfunction was noted in two-thirds of the 48. Eight patients had ocular muscle paresis, indicating involvement of their nerves in the retrobulbar region, cavernous sinus, or tentorial notch. Of these large tumors 30% produced occlusive hydrocephalus (Table 3).

Table 3. Symptoms in large sellar region tumors ($n = 48$)

Visual	43
Hormonal	32
Increased ICP	14
Ocular muscle deficits	8
Limb paresis	2

Preoperative Investigations

Radiological tests are aimed not merely at determining the diagnosis but also at helping to outline tumor extension and establishing its relationship to the important neighboring structures. This helps decide the type of exposure and provides guidance at operation. Plain X-rays, computed tomography (CT), and magnetic resonance imaging (MRI) all have their assets and are used routinely for obtaining varying sorts of information. Cerebral angiography is also helpful when dealing with such large tumors as it provides a clue to the extent and source of tumor vascularity and identifies the involvement and displacement of the major cerebral vessels and the circle of Willis.

However, it must be noted that despite advances in neuroradiology the preoperative diagnosis of the location of the chiasma, whether pre-, post-, or normofixed, is generally lacking. This vital anatomic information is usually defined only at surgery, and some of the operative techniques have been especially developed to deal with the anomalous positions of the chiasma (Rand 1978).

Neuroendocrinal investigations are carried out to assess the hormonal imbalance caused by these tumors and, on suitable occasions, after surgery (and radiotherapy, if it has been offered) to evaluate the change in their status, and to regulate the disturbed values.

Surgical Approach

A sound knowledge and experience of the microsurgical anatomy of this intricate region is an essential prerequisite to surgery (Rhoton et al. 1978).

The arterial anatomy of the chiasmatic cistern is very important. Complicated series of perforators perfuse this region after arising from the carotid artery in the front and from basilar artery behind. As the sellar tumor grows upward, the entire circle of Willis becomes stretched, the perforators being most susceptible to the stretching. A further surgical insult must be avoided.

Most of these tumors growing upward from the sella are in direct contact with the hypothalamus. Although it compresses the hypothalamus, the tumor rarely assumes the kind of intimacy which makes it really adhere to the hypothalamus. Separation of the tumor from this sensitive neural structure is one of the highlights of this surgery, and the care and ease with which the separation is achieved dictates the postoperative course. The larger the tumor is the greater are its effects on hypothalamic functions. Hence it may be apt to say that the prognosis of surgical removal of tumors in this region is commensurate with the degree of involvement of the third ventricle and hypothalamus. It is for this reason that one must sometimes be content with a subtotal removal. In this context there is no doubt that the neurosurgeon's experience in dealing with tumors of this region is an important decisive factor – the experienced surgeon is not only capable of doing more total removals with success, but he also knows to restrain himself from dangerously going further.

The pituitary stalk is often compressed beyond the point of recognition. However, there are times when it is stretched and displaced, and it may be possible to salvage it by carefully taking the tumor away from it. In some instances, such as in a craniopharyngioma, it may be infiltrated or may not be feasible to dissect out without leaving some tumor behind. The inseparably involved stalk may be best sectioned at its proximal end, for the attempt to separate tumor from stalk may result in traction damage to the hypothalamus.

Fundamentally, any surgical approach while aiming at effectively dealing with the pathology should be based on two factors: the least surgical trauma and the shortest route to the disease process. Such large suprasellar tumors are generally not suited to transsphenoidal removal. Of the transcranial approaches an extracerebral route is preferred to the relatively traumatic transcortical or transcallosal routes. For most tumors a right frontal craniotomy is made. For a very large mass a bifrontal approach is preferred for the wider access it provides. Since the majority of tumors are directly beneath the chiasma, hypothalamus, and third ventricle, and at times subfrontal and prepontine, a subfrontal route would lead us directly to the top of the lesion. A careful, patient, and gentle retraction of the frontal lobe carries virtually no risk of postoperative brain swelling or infarction and provides excellent visibility and a direct access to the tumor by the shortest and least obstructed route. We have not found it necessary to resect the undersurface of the frontal lobe for improving visualization, as has been suggested by other authors (Patterson and Danylevich 1980).

On retracting the right frontal lobe obliquely over the roof of the orbit, the right carotid artery and optic nerve are first visualized. The olfactory tract appears near midline and may be sectioned (or becomes avulsed) when a large pituitary tumor demands further frontal lobe retraction. A large tumor may obscure most other structures when initially exposed. Hence no attempt is made to visualize them at this stage. Three fundamental principles need to be observed: (a) The tumor is excised piecemeal by performing an intracapsular reduction. As the capsule relaxes, it tends automatically to separate itself from its delicate neighbors, thus minimizing the need to handle the latter. (b) The optic apparatus is never retracted, whatever seems the need or temptation. (c) The tiny blood vessels entering the optic apparatus, including those coursing over the tumor to reach the latter, are never coagulated, except those branches actually entering and feeding the tumor.

Case Reports

The operative strategy naturally varies with each case but a few representative cases will be demonstrated to highlight some of the important points.

Case 1. The CT scans of an elderly man depict an enormous and lobulated pituitary adenoma growing substantially (Fig. 1a) and with an asymmetrical left-sided extension (Fig. 1b,c). The tumor is compressing not only the third

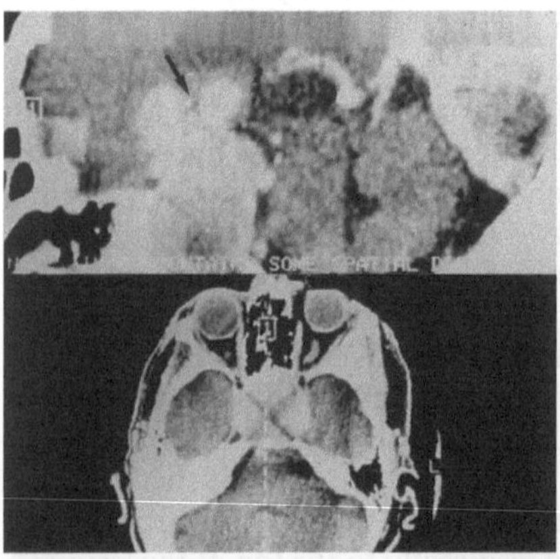

a

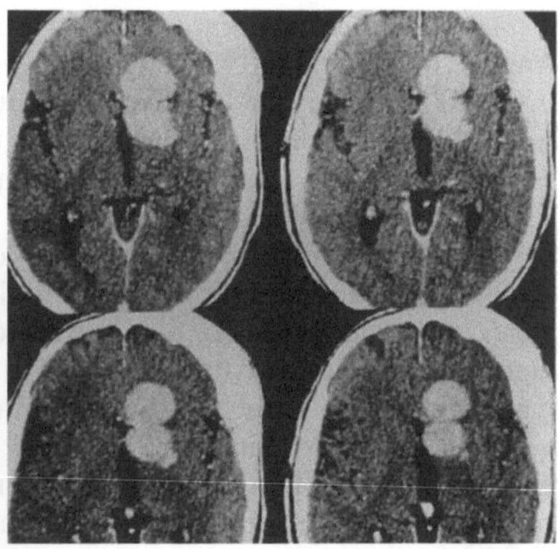

c

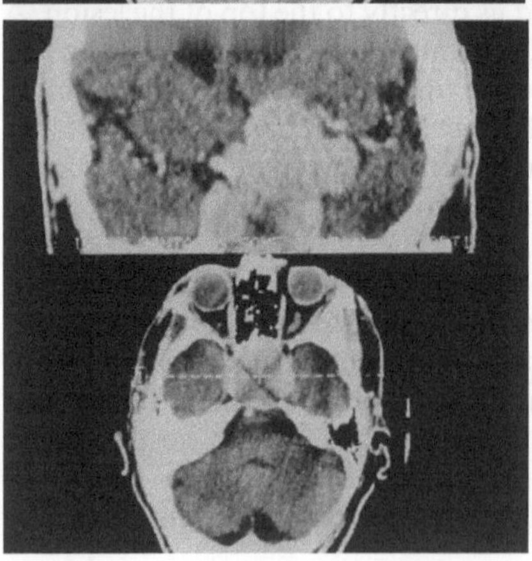

b

Fig. 1. a CT scan appearance of an elderly man with a giant, invasive pituitary adenoma assuming a high dimension and a lobulated appearance. The third ventricle is pushed upward and is reduced to a small chink. The floor of the lateral ventricle is also pushed up and the frontal horn truncated. The anterior cerebral artery (ACA) saddles in the niche or rather grooves the tumor which appears to be thereby lobulated (*arrow*). **b, c** Coronal reconstruction and axial views of the CT show the lop-sided growth remaining preferentially to the left

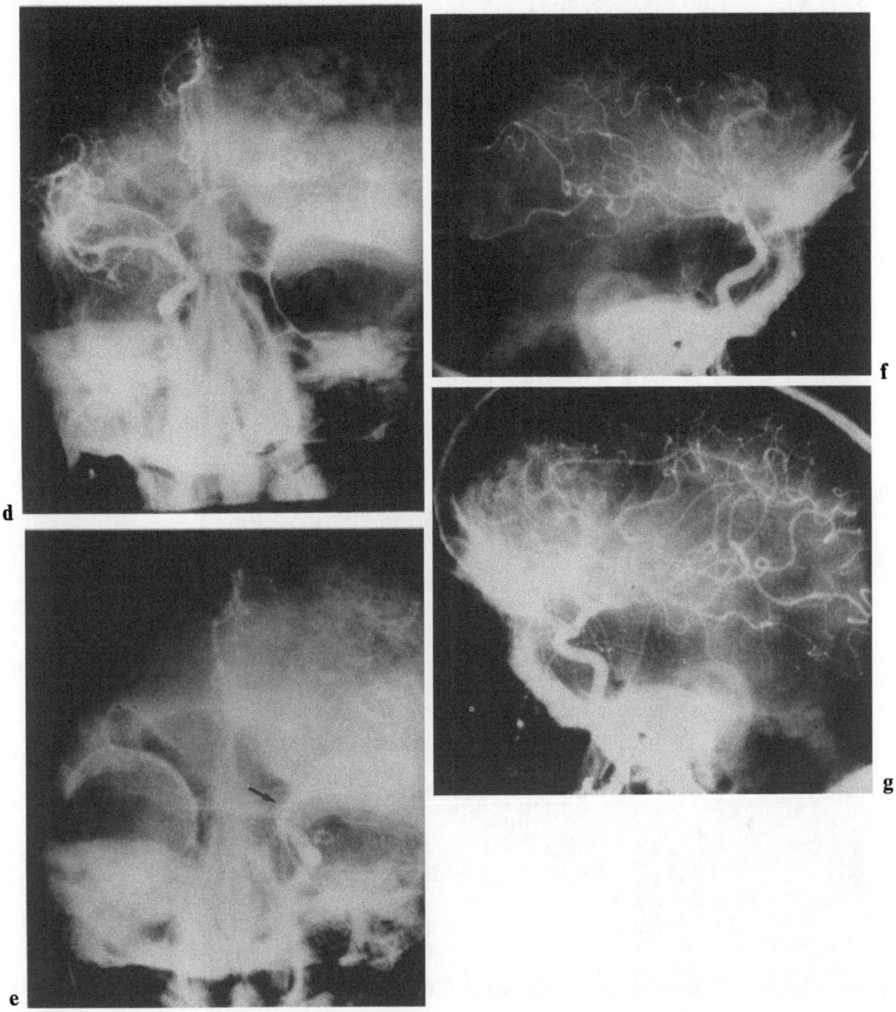

Fig. 1. d–g Carotid angiography is performed routinely by us despite the information provided by the modern generation of CT scans. Angiograms demonstrate the tumor blush, source of vascularity, and displacement or compromise of major vessels, which are a useful guide at surgery. The right and left anteroposterior views show marked lifting up of the A1 segments on both sides, the left side appearing much thinner due to a greater stretch. By the same token, the left internal carotid artery (ICA) is also deviated more laterally by this greater left-sided growth (*arrow*). The lateral views show unwinding of the carotid siphon and the pushing backwards of the ACAs indicating subfrontal tumor extension

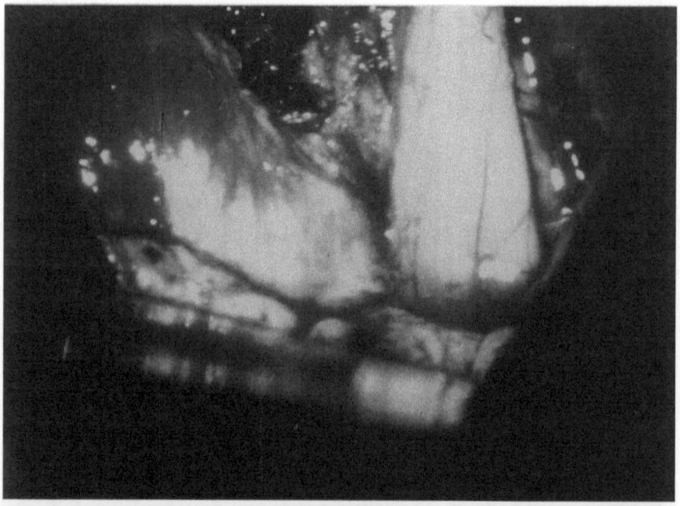

h

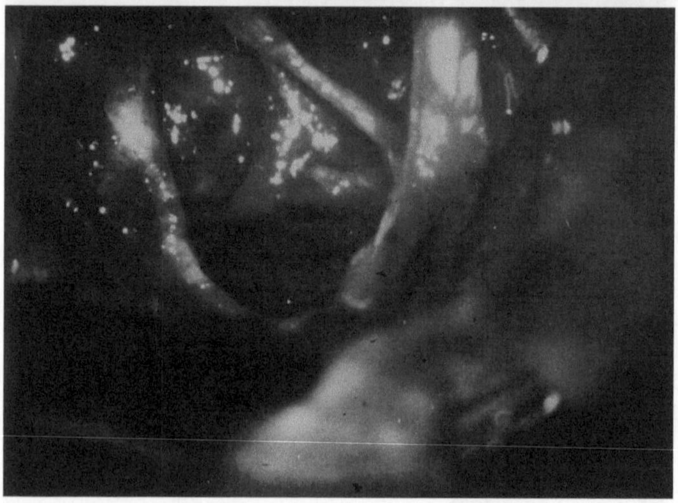

i

Fig. 1. h Appearance of the tumor as seen through the surgical microscope after retraction of both frontal lobes. The broad white vertical band to the right is the right optic nerve, and at far right the right ICA. The tumor occupies the rest of the operative field, totally obscuring visualization of the left optic nerve. **i** After removal of the tumor the extremely thinned out left optic nerve and the deformed chiasma are seen. The left ICA is seen to bifurcate, the A1 segment appearing rather elongated due to the tumor stretch

ventricle but even the lateral ventricles. The left anterior cerebral artery (ACA) is trapped between the two lobules (Fig. 1a, arrow). Right and left carotid angiography anteroposterior views demonstrate elevation of A1 segments of both ACAs (Fig. 1d, e) and lateral displacement of left internal carotid artery (ICA) (Fig. 1e, arrow). The lateral views (Fig. 1f, g) show a marked opening out of the carotid siphon.

A bifrontal craniotomy was performed. Such a craniotomy, is preferred for a bilaterally extensive tumor, as retraction of both frontal lobes provides a view of both optic nerves, both ICAs and ACAs, the under surface of the chiasma and structures at the tentorial margin. The tumor is visualized between the two optic nerves. The right optic nerve, and lateral to it the right ICA, are seen while the tumor (seen in the left half of the picture) has completely obscured the left optic nerve (Fig. 1 h). The tumor is removed by making an entry into its anterial portion and debulking it. The opposite optic nerve is generally decompressed earlier than the ipsilateral one. In other words, while operating from the right side, the left optic nerve is decompressed first. The reason is that it is easier to see and free the contralateral side, especially the more critical undersurface. Since retraction of the optic apparatus is forbidden, tumor bits may remain concealed or resist excision under the ipsilateral nerve. After careful and total tumor removal the left optic nerve can be visualized, appearing as a thin twine (Fig. 1 i). The left ICA and A1 are also visualized.

Case 2. This large pituitary adenoma grows upward from the right side and is also seen to compress the third ventricle and extend into the lateral ventricles (Fig. 2 a, b). The lateral reconstruction scan shows the compressed lamina terminalis (Fig. 2 c, arrow), the optic nerve, and basilar artery. The tumor extends asymmetrically below and behind the chiasma and upward into the third and lateral ventricles. At operation the tumor was at first visualized between the optic nerves and between the right side of the optic apparatus and the right ICA (Fig. 2 d). This part of the tumor was first excised. The contiguous postchiasmatic component was then chased and delivered piecemeal with microtumor forceps out of the anterior third ventricle whose cavity appeared fully exposed (Fig. 2 f). The lateral ventricular component was also mobilized, delivered downward, and excised. Finally, the last bit that lay prechiasmatic (Fig. 2 g) was tackled, carefully preserving all blood vessels to the optic apparatus while decompressing it (Fig. 2 h). Thus surgery in this region requires extreme caution and care at preservation of the optic apparatus, the circle of Willis and its profuse outflow, and the hypothalamus.

Case 3. The third case is that of an adult with a craniopharyngioma. The CT scan demonstrated a well-circumscribed, rounded mass, rather posteriorly placed and causing obliteration of the third ventricle and foramen of Monro and resulting obstructive hydrocephalus (Fig. 3 a). At surgery, the chiasma appeared relatively anterior, perhaps dislocated forward to an extent by the posteriorly located mass. Although a bit of tumor could be visualized in the prechiasmatic region, with this very little room available in the front it was unwise trying to deliver it from the front for the risk of scraping against the optic chiasma, as well as imposing a traction on the unvisualized hypothalamus and stalk (Fig. 3 b, c). A further, gentle and backward retraction of the frontal lobes led us to the postchiasmatic region. The A1 segments, lying in the cistern of lamina terminalis was thus exposed (Fig. 3 d). On opening it the tumor could be visualized directly in the third ventricular cavity (Fig. 3 e).

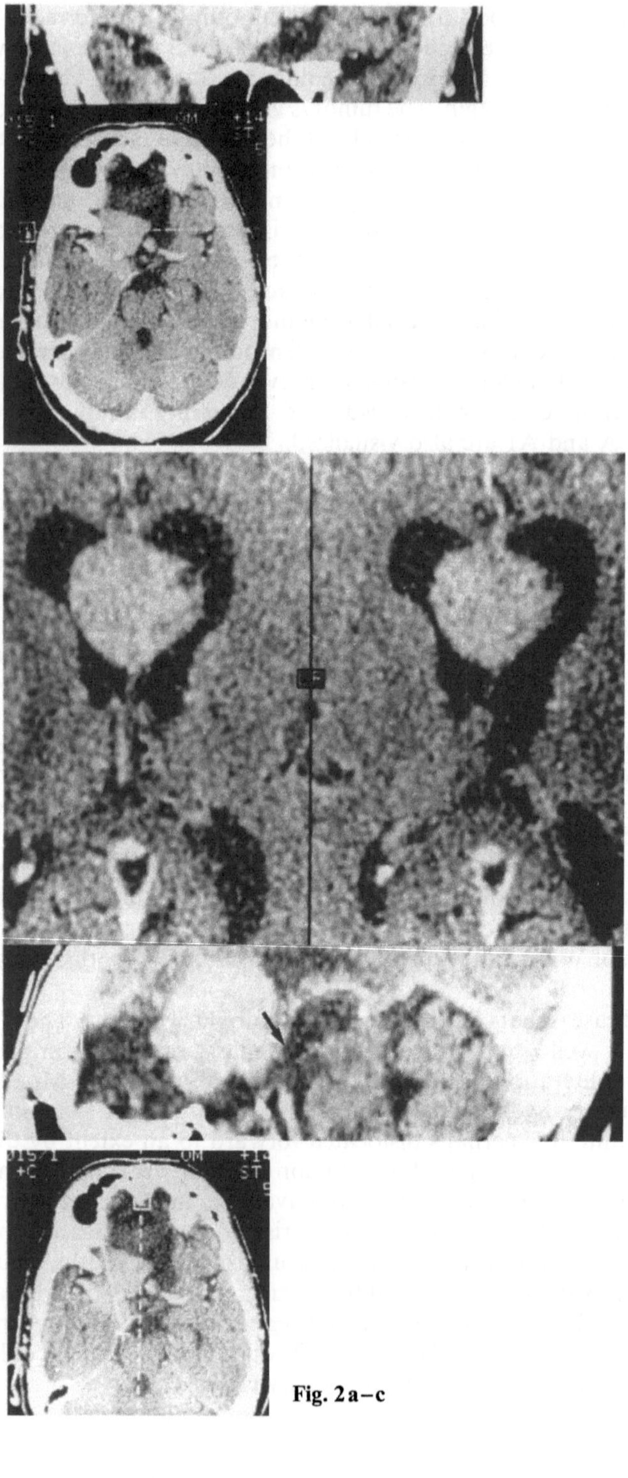

Fig. 2a–c

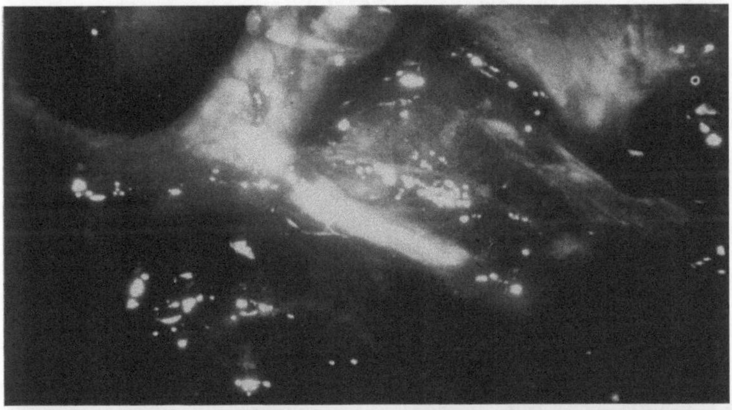

d

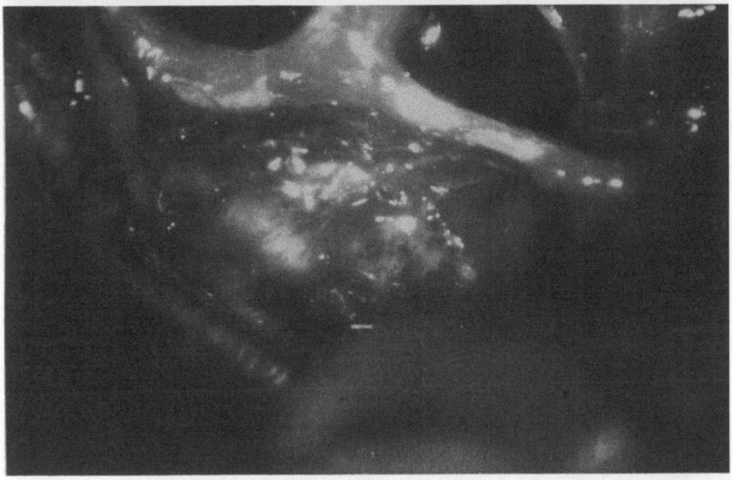

e

Fig. 2. **d** The presence of tumor in the right opticocarotid space (lateral to the right optic nerve, chiasma, and tract). The tumor extends further backward in the post-chiasmatic region seen in the dark foreground. **e** Moving backward in the retrochiasmatic region one can discern the solid mass of tumor between the optic apparatus and the third ventricle. Note the absence of tumor now between the right optic nerve and ICA. While removing this bit one must be careful to preserve all blood vessels going to the optic apparatus and diencephalon. Those vessels running over the tumor are spared; only those actually entering the tumor are coagulated and cut

◄

Fig. 2. **a** CT scans of a middle-aged woman demonstrating another large pituitary adenoma with an asymmetrical extension to the right para- and suprasellar regions, and upward into the third ventricle. **b** The tumor extending further upward through the floor of the lateral ventricles. **c** The tumor component extending into the third ventricle is seen to flatten the lamina terminalis (*arrow*). Lower, one can visualize the optic nerve and basilar artery

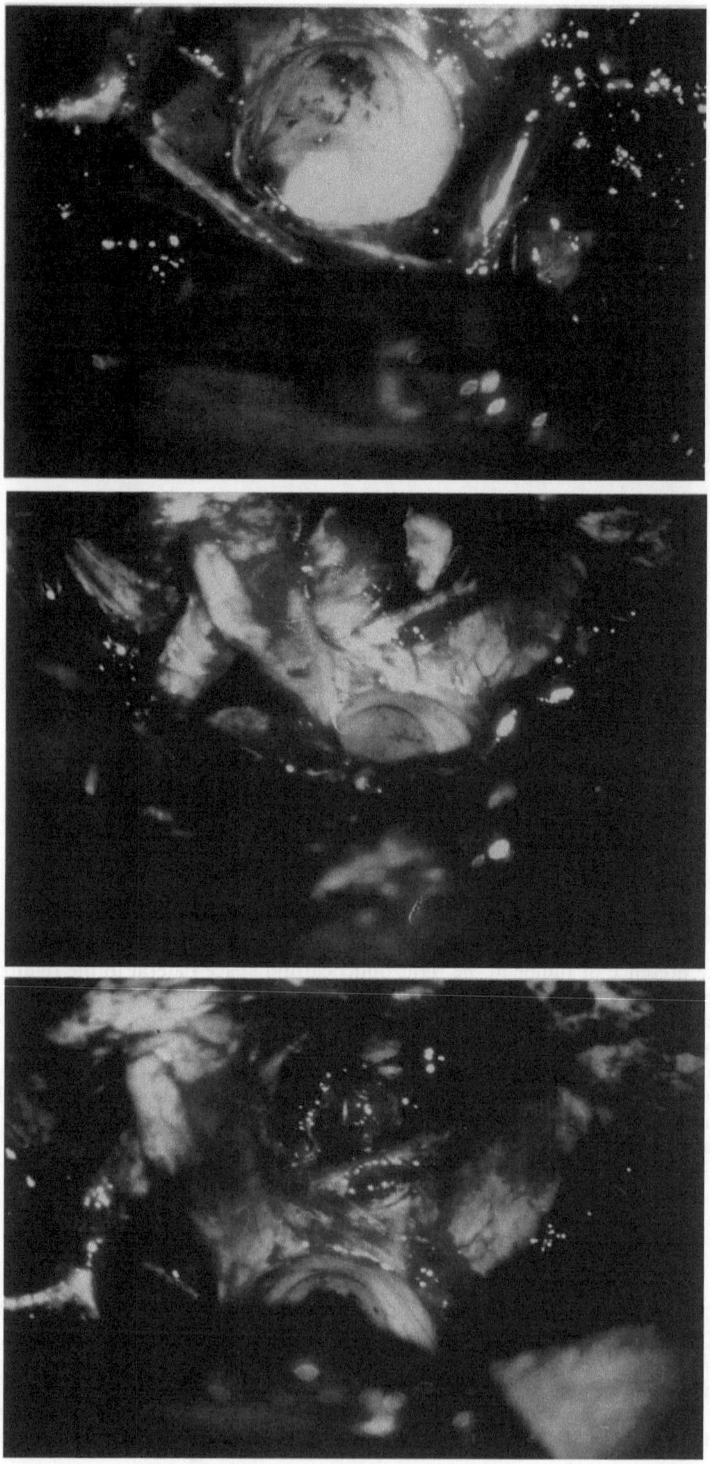
f
g
h

Fig. 2f–h

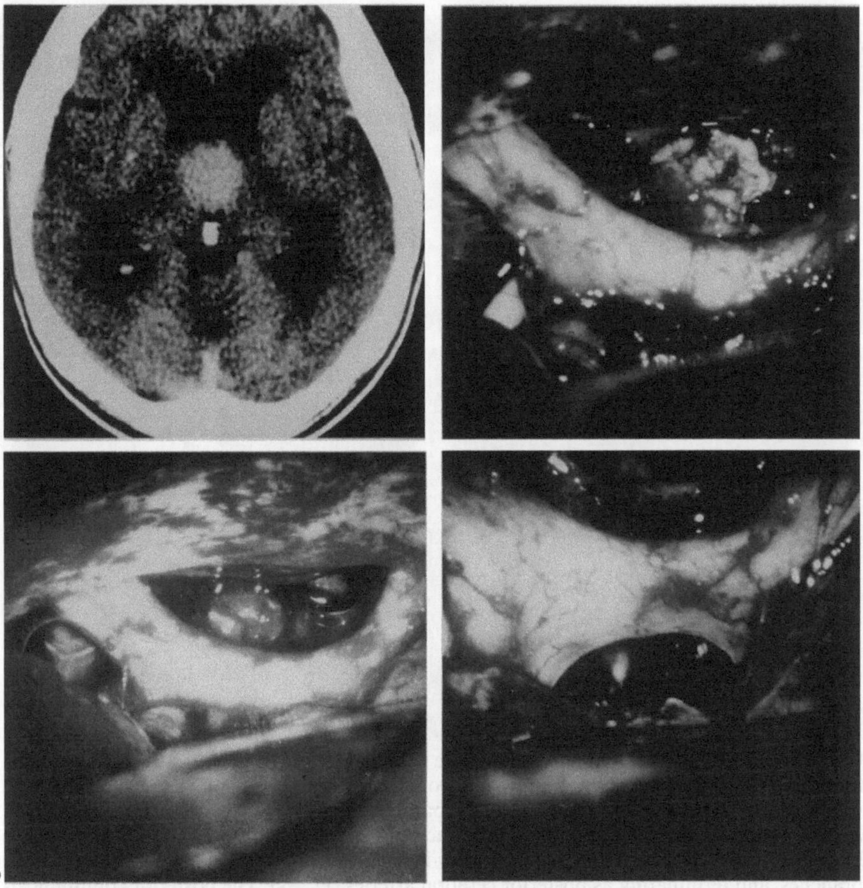

a c b d

Fig. 3. **a** CT scans of a middle-aged man who presented with progressive dementia. The CT scan demonstrates the hydrocephalus to be the cause of his mental deterioration, on account of a large rounded "intra" third ventricular tumor which is occluding both the interventricular foramina. **b** The standard bifrontal craniotomy has been carried out. The optic nerves and chiasma are identified. The chiasma appears to be dislocated forward by the push of the retrochiasmatic space-occupying lesion. A small bit of tumor can be seen through a 1–2 mm gap between the chiasma and diaphragma sella. However, this distance is not enough to permit a safe or substantial tumor removal from there. **c** An attempt made at biting the tumor. The left oculomotor nerve is seen running close to the left optic tract. **d** Since the chiasma is relatively prefixed and most of the tumor sub- and retrochiasmatic, it was necessary to attack the lesion directly from the retrochiasmatic region. The frontal lobes are retracted, and along with it both the A1 segments. The lamina terminalis appears here as a glassy curtain

←

Fig. 2. **f** After removal of the large portion of tumor impacting against the lamina terminalis and creating a craterlike effect in the third ventricle. The two ICAs with their respective A1 segments are seen "joining hands" as they course over the optic tracts. **g, h** The smaller prechiasmatic element is visualized and excised

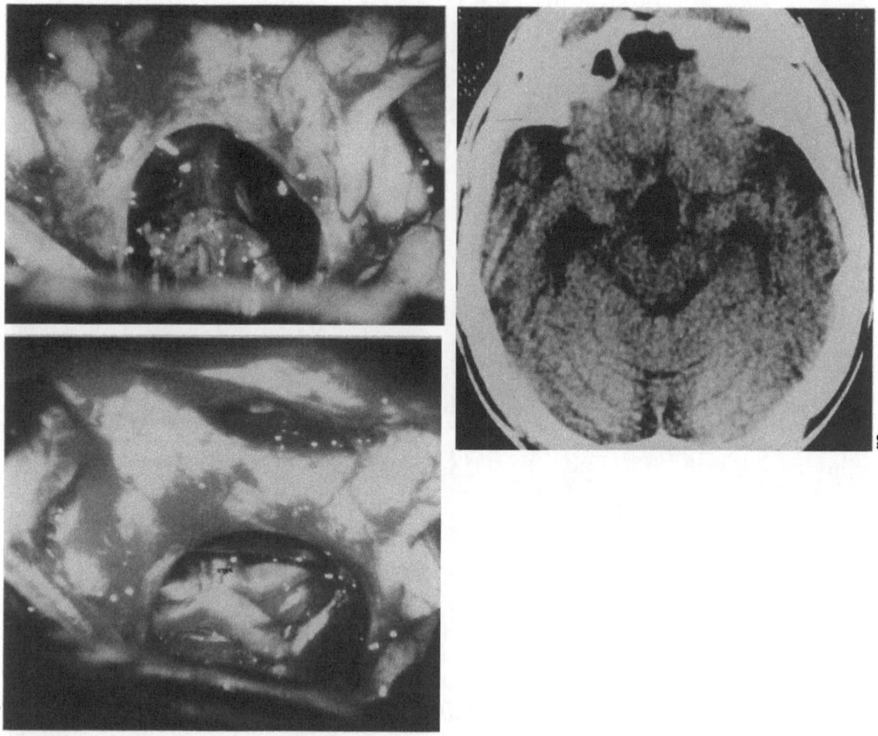

e

f

g

Fig. 3. **e** On opening the lamina terminalis we entered the third ventricle and confronted the tumor directly as it made its way upward through the floor and then the cavity of the third ventricle. The tumor was very carefully separated from the walls of the third ventricle containing the fornix and the hypothalamic nuclei. **f** Having achieved total tumor removal, we have this incredible sight of the basilar artery in the prepontine space, giving out its various branches, and terminating as posterior cerebral arteries (PCAs). The posterior communicating artery can be seen to communicate the right PCA with the right ICA. The ICAs and A1 segments are also seen, and full formation of the circle of Willis can now be appreciated. **g** Postoperative CT scan shows total tumor removal

Craniopharyngiomas arise from the remnants of Rathke's pouch, and its point of origin is the hypothalamic stalk. As the tumor grows upward it invaginates its way into the floor of the third ventricle, and while doing so, the lateral walls of the third ventricle in turn become the cap covering the dome of the tumor. With persistent pressure this cap may thin out, and the tumor may be seen through its transparency or may directly break through it and appear "bare" inside the third ventricular cavity. The gliotic tumor capsule is very delicately separated from the walls of the third ventricle which contain, in this region, the pillars of the fornix and the hypothalamic nuclei. Any traction on the diencephalon would be signaled by changes in the vital parameters and would also have serious repercussions on the patient's health. On total tumor removal, and with the third ventricle opened out all the way through by the

Table 4. Pre- and postoperative symptoms in large sellar region tumors ($n = 48$)

Symptoms	Preoperative	Postoperative	
		Improvement	Worsening
Visual	43	26	3
Hormonal	32	10	6
Increased ICP (hydrocephalus)	14	11	6
Ocular nerve deficits	8	3	2
Limb paresis	2	1	3

Other postoperative complications include: diabetes insipidus, 8 (6 controlled with drugs, 2 uncontrolled); CSF leakage, 6; death, 5 (11%). Causes of death: cardiopulmonary problems, 3; pulmonary embolism, 1; central brain damage, 1.

tumor we are directly in the prepontine region and can visualize the basilar artery and its outflow (Fig. 3f).

Results

Of the 48 large sellar region tumors, 38 were removed transcranially and 7 transsphenoidally. Three of the patients were provided with a shunt and did not have a definitive operation for various reasons. Of the 45 large tumors thus operated upon, 75% could be totally excised. In the remainder total excision was not possible owing to penetration of cavernous sinus (8), dense adhesion to hypothalamus (1), and/or malignancy (2).

Postoperatively, visual improvement occurred in 60% and normalization of hormonal status in 30% of those already having dysfunction. Ocular muscle palsies also recovered in three out of eight patients, and intracranial pressure dropped promptly after tumor removal in nearly 80% of them (Table 4). Postoperatively, worsening occurred in the form of visual deterioration in three, hormonal dysfunction in six, ocular palsies in two, hemiparesis in three, and hydrocephalus in six. CSF leakage that did not remit spontaneously occurred in six patients and ceased in all of them after they were repaired. Diabetes insipidus occurred in one of every six patients, most of them being controlled by drugs. Five out of 45 operated patients died. All of them were over 60 years old, and most hat preexisting cardiopulmonary problems. One death occurred due to hypothalamic damage.

Conclusion

Despite the general belief that most sellar region tumors, particularly the pituitary adenomas, are now diagnosed relatively early our experience has shown that every second tumor is large enough to involve the third ventricle. Their removal is challenging due to the important anatomic and physiologic structures in its vicinity. There is no definitive strategy that can apply uniformly to all cases; however, some fundamental surgical principles need to be observed irrespective of the lesion, in this sensitive area. The subfrontal approach (uni- or bilateral) is ideally suited for removal of most of these tumors. Since most of these lesions are benign, their total removal leads to a cure. However, in view of the intricate anatomic location, the experience of the surgeon in the planning and execution of such operations holds the key. Indeed, microsurgical techniques have proven that surgery in this region can be radical without being unsafe.

References

Patterson RH, Danylevich A (1980) Surgical removal of craniopharyngiomas by a transcranial approach through the lamina terminalis and sphenoid sinus. Neurosurgery 7:111–117

Rand RW (1978) Transfrontal transphenoidal craniotomy in pituitary and related tumors. In: Rand RW (ed) Microneurosurgery. Mosby, St. Louis, pp 93–104

Rhoton AL, Renn WH, Harris FS (1978) Microsurgical anatomy of the sellar region and cavernous sinus. In: Rand RW (ed) Microneurosurgery. Mosby, St. Louis, pp 71–92

The Use of CO_2 Laser in Transsphenoidal Surgery of Sellar Region Tumors

Ž. Gnjidić, L. Negovetić, D. Vidović, and D. Lovrić, Zagreb/Yugoslavia

Due to its complex anatomic and topographic relationships, surgery of tumors in the sellar region confronts the surgeon with very complex technical problems. Nowadays, over 90% of tumors of this region are removed by the transsphenoidal microsurgical approach. Surgical microscope, bipolar coagulator, special microsurgical instruments, irrigation and suction instruments, X-ray instrument with a special electronic amplifier, and the intraoperative measuring of the hyperproduced hormone enable the neurosurgeon to achieve an enviable success. But some problems of a technical nature still remain to be solved.

Sellar region tumors by their growth usually enlarge the sella turcica and erode its front wall. Still, in 4% of cases the front wall and the sellar floor are thicker than 2 mm (Fig. 1). In such cases up to now a high-speed diamond drill has been used for the opening of the front sellar wall. Although its construction is maximally adapted to its function, this instrument obstructs the field of sight. If we use a bayonet chisel, visibility is good and the opening of the thick bone efficient, but we have learned through our experience with over 200 operated patients that in such cases there are almost regular intensive bleedings that

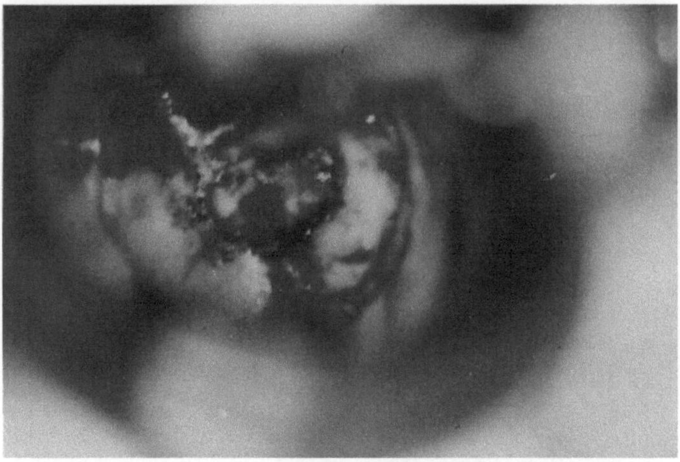

Fig. 1. Pituitary microadenoma. Hypoplastic sphenoid sinus with a thick front sellar wall and sellar floor

occur from the front intracavernous sinus whose diameter ranges from 0.5 to 4 mm. Without a special bipolar forceps or the application of hemostatic clips it is very difficult to achieve hemostasis. Due to the solid blast and water blast effect caused by the chisel opening, cavernous sinus venous plexus tends to become ruptured, and profuse bleedings often occur from the lateral sellar walls during the tumor extirpation. These bleedings interfere with the microsurgical procedures so that, besides other complications, there is a possibility of more frequent cases of residual tissue and recurrences.

Up to now the CO_2 laser has been used in surgery and microsurgery for the coagulation and vaporization of tissue. We have tried to solve the problem of bone opening by the CO_2 laser with the wave length of 10.6 mm (Takizawa et al. 1980; Ascher 1982). Focused CO_2 laser beam with the power of 2.5–5 W provides the energy sufficient to perform the incision of the bone layer of the front sellar wall without the absorption of heat into the surrounding tissue and with no possibility for detrimental effects. So by means of the "no-touch technique," without the use of a direct mechanical force and with the maximum visibility of the field of sight, the front sellar wall is fenestrated, and we then proceed with the conventional microsurgical opening technique (Fig. 2). After the bone has been opened, we encounter the front sellar wall periostal layer saturated by the intracavernous sinus and numerous smaller blood vessels. By diminishing the laser beam power to 0.5 W and defocusing it to 1.5 mm we incite the coagulation of blood vessels. The beam is then focused to 0.1 mm, and by the power of 1 W we perform the crosslike incision of the formerly coagulated periostal layer and approach the tumor (Beck 1980).

Pituitary adenomas, the most frequent tumors of the sellar region, being soft in consistency, are easily removed by the classical microsurgical technique without affecting the surrounding pituitary tissue. However, sometimes these

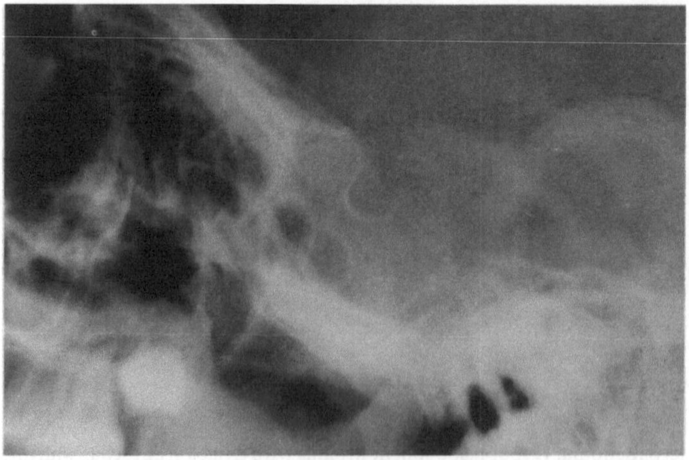

Fig. 2. Front sellar wall fenestrated by the CO_2 laser

tumors tend to be fibrously altered, of hard consistency, containing calcificates, and can be extremely adherent to the surrounding anatomical structures. Such alterations are sometimes encountered in large adenomas in which micronecrosis and fibrous hyaline changes took place after the incomplete surgical interventions. In such cases the removal of tumor by a classical microsurgical technique becomes very difficult and hazardous because of a possibility of damage to the remaining unaffected pituitary tissue, possible infundibulum lesion, diaphragm lesion, a more intensive intraoperative CSF leak as well as a possibility of bleeding from the cavernous sinus. By changing the laser beam focus from 0.5 to 2 mm and its power from 1 to 5 W we can, without any risk to the surrounding unaffected tissue, perform either the coagulation or the vaporization of the tumor, thus avoiding most unpleasant difficulties. (Tew and Tobler 1983).

The CO_2 laser can also be used for opening the front sellar wall in cases of reoperation. The bone layer, which served for the reconstruction of the front wall, has grown into both the surrounding bone and the implantation after the first surgery and it has formed a strong scar. By means of the technique described above the scar can be vaporized, the bone preserved for the new, repeated closure, intrasellar scar tissue can be vaporized, and then surgery is continued by the conventional microsurgical technique.

In conclusion we can say that the conventional microsurgical technique still predominates in the removal of sellar region tumors. However, as in the above cases, its advantages can be used for achieving better surgical results if we consider the CO_2 laser a complementary technical means.

References

Ascher PW (1982) Neurosurgery. In: Andrews AH, Polany TG (eds) Microscopic and endoscopic surgery with the CO_2 laser. Wright, Bristol, pp 298–314

Beck OJ (1980) The use of the Nd: YAG and the CO_2 laser in neurosurgery. Neurosurg Rev 3:261–266

Takizawa T, Yamazaki T, Miura N et al. (1980) Laser surgery of basal, orbital and ventricular meningiomas which are difficult to extirpate by conventional methods. Neurol Med Chir (Tokyo) 20:729–737

Tew JM, Tobler WD (1983) The laser: history, biophysics and neurosurgical application. Clin Neurosurg 31:506–549

Practical Experiences with the Nd-YAG Laser in Pituitary Surgery

R. Oeckler, München/FRG

Introduction

The introduction of microsurgical techniques led to a great advance in the surgery of sellar and circumsellar lesions. It favored not only the renascence of the transsphenoidal approach and the possibility of selective adenomectomy, (Guiot and Thibaut 1959) but also improved the results in transcranial procedures by facilitating the preparation of the tumor from the important and sensitive surrounding structures, i.e., the vessels of the circle of Willis and the optic pathways (Fahlbusch and Marguth 1978). Nevertheless, some critical conditions may be harmful for complete tumor removal. This may be especially true in recurrent cases, where tumor remnants often adhere to the chiasm or vessels. Furthermore, adenomatous tissue can also be found fixed in scar formation in transsphenoidal reoperations. Therefore, second procedures are burdened with a significantly higher incidence of complications and a lower functional success rate (Laws et al. 1985). Following these considerations, the search for new technical methods to improve the results was an obvious one.

For nearly one decade, laser systems have been used in experimental and clinical examinations for different surgical purposes. In neurosurgery, however, the CO_2 as well as the Nd-YAG laser has been preferably used and investigated (Ascher 1978; Beck 1980; Saunders et al. 1980). For reasons discussed below, we focused on the properties and advantages of the Nd-YAG laser system.

Physical and Technical Remarks

The neodymium-YAG laser (wavelength 1.32 mm) which we used in our experimental and clinical studies is primarily a coagulation device. The light energy of the laser beam is taken up by scattering in the irradiated tissue. The lesion produced is a thermic one. In biological tissue, which is principally a protein solution with a high water content, the lesion borderline runs along the region where the tissue temperature falls below 55 °C, i.e., the critical temperature of protein coagulation. Due to the energy uptake by scattering and the energy distribution in the light beam itself (following the Gaussian distribution), a lesion of parabolic profile occurs in standard conditions. In biological tissue, however, this profile is changed by its physical properties,

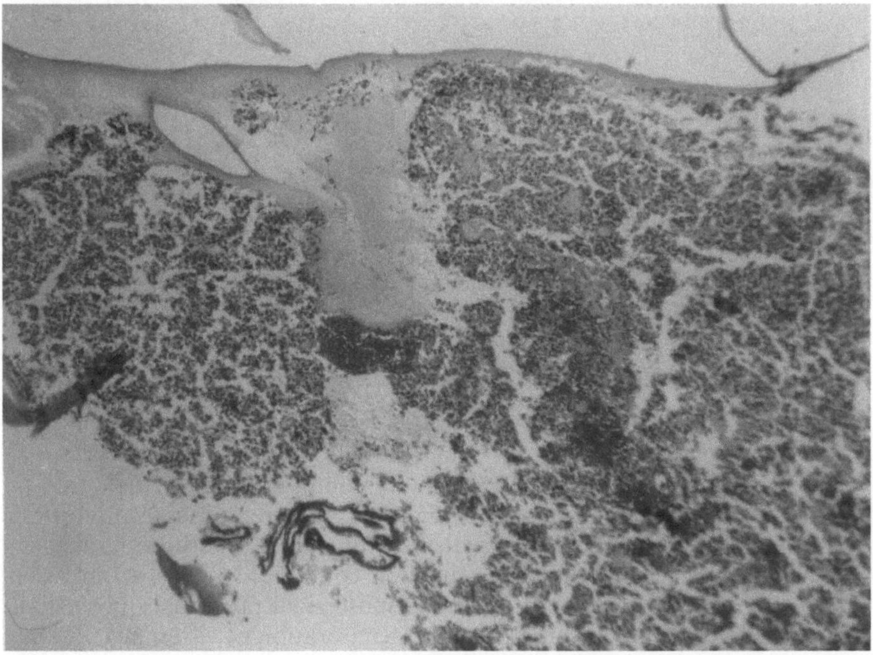

Fig. 1. Histological specimen of a nonfunctioning pituitary adenoma (H&E staining) following defocused Nd-YAG laser coagulation with 30 W per 2 s. The thermic lesion is sharply demarked and has a parabolic profile. Penetration depth is about 1 mm

namely its water content and homogeneity. All these findings could be confirmed also in specimens of pituitary adenoma tissue (Fig. 1). Additionally, the depth shrinkage effect of Nd-YAG laser coagulation leads to a compression effect on the vessels in the coagulated area so that hemostasis occurs.

Methods and Materials

In all surgical procedures, we used the Nd-YAG laser equipment produced by Messerschmitt Medizintechnik (wavelength 1.32 mm). In view of the deep and narrow operative fields, we designed a microsucker-like, flexible tube, containing a flexible optic fiber. This is cooled by CO_2 gas blown along the tube (Fig. 2).

Our experiences were based on a total of 200 surgical procedures with the additional use of the laser. Most of the lesions were pituitary adenomas, followed by craniopharyngeomas and suprasellar meningeomas. In 171 cases, we used the transsphenoidal approach; 29 patients were operated on transcranially; 32 surgical procedures had to be performed on recurrent cases.

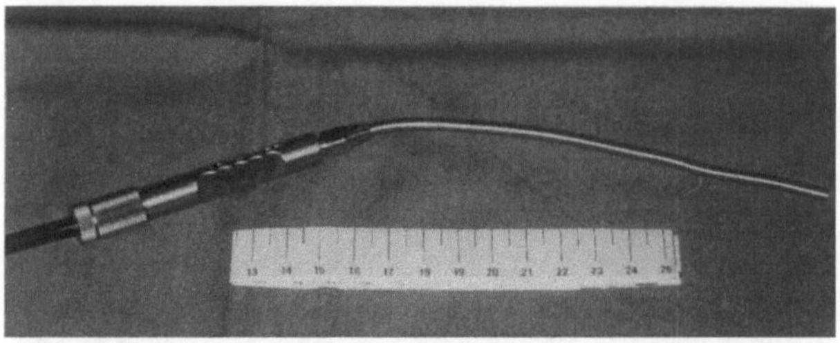

Fig. 2. Microsurgical handpiece for application of the Nd-YAG laser, containing the flexible fiber optic of 1.2 mm. External diameter is 2.4 mm. The light fiber is cooled using CO_2 gas

Results

When analyzing the functional results, we found a normalization rate in hypersecreting adenomas of 80% (55% in recurrent cases). Following transsphenoidal decompression in patients with chiasm syndrome, normalization or improvement of the visual function could be documented in 93%; worsening or blindness did not occur. In transcranial procedures, normalization or improvement was found in 58%, deterioration of vision in 15%, but none of the patients became completely blind. The postoperative worsening rate of the anterior lobe functions was 10% following transsphenoidal procedures, which was found both in primary and secondary procedures. These results correlate very well with the findings described in conventional series (Fahlbusch and Marguth 1980; Findlay et al. 1983).

The rate of postoperative posterior lobe dysfunction, however, was relatively high. An explanation for this finding is given below. Generally, the complication rate was low when compared with other reported series using conventional techniques. There was no mortality, and, importantly, no clinically apparent rebleeding episodes occurred.

Discussion

Sellar lesions, especially pituitary adenomas, seemed a very suitable object to verify the advantages of the Nd-YAG laser as well as possible side effects by objective and sensitive investigation parameters, ranging from skillful endocrinological and ophthalmological function tests to high-resolution imaging methods such as computed tomography and magnetic resonance imaging.

In practice, the application of laser light using the microsurgical handpiece was very comfortable and superior to the more space-occupying action of the bipolar forceps, especially in deep and narrow operative fields of transsphenoidal or transventricular approaches. The free three-dimensional handling of the handpiece with the additional possibility of flexing the tube permitted an unhindered access even to more hidden parts of the operative area.

The well-definable and controllable lesion size (in contrary to the bizarre and unpredictable shape following bipolar electric coagulation) and its smooth shrinkage effect facilitated a safer preparation of the tumor capsule from vessels and the optic pathways, which could be objectified by the follow-up, when focusing on the ophthalmological results. The reliable hemostatic effect could also be documented by the observation that no clinically apparent postoperative hemorrhage occurred in this series.

There seemed to be only one side effect attributable to the use of the laser, the strikingly high incidence of postoperative posterior lobe dysfunction. Our explanation for this is that the posterior lobe is localized in the straight main direction of the laser in transsphenoidal procedure and therefore may be easily damaged. We recognized this condition following our very first follow-ups and thereafter covered the region of the posterior lobe using a cotton pledget during intrasellar laser application. Consecutively, the incidence of this complication decreased promptly. The safety and selectivity of laser coagulation could be further verified when analyzing the postoperative worsening rate of anterior lobe functions following selective transsphenoidal adenomectomy. This was 10% in primary as well as in secondary procedures, which seems a good result when compared with conventional series (McLanahan et al. 1978; Nelson et al. 1984). Especially in recurrent cases, the use of the laser was very effective and gentle, as reflected by the functional endocrinological and ophthalmological results.

In summary, the Nd-YAG laser fulfilled the expectations which were made in view of the physical properties revealed by experimental studies. Its use is more comfortable, more effective, and safer than electric bipolar coagulation, which was demonstrable by a skillful follow-up using very sensitive functional parameters in sellar lesions. These advantages became especially apparent during the microsurgical preparation of the tumor capsule from the adjacent structures and in recurrent cases. Whereas ultrasonic aspiration units are very helpful for removal of the tumor mass, the Nd-YAG laser appeared to be an effective supplemental device for microsurgical preparation procedures and therefore presents another useful microsurgical instrumentation.

References

1. Ascher PW (1978) The use of CO_2-laser in neurosurgery. In: Kaplan J (ed) Laser surgery II. Jerusalem Academic Press, Jerusalem, pp 76–78
2. Beck OJ (1980) The use of the Nd-YAG-laser and the CO_2-laser in neurosurgery. Neurosurg Rev 3:216–266
3. Fahlbusch R, Marguth F (1978) Development in surgical treatment of pituitary adenomas. Neurosurg Rev 1:5–13
4. Fahlbusch R, Marguth F (1980) Optic nerve compression by pituitary adenomas. In: Samii M, Jannetta PJ (eds) The cranial nerves: anatomy, pathology, pathophysiology, diagnosis, treatment. Springer, Berlin Heidelberg New York, pp 140–147
5. Findlay G, Fadzean RMMc, Teasdale G (1983) Recovery of vision following treatment of pituitary tumors, application of a new system of assessment to patients treated by transsphenoidal operation. Acta Neurochir 68:175–186

6. Guiot G, Thibaut B (1959) L'extirpation des adenomes hypophysaires par voie transsphenoidale. Neurochirurgica 1:133–150
7. Laws ER, Nicolee CF, Redmond MJ (1985) Transsphenoidal surgery following unsuccessful prior therapy. J Neurosurg 63:823–829
8. McLanahan CS, Christy JM, Tindall GT (1978) Anterior pituitary function before and after transspenoidal microsurgical resection of pituitary tumors. Neurosurgery 3:142–145
9. Nelson AR, Tucker SG, Becker DP (1984) Residual anterior pituitary function following resection of pituitary macroadenomas. J Neurosurg 61:577–580
10. Saunders ML, Young MF, Becker DP (1980) The use of the laser in neurological surgery. Surg Neurol 14:1–10

Pituitary Interstitial Irradiation for Cushing's Disease and Acromegaly

L. M. SANDLER, High Wycombe/U.K., A. JADRESIC, D. H. CARR, J. M. BURRIN, R. ARNOT, and G. F. JOPLIN, London/U.K.

Introduction

For 30 years interstitial irradiation (by needle implantation) with either gold-198 (^{198}Au) or yttrium-90 (^{90}Y) has been used as a method of treating pituitary tumors at Hammersmith Hospital, London. We have recently published our long-term results of pituitary implantation for both Cushing's disease (Sandler et al. 1987) and acromegaly (Jadresic et al. 1987), and these data form the basis of this report. Our long-term follow-up of patients with macroprolactinomas treated with ^{90}Y is currently being analysed.

Methods

The planning of dosimetry (Duggan et al. 1963) and the operative technique (Joplin et al. 1978) have previously been described. Initially, ^{198}Au alone was used (a γ-emitter), but this was soon replaced by ^{90}Y (a β-emitter). The selection of β-emitter resulted primarily from the necessity to localise the irradiation as much as possible within the small volume of gland so as to enable a large dose to be given, yet not to over-irradiate adjacent nervous tissues. The isodose curves indicate that the dose of 90Y is halved at each millimeter distance from the rod (Fig. 1). The dose of ^{198}Au used was usually 10 Gy, but that of ^{90}Y varied from 20 to 150 Gy according to the clinical objective, being the planned maximum doses at the gland surface.

Long-Term Results

Cushing's Disease. Between May 1959 and June 1982, 86 patients with pituitary-dependent Cushing's disease were treated with interstitial irradiation using ^{198}Au and ^{90}Y or both. None had previously received external beam radiotherapy or pituitary surgery. Four patients returned overseas immediately after the implant and were lost to follow-up. In the 82 patients who were reassessed 1 year after treatment, 63 (77%) achieved remission (although this has more recently risen to 85% of 52 patients; Table 1). Remission is defined by (a) clinical and biochemical evidence of corticosteroid deficiency and (b) normal urinary free cortisol or 17-oxogenic steroid excretion, which remained

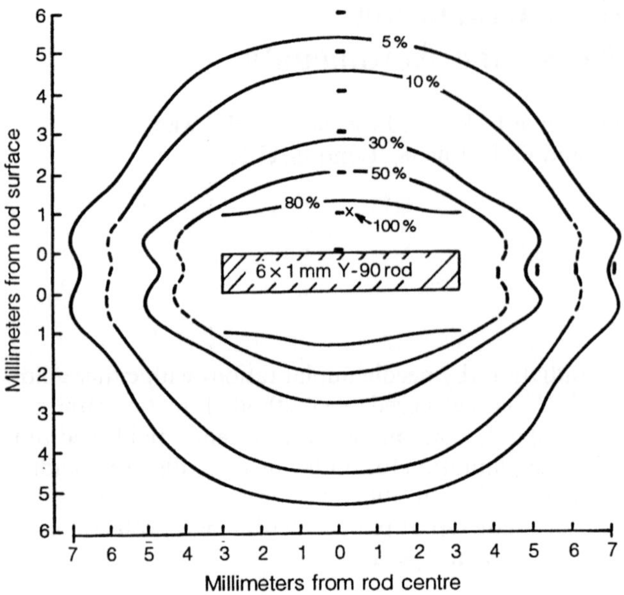

Fig. 1. Isodose lines around a ^{90}Y rod. (Adapted from Duggan et al. 1963)

so during the follow-up period. The present study comprises the outcome and complications in the 54 patients (48 women) who had a remission, and whom we were able to follow up. The follow-up period ranged from 3 to 26 years (mean 10.5) from the time of remission. No instance of clinical or radiological relapse has occurred in 569 patient-years follow-up (Table 1). Of these 54 patients, ^{90}Y alone was used in 32, of whom 12 (37%) required corticosteroid or T_4 replacement therapy in a mean time of 3.5 months; in 7 of these 12 we elected to give an ablative dose. Hypopituitarism was defined as the clinical need for replacement therapy with a glucocorticoid or T_4 combined with biochemical deficiency of the hormone(s). The number of patients requiring hormone replacement after ^{90}Y is shown in Fig. 2.

^{198}Au alone was used in 15 patients, of whom 7 (47%) developed hypopituitarism in a mean of 76 months. Both isotopes were used in 7 patients. A diurnal serum cortisol rhythm was found in 28 of the 31 patients who were

Table 1. Overview of pituitary interstitial irradiation

	Remission	Hypopituitary	Recurrence
Cushing's disease	1957–1982: 77% 1977–1987: 85%	All doses: 35% Single low dose: 10%	Nil 569 patient-years
Acromegaly	1965–1973: 52% 1977–1987: 65%	39%	Nil 588 patient-years

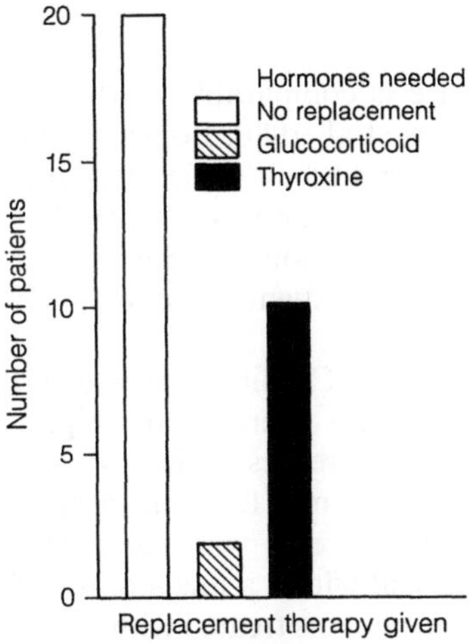

Fig. 2. Replacement hormones required in patients with Cushing's disease treated with ^{90}Y alone ($n = 32$)

not receiving corticosteroid therapy. In five of the seven patients with an initially abnormal pituitary fossa, serial radiological studies revealed remodelling in three. There have been no complications in the last 17 years.

Acromegaly. Between 1 January 1965 and 31 July 1973 a total of 155 acromegalic patients (79 women) were treated by implantation with ^{90}Y in the Endocrine Unit at Hammersmith Hospital, London. In 100 patients 90Y was the sole treatment for the macroadenoma, and in 52 of them growth hormone (GH) level at 60, 90 and 120 min of an oral glucose tolerance test was under 5 mIU/l after pituitary implant. More recently (1977–1987) the remission rate has increased to 65%. From the parent group of 100 patients, 38 were selected insofar as we were able to follow them up for more than 10 years after a single dose of ^{90}Y interstitial irradiation as the sole treatment. Mean serum GH had fallen from 106 to 24 mIU/l within 3–6 months and then slowly declined to 4 mIU/l after 10 years. GH levels of 5 mIU/l or lower were obtained in 8% of patients at 3–6 months and in 18% at 1 year, the cumulative proportion increasing to 53% at 10, and 76% at 14 years. The proportion of patients requiring hormone replacement therapy rose from nil preceding implant to 16% by 3–6 months and then slowly increased to 39% by 14 years.

Serial coned radiographs of the pituitary fossa were available for 32 of the 38 patients for at least 10 years of follow-up; 16 showed thickening of the dorsum sellae and/or reduction of at least one diameter by 3 mm.

Concerning symptoms, all 29 patients whose GH level fell to 5 mIU/l or below showed improvement, 22 becoming asymptomatic. Seven patients with lesser falls in GH levels (from a mean of 193 to a mean of 15 mIU/l) all improved, one becoming asymptomatic. Two showed no variation. There has been no recurrence in 588 patient-years of follow-up.

Discussion

The remission rate achieved in Cushing's disease is comparable to that of transsphenoidal surgery (Salassa et al. 1978; Bigos et al. 1980; Boggan et al. 1983). In acromegaly biochemical remission is more difficult to define: some groups rate as satisfactory a drop in serum GH concentrations to 20 mIU/l or below (Christy 1982), whilst others regard "normalisation" as 5 mIU/l or below (Teasdale et al. 1982). In our study 76% of patients achieved GH concentrations of 5 mIU/l or lower.

Preservation of pituitary function after treatment of Cushing's disease and acromegaly and avoidance of long-term replacement therapy are highly desirable. In Cushing's disease, there were marked differences in the response to ^{90}Y and ^{198}Au. Overall, ^{90}Y resulted in hypopituitarism in less than 40% of patients (usually in those electively given a higher dose or who required a supplementary implant). Pituitary failure was evident within a few months of the ^{90}Y-implanted patients; in contrast, 50% of the patients treated with ^{198}Au are presently receiving replacement therapy, and pituitary failure was usually of late onset.

In acromegaly, hypopituitarism after ^{90}Y was cumulative up to 14 years after implant, although beyond 5 years only 4 of 27 patients showed subsequent deterioration. The large size of the tumors that were frequently found in acromegaly probably resulted in the occurrence of central tumor necrosis but only cessation of mitoses at the periphery after interstitial irradiation. This may account for the ongoing increase in biochemical remission and pituitary failure which is not evident in ^{90}Y-treated Cushing's disease.

The total lack of recurrence after long-term follow-up of more than 1000 patient-years in both Cushing's disease and acromegaly appears to be unique.

Pituitary implantation of ^{90}Y results in good remission rates in both Cushing's disease and acromegaly. Around one-third of patients require hormone replacement therapy, but the procedure has had no other complications for nearly two decades. As yet, there has been no instance of tumor recurrence once remission has been achieved.

References

Bigos ST, Somma M, Rasio E, Eastman RC, Latnthier A, Johnston HH, Hardy J (1980) Cushing's disease: management by transsphenoidal microsurgery. J Clin Endocrinal Metab 50:348

Boggan JE, Tyrell JB, Wilson CB (1983) Transsphenoidal microsurgical management of Cushing's disease: report of 100 cases. J Neurosurg 59:195

Christy NP (1982) Chosing the best treatment for acromegaly (editorial) JAMA 247:1320

Duggan MH, Jones E, Mallard JR, Joplin GF (1963) The technique and dosimetry of pituitary implantation using sources of Y^{90}. In: Production and use of short-lived radioisotopes from reactors, vol II. International Atomic Energy Agency, Vienna, p. 213

Jadresic A, Jimenez LE, Joplin GF (1987) Long-term effect of ^{90}Y pituitary implantation in acromegaly. Acta Endocrinol 115:301

Joplin GF, Banks L, Child DF, Diamant L, Doyle FH, Jadresic A, Koochek MH (1978) In: Treatment of pituitary adenomas. Thieme, Stuttgart, p 261

Salassa RM, Laws ER, Carpenter PC, Northcutt RC (1978) Transsphenoidal removal of pituitary microadenoma in Cushing's disease. Mayo Clin Proc 53:24

Sandler LM, Richards NT, Carr DH, Mashiter K, Joplin GF (1987) Long term follow-up of patients with Cushing's disease treated by interstitial irradiation. J Clin Endocrinol Metab 65:441

Teasdale GM, Hay ID, Beasttall GH et al. (1982) Cryosurgery or microsurgery in the management of acromegaly. Am J Med 247:1289

Complications of Transsphenoidal Surgery

E. R. Laws, Washington, D.C./USA

The complications of transsphenoidal surgery may be divided into a number of categories, based both on anatomic structures and on systems that may be involved.

Intracranial complications are probably the most devastating and have frequently been reported as the cause of fatalities. Hypothalamic injury may result both from direct trauma and from secondary vascular effects. The clinical course, marked by high fever and coma, is characteristic and is usually not reversible. Direct intracranial hemorrhage may occur, and subarachnoid bleeding or organized intracranial hematoma may be the cause of complications or death. Branches of the circle of Willis may be torn or injured.

Meningitis may result from the introduction of organisms into the intracranial space at operation or later if postoperative cerebrospinal rhinorrhea occurs. Other infectious complications have included subdural empyema and brain abscess, usually secondary to meningitis. Large amounts of air may enter the head when a CSF leak occurs and may produce cerebral toxicity or tension pneumocephalus. A persistent CSF leak requiring a lumboperitoneal shunt occurred in a patient with a pituitary adenoma and a pseudotumor syndrome with increased intracranial pressure.

One of the contraindications to the transsphenoidal approach is extension of the tumor into the anterior or middle cranial fossa. When unusually large intracranial tumor extensions are present, transsphenoidal surgery frequently results in complications related to the part of the tumor which is not accessible by the approach selected.

Carotid artery complications are equally devastating and commonly result in death or serious disability. Arterial hemorrhage may occur when the carotid artery, as it lies in the cavernous sinus, is lacerated, perforated, or avulsed. Hemorrhage may be difficult to control without interrupting distal blood flow to the brain. Occlusion of a more spontaneous nature has been reported to follow procedures that were otherwise uncomplicated. The cavernous carotid artery may develop spasm as a result of manipulation within the sella turcica, and thrombosis and distal embolism also have been encountered. A few patients have tortuous or redundant carotid arteries that may be visible within the sella and may be so close together as to occlude the sella, making the transsphenoidal approach impractical. Intrasellar aneurysms may simulate or coexist with pituitary lesions, but their presence should be recognized if preoperative angiography is performed. False aneurysms may occur after injury to the carotid artery and presumed satisfactory control; under the same circumstances, a carotid cavernous fistula might also be provoked. False aneurysms, if untreated, nearly always result in disastrous secondary hemorrhage. Epidural hematomas following surgery have been reported, perhaps related to decrease in intracranial pressure from intraoperative loss of CSF.

Complications involving the visual system are rarely fatal, but obviously they are capable of causing severe disability. Direct damage to the optic nerves or chiasm may occur while a tumor is being removed from the sella or from the suprasellar space. Such damage is more likely to occur when the capsule of the tumor is disrupted, or when the capsule is adherent to the visual apparatus, as in patients with prior craniotomy. Dissection and removal of pituitary tumors may interfere with the blood supply to the optic nerve or chiasm and may produce visual deficits related to ischemia. Loss of vision has developed after removal of a pituitary tumor when the chiasm has been compressed by the muscle, fat, or gelfoam plug commonly used to obliterate the empty sella. One of the reasons for using such a plug was the report of visual loss caused by postoperative prolapse of the optic nerves and chiasm into an empty sella after removal of a tumor. The optic nerves have been damaged by fractures of the optic foramen and orbit produced by the retractors used for the transsphenoidal approach.

The cavernous sinus, together with its contents other than the carotid artery, also serves as a source for complications during transsphenoidal surgery. Venous hemorrhage may occur directly from the sinus or from anomalies of connections between the two cavernous sinuses. Laceration or disruption of these venous structures may be accompanied by venous bleeding. In addition, the removal of a tumor or a normal pituitary gland from the lateral wall of the sella may be associated with damage to the cranial nerves that run within the cavernous sinus, and involvement of the III, IV, V, and VI cranial nerves subsequent to operation has developed as either an immediate or a more delayed complication. These cranial nerve injuries usually are produced by trauma or stretching and are almost always transient. Venous air embolism may occur into the cavernous sinus, particularly when the patient is positioned with the head higher than the right atrium of the heart.

The sella turcica itself may be the source of complications. The anatomy of the sella and its relation to the sphenoid sinus are of primary importance, and the direction of approach should be carefully monitored to avoid misdirection to the floor of the anterior fossa or the clivus. In patients with a shallow sella and a poorly pneumatized sphenoid sinus, the transsphenoidal approach may require excavation of bone with a drill or curette which may be so difficult technically as to be a contraindication to the approach. Postoperative cerebrospinal rhinorrhea most commonly occurs through the sella and its diaphragm, either because of iatrogenic injury to the roof of the sella or because of anatomic variations such as prolapse of the arachnoid into the sella or unusually large infundibular apertures in the diaphragm. Excessive enlargement or erosion of the sella from a very large intrasellar lesion has been recognized as a predisposing factor to complications, generally of a hemorrhagic nature. CSF leaks are also more likely to occur with tumors that invade or erode the bone at the base of the skull.

Sphenoid sinus and nasofacial complications are rarely fatal but can be troublesome and may persist for some time after surgery. Imprudent use of the retractor may result in diastasis or fracture of the hard palate. Injury to or fracture of the cribriform plate may cause cerebrospinal rhinorrhea. In the postoperative period, the mucosa of the sphenoid sinus may produce symptomatic sinusitis and eventual development of a mucocele. Technical errors in handling the nasal mucosa, the nasal septum, and the nasal spine may result in postoperative external nasal deformity. Inadequate control of bleeding from the nasal portion of the procedure may lead to superficial wound hemorrhage and swelling. Toxic shock syndrome has been reported, related to the nasal packing.

Finally, endocrinological complications that are related both to the preoperative endocrine state of the patient and to postoperative alterations in endocrine function may occur. Most common of these is hypopituitarism, which should be quickly recognized and rapidly corrected with the use of appropriate steroid, thyroid and gonadotropic medications. Diabetes insipidus may develop because of damage to the posterior pituitary gland or the infundibulum; water and electrolyte imbalance may follow and may be life threatening if unrecognized or inappropriately treated. Later in the postopera-

Table 1. Transsphenoidal surgery 1972–1988

Diagnosis		Number of patients	Number of procedures
Pituitary adenoma		1767	1826
Acromegaly	315		
Prolactin adenoma	668		
Cushing's disease	217		
Nelson's syndrome	46		
Nonfunctioning and miscellaneous	521		
Craniopharyngioma		77	88
Empty Sella/CSF rhinorrhea		59	63
Hypophysectomy		25	26
Chordoma		18	20
Miscellaneous lesions		73	76
Total		2019	2099

Table 2. Fatal complications in 2000 transsphenoidal operations

Hypothalamic injury	1
Intracranial hemorrhage	2
Vascular occlusion	2
Meningitis	3 (1 delayed)
Carotid injury	1

tive period (usually 5–10 days), symptomatic SIADH with hyponatremia may develop, and it must be recognized and appropriately treated. In patients who present for operation with pituitary tumors producing endocrine hyperactivity, persistence of such hyperactivity subsequent to surgery must be regarded as an unsatisfactory result.

After establishing these categories by analysis of the published experience with transsphenoidal surgery, our experience was reviewed.

Between September 1972 and March 1988, 2000 transsphenoidal procedures were performed. A modification of the sublabial, transseptal approach utilized by Cushing, Guiot, and Hardy was used. The pathological findings are presented in Table 1. A recapitulation of the complications is seen in Tables 2–4. There were eight fatalities, giving a mortality rate of 0.4%. Both the nature

Table 3. Serious nonfatal complications in 2000 transsphenoidal operations

Stroke	5
(vascular occlusion 1, spasm 4)	
Meningitis	5
(+ subdural empyema 1)	
(+ brain abscess 1)	
Visual loss, permanent	7
Hemorrhage	4
(requiring reoperation or abandonment of procedure)	
CSF rhinorrhea	18
Vascular injury	4
(+ false aneurysm 1)	
Cranial nerve paralysis (6th)	2
Total	45 (2.2%)

Table 4. Less serious complications in 2000 transsphenoidal procedures

Transient visual loss	5
Transient cranial nerve palsy	
Sixth nerve	2
Third nerve	3
Excessive bleeding (intraoperative)	7
Symptomatic nasal septal perforation	7
Severe postoperative sinusitis	3
Permanent diabetes insipidus	24
Symptomatic SIADH	14
Palatal diastasis	2
Gingival resorption	1
Tooth denervation	1
Symptomatic pneumocephalus	2
Total	71 (3.6%)
Total complications	124 (6.2%)

and the incidence of these complications are compatible with other large series of transsphenoidal procedures.

Although some complications may be anticipated and avoided, others must be recognized promptly and treated appropriately. A review of our experience would indicate that the patients most likely to suffer serious complications are those who have been subjected to previous craniotomy. Additional factors that increase risk include excessive enlargement of the sella, erosion of parasellar bony structures, very large suprasellar extensions of pituitary tumors, and poor general medical condition because of other disease factors. Participation by ear, nose, and throat specialists in the transsphenoidal procedure has been a major factor in reducing postoperative nasal complications.

Diligent preoperative assessment of the anatomy is essential and should be accomplished in each case with appropriate imaging studies. Prophylactic antibiotic therapy is routinely employed.

Undoubtedly the nature of complications of transsphenoidal surgery will continue to change as the approach is utilized for the treatment of new conditions, and it is our hope that in the future this procedure may be applied even more effectively and with greater safety than it is at present.

References

1. Guiot G, Derome P (1976) Surgical problems of pituitary adenomas. In: Krayenbuhl H (ed) Advances and technical standards in neurosurgery, vol 3. Springer, Vienna New York pp 3–33
2. Landolt AM (1986) Complications and pitfalls of transsphenoidal pituitary surgery. Neurosurgeons 4:395–404
3. Laws ER Jr (1982) Complications of transsphenoidal microsurgery for pituitary adenoma. In: Brock M (ed) Modern neurosurgery. Springer, Berlin Heidelberg New York, pp 181–186
4. Laws ER Jr, Fode NC, Redmond MJ (1985) Transsphenoidal surgery following unsuccessful prior therapy. J Neurosurg 63:823–829
5. Nicola GC, Tonarelli GP, Griner AP (1980) Complications of transsphenoidal surgery in pituitary adenomas. In: Derome PJ, Jedynak CP, Peillon F (eds) Pituitary adenomas: biology, physiopathology and treatment. Asclepios, Paris, pp 237–240

The Effects of Pituitary Adenoma on the Facial Skeleton in Cases of Acromegaly

M. Farmand, Erlangen/FRG, A. Künzler, and B. de Giacomi, Zürich/Switzerland

Introduction

Typical clinical signs of long-lasting overproduction of growth hormone in adults such as enlargement of the feet, hands, mandible, and nose allow the diagnosis of a pituitary adenoma in cases of acromegaly at first sight. Since most of the bony changes are not reversible (Kinnmann 1976), it would be of great value if an acromegaly could be diagnosed before visible changes are present. In addition, not all patients develop the typical deformations, and therefore the time between the first manifestation of the disease and the beginning of treatment may be quite long (Landolt 1986). Nonspecific symptoms such as headache and amenorrhea generally lead the patient with an acromegaly first to a dentist, general practitioner, or gynecologist rather than to a neurosurgeon.

To determine whether there is a possibility to verify the diagnosis of an acromegaly on the basis of bony changes, we examined X-rays for objective pathological signs in the face and assessed their valency for the diagnosis.

Patient Material and Investigation Methods

A total of 30 patients with GH-secreting pituitary adenoma were investigated. Most of them were treated by Prof. Landolt (Neurosurgical Clinic, University Hospital, Zürich). The operation had been performed at least some months earlier, so that the soft tissue changes had been reduced. Nevertheless, all visible pathological signs of the face were registered. In a standardized lateral cephalometric X-ray of the skull all roentgenological changes of the bony structure of the facial skeleton and the neurocranium were recorded objectively. The obtained data were compared with the same method of measurement to a control group of 21 volunteers in the same age range. The viscerocranium was measured according to the Zürich planning method for orthodontic patients (Rutz 1977) (Fig. 1). In the neurocranium the following parameters were metrically evaluated: the length and depth of the sella, thickness of the calvaria, prominence of the supraorbital rim, surface of the pneumatization of the mastoid cells and the frontal sinus (Fig. 2). Differences between the acromegaly group and the control group were analyzed statistically.

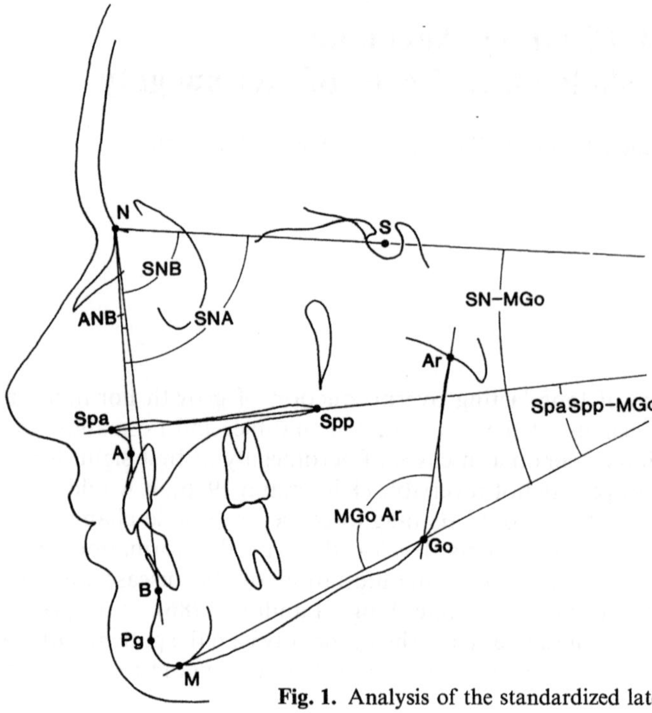

Fig. 1. Analysis of the standardized lateral X-ray of the skull

Results and Discussion

IGF and growth hormone levels were normal in all patients after treatment of the adenoma. It was noticed that incidence and valency of the bony changes varied. Not all the measured values were pathological in all patients. The changes in the jaws are described in detail in the literature (Korkhaus 1955; Markovic and Trisovic 1978; Bernsmann 1980). There were no significant changes in the maxilla and the anterior cranial base, in contrast to the mandible. There was a highly significant elongation of the mandibular ramus and body and a change of the ANB angle resulting in a prognathism, although not in all patients (Fig. 3). These parameters are of special interest to dentists and orthodontists.

In the neurocranium the changes were not constant. Comparing each bony change with the control group, it seemed that there exist parameters with very good discrimination. We found that the depth and length of the sella and the thickness of the calvaria had a discrimination of over 70% between real-positive changes in acromegaly patients and false-positive changes in the control group. The other three parameters (supraorbital rim enlargement, mastoid pneumatization, length of the mastoid process) had a discrimination of only around 50%. The enlargement of the frontal sinus had practically no discrimination (Table 1).

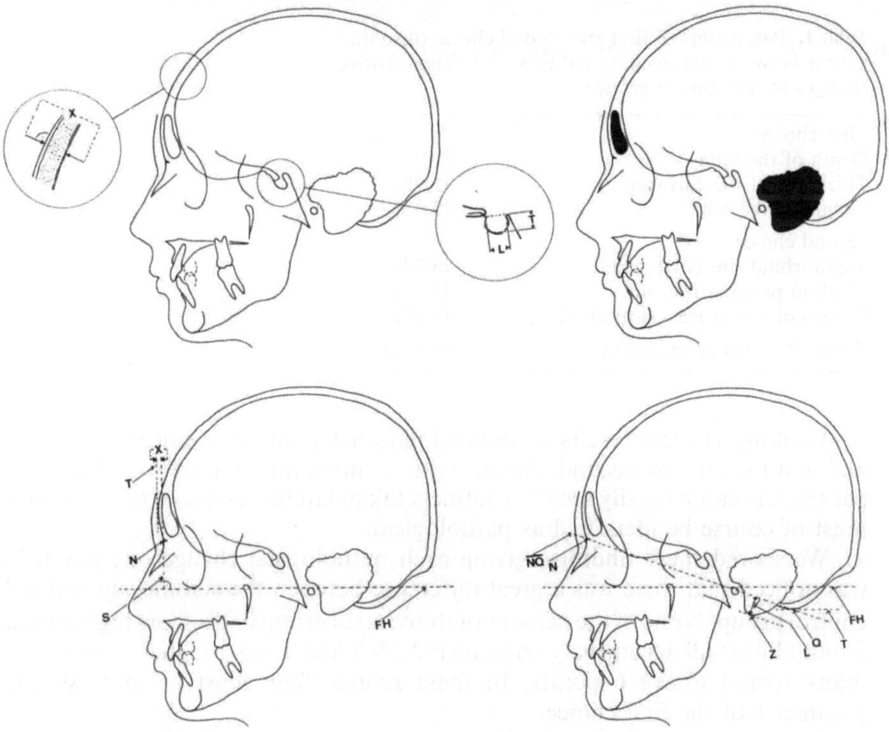

Fig. 2. Analysis of pathological changes in the neurocranium

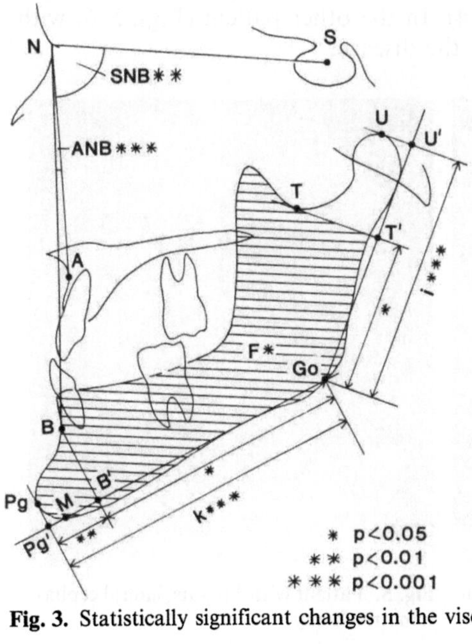

Fig. 3. Statistically significant changes in the viscerocranium in patients with acromegaly

Table 1. Parameters of first and second choice (discrimination between acromegaly patients and false-positive changes in the control group)

First choice	
Depth of the sella	71.9%
Thickness of the calvaria	71.9%
Length of the sella	70.0%
Second choice	
Supraorbital rim enlargement	56.7%
Mastoid pneumatization	49.5%
Length of the processus mastoideus	43.4%
(Sinus frontalis enlargement)	(0.4%)

Looking at these results we defined parameters of first choice with a high and parameters of second choice with a moderate sensibility. All these parameters can be easily seen in routinely taken lateral X-rays of the skull and must of course be identified as pathological.

We scored these findings, giving each pathological change one point. It was noticed that there was a great difference between the acromegaly and the control group. None of the persons of the control group had a score higher than 2, but almost all acromegaly patients (93.3%) had a score higher than 3, and many scored above 6 points. In these scores there must be at least one parameter of the first choice.

Figures 4–6 present two patients with acromegaly having different scores. In the patient with 6 points the acromegaly is clinically obvious and can be diagnosed without difficulty (Fig. 4). In the other patient (Figs. 5, 6) with 3 points it took 19 years to discover the disease.

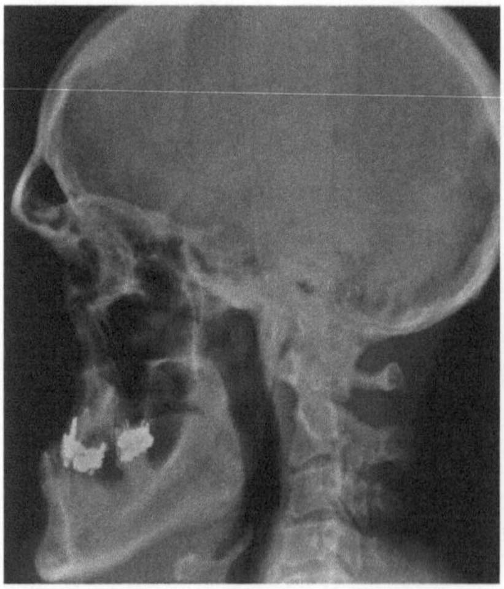

Fig. 4. Patient with 6 points, lateral cephalometric X-ray

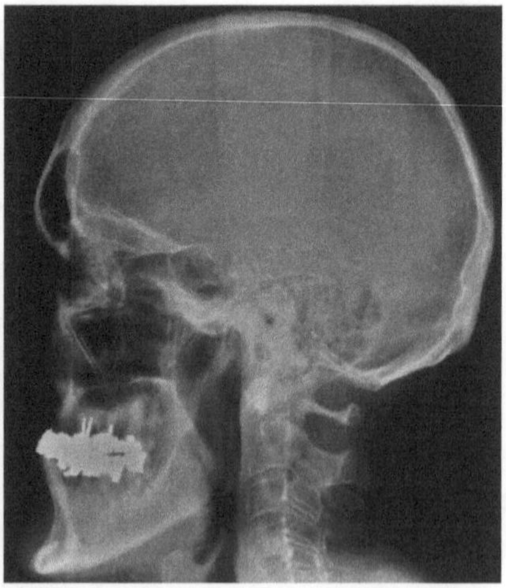

Fig. 5. Patient with 3 points, lateral cephalometric X-ray

Fig. 6. Patient of Fig. 5 with 3 points

In these cases our score may serve as an indicator for undertaking further extensive investigations to clarify the diagnosis of a suspected growth hormone producing adenoma. Patients must, of course, then be referred for an examination of growth hormone and IGF levels (Froesch 1986).

References

Bernsmann JF (1980) Das Fernröntgenbild bei Akromegalie. Fortschr Kieferorthop 41:513–523

Froesch EZ (1986) Growth-hormone dependent and Growth-hormone independent growth factors. In: Labhart (ed) Clinical endocrinology, therapy and practice, 2nd ed. Springer, Berlin Heidelberg New York Tokyo, pp 101–112

Kinnmann JEG (1976) The prognosis in acromegaly treated by transantero-sphenoidal operation. Acta otolaryngol 82:420–430

Korkhaus G (1955) Veränderungen im Gebiss und Gesichtsschädel bei Akromegalie. Dtsch Zahn Mund Kieferheilkd 22:93–128

Landolt AM (1986) Acromegaly. In: Labhart (ed) Clinical endocrinology, theory and practice, 2nd ed. Springer, Berlin Heidelberg New York Tokyo, pp 135–154

Markovic M, Trisovic D (1978) Some results of occlusal and metric analysis of acromegalic cases. Int J Orthod 16:13–20

Rutz A (1977) Zürcher Analyse der seitlichen Schädelaufnahme (unveröffentlicht)

Some Rare Tumors of the Sella Turcica and Paranasal Sinuses

A. El Khamlichi, A. M. Khamlichi, D. Moussaif, A. El Ouahabi, and A. Aghzadi, Rabat/Morocco

The most frequent tumors of the sella turcica region are certainly the pituitary adenomas, craniopharyngiomas, and meningiomas. However, while reviewing our 101 patients operated on during the 5 years from 1983 to 1987, we found some histological forms which are rarely reported in the literature: the epidermoid cyst, chordoma, eosinophilic granuloma, myeloplax tumor, and rhabdomyosarcoma. The following observations regarding these rare tumors illustrate the anatomoclinical aspects and the therapeutic problems.

The *first observation* concerns the epidermoid cyst in a 22-year-old woman admitted in April 1987 for secondary amenorrhea without galactorrhoea and ICH with left hemiparesis. The endocrinological examinations were normal with normal prolactin level. Skull X-ray showed clear modifications of the sella turcica; computed tomography (CT) showed an expansive process intra- and suprasellarly invading the sphenoidal sinus, isodense, with an enhancement of the capsula after contrast injection; and angiography showed a clear lifting of the A_1 segment of the anterior cerebral arteries without tumoral vascularization. A total surgical excision was performed by a transsphenoidal approach. The diagnosis of epidermoid cyst was confirmed by the macroscopic and histological aspects showing an epithelium with uniform layers of keratinizing squamous cells. The outcome was good with recovery of the hemiparesis, and the CT scan control confirmed total absence of the tumoral process.

The *second observation* is the most typical of the three cases of chordoma. This was a 47-year-old man admitted in July 1986 for headache starting 5 years earlier. At admission the examinations showed: on skull X-ray, an enlargement of the sella turcica with posterior destruction; on CT scan, a sphenoidal expansive process with supra- and intrasellar development; and on angiography, a lifting and stretching of the terminal parts of the carotid arteries. We carried out a subtotal excision by transsphenoidal approach. The histological examination showed a typical aspect of chordoma with eosinophilic cells and clear cells (called vegetal cells) drowned in the mucus. The patient was submitted to a complementary radiotherapy focused on the sella turcica region. The outcome was favorable without recurrence within 2 years, as shown on the CT scan. Nevertheless, this kind of tumor generally recurs within an average period of 5–8 years.

The *third observation* concerns a particular type of rhabdomyosarcoma, the embryonal rhabdomyosarcoma, in a 22-year-old woman admitted in May 1983 for ICH, bilateral anosmia, and left exophthalmos. The examinations at admission showed: on skull X-ray, dismineralization and filling of ethmoidal and sphenoid sinuses; on CT scan, a tumoral process of the anterior fossa filling the ethmoidal and sphenoidal sinuses; and on angiography, a tumoral vascularization with arterial displacement. A subtotal excision was carried out. Histology confirmed the diagnosis of an embryonal rhabdomyosarcoma of the alveolar type showing a very high cellular proliferation density with an alveolar architecture made of rhabdomyoblastes. The evolution was bad, with postoperative hydrocephalus and tumoral recurrence followed by death 3 months later. This is a rare tumor. Twenty cases were reported before 1983 when we encountered our case. The evolution is fatal, with local recurrence and metastases. The average survival time in the published cases was 7.5 months.

The *fourth observation* concerns a myeloplax tumor in a 46-year-old woman admitted in May 1984 for left temporobasal tumefaction starting 15 months before. The examinations at admission showed: on skull X-ray, a destruction of temporobasal and sphenoidal bones; on CT scan, a huge heterogenous tumor with a high enhancement localized in the middle fossa and sphenoid sinus; and on angiography, a high tumoral vascularization, especially of external carotid artery branches. The surgical excision was total, and the histological diagnosis was a giant cell tumor with a high density of multinuclear giant cells without any malignant signs. The evolution was good without recurrence within 3 years, as shown by the last CT scan.

The *fifth observation* concerns histiocytosis X in a 24-year-old woman admitted in October 1986 for ICH with optical atrophy and right sixth nerve paralysis. The examinations at admission showed: on skull X-ray, destruction of the sella turcica and sphenoidal sinus with suprasellar calcifications; on CT scan, a huge intrasellar expansive process with expansion into the sphenoidal and ethmoidal sinuses; and on angiography, no tumoral vascularization. We carried out subtotal excision by a transsphenoidal approach. The histology showed an eosinophilic granuloma with abnormal histiocytes (sometimes on mitosis) and inflammatory cells (polynuclears and plasmocytes). The patient was submitted to radiotherapy, and the outcome was favorable within 21 months. This case of an isolated eosinophilic granuloma in the sellar region is the fourth reported in the literature. The other three cases were published, respectively, by Goodman et al. in 1979 (adult), Belloir et al. in 1984 (infant), and Thiebot et al. in 1987 (infant).

Conclusion

These observations show that the pituitary adenomas, craniopharyngiomas, and meningiomas do not represent all the tumors of the sella region. Some other rare tumors generally localized in the cranial vault or in the other part of the skeleton may also be encountered here. We draw the following conclusions from our experience:

- The CT scan helps in diagnosis.
- The transsphenoidal approach remains the best for the biopsy and surgical excision of these lesions.
- The prognosis depends on the histological malignancy and the complete or nonsurgical excision. Recurrence is possible in most of the cases.

References

Belloir C, Chavane G, Mazabrand A (1984) Tumeurs osseuses. Observation 18 – hystiocytose. Ann Radiol 27:537–539

Goodman RH, Post KD, Molitch ME, Adelman LS, Altemus LR, Johnston H (1979) Eosinophilic granuloma mimicking a pituitary tumor. Neurosurgery 5:723–725

Thiebot J, Clavier E, Challine B, Henry J (1987) Localisation tumorale sphénoidale isolée chez un enfant révélatrice d'une histiocytose X. J Radiol 68:309–311

The Empty Sella and the Sellar Arachnoidal Cyst

G. Hildebrandt, H. Stracke, H. Schatz, A. L. Agnoli, N. Klug, and P. Christophis, Giessen/FRG

Introduction

Busch first used the term empty sella in 1951. This was follwed by the classification into primary empty sella, due to a congenitally deficient sella diaphragm and secondary empty sella, which refers to those in which extension of the subarachnoid space into the sella is caused by operative procedures or radiation (Tindall and Barrow 1986; Bonneville et al. 1986). When the cases referred to the Neurosurgical Department at the University of Giessen during a 6 year period were studied, it was found that while this subdivision is still useful, there are pathological features termed empty sella which do not fit into these classical categories.

The aim of this study was to demonstrate the different causes of the so-called empty sella and the necessity to use the whole diagnostic armamentarium in order to clarify the etiology.

Material and Methods

From a group of 220 patients harboring lesions within the sella or the suprasellar area, 21 cases were studied which were thought to have empty sella. In each of the 21 cases a full clinical and ophthalmological work up was performed including assessment of visus, performing perimetry, and recording visual evoked potentials (VEP). Endocrine studies included a combined anterior pituitary stimulation test by intravenous injection of the four releasing hormones GHRH (GH-releasing hormone), TRH (TSH-releasing hormone), GnRH (FSH- and LH-releasing hormone), and CRH (ACTH-releasing hormone). Basal and stimulated levels of anterior pituitary hormones, basal levels of plasma antidiuretic hormone (PADH), and serum aldosterone (Aldo) were measured by commercially available radioimmunoassays. Water electrolyte balance was investigated by measurement of osmolality (serum and urine), sodium and potassium (serum), and renal fractional sodium excretion. The patients received X-ray studies (frontal and lateral tomograms of the sella turcica), intrathecally and intravenously enhanced computed tomography (CT; Somatom 2) and intravenously enhanced (gadolinium-DTPA) magnetic resonance imaging (MRI) studies (Picker International, Dr. Halbsguth, Dr. Lochner, Frankfurt).

Results

Primary Empty Sella. In 9 out of 21 patients no cause of the empty sella could be established, and they were thus classified as real primary empty sellas. Clinical findings, signs, and symptoms were rather nonspecific. The majority (six out of nine patients) were multiparous, obese, and hypertensive women complaining of headache. VEP revealed no abnormality in any of these cases. The endocrine findings in this subgroup disclosed a decrease of basal as well as maximal serum GH and cortisol secretion and of FSH/LH in postmenopausal women (Table 1). Posterior pituitary function and water electrolyte balance were not compromised (Table 1). Radiological findings disclosed unique

Table 1. Primary empty sella: results of endocrine studies (6 women, 3 men)

Pituitary stimulation test			Controls ($n = 23$)		
	M	SD	M	SD	WLX
GH (ng/ml)					
Basal	0.6	1.0	1.0	1.1	+
Maximum	8.5	7.4	14.4	6.7	+
ACTH (pg/ml)					
Basal	24.0	10.6	27.2	20.1	
Maximum	38.0	10.3	54.1	27.3	
Cort (μg/dl)					
Basal	13.8	2.9	16.6	4.9	+
Maximum	21.4	4.9	29.1	7.6	*
Postmenopausal women ($n = 3$)			Controls ($n = 10$)		
LH (mU/ml)					
Basal	12.6	12.1	28.2	10.0	
Maximum	54.0	46.0	67.8	19.7	
FSH (mU/ml)					
Basal	20.5	20.2	37.1	16.8	
Maximum	40.5	29.6	48.4	17.7	

Water electrolyte balance		M	SD
PADH	(pg/ml)	1.8	0.6
S_{osmo}	(mosmol/kg)	288	9.3
S^+_{Na}	(mmol/l)	140	2.3
S^+_{k}	(mmol/l)	4.2	0.2
Aldo	(ng/dl)	5.3	1.8
U_{osmo}	(mosmol/kg)	873	181
FE^+_{Na}	(%)	0.5	0.2

GH, Growth hormone; ACTH, adrenocorticotrope hormone; Cort, cortisol; LH, luteinizing hormone; FSH, follicle stimulating hormone; WLX, Wilcoxon test; PADH, plasma antidiuretic hormone; S_{osmo}, serum osmolality; S^+_{Na}, serum sodium; S^+_{K}, serum potassium; Aldo, serum aldosterone; U_{osmo}, urine osmolality; FE^+_{Na}, renal fractional sodium excretion; +, $p < 0.05$; *, $p < 0.01$; M, mean; SD, standard deviation.

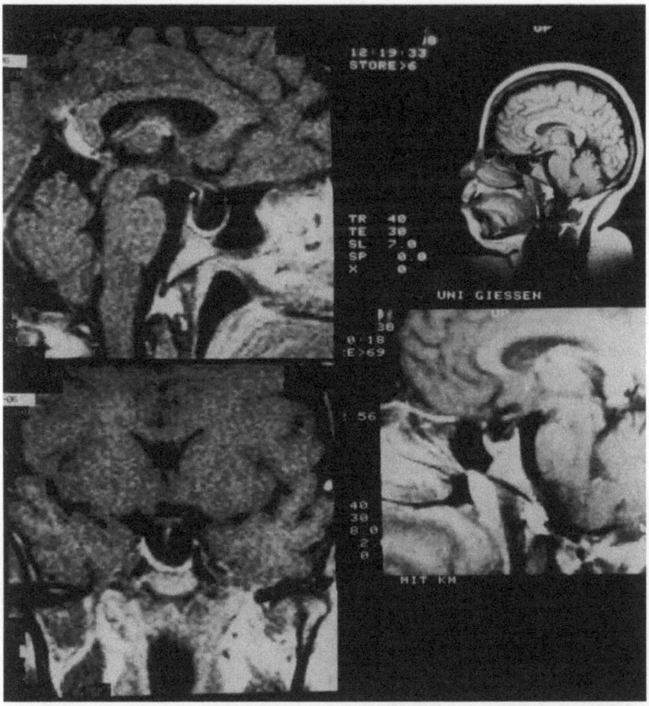

a b c d

Fig. 1 a–d. Sagittal (**a**) and coronal (**b**) MRI sections (T_1-weighted sequences) in a primary empty sella, demonstrating endosellar extension of the subarachnoid space, flattened pituitary gland stretched along the floor of the sella, dislocation of the pituitary stalk towards the dorsum sellae (increase of signal intensity after gadolinium injection). Sagittal MRI sections (T_1-weighted sequences) before (**c**) and after (**d**) gadolinium injection. **d** Increase of signal intensity of the pituitary at the sellar floor

features: generally a nonprogressive enlargement of the sella (eight out of nine patients), endosellar extension of the subarachnoid space, flattened pituitary gland stretched along the floor of the sella, elongation of the pituitary stalk, dislocation of the pituitary stalk towards the dorsum sellae, increase in signal intensity within the hypophysis at the sellar floor after gadolinium injection (Fig. 1) and CSF-like signal intensity of the sellar space in T_1- and T_2-weighted MRI sequences (Fig. 2).

Secondary Empty Sella. In 151 transnasal or transcranial operative procedures there were six secondary empty sellas (Fig. 3). A tendency was observed toward increased basal prolactin (PRL) secretion in these cases compared to those after total tumor removal without empty sella (Table 2), indicating operative damage to venous portal vessels and/or stretching of the pituitary stalk due to cisternal herniation. In one case combined operative and medical treatment (dopamine agonist) caused a secondary empty sella and a

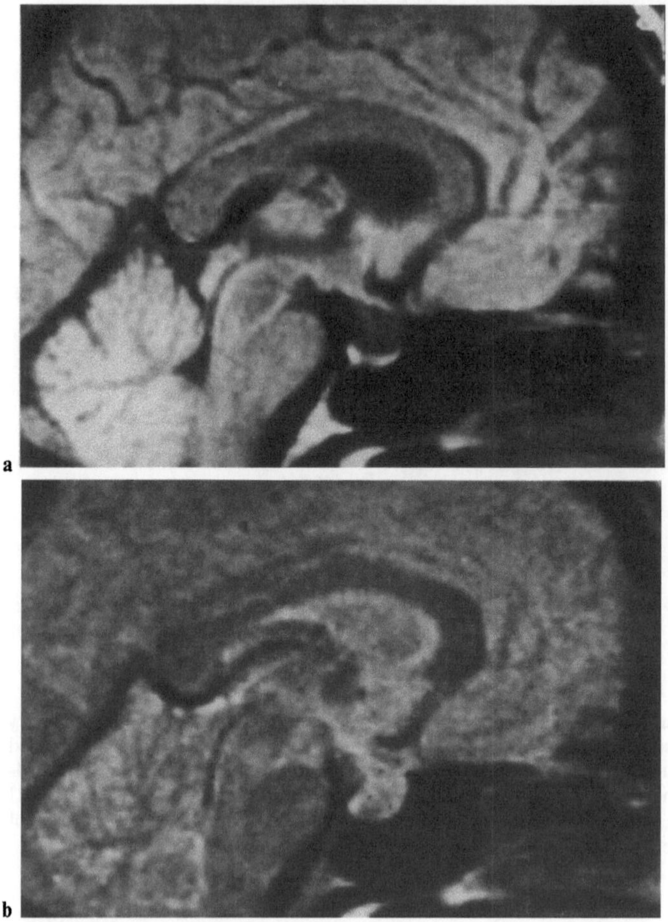

Fig. 2a, b. Sagittal MRI sections in a case of primary empty sella. T_1- (**a**) and T_2-weighted (**b**) MRI sequences demonstrate CSF-like signal intensity of the sellar space

Table 2. Secondary empty sella ($n = 6$, after 151 transnasal/transcranial operations)

			Inactive adenomas total removal ($n = 15$)	
	M	SD	M	SD
PRL (ng/ml)				
Basal	28.3	35.3	11.2	8.8
Maximum	38.1	57.0	18.2	14.5

PRL, prolactin.

a

b

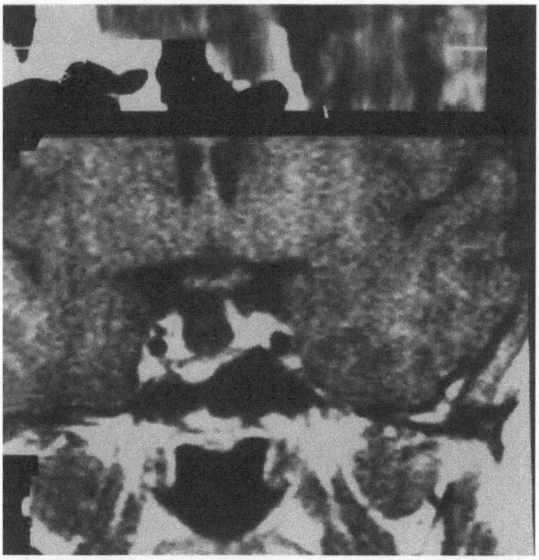

Fig. 3 a, b. Sagittal CT section before (**a**) and coronal MRI view (T_1-weighted sequence) after (**b**) total removal of an inactive macroadenoma. Postoperative secondary empty sella with persistent dislocation of the pituitary stalk and the normal pituitary (increase of signal intensity after gadolinium injection)

CSF fistula due to complete shrinkage of an invasive macroprolactinoma. Deterioration of vision never occurred in cases harboring secondary empty sella.

Partial Empty Sella. In 2 out of 21 patients partial empty sella was detected. One patient presented with very slight acromegalic features and a GH level of 6 ng/ml, which was not suppressible in the oral glucose tolerance test; the other one was diagnosed as primary empty sella during routine X-ray studies at first glance. Intrathecally and intravenously enhanced CT and intravenously enhanced MRI studies revealed the exact nature of the lesion (Fig. 4), because empty sella plus tumor remnants could be outlined clearly.

Various Entities. There were other entities in the group of the 21 patients which had to be excluded before the diagnosis of a real primary or secondary empty sella could be established: one due to suprasellar pathology (aqueductal stenosis and hydrocephalus, Fig. 5 a) and two due to infrasellar pathology (transsphenoidal meningocele, primary invasive sphenoidal macroprolactinoma, Fig. 5 b). Most of the criteria for establishing the diagnosis of a primary empty sella could have been misleading in the patient presented in the following case report.

Case Report. A CT scan of a young woman who presented with a history of headache and dizziness and a recent acute worsening of the symptoms

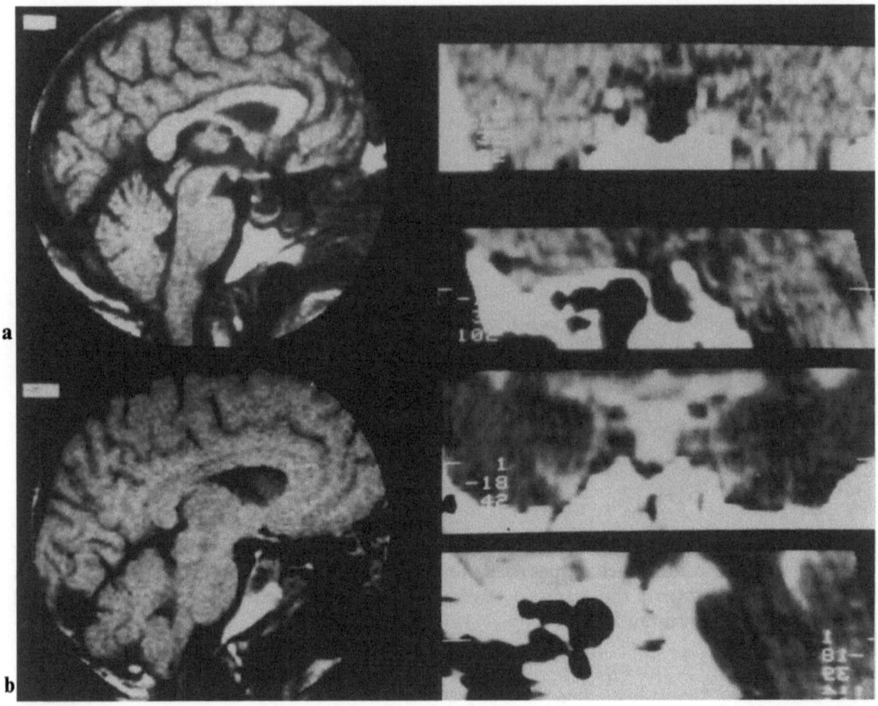

Fig. 4a–f. The midline sagittal MRI view (T_1-weighted sequence) (**a**) reveals a partial empty sella, the paramidline section discloses a small GH-secreting adenoma after gadolinium injection (**b**). **c, d** Coronal and sagittal CT section in the same patient showing an endosellar extension of the subarachnoid space. **e, f** Cisternogram of the same patient outlining the hypodense adenoma at the sellar floor in coronal (**e**) and sagittal (**f**) views

(Fig. 6a, b). The endocrine findings in this patient were normal. CT suggested first an empty sella, but it was not possible to identify the hypophyseal stalk on coronal and sagittal CT sections. Not until intrathecal contrast medium had been injected (Fig. 6c, d) was a noncommunication of the intra/suprasellar and the subarachnoidal space demonstrated. An MRI study (Fig. 7) revealed a CSF-like signal intensity of the lesion in T_1- and T_2-weighted sequences. Upon opening the cyst via a transnasal route to establish a communication between the cyst and the subarachnoid space, the lining of the cyst resembled very much that of an arachnoid membrane which extended into the subchiasmal region.

Discussion

In a series of 21 patients it was possible to determine the exact nature of an empty sella in each case by using endocrine and radiological methods including intrathecally and intravenously enhanced CT and intravenously enhanced

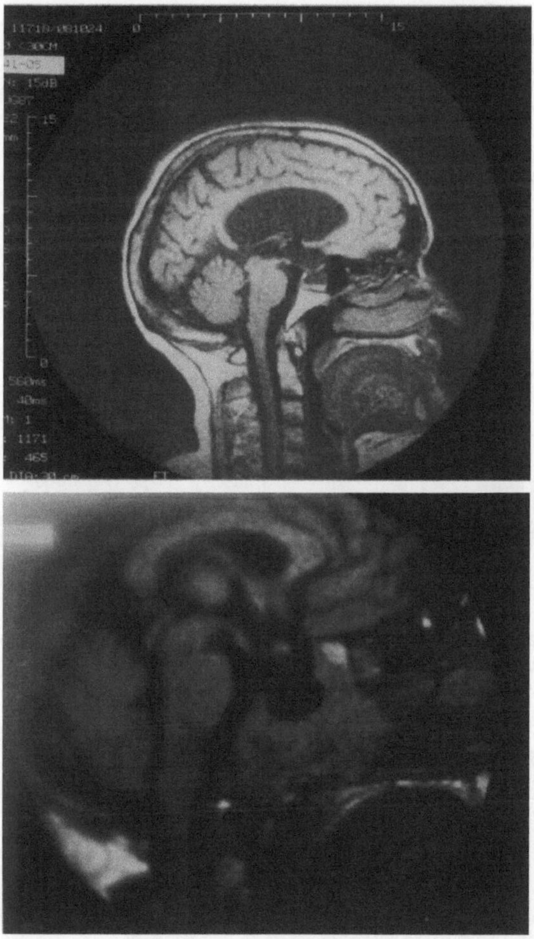

a

b

Fig. 5. **a** Sagittal MRI view (T_1-weighted sequence) in a case of empty sella which is caused by hydrocephalus and aqueductal stenosis. **b** Sagittal MRI section (T_1-weighted sequence) in a case of an invasive sphenoidal macroprolactinoma with a concurrent empty sella

MRI studies. Cystic lesions mimicking an empty sella turcica appear to be the most important entity which have to be excluded by sophisticated radiological investigations. Benign cysts of the sellar area are subdivided into those which are lined by epithelium (Rathke's cleft cyst, dermoid/epidermoid cyst, neuroepithelial cyst, cystic craniopharyngeoma and mucocele) and into those which are lined by an arachnoid membrane (Shaw and Alvord 1977).

According to Kasdon et al. (1977) and Fox and Al-Mefty (1980) suprasellar arachnoidal cysts are interpreted as a consequence of traumatic leptomeningeal cysts, cystic adhesive arachnoiditis, maldevelopment, or extension of the membrane of Liliequist. As congenital maldevelopments arachnoidal cysts of the sellar area may result from a so-called sealed-off arachnoidal diverticulum or an intraarachnoid maldevelopment (Fox and Al-Mefty 1980).

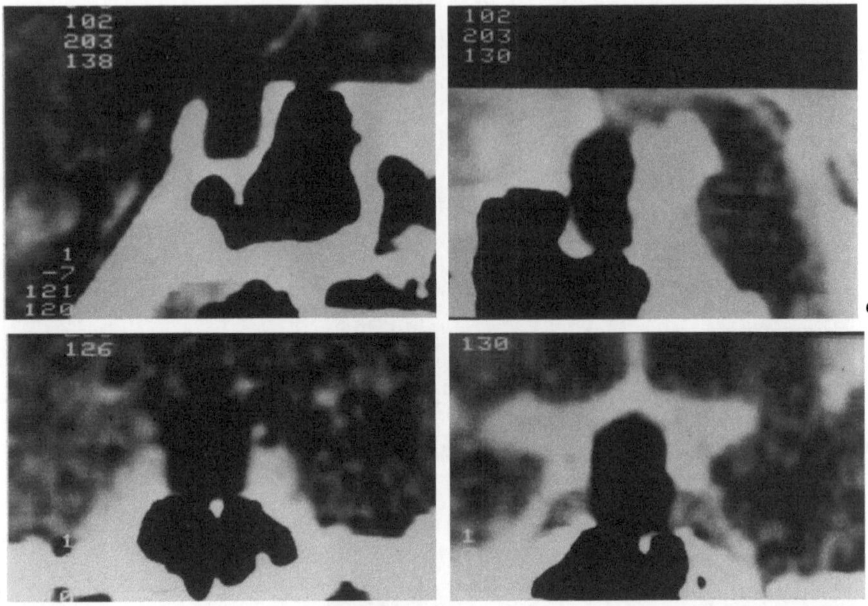

a c b d

Fig. 6a–d. Sagittal (**a**) and coronal (**b**) CT view in a case of suspected primary empty sella. Sagittal (**c**) and coronal (**d**) cisternogram of the same patient demonstrating a noncommunication between the sellar and the subarachnoidal space

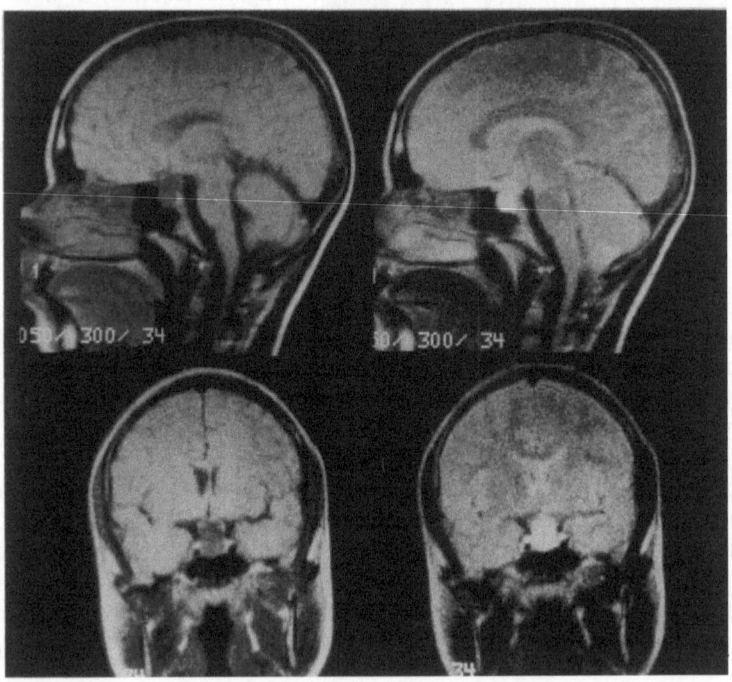

a b

Fig. 7a–d

Calkins et al. (1968) made no distinction between the terms empty sella, intrasellar arachnoid diverticulum, intrasellar subarachnoid space, and intrasellar/suprasellar cyst whether communicating or noncommunicating, so that one may assume a continuum between empty sella or an intrasellar arachnoid diverticulum and a so-called sealed-off arachnoid diverticulum possibly causing an endosellar arachnoidal cyst which may rupture and form an empty sella again (Bonneville et al. 1986). This is in line with the fact that the existence of a true primary intrasellar arachnoidal cyst is still highly doubtful since the leptomeninges surround only the infundibulum and do not normally extend into the sella turcica (Shaw and Alvord 1977).

Whether the endocrine findings in primary empty sella represent some damage to the hypophysis in the previous history of the patients [the results of Bjerre et al. (1986) suggested that partial or complete disappearance of a GH-secreting pituitary adenoma occurs rather frequently, leaving the sella empty or partially empty] indicating that the incidence of real primary sella is even less than expected, or whether decreased GH responses to GHRH are related to obesity (Kelijman and Frohman 1988) independently of the pituitary lesion, could not be settled in this study.

Conclusion

For a neurosurgeon it is not enough to confirm the diagnosis of an empty sella. To determine the etiology and not to miss the chance of an operative or medical treatment in these cases one should use the whole armamentarium of diagnostic procedures including dynamic testing for anterior pituitary function, intrathecally enhanced CT and intravenously enhanced CT and MRI studies. In this way lesions such as an endosellar arachnoid cyst and special types of the so-called empty sella are not confused with a primary empty sella. To avoid diagnostic and therapeutic confusion we propose a new classification for the term empty sella according to modern imaging technologies and endocrine diagnostic procedures.

References

1. Bjerre P, Lindholm J, Vidbaek H (1986) The spontaneous course of pituitary adenomas and occurrence of an empty sella in untreated acromegaly. J Clin Endocrinol Metab 63:287–291
2. Bonneville JF, Cattin F, Dietmann JL (1986) Computed tomography of the pituitary gland. Springer, Berlin Heidelberg New York Tokyo, pp 167–174

◄

Fig. 7a–d. Sagittal (**a**) and coronal (**b**) MRI view (T_1-weighted sequence) in a patient harboring an endosellar arachnoid cyst with extension towards the subchiasmal region. Sagittal (**c**) and coronal (**d**) MRI section in the same patient demonstrating CSF-like signal intensity in the endo- and suprasellar area (T_2-weighted sequences)

3. Busch W (1951) Die Morphologie der Sella turcica und ihre Beziehungen zur Hypophyse. Virchows Arch 320:437–458
4. Calkins RA, Pribham HFW, Joynt RJ (1968) Intrasellar arachnoid diverticulum, a case report. Neurology 18:1037–1040
5. Fox JL, Al-Mefty O (1980) Suprasellar arachnoid cysts: an extension of the membrane of Liliequist. Neurosurgery 7:615–618
6. Kasdon DL, Douglas EA, Broughman MF (1977) Suprasellar arachnoid cyst diagnosed preoperatively by computerized tomographic scanning. Surg Neurol 7:299–303
7. Kelijman M, Frohman LA (1988) Enhanced growth hormone (GH) responsiveness to GH-releasing hormone after diatry manipulation in obese and nonobese subjects. J Clin Endocrinol Metab 66:489–499
8. Shaw CM, Alvord EC (1977) Congenital arachnoid cysts and their differential diagnosis. In: Vinken PJ, Bruyn GW (eds) Handbook of clinical neurology, vol 31, part II, North Holland, Amsterdam New York Oxford, pp 75–135
9. Tindall GT, Barrow DL (1986) Disorders of the pituitary. Mosby, St. Louis Toronto Princeton, pp 473–475

C. Craniopharyngiomas

Craniopharyngioma: A Puzzling Neurosurgical Problem

B. KLUN, Ljubljana/Yugoslavia

Craniopharyngioma is a benign tumor which may ultimately lead to death, a tumor which may relapse after months, decades, or not at all, and one which before being declared cured probably requires the longest follow-up postoperative period, indeed, a maverick among brain neoplasms.

Few surgeons have been able to gain extensive experience in treating these tumors, since large series of 100 personal cases or more are rather exceptional. The period of 10 years covered by this study seems to allow a more realistic evaluation of results than comparative studies conducted over a longer time span, as it takes into consideration more effectively the substantial changes in philosophy of treatment and advances in operative techniques in recent years.

The studied series included 32 patients, of whom 7 were lost to follow-up and therefore excluded from the study. For the remaining 25 only the basic data are given as more space is dedicated to the conclusions. There were 17 adults and 8 children in the series. The mean follow-up period was 54 months. The following operative approaches were used: pterional in 13 cases, subfrontal in 4, interhemispheric or transcallosal in 4, and transsphenoid in 4. The surgeon felt the procedure to be radical in 13 cases and not radical in the remaining 12. In no case did the intervention consist of evacuation of the cyst(s) only. Fifteen patients were treated by surgery alone and 10 by a combination of surgery and radiotherapy. There were six recurrences: two in the group having undergone radical removal and four in the group with incomplete removal. Four of these recurrences were patients who had not received radiotherapy.

The results are as follows:

- Three patients died, two postoperatively and one in the following years due to recurrence.
- Nine patients are well, free of neurological and hormonal deficit and fully indepedent.
- Twelve patients – all of the children fall into this group – are independent, yet showing some, mainly endocrinological deficits.
- Two patients are severely visually impaired.

On the basis of the presented results and our experience gained in the previous years, the following conclusions have been made.

The first operative procedure determines whether removal is radical or not. A nonradical procedure is rarely followed by a total removal. This of course does not apply to planned two-stage procedures.

Total removal remains a goal, but not at all costs. The anatomical peculiarities as well as the size of the tumor may be the cause of serious dilemmas. The decision whether to accept a significant deficit is often difficult during surgery. The tumor attachment to the optic nerves or tracts is usually firm and even a gentle dissection may, in cases with previous poor vision, cause deleterious visual deficits. The same is true for the dissection from the hypothalamus. It is questionable whether the glial layer described by Sweet (1980) between the tumor and the hypothalamus is indeed always present and a reliable guide against neural damage. Surprisingly, even cystic tumors may present serious technical problems; the tiny capsule may easily be torn and is then difficult to be traced in the hypothalamus into which it retracts.

Stalk preservation is imperative. Although stalk preservation, especially in children, is not a guarantee against diabetes insipidus, this possibility, rendered today by modern microsurgical techniques, does at least facilitate faster and more complete regaining of endocrine function than in cases in which it is not preserved.

Transsphenoidal approach – if possible – gives the best functional results. The use of a variety of surgical approaches demonstrates that none is ideal in all cases. In view of large differences in size and extension of tumors, a universally ideal approach is not attainable. In the relatively small number of the cases treated, the transsphenoidal approach seems to offer the best prognosis not only because it usually allows radical removal of the tumor, but also because it renders possible the preservation of the pituitary gland in most cases (Laws 1980; König et al. 1986). In our series, all the tumors of intrasellar localization and/or with moderate suprasellar extension were totally removed by this approach.

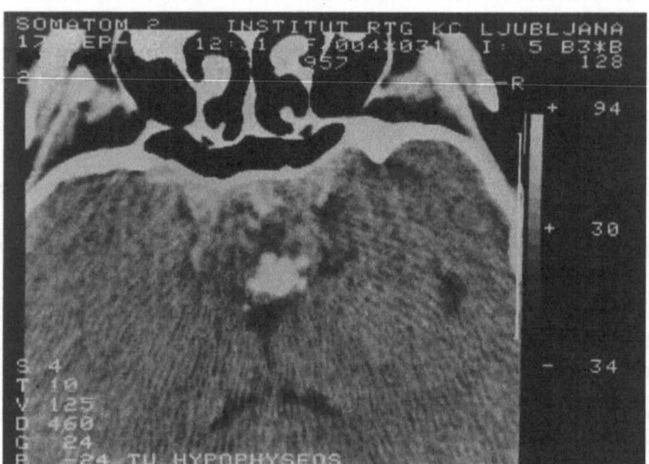

Fig. 1. The diagnosis of this craniopharyngioma was made 13 years previously, but the family refused the operation. During this period, the clinical picture remained virtually unchanged

Radical removal does not guarantee cure. This opinion is not shared by all neurosurgeons. The author is often in doubt whether to consider a removal radical or not when operating on large tumors. The remaining cavity is difficult to inspect, and small remnants could be left behind, especially in large, solid tumors removed by piecemeal techniques (Carmel 1985; Amacher 1980). Generally, the surgeon's judgement is based on his intraoperative appraisal and on the postoperative computed tomography scan. The latter, however, only shows tumor remnants larger than 4–5 mm, and recurrence may start from cell clusters much smaller than that. Two of our patients whom we considered radically operated upon later developed recurrences.

Nonradical removal does not necessarily lead to recurrence. Although the majority of incompletely removed tumors will relapse, some show extremely long quiescent intervals, as illustrated by the Fig. 1 and 2. Maier (1985) reports on his own craniopharyngioma which after two sessions of radiotherapy, inadequate by present standards, remained well under control for 66 years.

Incomplete removal should be followed by radiotherapy. There is enough evidence that radiation therapy is successful and delays or even prevents recurrence (Manaka et al. 1985). The modern radiation techniques have considerably diminished the danger of damage to the optic nerves and hypothalamic structures. However, the decision about X-ray treatment in children under the age of 5 years should be deliberate and taken with great caution.

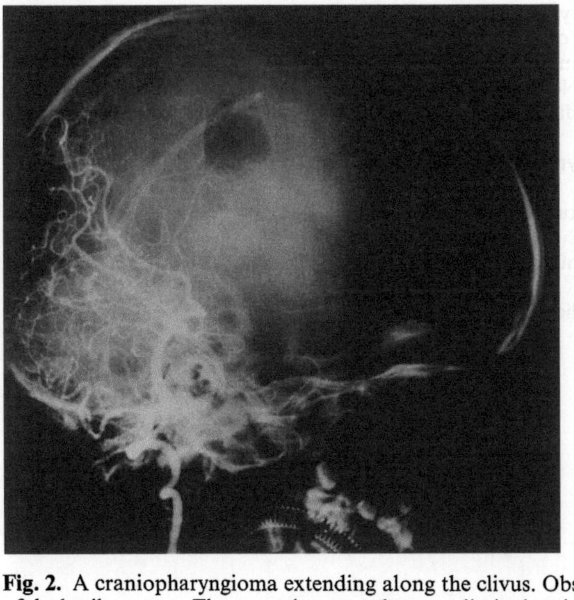

Fig. 2. A craniopharyngioma extending along the clivus. Observe the enormous dislocation of the basilar artery. The operative procedure was limited to drainage of cysts. The child made excellent recovery and was, besides retarded growth, without serious deficits. A recurrence developed after 15 years. (This patient is not included in the presented series)

Independent social and behavioral function should be the goal, not the survival rate. Our statistic evaluations are still expressed mostly in terms of mere survival. The points of view of the surgeon and the patient as to the postoperative success, however, may be quite different. To the surgeon, who may have spent hours in resecting a most difficult tumor, the final deficit may seem perfectly acceptable. The opinion of the patient could be completely different. He is not aware of the fact that his life might have been saved but sees insurmountable obstacles to his returning to an independent and competitive life. The operation leaving the patient with severe mental or endocrine handicaps can hardly be considered successful, even if the surgeon did a remarkable technical performance.

To consider a patient cured, a time span of two decades or more is necessary. The odd behavior of the tumor does not allow a cure to be declared before at least two decades following surgery (Symon and Sprich 1985). Occasionally, recurrences are seen which are due to no obvious reason, in cases in which tumors have been quiescent for very long time spans. This includes tumors which have been radically removed as well. Therefore, cure must always be declared with caution.

References

Amacher AL (1980) Craniopharyngioma: the controversy regarding radiotherapy. Child Brain 6:57–64

Carmel PW (1985) Craniopharyngiomas. In: Wilkins RH, Rengachary SS (eds) Neurosurgery. McGraw-Hill, New York, pp 905–916

König A, Lüdecke DK, Herrmann HD (1986) Transnasal surgery in the treatment of craniopharyngioma. Acta Neurochir (Wien) 83:1–7

Laws ER (1980) Transsphenoidal microsurgery in the management of craniopharyngioma. J Neurosurg 52:661–666

Maier HC (1985) Craniopharyngioma with erosion and drainage into the nasopharynx. J Neurosurg 62:132–134

Manaka S, Teramoto A, Takakura K (1985) The efficacy of radiotherapy for craniopharyngioma. J Neurosurg 62:648–666

Sweet WH (1980) Recurrent craniopharyngiomas: therapeutic alternatives. Clin Neurosurg 27:206–229

Symon L, Sprich W (1985) Radical excision of craniopharyngioma. J Neurosurg 62:174–181

The Radical Operation for Removal of Craniopharyngioma

W. H. Sweet, Boston/USA

Before 1950, the clinical course of patients with craniopharyngioma was usually a dismal succession of brief improvements followed by relentless deteriorations after partial removals and aspirations of cystic components or perhaps an even more rapid death following an attempted total removal. For example, Northfield (1957) summarized the treatment of these patients as "fraught with difficulty and disappointment to a degree probably not offered by any other intracranial tumor with the exception of glioblastoma." In the same vein is the experience of Olivecrona who from 1924 to 1958 ventured a radical extirpation in only 25 of 106 patients (Svolos 1969); the early postoperative mortality was 24.5%, and survival in the remaining 80 patients was described as "useful" in only 35.

Rationale for Initiating Radical Removal

By 1950 pituitary replacement therapy had reached a sufficiently promising level to induce Matson (1969, p. 556) and me independently to attempt total removals as a routine policy. I was led finally to this decision by a distinguished neurologist who advocated a "kill or cure" policy in this disorder. He was willing to refer to me a relative to whom he was deeply devoted only if I would agree to remove her whole craniopharyngioma even if I estimated the operative mortality of so doing at over 95% – so convinced was he that life was not worth leading after a partial removal. (At that time the erroneous opinion that radiation therapy was of no value for this tumor was widely held). The successful removal of her large solid tumor provided a major impetus to continue the series.

Observations Regarding the Surgery

This is a report of some of the lessons learned and the results in 43 patients with craniopharyngiomas on whom I operated between 1950 and July 1979, my last procedure on such a tumor. The present meager set of data suggests that late direct sequelae of radiation or its late failure to prevent regrowth of tumor may be more likely to occur after 10 years than is late recurrence of a tumor not quite totally removed at radical operation. The studies in which percentages of deaths at various intervals are compared for various forms of treatment with no indication as to the cause of death have dubious value. They become progressively more worthless the longer the time after the treatment because of the increasing likelihood that the death may be due to some unrelated cause. In the important publication of Manaka et al. (1985) their best results were in the group in whom a subtotal removal was combined with radiation therapy. In those, 87% of 18 patients were surviving at 10 years, as compared with 81% of 21 patients likewise surviving at 10 years following presumed total removal (without radiation). The crucial datum of the cause of the deaths was not given, but one notes in their figures that 100% of the subtotal removal – radiation group of 18 had survived for 5 years, whereas in the total removal group all of the deaths had occurred in the first 5 years with the 80% who survived 5 years still being alive 5 years later.

Extended accounts have already been published of my appraisal of my experience and that of others (1976, 1980, 1988). I present here a cursory summary. The principal point that I have sought to make is that these tumor cells tend to provoke a glial reaction at their periphery especially in that portion of the tumor within the third ventricle. To the naked eye this reactive gliosis can look remarkably and alarmingly similar to brain tissue, for which I mistook it in my first cases. However, histologic examination in an effort to discern which hypothalamic nuclei I had removed, revealed the reassuring spectacle of pure glia with an occasional epithelial island of cells. Indeed in two of my cases the glial reaction was so extensive that my initial impression that the tumor was a glioma of the anterior visual pathway or hypothalamus was corrected only by the pathologist's study of my biopsy. I would agree with Pertuiset (1975) that histologically there is no evidence of a cleavage plane between this glial tissue and the normal brain. It has seemed to Pertuiset (1975) "obvious that any dissection can damage the normal tissue," a view held earlier by Bailey et al. (1939) and by Van den Bergh and Brucher (1970). However, the de facto situation is that one can gently – very gently – pull this capsulelike tissue away from the vitally important adjoining hypothalamic, thalamic and visual structures in patients in whom no previous operative dissection has removed this somewhat protective barrier.

In particular those portions of the tumor extending upward into middle and upper third ventricle can be blindly but cautiously drawn down into the operative field of a subfrontal approach. The advent of the operative microscope has permitted two surgeons to see and work in the small field of an open lamina terminalis behind chiasm in front of the anterior cerebral arteries. With at times the hand of the assistant surgeon using a mouse-toothed long-shafted micro-alligator forceps to prevent the tumor capsule from slipping back up into the third ventricle, the surgeon may be able to deploy his two similar forceps or dissectors more effectively to draw tumor out above the chiasm. In 25 of my 43 patients, the tumor had either presumably grown up from below elevating hypothalamic floor and then breaking into third ventricle or had arisen from the wall of that ventricle so that following removal there was no hole in the ventral hypothalamus. In the latter situation the presumed congenital rests must of course have lain separated by cerebral tissue from their usual sites around pituitary stalk and tuber cinereum. The really tenacious attachments in the tumors at primary operation appear to be only anteroinferiorly mainly in the general region of the chiasm, hypothalamus, and major arteries. Such attachments are sufficiently frequent to lead me to question the advisability of a transsphenoidal approach to a large suprasellar mass even if the tumor also enlarges the intrasellar volume. Above the sella, separation of tumor from adjoining normal structures is better done under direct vision. For the same reason I have not used the transcallosal or transcorticoventricular approach even if much of the third ventricle is filled or displaced by tumor.

Although a glial barrier was present in critical areas in 40 of my cases, it is not an invariable finding. Thus Landolt (1972), in electron microscopic studies of five biopsy specimens, identified no glial tissue in one patient whose tumor lay entirely in the third ventricle. Shillito (1976) finds that the layer of gliosis around the tumor may be either too dense and adherent to the brain or too nonexistent to be of value.

Although three of Matson's patients with radical removal at his first operation died at 1, 2, and 6 years postoperatively of hypothalamo-hypophyseal disorders (Katz 1975), such late hypothalamic deaths have been less of a problem as management of associated endocrine deficiencies has improved. However, failure to increase their secretion of cortisone in response to such stresses as infection is a common major hazard in these patients as emphasized by Till (1982). Patients or their relatives should be instructed to double or triple the maintenance dose of steroid with any illness or other significant stress.

Those cystic portions of tumor near its surface may have such thin walls as to be readily torn or shredded with the obvious increased danger of leaving tumor behind. This is also true of any extracerebral parts of the tumor which do not excite a glial response as they expand in the subarachnoid or commonly in the subdural space. This usually permits drawing of tumor blindly up into the operative field from prepontine or cerebellopontine angle regions leaving behind an intact arachnoid over these areas.

The confidence with which one can remove the walls of large cystic tumors appears to depend upon their locus, fragility, and adherence. The four almost wholly cystic tumors in my series contained 20–56 ml fluid. I thought that I had removed them totally, and they have not recurred. The gigantic cysts of Djordjevic et al. (1979) of 100, 150, and 200 ml and of Gelabert et al. (1985) of 620 ml were all bilateral frontal or subfrontal and were all totally removed. Symon et al. (1981), on the other hand, encountering some almost completely cystic tumors within the third ventricle found them "unsatisfactory for radical excision . . . since every movement of dissection of the walls is accompanied by considerable disturbances of the third ventricular structures." Kobayashi et al. (1981) in autopsy studies of seven patients who had no major operations on their tumors found in the cystic parts of the tumors a layer of fibrillary astrocytes and Rosenthal fibers varying in thickness from several hundred microns to a few millimeters. This layer lay between the cyst wall and viable ganglion cells in the thalamus or hypothalamus.

Results of Personal Series

Of the 43 patients on whom I have operated since 1950, 21 were under 17 years of age at the time of my first operation and 22 were 17 years or older. In only three of these did I stop with a partial removal. Radiation doses of 34 Gy in one and 45 Gy to another, respectively, 24 and 17 years earlier, and a subtotal removal 14 years earlier by me in a third patient had been followed by an intense fibrosis. The operative specimens of the third patient included one showing hypothalamic neurones adherent to the dense astroglia. Here the cleavage plane had lain within the hypothalamus. My operation in the remaining 40 patients was the patient's first attempt at radical extirpation.

There were three postoperative deaths. In two patients moderate bleeding from a major artery during the dissection was readily controlled and I went on to complete the removal. Early postoperative favorable status soon deteriorated and at postmortem extensive thrombosis in the anterior part of the circle of Willis was found, having spread presumably from the intraarterial clot at the site of the hemorrhage. Since then I have stopped the operation if I have provoked significant arterial bleeding in order to minimize intra- or extraarterial pressure changes at the locus of hemorrhage which might ensue upon further local manipulation. This policy in two subsequent cases was followed by absence of postoperative sequelae, and the removal was completed at a second stage some weeks later. The third death followed a succession of several minor hemorrhages shown by frank red cells in the lumbar spinal fluid. A somewhat larger hemorrhage proved at autopsy to emanate from the wall of the third ventricle, the site of the previous tumor and not from the expected aneurysm. The bleeding point was not identified. I find no similar recorded case.

The assiduous efforts of my research assistant, Mrs. D. Wallace, have yielded a follow-up on all patients including information on the probable cause of death in all who have died. I emphasize the importance of determining whether or not the late deaths are related to tumor recurrence, a late sequela of cerebral operative damage, radiation necrosis, radiation induced new tumor or unrelated cause. I thought my removal was in fact total in 31, but that I had left a minimal soft tissue remnant in 7. In two I was aware only of a large irregular chunk of calcium being left behind. There have been clinical recurrences in only three of the seven patients in whom I thought I had left some soft tumor. Of the other four of the seven, three are still living at 22, 24, and 21 years after operation and the fourth died 11 years postoperatively from a myocardial infarct.

The danger of treating by radiation an asymptomatic recurrence is demonstrated by the outcome in an 8-year-old patient in whom 1 year after my presumed total removal routine follow-up computed tomography (CT) scan checked by a pneumoencephalogram were both thought to show a small asymptomatic recurrence. With elaborate precautions as to precise placement of portals and dosimetry our radiation therapists delivered 55.2 Gy in 46 elapsed days [in accordance with the recommendation of Kramer (1976) after presumed total removal]. A rather rapidly developing massive radiation necrosis was found at autopsy 5.5 months later. This experience has led me to a much more conservative approach to small presumed recurrences, subsequently seen in magnetic resonance imaging and CT scan in one patient. This was not treated by operation, aspiration, or radiation; it remains asymptomatic with unchanged radiographic images 7 years later.

Both Matson (1969, p. 562) and I have emphasized the much greater operative mortality of a second attempt at radical removal following a recurrence after an initial radical effort. In every one of the five patients in whom I essayed a second radical operation in this situation death occurred within a maximum of 6 months. Chastened by these results, I carried out a less extensive removal in this situation in my last two clinical recurrences without mortality but with continuing deterioration.

A less extensive resection had been carried out elsewhere before my attempted total removal in eight patients. Of these, two had radiation therapy – 50 Gy concluded 11 months before my operation in one and 40 Gy in 29 days after a partial removal in the other. Vision continued to fail, and I then performed the radical removal. No major technical problems were met in these eight cases, and I decided that a radical attempt may succeed after up to a subtotal removal, radiation not over 1 year earlier, or both. However, my worst result in my last decade of 12 radical operations was in a young man who had a "grossly total removal" and then cyst aspiration by another neurosurgeon in the year previous to my operation. After my procedure he was at first "fair" but gradually deteriorated despite no recurrence to live for 15 years, left merely with the ability to feed and dress himself.

The condition and causes of death at various time intervals in the remaining 37 cases are summarized in Tables 2 and 3. I use the classifications as to quality of survival of Katz (1975) and the radiation therapists Bloom (1979) and Danoff et al. (1983) which are roughly the same (Table 1).

In the 7-year interval between 1981 and 1988 the only two deaths were from causes unrelated to the tumor. This result continues to validate our earlier report (Sweet 1982) that no patient who has lived more than 10 years after

Table 1. Performance grades

	Katz	Radiotherapists
"Excellent"	"Independent, functioning normally"	"No disability"
"Good"	"Functioning well with some deficiency"	"Mild handicap"
"Fair"	"Functioning but dependent"	"Major disability, capable of self-care"
"Poor"	"Nonfunctional"	"Incapable of self-care"

Table 2. Results of radical excision (19 children, 21 adults)

Years since first radical operation	Number operated on who might have lived for stated interval (number "at risk")	Number alive for stated interval	Death caused by surgery and/or recurrent tumor	Death from unrelated disease
3	40	32 (80%)	0–3 years 3 (8%)	4 (10%)
5	32	31 (97%)	3–5 years 2 (5%)	0
10	31	27 (87%)	5–10 years 2 (5%)	2 (5%)
10–15	24	22 (92%)	10–15 years 0 (0%)	2 (5%)
15–38	15	14 (93%)	15–38 years 0 (0%)	1 (3%)

There have been three (8%) postoperative deaths, four (10%) from recurrent tumor and nine (23%) from disease unrelated to surgery or recurrence.

Table 3. Quality of survival

	Status early after operative recovery in patients surviving operation but who died by 1981 either from recurrent tumor or unrelated cause	Patients alive in 1981	Patients alive in 1988
Excellent	6	13	12
Good	3	6	6
Fair	1	4	4
Poor	1	1	0
Inadequate data	2	0	0

radical removal has died of the tumor or operation, but also that no such patient has had a clinically significant recurrence. This is perhaps the most important fact to emerge from the long follow-ups. Since so many of these patients are children or young adults, it is especially important to provide a good result for several decades. A similar conclusion was reached by Pierre-Kahn et al. (1986). Half of their 52 children had a radical removal and half a lesser removal plus radiation. At 9–10 year follow-ups the radical removal group had a 24% recurrence rate whereas in the other group with radiation this rate was only 20%. However, at the 15 year follow-up there was a dramatic difference in the two groups. As in our series there were no more recurrences among those with radical removals. In the French series the rate remained at 24% whereas in those with lesser removals and radiation the recurrence rate had jumped to 50%. In the series of Sorva and Heiskanen (1986) the recurrence-free survival rate at 10 years in the group of 27 patients with radical operation had dropped to 60% from 72% at 5 years. Their figures in the eight patients with "nonradical operation and radiotherapy" show a much greater drop to 25% at 10 years compared with 62% at 5 years. The data from the radiotherapists are those of Kramer (1976) who reports that of nine children treated over 10 years earlier only two had died of recurrent disease (at 3 years 8 months and at 5 years).

Conclusion

Although the results of radiation plus less than radical removal may perhaps be better than radical removal without radiation at 5 or 10 years, the results at time intervals longer than 10 years may well prove to be superior. We need more data including more from radical removals in the era of microsurgery and improved imaging.

Acknowledgement. The author wishes to express his gratitude to the Neuro-Research Foundation for its support during the preparation of this manuscript.

References

Bailey P, Buchanan DN, Bucy PC (1939) Intracranial tumors of infancy and childhood. University of Chicago Press, Chicago, pp 349–375

Bloom HJG (1979) Recent concepts in the conservative treatment of intracranial tumors in children. Acta Neurochir 50:103–116

Danoff BF, Cowchock FS, Kramer S (1983) Childhood craniopharyngioma: survival, local control, endocrine and neurologic function following radiotherapy. Int J Radiat Oncol Biol Phys 9:171–175

Djordjevic M, Djordjevic Z, Janicijevic M et al. (1979) Surgical treatment of craniopharyngiomas in children. Acta Neurochir [Suppl] 28:344–347

Gelabert M, Reyes F, Bollar A et al. (1985) Successful treatment of a giant cystic craniopharyngioma. J Neurosurg Sci 29:263–265

Katz EL (1975) Late results of radical excision of craniopharyngiomas in children. J Neurosurg 42:86–90

Kobayashi T, Kageyama N, Yoshida J et al. (1981) Pathological and clinical basis of the indications for treatment of craniopharyngiomas. Neurol Med Chir (Tokyo) 21:39–47

Kramer S (1976) Craniopharyngioma: the best treatment is conservative surgery and postoperative radiation therapy. In: Morley TP (ed) Current controversies in neurosurgery. Saunders, Philadelphia, pp 336–343

Landolt A (1972) 8: Die Ultrastruktur des Kraniopharyngeoms. Schweiz Arch Neurol Neurochir Psychiatr 111:313–329

Manaka S, Teramoto A, Takakura K (1985) The efficacy of radiotherapy for craniopharyngioma. J Neurosurg 62:648–656

Matson DD (1969) Neurosurgery of infancy and childhood, 2nd ed. Thomas, Springfield

Northfield DWC (1957) Rathke-pouch tumors. Brain 80:293–312

Pertuiset B (1975) Craniopharyngiomas. In: Vinken PJ, Bruyn GW (eds) Handbook of clinical neurology, vol 18, Elsevier, Amsterdam pp 531–572

Pierre-Kahn A, Hirsch JF, Renier D et al. (1986) Craniopharyngiome de l'enfant – indication therapeutique et devenir a long terme. Analyse retrospective de 52 observations. Abstract for the meeting of the Société de Neurochirurgie de Langue Francaise, Paris, France, December 8–11

Shillito J (1976) The treatment of craniopharyngiomas of childhood. In: Morley TP (ed) Current controversies in neurosurgery. Saunders, Philadelphia, pp 332–335

Sorva R, Heiskanen O (1986) Craniopharyngioma in Finland. Acta Neurochir 81:85–89

Svolos DG (1969) Craniopharyngiomas. A study based on 108 verified cases. Acta Chir Scand [Suppl] 403:1–44

Sweet WH (1976) Radical surgical treatment of craniopharyngioma. Clin Neurosurg 23:52–79

Sweet WH (1980) Recurrent craniopharyngiomas: therapeutic alternatives. Clin Neurosurg 27:206–229

Sweet WH (1982) Craniopharyngiomas, with a note on Rathke's cleft or epithelial cysts and on suprasellar cysts. In: Schmidek HH, Sweet WH (eds) Operative neurosurgical techniques. Grune & Stratton, New York, pp 291–325

Sweet WH (1988) Craniopharyngiomas (with a note on Rathke's cleft or epithelial cysts and on suprasellar cysts). In: Schmidek HH, Sweet WH (eds) Operative neurosurgical techniques. Grune & Stratton, New York, pp 349–379

Symon L, Jakubowski J, Logue V (1981) The surgical treatment of craniopharyngioma. Neurochirurgia [Suppl]: 262

Till K (1982) Craniopharyngioma. Child Brain 9:179–187

Van den Bergh R, Brucher JM (1970) L'abord transventriculaire dans les craniopharyngiomes du troisième ventricule. Aspects neurochirurgicaux et neuropathologiques. Neurochirurgie 16:51–65

Experiences with Radical Excision of Craniopharyngioma

L. Symon, London/U.K.

Introduction

As Cushing [2] pointed out in 1932, of all the tumors of the suprasellar region, few are as difficult to deal with as craniopharyngiomas. As a result widely varying opinions have been expressed about their management, and although it is well known that the lesions are almost invariably "benign", most surgical series quoted over the past 20 years attest to the difficult management characteristics of these tumors and their tendency to recur. They carry in general a miserable prognosis. Early surgical experience without the operating microscope showed that as the tumors were followed over a long time recurrence was inevitable despite apparent initial complete resection [6, 8, 10, 11]. Not surprisingly, the poor results of surgery led to aggressive radiotherapy, and following the series reported by Kramer et al. [5] many surgeons inclined to the view that minimal surgical intervention with the aspiration of any associated cyst in establishment of the diagnosis followed by aggressive radiotherapy had better results. Nevertheless, isolated reports [4, 11] have suggested that an aggressive approach remains justified, and our own experience with a total of 36 radically excised cases suggests that radical excision, if possible, is undoubtedly the best method of management. Our series is almost exclusively adult, only 4 of the 36 were below the age of 20 years, and only one was younger than 10.

They were almost all properly classified as retrochiasmatic and almost invariably operated upon by a sub-temporal approach. This was not a consecutive series, radical excision being attempted only when it appeared a favourable prospect.

Presenting Features

Over 70% of the patients presented with failure of vision and about one-fifth of them with raised intracranial pressure. The usual visual disturbance was a bitemporal haemianopia although a predominantly unilateral abnormality could be present where the tumor was conspicuously offset to one side. Loss of libido secondary to decreased follicle-stimulating hormone (FSH) and luteinizing hormone (LH) and fatigue secondary to decreased thyroid-stimulating hormone (TSH) [3, 7] were common and often confirmable by endocrine assessment, which was not invariably detailed before surgery.

Radiological Investigations

Plain skull X-rays, computed tomography with and without enhancement [9], four-vessel angiography and occasionally pneumoencephalography with tomography of the anterior third ventricular region [12] were routine methods of assessment in the early part of the series, and more recently as magnetic resonance imaging has become available, it has become clear that this is the ideal method of investigation for these cases. Not since the abandonment of pneumoencephalography has such clear demonstration of the relationship of these tumors to the ventricular system been available as with a high field strength magnetic resonance scan.

We used angiography to delineate the vessels of the circle of Willis and to plan the surgical attack. The usual plain skull changes with truncation of the dorsum sellae were common, occurring in about 40% of cases and pathological calcification was present in 25% as in Banna's analysis [1]. It is worthwhile pointing out that calcification was visible in over 40% of cases assessed by CT but in only about a quarter in plain X-ray.

Follow-Up

The 34 patients have been followed up for a minimum period of 2 years, and the average follow-up period is now over 5 years.

Surgical Philosophy

The great majority of cases considered suitable for radical excision in this group were the retrochiasmatic group of cases. These lie partly within the cavity of the third ventricle, partly in its floor, and partly in the interpeduncular cistern behind the chiasm. The relative proportion of these three sites varies from tumor to tumor. Sweet [11] was the first to point out that where the tumor lay in contact with brain, it was surrounded by a reactive gliosis of long-standing as evidenced by the presence of Rosenthal fibres. Encephalography as shown in Fig. 1 demonstrates that the tumor separates the walls of the third ventricle, and from what we know of the histology its apposition to the floor of the third ventricle is accompanied by this extensive gliosis. The philosophy of excision, therefore, is to dissect through this plane of gliosis carrying the process of morcellement through the centre of the tumor initially and developing a plane of excision first attacking the free portion of the tumor within the interpeduncular cistern. Detailed knowledge of the vascular anatomy enables one to avoid the perforating vessels of the hypothalamus which are usually in two groups from the posterior communicating artery, and enables one to avoid also the anterior and postero-lateral choroidal arteries which may be considerably stretched by the tumor (Fig. 2).

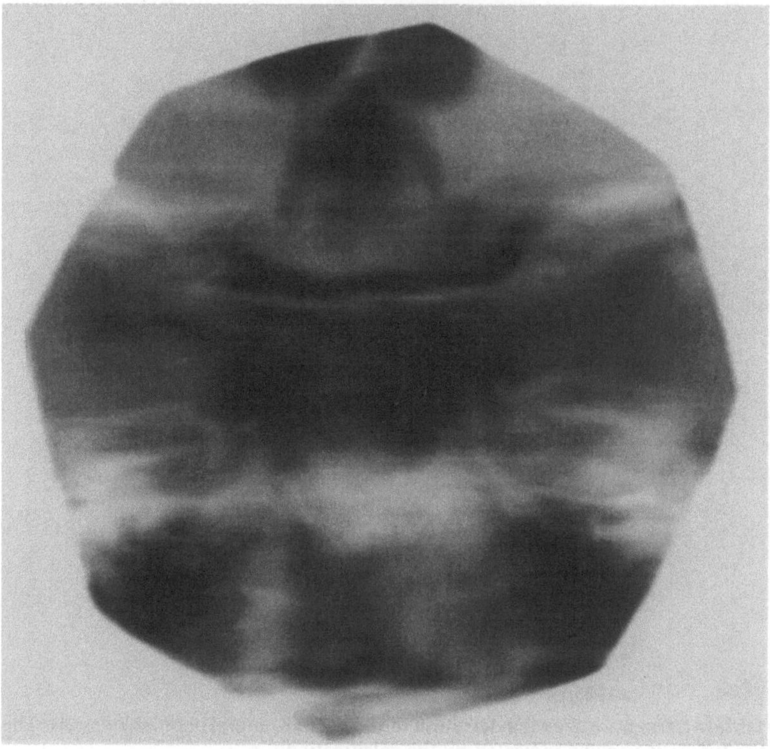

Fig. 1. AP encephalogram showing the third ventricle draped over the surface of the tumor

Surgical Approach

Where the tumors are pre-chiasmatic, a standard sub-frontal approach may be appropriate. Where the tumors are almost completely within the third ventricle (they are never completely within the third ventricle, but also always partly within its floor), a transventricular approach through the lateral ventricle and foramen of Monroe, may provide an adequate excision.

Classically, however, the retrochiasmatic approach is best performed by a fronto-temporal craniotomy carrying the craniotomy well to the floor of the middle temporal fossa and followed by a small one inch excision of the anterior temporal pole to lay the sub-frontal and temporal regions in continuity. Self-retaining retraction and the microscope are essential, and the dissection is initially in the interpeduncular cistern although if the tumor is small and has little posterior extension, it may be necessary to dissect in the triangular space between the carotid artery and the anterior choroidal artery anteriorly or between the posterior communicating artery and its anterior thalamo-perforating branches inferiorly and posteriorly. All these vessels must be carefully preserved.

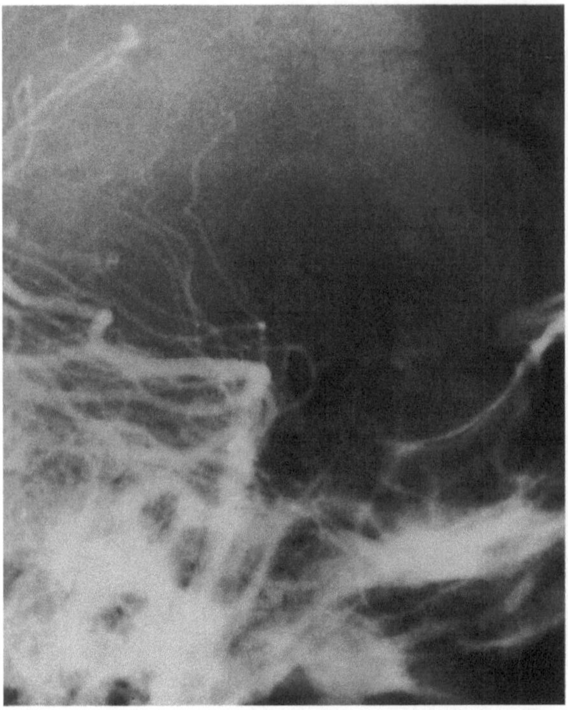

Fig. 2. Vertebral arteriogram showing deviation of the postero-lateral choroidal vessels. The outline of the tumor is vaguely indicated by the calcification

Incision of the capsule of the tumor now allows decompression of the interior of the mass which may be considerably accelerated by the use of the ultrasonic dissector and is possible with gentle controlled suction, monopolar coagulation and dissection with sharp and blunt hooks. Strong retraction on portions of the capsule anterior to the third ventricle or major vascular structures should be avoided and the tumor dissected progressively from the vessels on the side of the exposure and on the contralateral side. Some blind dissection may be necessary anteriorly, but if adequately debulked the capsule comes down with gentle traction and under direct microscopic vision. Sharp dissection is usually necessary to free the tumor from the posterior aspect of the chiasm and the approach enables excellent visualization of the ipsilateral optic tract and nerve and if necessary visualization of the opposite optic nerve allowing adequate calculation of the line of the chiasm. Retraction on the chiasm itself is dangerous and should be avoided.

Densely calcified tumors may be adherent to the medial aspect of the ipsilateral or contralateral carotid artery and may occasion abandonment of the procedure since damage to the carotid is extremely difficult to repair.

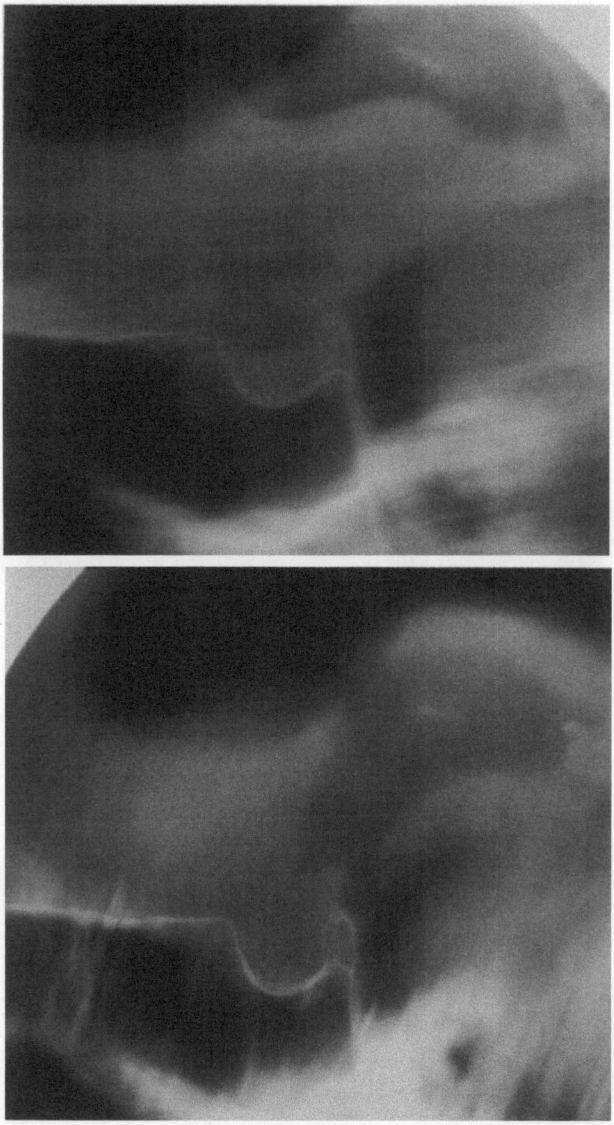

a

b

Fig. 3. **a** Pre-operative air encephalogram in a typical retrochiasmal craniopharyngioma. **b** Post-operative air encephalogram showing communication between the pre-pontine cistern and the third ventricle. The third ventricle has returned to normal position

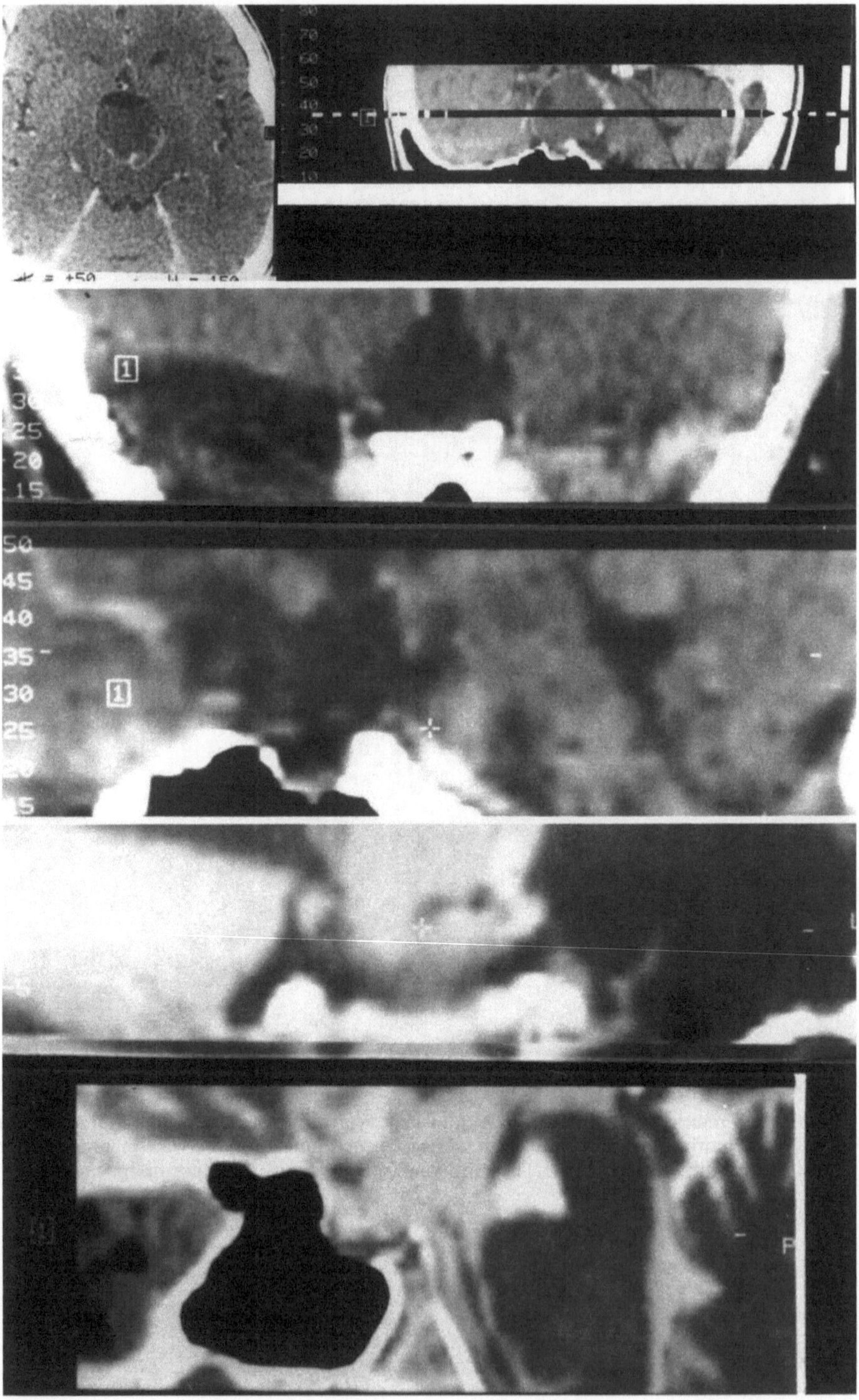

Fig. 4a–c

Table 1. Recurrence rate

	n	Known recurrences
Primary radical excision	34	10
Primary sub-total excision	45	17 (38%)
Partial excision	25	11 (44%)
Aspiration/biopsy/shunt	21	16 (76%)

General Management

The use of high-dose steroids and of synthetic DDAVP (desamino-*cis*-1-8-D-arginine vasopressin) have revolutionized the capacity to manage the electrolyte and hormonal disturbances associated with disturbance of the hypothalamus almost inevitable in such excisions. However, the hypothalamus is a paired organ and curiously the contralateral dissection can be done extremely gently by this approach and profound endocrine disturbance has been progressively less common.

Assessment of Completeness of Excision

Early in the series completeness of excision was assessed by encephalography following surgery, the characteristic return of the anterior third ventricle anatomy and the visualization of a hole in the floor of the third ventricle being adequate evidence that the tumor had been radically removed. (Fig. 3). More recently, computed cisternography with iohexol has been employed (Fig. 4), and in the most recent cases post-operative magnetic resonance imaging will clearly be the key. One should note that swelling immediately post-operatively may produce small defects in ventricular anatomy, and that all the subtle changes of magnetic resonance images are as yet incompletely understood, and the differentiation between the residual oedema and tumor may be quite difficult in the early stages.

Results

There have been two deaths in the series, one with typical hypothalamic damage, and the other from Stevens-Johnson syndrome probably related to anticonvulsant medication in the early post-operative period. Of the 34 cases available for follow-up none has recurred within 2 years of surgery (Table 1).

◄

Fig. 4. a Pre-operative computed tomogram of a partially cystic craniopharyngioma. **b** Post-operative unenhanced computed tomogram showing the temporal lobectomy and the space left by removal of the tumor. **c** Iohexol cisternogram showing the empty sella and the space previously occupied by the craniopharyngioma

Table 2. Surgical management of craniopharyngioma: 1958–1986

	n	Operative death	Mortality rate
Radical excision	36 (28%)	2	(6%)
Sub-total excision	45 (35%)	0	(0%)
Partial excision	25 (20%)	5	(20%)
Aspiration/biopsy/shunt	21 (17%)	0	(0%)
Total	127 (100%)	7	(6%)

This contrasts with the general experience of the clinic in over 100 cases (Table 2) which indicates clearly that lesser operations on craniopharyngioma show a higher recurrence rate.

References

1. Banna M (1976) Craniopharyngioma: based on 160 cases. Br J Radiol 49:206–223
2. Cushing H (1932) Intracranial tumors. Notes upon a series of two thousand verified cases with surgical-mortality percentages pertaining thereto. Thomas, Springfield, IL, pp 93–98
3. Kahn EA, Gosch HH, Seeger JF, et al. (1973) Forty-five years experience with the craniopharyngiomas. Surg Neurol 1:5–12
4. Katz EL (1975) Late results of radical excision of craniopharyngiomas in children. J Neurosurg 42:86–90
5. Kramer S, McKissock W, Concannon JP (1961) Craniopharyngiomas. Treatment by combined surgery and radiation therapy. J Neurosurg 18:217–226
6. Matson DD (1964) Craniopharyngioma. Clin Neurosurg 10:116–129
7. Matson DD, Crigler JF Jr (1966) Management of craniopharyngioma. Am Surg 152:699–704
8. Matson DD, Crigler JF Jr (1969) Management of craniopharyngiomas in childhood. J Neurosurg 30:377–390
9. Naidich TP, Pinto RS, Kushner MJ et al. (1976) Evaluation of sellar and parasellar masses by computed tomography. Radiology 120:91–99
10. Shillito J Jr (1980) Craniopharyngiomas: the subfrontal approach, or none at all? Clin Neurosurg 27:188–205
11. Sweet WH (1976) Radical surgical treatment of craniopharyngioma. Clin Neurosurg 23:52–79
12. Symon L (1982) Operative surgery in craniopharyngiomas. In: Edwards JMR (ed) Topical reviews in neurosurgery, vol 1. Wright-PSG, Bristol London Boston, pp 134–160

Some Technical Considerations Regarding Craniopharyngioma Surgery: The Bifrontal Approach

W. Bini, A. Sepehrnia, and M. Samii, Hannover/FRG

Introduction

The sellar region is a vital anatomical and functional landmark in the skull base. This region is a crossroad for the major blood vessels supplying the brain, several cranial nerves concerned with vision and eye movement, the centers of both endocrinal and autonomic system, and those concerned with the emotional sphere of our being. These structures or centers can be affected by pathological processes in this region or neighboring areas as well as by the therapy directed to eradicating them. The vast majority of the processes are benign and slowly growing tumors, such as pituitary adenomas, craniopharyngiomas, tuberculum sellae meningiomas, and epidermoid-dermoid tumors. Surgery in this area, irrespective of the surgical process, is not without risk. Microtechniques have contributed to make such surgery safer and have enhanced the surgeon's ability to approach this landmark through various routes, improving the results. Nevertheless, the surgical treatment or operative strategy of craniopharyngiomas is a matter of ongoing discussion, and to use Cushing's words "constitutes a baffling problem to neurosurgeons" still today, and the management of such a tumor in a patient is complex.

While there is general agreement on the fact that surgery should play a major role in the management of craniopharyngiomas, there are still questions regarding the exact nature and limits of it, and regarding which approach is the optimal one. The primary reasons for the substantial difficulty stem from the close relationship between the tumor and the floor of the third ventricle (hypothalamus) and the perforating vessels which nourish it. Of course the extension and size of the tumor determine the problem of complete removal and operative morbidity, which may be high. From 1981 to June 1988, 24 patients with craniopharyngiomas were operated on at the Neurosurgical Clinic of the Nordstadt Hospital in Hannover.

Based on this experience we have come to prefer the subfrontal route via a bifrontal approach. Fundamental for this is that due to the often large suprasellar mass (especially in cystic tumors) the optic pathway or chiasm and the hypothalamus are completely elevated and displaced from the skull base, and thus we acquire a large space for preparation and dissection of the important structures by gradual tumor debulkment and can even reach the whole prepontine area visualizing the basilar artery and its branches (Fig. 1).

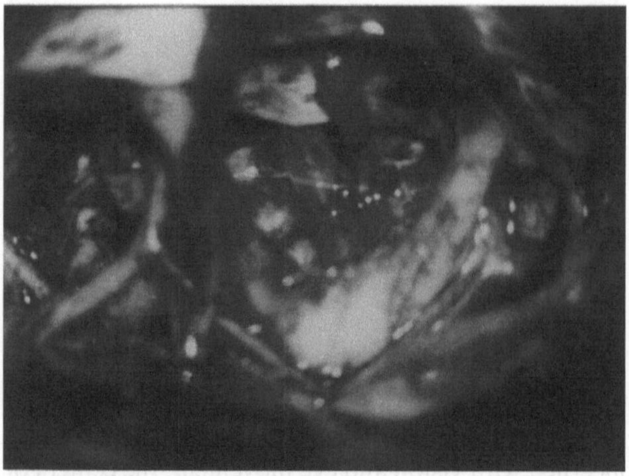

Fig. 1. Intraoperative view of the basilar tree and prepontine area after complete removal of a craniopharyngioma via a bifrontal (subfrontal) approach

Materials and Methods

Our 24 patients icluded 11 women and 13 men. The medium age was 39 years (2–64 years); ten patients were aged 20 or younger. All patients were preoperatively evaluated with nonenhanced and enhanced high-resolution computed tomography (CT) scans and bilateral carotid angiograms. The major tumor diameter was 4 cm or more in three patients and 3–4 cm in one. All tumors were prevalently retrochiasmatic; three of these extended to the level of the middle clivus, invading the prepontine cistern. There was no operative mortality, no patient experienced a neurological worsening after the operation, but all patients required hormonal replacement. We had one case of personal recurrence (4.2%). Two other cases were recurrences with primary operation elsewhere. Adjuvant therapy was not given in any case. All patients are clinically and CT recurrence-free over a follow-up period 5 months to 8 years after operation. Although in two of the large craniopharyngiomas we used a subfrontotemporal approach, we performed a bifrontal craniotomy in the majority of our cases. Via this approach the pituitary stalk, visual pathway and adjacent arteries of the anterior part of the circle of Willis can be optimally visualized and kept under view during the operation. This approach also allows us to preserve both olfactory nerves preparing and dissecting them until exposure of both optici.

We begin with a bitemporal skin incision from zygoma to zygoma behind the frontal hairline. The burrholes are placed left and right on the orbital buttons, another on the midline over the frontal sinus, and the fourth above the glabella (Fig. 2). An osteoplastic craniotomy is performed, and then the frontal sinuses which may be opened are cleaned out and packed with antibiotic gauze (until the end of the operation). The sinus is ligated, the dura incised laterally,

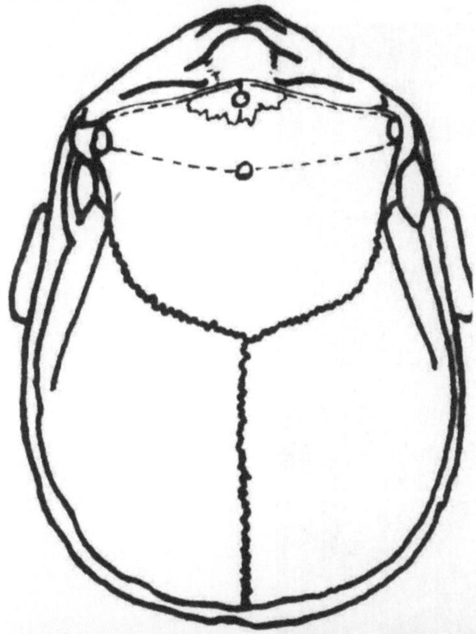

Fig. 2. Schematic representation of the bifrontal exposure. Note the placement of the midline burrholes

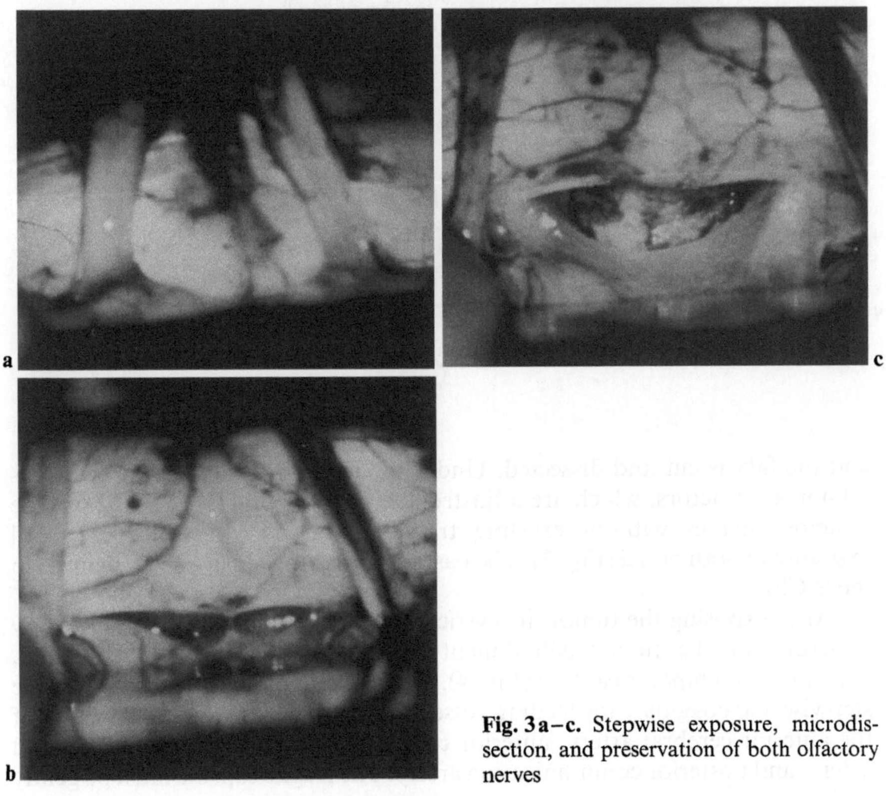

Fig. 3a–c. Stepwise exposure, microdissection, and preservation of both olfactory nerves

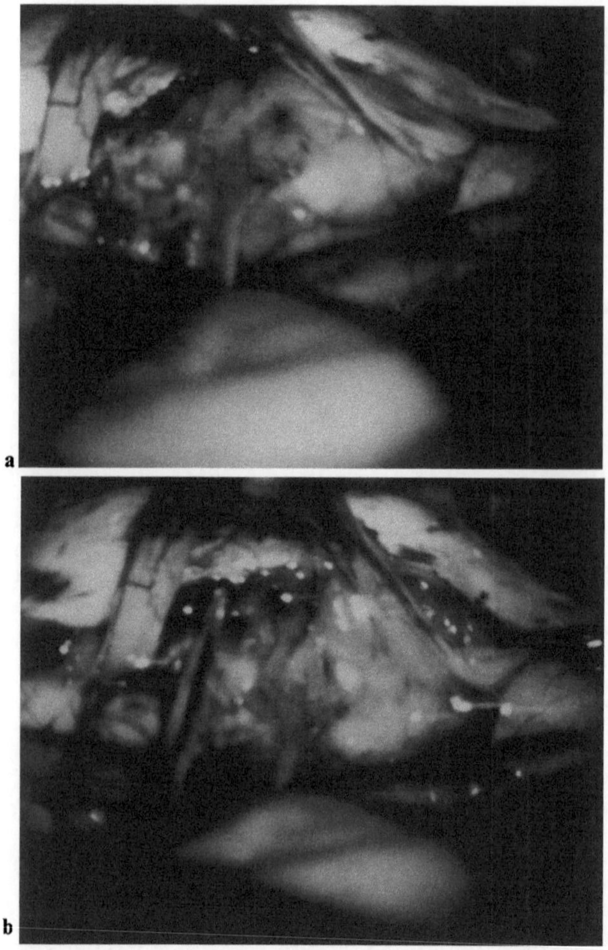

Fig. 4 a, b. Initial exposure and blunt preparation of tumor capsule

and the falx is cut and dissected. Under the microscope and using two self-retaining retractors, which are adjusted stepwise, we obtain CSF, expose the olfactory nerves without exerting traction, and microdissect these until exposure of both optici (Fig. 3). The basal cisterns are then opened to gain even more CSF.

After exposing the tumor, its cystic contents are drained when present by puncture, or the tumor debulkment begins under semiconstant suction, irrigation, and bipolar cautery (Fig. 4). We further mobilize the tumor capsule stepwise and coagulate the feeding vessels originating from the small feeders of the anterior cerebral artery, anterior communicating artery, internal carotid artery, and posterior communicating artery. The fragile capsule must be gently

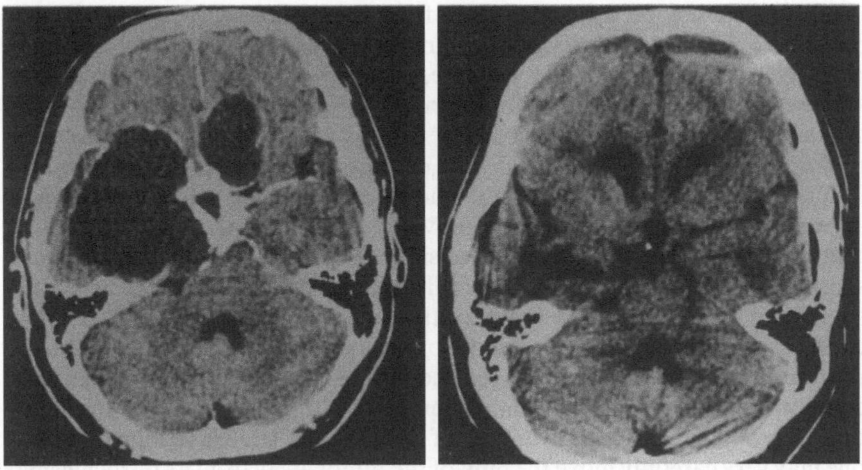

a b

Fig. 5a, b. Giant cystic craniopharyngioma with large middle fossa extension. Preoperative enhanced CCT and postoperative control study after complete tumor removal

handled throughout the operation. As the tumor is removed and the nearby structures relaxed and viewed, careful dissection (sharp or blunt) can proceed.

In cases of "postfixed" chiasma, the tumor removal is first performed between the optic nerves; after tumor debulkment the optic pathway is relaxed allowing us to proceed, visualizing the pituitary stalk. In cases of "prefixed" chiasmas, a situation with which we are confronted in large tumors (prevalently retrochiasmatic), the tumor debulkment is performed through the lamina terminalis. In these cases the walls of the third ventricle appear after relaxing the tumor capsule. By this approach even extensive growths, affecting the middle fossa, can be tackled lateral to the carotid artery and corresponding optic nerve (Fig. 5).

In some cases the tuberculum sellae can be removed with diamond drill to improve the view into the sella. Once this procedure is completed, the opened ethmoid cells must be covered; for this we utilize the galea periost flap to close the frontal sinuses.

Discussion

Among the different topographical locations of these tumors the retrochiasmatic one is considered the most difficult to approach surgically, and it so happens that large cystic craniopharyngiomas are usually retrochiasmatic – in our series this was the case in all of them.

The difficulties stem from the close relationship between the tumor, the floor of the third ventricle (hypothalamus), and the perforating vessels. Nevertheless, these tumors usually at primary operation do not transgress

Lilequist's membrane separating them from the basilar artery system. Retrochiasmatic craniopharyngiomas may grow superiorly, invaginating and thinning out the floor of the third ventricle, posteriorly toward the interpedicular and prepontine cistern, and anterioinferiorly to the chiasma widening the space between the optic nerve and internal carotid artery. The optic chiasma is pushed forward or prefixed. Besides the ensuing vascular displacements and the frequent coexistence of subfrontal and/or subtemporal parts, the surgical approach should allow for tumor removal through the real or potential spaces that are created by the tumor expansion. A bilateral subfrontal approach sparing the olfactory nerves bilaterally allows unobstructed access to the subfrontal extension, many times cystic, and in addition one has complete control as mentioned earlier over both optici and both ICA. The prechiasmatic portion may be removed in between the optic nerves and the retrochiasmatic component by opening the lamina terminalis or by working in the interopticocarotid space. The advantage of opening the lamina terminalis is that one may very carefully move the bulk of the tumor away from its retrochiasmatic location at the level of the third ventricle by pushing it downward and anteriorly to be removed between both optici or downward and laterally to be delivered through the interopticocarotid space.

A pterional approach allows better exposure of the interopticocarotid space, and if extended posteriorly and combined with superior retraction of the temporal lobe, it may somewhat better expose the lateral tumor portion in the space between the third nerve and posterior communicating artery inferiorly and the optic tract superiorly. However, the exposure of the contralateral optic nerve and ICA and of the contralateral frontal fossa is worse.

Based on our experience no increased morbidity or postoperative complications have resulted, even when giant tumors have been so approached and a complete tumor removal performed, as this bilateral subfrontal approach due to the surgical anatomy allows us a good overview of the area and permits us a complete tumor removal through the different portals. Other approaches that may be useful in special situations combined in part or in toto are the transventricular one for pure intraventricular tumors, the subtemporal one for small retrochiasmatic tumors, the transpetrosal-transtentorial Hakuba approach for large retrochiasmatic tumors, and the transsphenoidal route preferably for tumors confined to the sella.

The best surgical approach, nevertheless, may only be selected after careful evaluation and study of the topography of the tumor by neuroradiological examination, angiography, CT and MRI and based on the experience of the surgeon.

Two Rare Craniopharyngiomas

J. P. TSIRIMBAS, Athens/Greece

We present here two cases of rare craniopharyngiomas. The first is a solid and cystic tumor that developed in the sphenoid air sinus and extended rapidly as far as the left optic nerve. Over the course of 6 days the otherwise healthy patient K. P., a-24-year-old soldier, had successive headaches, dizziness, vomiting, left third cranial nerve paresis, impairment of vision in the left eye, and a visual field defect in the left eye, and he became drowsy and semicomatose. The investigation included X-ray, air encephalography, and a left carotid angiography (Figs. 1–4). A left subfrontal approach was performed. The gray solid part of the tumor was situated in front of the hypophysis and compressed the left optic nerve from below. The cystic part of the tumor was found occupying the sphenoidal cavity and contained yellow fluid with crystals of cholesterol. The tumor was totally excised through an opening on the planum sphenoidale, which was closed by muscles, gelfoam, and fascia. The postoperative course was complicated by a meningitis which kept the patient in the state of coma for approximately 2 months. On the 3rd postoperative day the patient underwent an exploratory operation because of suspicion of a bleeding, which was not detected (Fig. 5). The patient was discharged from hospital with a slight right hemiparesis, loss of vision in the left eye due to optic disc atrophy, and a right elbow ossification. In the intervening 11 years the patient has suffered no recurrence of the tumor.

The second case is that of a woman, K. T., 56 years old, who appeared to have two separated craniopharyngiomas in different places, in a period of 3 months. The first was located in the region of the sella turcica and the other in the hypothalamus – third ventricle. The patient had been investigated in another hospital and referred to us in good condition for surgery. She was a little confused, drowsy, and hemianopic. On computed tomography (CT) a typical cystic tumor in the sella turcica was evident. The operation was performed by a subfrontal approach. The tumor was found connected with the posterior hypophysis and the under surface of the optic chiasma and tracts. It consisted of solid and cystic parts. The solid tumor was gray and contained a mass of small crystallic yellow grains resembling sand. The cystic tumor extended posteriorly and somewhat upward to the solid part between the optic tracts and contained yellow fluid. The excision of the mass-cyst tumor was total, but small remnants conncected with the undersurface of the optic tracts remained. The operation was bloodless, and the patient had an excellent recovery. A new CT examination showed an area of calcified grains in the

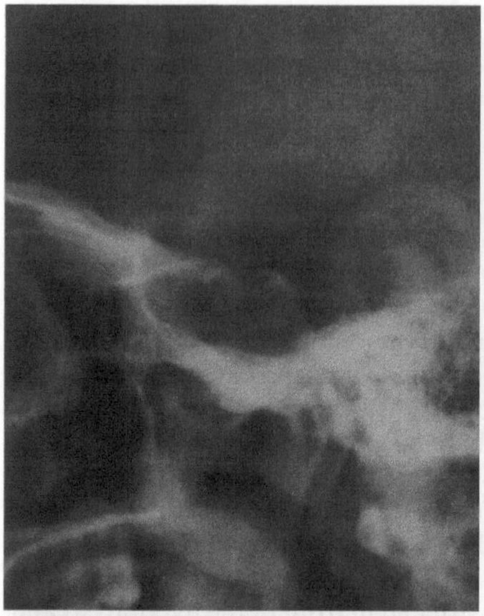

1

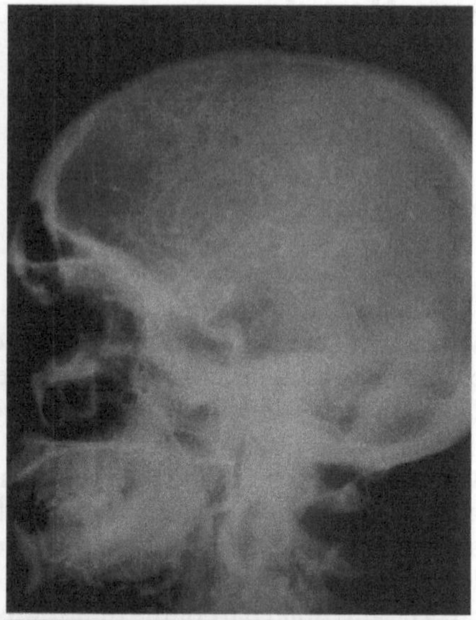

3

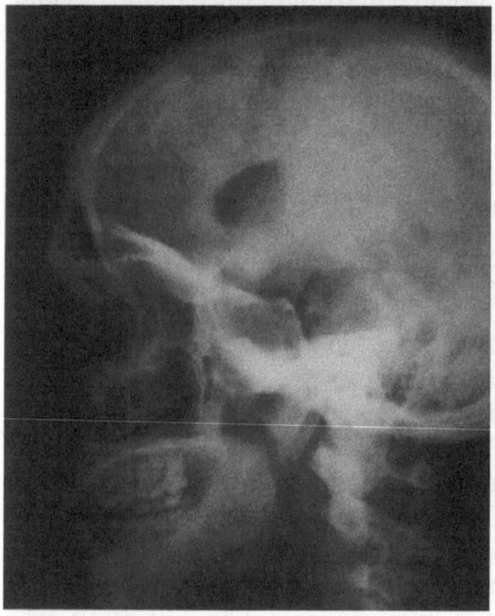

2

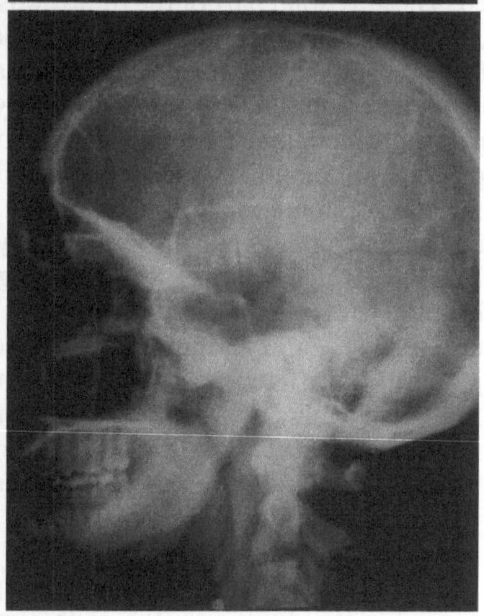

4

Fig. 1. Case 1: X-ray film of intrasphenoidal craniopharyngioma. The enlargement of sphenoidal sinus is evident

Fig. 2. Case 1: air encephalography

Fig. 3. Case 1: arterial phase of left carotid angiography. Avascular parasellar tumor

Fig. 4. Case 1: venous phase of left carotid angiography. Avascular parasellar tumor

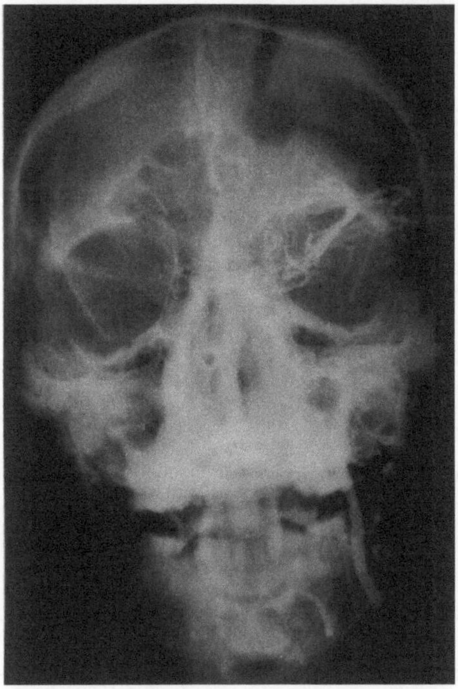

Fig. 5. Case 1: postoperative left carotid angiogram

6 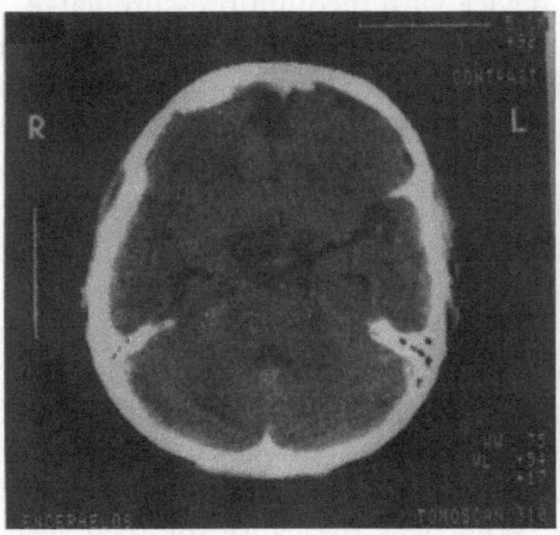

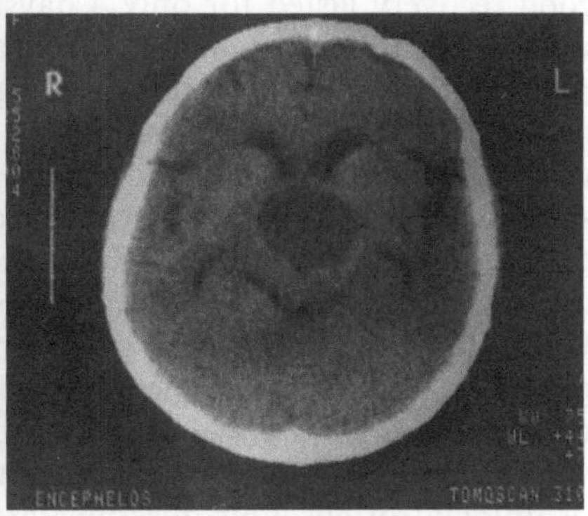 7

Fig. 6. Case 2: CT scan showing the solid and cystic part of the second craniopharyngioma

Fig. 7. Case 2: CT scan, cystic part in the region of the third ventricle

8 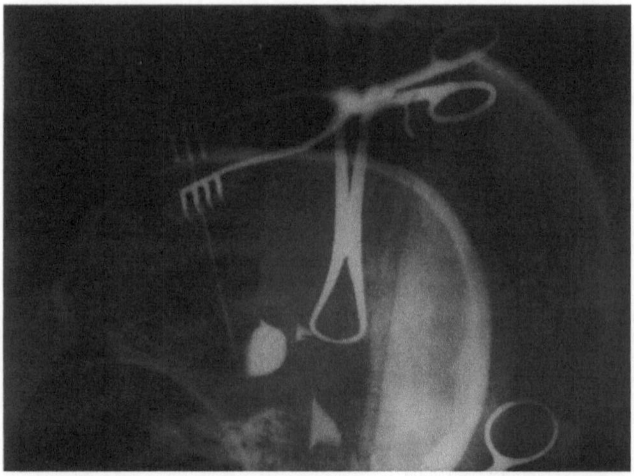

Fig. 8. Case 2: air and contrast cystography of the craniopharyngioma which communicates with the aqueduct and fourth ventricle

hypothalamus which in our opinion became the solid part of the second craniopharyngioma.

After an excellent condition for about 2.5 months, she repeatedly felt confused, drowsy, and demented, and in 21 days she became semicomatose. A new CT examination revealed another cystic tumor in the hypothalamic region (Figs. 6, 7). Following air ventriculography the cyst was located and aspirated. It contained yellow fluid. Air and contrast cystography was performed. The cyst in the region of the third ventricle communicated with the fourth ventricle (Fig. 8). For radiotherapy of the patient we drained the cyst into the gallbladder. We used for it a shunting device with medium pressure valve. A good recovery lasted for only 4 days. Hypotonia, mild fever, and cardiorespiratory disturbances then occurred, and the radiotherapy was stopped at the beginning. For fear that the shunting into the gallbladder might have been responsible for these symptoms, the peritoneal catheter of the shunt was withdrawn from the gallbladder and connected to a closed micro Red-o-pack system, put into a colostomy bag filled with antiseptic povidone-iodine. Both the above means of drainage have proven successful. Despite the communication of the cyst with the aqueduct, no CSF was drained in the vacuum system. Unfortunately, however, every effort of ours to cope with the above syndrome proved unsuccessful, and the patient died 15 days later from cardiac arrest. In conclusion, therefore, we would say that undoubtedly the two cystic tumors were separated, in different places of the brain and were craniopharyngiomas of different origin. Our questions are: What was the cause for their nearly parallel development? Is there any predisposing factor which can influence the increase of craniopharyngiomas, or was it an incidental coincidence?

References

1. Guidetti B (1979) Craniopharyngiomas. Results of surgical treatment. Acta Neurochir [Suppl] 28:349–351
2. Metzger J (1979) CT follow up studies in postoperative craniopharyngiomas. Acta Neurochir [Suppl] 28:357–361
3. Northfield DWC (1973) The surgery of the central nervous system, chapter II. The pituitary adenomas: craniopharyngioma. Blackwell, Oxford London Edinburgh Melbourne, pp 280–330

Chapter 3: Other Space-Occupying Lesions

Thermoregulation in Patients with Skull Base Tumors*

R. BEHR, C. DIETRICH, G. HILDEBRANDT, N. KLUG, and K. BRÜCK, Giessen/FRG, and Köln/FRG

Introduction

Severe disturbances of temperature regulation are well known in acute traumatic brain stem lesions. In chronic hypothalamus lesions due to space-occupying processes spontaneous hyperthermia was described by Strauss and Globus in 1931, by Alpers in 1936, by Globus and Kuhlenbeck in 1942, and recently by Lipton et al. in 1981. Hypothermic temperature courses were reported by Davison and Selby in 1935, Cooper in 1970, and Fox et al. in 1970. A comprehensive study of temperature courses in patients with different hypothalamic and brain stem lesions was performed by Lausberg in 1972.

The goal of this study was to describe modifications of thermoregulatory mechanisms in patients with basal midline tumors and clinically normal body temperature by means of external thermal loads and measurement of thermoregulatory effector mechanisms in relation to body temperature changes.

Methods

The patients and healthy control subjects were exposed to cold and heat in a climatic chamber, dressed in bathing clothes and resting in supine position in a chair. After an initial period of about 30 min at an ambient temperature (T_a) of 30 °C (wall and air temperature equal; relative humidity 50 %), T_a was reduced to 10 °C until the patients and controls exhibited shivering (electric muscle activity, EMA) and increased metabolic rate (MR). They were then rewarmed by increasing T_a to 55 °C (relative humidity 50 %). This heat load was continued until distinct heat dissipation reactions (vasodilatation, sweating) were observed. During the whole thermoregulatory examination the following variables were continuously measured and recorded by an electronic data acquisition and processing system.

Temperature

All temperatures were measured using copper-constantan thermocouples. Air temperature was measured in the middle of the climatic chamber (T_a). The

* This research was supported by the Deutsche Forschungsgemeinschaft.

temperature in the external auditory canal close to the tympanic membrane was measured; this is referred to as the tympanic temperature (T_{ty}). To insure the measurement of head core temperature the auditory canal was plugged with absorbent cotton, and the ear was covered by a capsule filled with absorbent cotton (Hessemer and Brück 1985). Skin temperature was measured at four different sites (upper arm, chest, thigh, lower leg). Mean skin temperature ($\bar{T}_{sk}$) was determined using the equation of Ramanathan (1964). The mean weighted body temperature represents the thermoregulatory responses influenced by both cutaneous and internal thermal receptors. Therefore these responses must be described as a function of both skin and core temperatures (Brück 1981; Brück and Zeisberger 1987). The weighted mean body temperature ($\bar{T}_b$) is calculated according to the following equation:

$$\bar{T}_b = T_{ty} \times a + \bar{T}_{sk} \times b$$

The coefficients a and b were determined to 0.87 and 0.13 (Brück et al. 1976, 1981, 1987; Hessemer and Brück 1985; Hessemer et al. 1986). These weighting factors are comparable to those determined by Nadel et al. (1971) for the description of evaporative heat loss. It is thus justified to describe modifications in regulatory characteristics of the thermoregulatory system by alteration of the threshold temperature for heat production and heat dissipation on the basis of mean weighted body temperature ($\bar{T}_b$). The coefficients denote the relative sensitivity of the thermoregulatory system against the changes of core and skin temperatures, respectively, but they cannot be used to calculate heat content of the body.

Measurement of Metabolic Rate

Oxygen consumption (STPD) was measured by an open system, consisting of a plastic hood, placed over the subjects' head, a pump, flow meter, and paramagnetic oxygen analyzer (Servomex OA 184), as previously described (Hessemer and Brück 1985). MR was calculated from $\dot{V}_{O_2}$ (STPT) using a caloric equivalent of 20.4 KJ/l, assuming a respiratory quotient of 0.86.

Measurement of Electric Muscle Activity

EMA was recorded by surface electrodes over the latissimus dorsi muscle and quadriceps femoris muscle. EMA was integrated, and arithmetic means of EMA values from the two muscles were calculated.

Measurement of Sweat Rate

Local cutaneous sweat rate (SR) was measured on the chest using a capsule which covered a skin area of 7 cm^2. Dried heated air of constant temperature was drawn through the capsule at a constant rate and led to an infrared water vapor analyzer (Uras 2t, Hartmann and Braun).

Measurement of Local Skin Blood Floow ($\lambda 1$, $\lambda 2$)

Local skin blood flow at forearm ($\lambda 1$) and thenar ($\lambda 2$) was measured using a heat clearance technique, previously described by Hessemer and Brück (1985).

Hormonal Findings

Analysis of hormones was performed routinely at the Department of Nuclear Medicine and the Department of Endocrinology of the University of Giessen, using the radioimmunoassay technique.

Statistics

Statistical tests included analysis of variance (ANOVA), Kruskal-Wallis test (KWT), Wilcoxon test, and Student's *t* test.

Results

Figure 1 outlines the typical temperature courses and thermoregulatory responses of a healthy subject during external cold and heat exposure. With decreasing T_a, mean skin temperature ($\bar{T}_{sk}$) decreases, and the core temperatures, esophageal and tympanic temperatures increase due to improvement in heat preservation. The mean body temperature ($\bar{T}_b$), however, is reduced and drives the central regulating system to activate heat production. This activation is registered by an enhancement of MR and EMA. During external heat load $\bar{T}_b$ rises, and at a critical body temperature vasodilatation and sweating are enforced, indicating the onset of heat dissipation. The critical temperatures at which heat production and heat loss are activated by the central thermostat are called thermoregulatory threshold temperatures. They are determined by relating heat production and heat dissipation mechanisms to changes of the mean weighted body temperatures ($\bar{T}_b$). Calculating linear regression lines of the resting and activated period of each mechanism provides the threshold temperatures at the points of intersection of each of regression lines. Figure 2 depicts the threshold temperatures calculated from the data represented in Fig. 1. The interthreshold zone has a width of 0.8 °C–1 °C and indicates the tolerance of the central controller to changes in $\bar{T}_b$. It is thus a measure for the accuracy of the regulation of body temperature.

In several studies, the normal values of threshold temperatures were established. In our study, five controls were examined; their threshold temperatures are comparable to the findings of other authors (see Table 1). Heat production is increased at a $\bar{T}_b$ of 35.96 °C and heat loss at 36.73 °C. The anthropometric data of controls and patients are compiled in Table 2.

Twelve patients with basal midline meningeomas extending over the plane of the optic chiasm and elevating the floor of the third ventricle were investigated (Fig. 3). The threshold temperatures were shifted to a higher

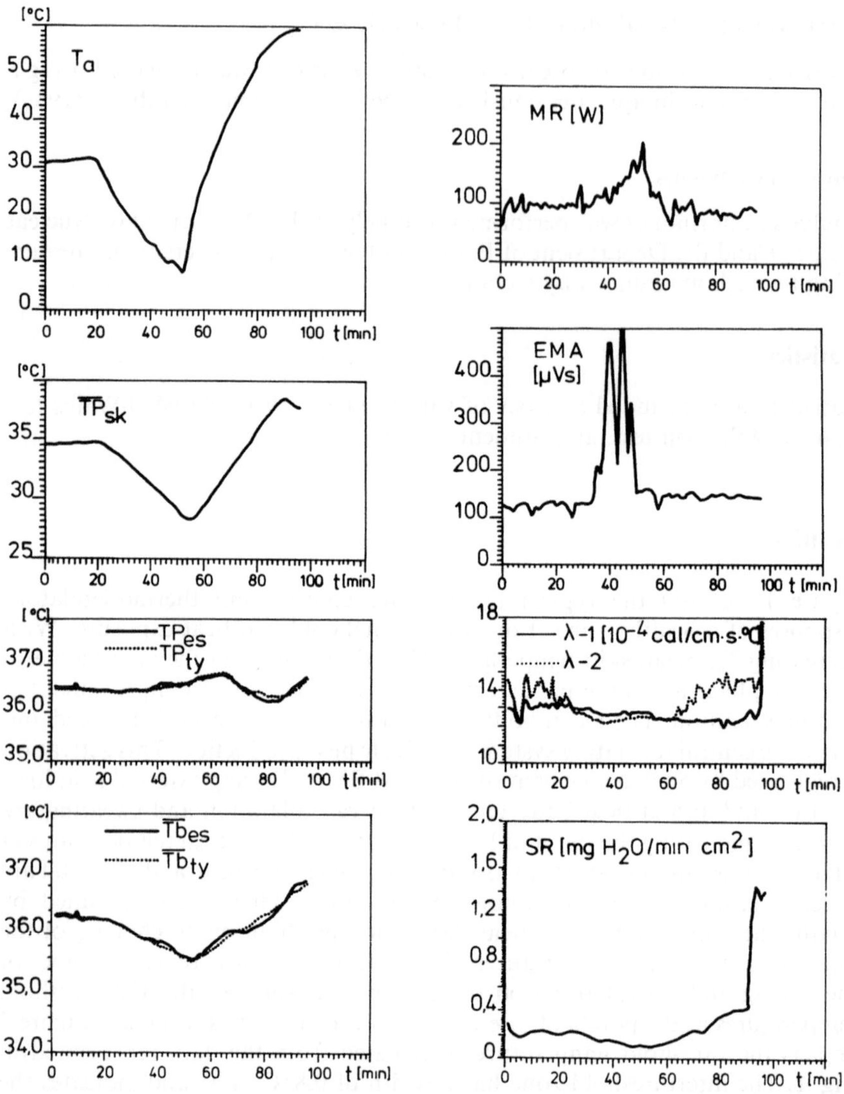

Fig. 1. Temperature courses (*left*) and thermoregulatory reactions (*right*) of a healthy subject over time. During cold load, provoked by reducing T_a, mean skin temperature decreases while body core temperatures rise slightly due to improvement in heat preservation by vasoconstriction. The decreasing $\bar{T}b_{es,ty}$ stimulates the central thermostat to activate metabolic rate and electric muscle activity, indicating heat production. During rewarming, there is a steep increase in $\bar{T}P_{sk}$ and a moderate rise in $\bar{T}b_{es,ty}$. The onset of heat loss is outlined by an enhancement of local skin blood flow and local sweat rate. T_a, ambient temperature; $\bar{T}P_{sk}$, mean skin temperature; TP_{es}, esophageal temperature; TP_{ty}, tympanal temperature; $\bar{T}b_{es}$, mean body temperature (calculated with esophageal temperature); $\bar{T}b_{ty}$, mean body temperature (calculated with tympanal temperature); *MR*, metabolic rate; *EMA*, electric muscle activity; λ_1, local cutaneous blood flow forearm; λ_2, local cutaneous blood flow thenar; *SR*, local sweat rate (chest)

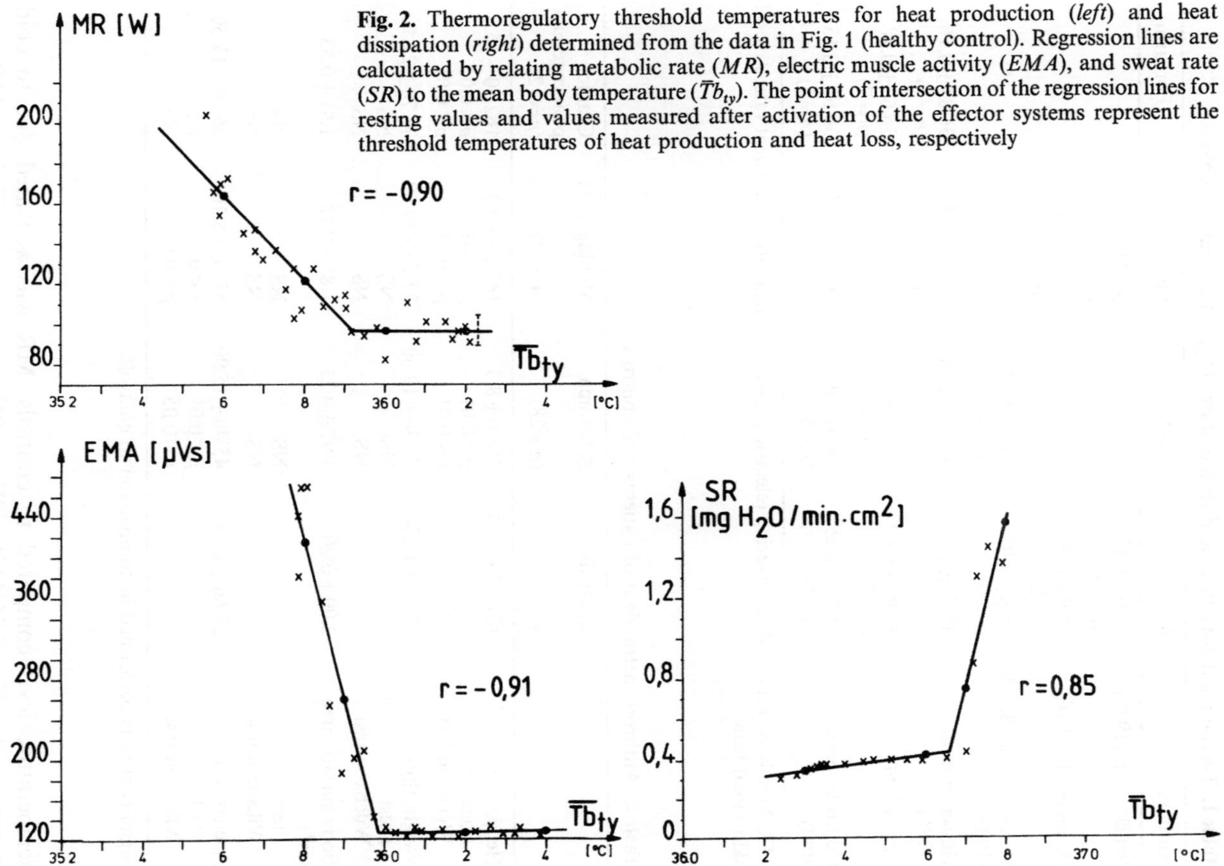

Fig. 2. Thermoregulatory threshold temperatures for heat production (*left*) and heat dissipation (*right*) determined from the data in Fig. 1 (healthy control). Regression lines are calculated by relating metabolic rate (*MR*), electric muscle activity (*EMA*), and sweat rate (*SR*) to the mean body temperature ($\overline{Tb}_{ty}$). The point of intersection of the regression lines for resting values and values measured after activation of the effector systems represent the threshold temperatures of heat production and heat loss, respectively

Table 1. Thermoregulatory threshold temperatures ($\bar{T}_b$, °C) in healthy subjects

Authors	MR	EMA	SR	λ_1 (Forearm)
Brück et al. (1976)	35.9±0.12 (*n*=15)		36.89±0.31 (*n*=4)	
Baum et al. (1976)	36.0±0.08 (*n*=20)		36.7±0.07 (*n*=11)	
Hessemer and Brück (1985)	35.8±0.07 (*n*=9)	35.8±0.07 (*n*=8)		
Hessemer et al. (1986)	35.82±0.13 (*n*=7)	35.70±0.14 (*n*=7)	36.78±0.18 (*n*=7)	36.87±0.16 (*n*=7)
Zeh (1985)	35.82±0.07 (*n*=7)	35.70±0.10 (*n*=7)	36.79±0.16 (*n*=7)	36.68±0.18 (*n*=7)
Controls, present study	35.96±0.09 (*n*=5)	36.04±0.12 (*n*=5)	36.73±0.15 (*n*=5)	36.65±0.13 (*n*=5)

MR, Metabolic rate; EMA, shivering (electric muscle activity); SR, local sweat rate; λ, local skin blood flow.

Table 2. Anthropometric data of patients and controls

	Controls (*n*=5)	Adenoma (*n*=28)	Meningeoma (*n*=12)	Cranio-pharyngeoma (*n*=12)
Height (cm)	181.00±7.14	168.6±8.3	163.80±7.3	162.90±16.0
t test		$p<0.01$	$p<0.01$	$p<0.05$
Wilcoxon test		$p<0.01$	$p<0.01$	$p<0.01$
Weight (kg)	73.20±12.38	79.30±19.60	77.70±19.60	73.70±21.10
t test		NS	NS	NS
Wilcoxon test		NS	NS	NS
Body surface area (m^2)	2.00±0.20	1.92±0.23	1.85±0.18	1.81±0.33
t test		NS	NS	NS
Wilcoxon test		NS	NS	NS
Age (years)	27.60±1.34	45.40±15.90	57.10±8.10	36.70±17.50
t test		$p<0.01$	$p<0.01$	NS
Wilcoxon test		$p<0.05$	$p<0.01$	NS

Significant tests performed in comparison to controls.

temperature level compared to controls. MR was activated due to cold exposure at a $\bar{T}_b$ of 36.44 °C (SE = 0.06; $n = 10$; in two patients MR was excluded from evaluation due to technical problems; $p = 0.0097$ by KWT). Heat loss, indicated by an enhancement of sweating, was increased at a $\bar{T}_b$ of 37.13 °C (SE = 0.08; $n = 12$; $p = 0.038$ by KWT; $p = 0.027$ by ANOVA). The interthreshold zone was unchanged. Hormonal findings of basal TSH, cortisol, and prolactin were in normal range (see Table 3).

Table 3. Basal hormonal findings

Hormone (normal range)	Controls	Meningeomas	Adenomas				Cranio-pharyngeomas
			Threshold shift	No threshold shift	Preoperative	Postoperative	
TSH (0.3–3.3 μU/ml)	1.2±0.14 (n=30)	1.93±0.49 (n=6)	1.46±0.25 (n=17)	0.82±0.19 (n=6)	1.12±0.49 (n=5)	1.60±0.65 (n=4)	1.54±0.37 (n=9)
Cortisol (7.2–26.5 μg/dl)	16.60±0.93 (n=27)	11.80±1.54 (n=5)	14.30±1.66 (n=17)	14.82±1.55 (n=6)	15.32±3.89 (n=5)	12.17±3.37 (n=4)	12.60±2.41 (n=9)
	$p<0.05$						
Prolactin (3–17 ng/ml)	7.56±0.60 (n=35)	8.85±1.93 (n=6)	1983±703 (n=18)	161±148 (n=6)	5219±2209 (n=5)	941±584 (n=4)	13.20±5.48 (n=9)
			$p<0.01$				

In all patients basal TSH and cortisol was in the normal range. Significant differences are indicated in the Table (Kruskal-Wallis test). Prolactin was increased due to ten patients with macroprolactinomas. Threshold shift: elevation of heat production and heat loss threshold temperatures; no threshold shift: thresholds not different from controls. (All data provided by G. Hildebrandt.)

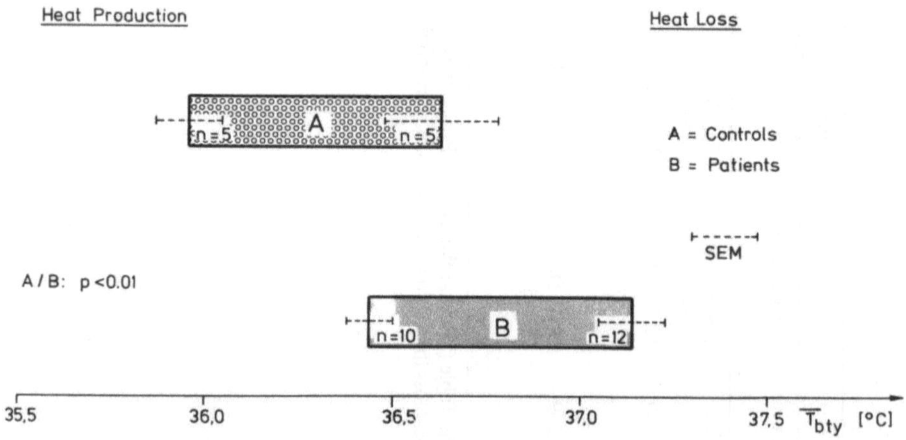

Fig. 3. Thermoregulatory threshold temperatures in patients with basal midline meningeomas. The *left and right edges of the bars* represent the threshold temperatures for heat production and heat loss, respectively; *bar width*, interthreshold zone. In patients with basal meningeomas both thresholds are significantly shifted to a higher temperature level. The width of the interthreshold zone remains unaltered. *SEM*, standard error from the mean; $\overline{Tb}_{ty}$, weighted mean body temperature

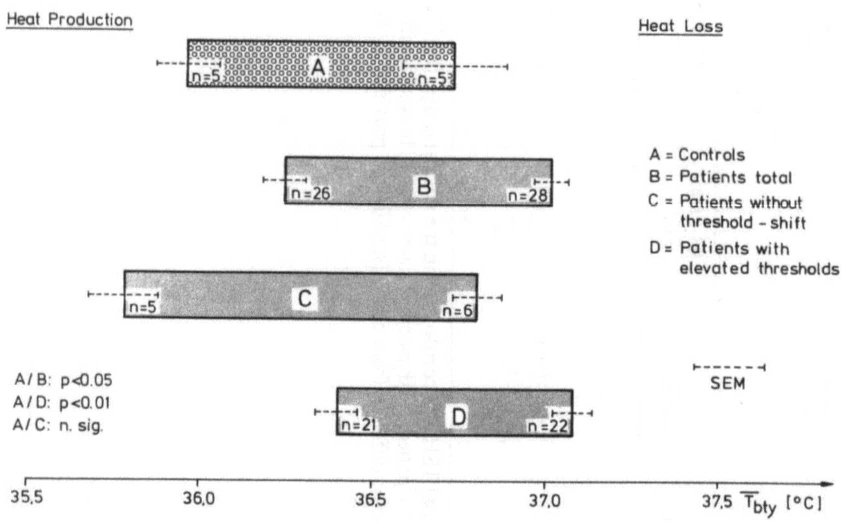

Fig. 4. Thermoregulatory threshold temperatures in patients with suprasellar pituitary adenomas. The majority of patients with suprasellar adenomas showed an elevation of the threshold temperatures (*D*) comparable to the findings in basal meningeomas (Fig. 3). In six patients no threshold shift was detectable (*C*). *B*, The thresholds of all adenoma patients (*C*+*D*), being significantly higher than controls

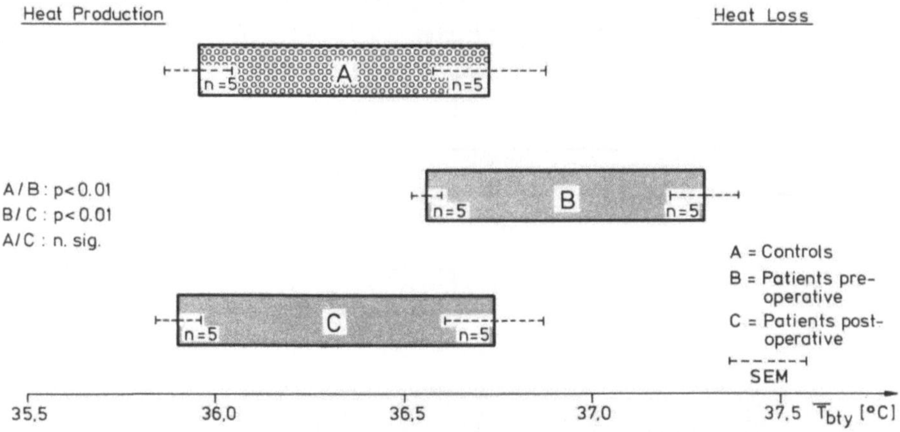

Fig. 5. Thermoregulatory threshold temperatures in patients with suprasellar pituitary adenomas: pre- and postoperative. In five patients the threshold temperatures and interthreshold zones were compared intraindividually before and in three cases 3–6 months, in two cases 1–2 weeks after surgery. After decompression of the hypothalamus the thermoregulatory thresholds were in normal range (*C*)

Similar results were obtained in 22 patients suffering from suprasellar pituitary adenomas, with compression of hypothalamic tissue (Fig. 4). MR was increased at an elevated threshold temperature of 36.40 °C (SE = 0.05), being 0.44 °C higher than control values ($p < 0.001$ by ANOVA and KWT). Heat loss by sweating was activated at 37.08 °C $\bar{T}_b$ (SE = 0.05; $p = 0.009$ by ANOVA; $p = 0.005$ by KWT). In six other patients, no significant threshold alteration was measured. For endocrine findings see Table 3. Intraindividual comparison of threshold temperatures of five patients pre- and postoperatively revealed a normalization of the elevated level of temperature regulation after surgery (Fig. 5).

Apart from an increased level of prolactin in ten patients with macroprolactinomas, the basal hormonal values of TSH and cortisol did not differ from controls (Table 3). Since the elevated threshold temperature coincided with an elevation of prolactin, and after surgery both parameters were normalized or reduced, a correlation between these parameters was sought. However, there was neither a significant linear nor logarhythmic correlation ($r \leq 0.21$; $p \geq 0.12$). Twelve patients with craniopharyngeomas extending towards the hypothalamus by elevating the floor of the third ventricle were investigated (Fig. 6). The observed pattern of temperature regulation in ten patients was different from that in patients with adenomas or meningeomas. In eight patients the interthreshold zone was markedly enlarged, measuring 1.59 °C. This was due to a reduction of the threshold temperature for heat production to 35.45 °C (SE = 0.09; $p < 0.001$ by ANOVA and KWT) and an elevation of the sweating threshold to 37.04 °C (SE = 0.08; $p = 0.01$ by ANOVA, and KWT). In two other patients it was not possible to determine a threshold temperature,

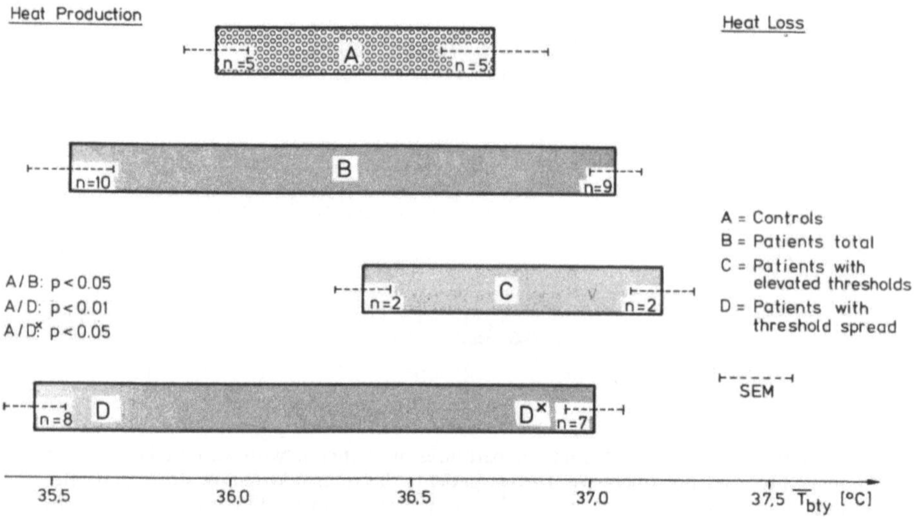

Fig. 6. Thermoregulatory threshold temperatures in patients with craniopharyngeomas. In most patients with craniopharyngeomas the interthreshold zone was enlarged (*D*) indicating a greater tolerance of the control system to body temperature changes. Two patients demonstrated a pattern of regulation similar to that of basal meningeomas and pituitary adenomas (*C*). Two other craniopharyngeoma patients are not represented since no threshold was detectable in a wide temperature range (see text)

since even at $\bar{T}_b$ below 33 °C and up to 38 °C neither heat production nor heat loss was activated. Both patients were on sufficient hormonal therapy. Two patients revealed a threshold elevation without change of the interthreshold zone, similar to the pattern of meningeomas or pituitary adenomas.

Discussion

This study demonstrated that in patients suffering from basal midline tumors, modifications of thermoregulatory characteristics are common. These alterations are often not detectable by routine clinical measurement of axillar or rectal temperature when patients stay in bed in a thermoneutral environment. However, by cold and heat exposure and measurement of thermoregulatory effectors and threshold temperatures even slight changes of the control of body temperature are seen (Brück and Zeisberger 1987).

Patients with basal meningeomas and pituitary adenomas revealed a very similar pattern of temperature regulation. In both groups the thresholds were shifted to a higher $\bar{T}_b$, indicating a setpoint elevation of the regulating system. The interthreshold zone was of normal width, pointing out that the tolerance of the hypothalamic thermostat to changes of $\bar{T}_b$ relative to the new setpoint was unaltered.

The setpoint elevation is interpreted as an irritation of hypothalamic structures by the local compression of the extracerebrally growing tumors. As demonstrated in five cases of pituitary adenomas being reinvestigated after surgery, the threshold temperatures were in normal range after hypothalamic decompression was achieved.

Hormonal abnormalities are unlikely to cause the modifications of the regulatory characteristics. Although endocrine data were not available in every patient, in the majority the basal values of TSH and cortisol were in normal range. There was no case of severe hypo- or hyperthyroidism. All patients investigated postoperatively were on sufficient hormonal therapy. Increased levels of prolactin were also not responsible for the threshold shifts, since there was no significant correlation between these two parameters. Furthermore, in meningeomas with normal prolactin levels, the setpoint elevation was even greater than in adenomas.

Age differences between patients and controls were not of significance in respect to temperature regulation. As reported by Mathew et al. (1986), the resting oxygen consumption is reduced over the age of 50 years compared to young subjects, but the cold induced increase of oxygen consumption was even greater in the elderly.

The pattern of thermoregulation in the majority of the patients with craniopharyngeomas was characterized by a widening of the interthreshold zone. This indicated that the central thermostat was less accurate in reacting to changes in $\bar{T}_b$. In two other patients this inaccurary of regulation was exaggerated since no threshold temperature was detectable. Even a reduction of $\bar{T}_b$ below 33 °C did not evoke an increase in metabolism to compensate heat loss. During heat exposure sweating was not observed at 38 °C $\bar{T}_b$, but there was a passive increase of metabolism due to the elevated body temperature. This pattern of regulation resembled poikilothermia although it is not known whether the patients would have regulated at even lower or higher temperatures. However, according to Bligh (1973), one can call this kind of regulation the "wide band control" of temperature regulation in contrary to the normal "narrow band control." It is difficult to answer why in craniopharyngeomas, with the exception of two cases, an enlargement of the interthreshold zone was observed although they are extracerebral tumors such as adenomas and meningeomas. Of importance is probably the ability of craniopharyngeomas to penetrate the adjacent nervous tissue by growing epitheloid pins and inducing a gliosis (Jänisch et al. 1988).

In conclusion, the fine control of body temperature was unaffected in basal meningeomas and suprasellar pituitary adenomas but was shifted to a higher temperature level. Patients with craniopharyngeomas exhibited an inaccuracy of regulation due to a widening of the interthreshold zone. In some cases the fine control of body temperature was abolished and replaced by a "wide band control," resembling poikilothermia.

References

Alpers BJ (1936) Hyperthermia due to lesions in the hypothalamus. Arch Neurol Psychiatry 35:30–42

Baum E, Brück K, Schwennicke HP (1976) Adaptive modifications in the thermoregulatory system of long-distance runners. J Appl Physiol 40:404–410

Bligh J (1973) Temperature regulation in mammals and other vertebrates. In: Neuberger A, Tatum EL (eds) Frontiers of biology, vol 30, pp 248–270

Brück K, Baum E, Schwennicke HP (1976) Cold-adaptive modifications in man induced by repeated short-term cold-exposures and during a 10-day and -night cold-exposure. Pflugers Arch 363:125–133

Brück K (1981) Basic mechanism in longtime thermal adaptation. In: Szelenyi Z, Szekely M (eds) Advances in physiological science, vol 32. Pergamon, Oxford New York

Brück K, Zeisberger E (1987) Adaptive changes in thermoregulation and their neuropharmacological basis. Pharmacol Ther 35:163–215

Cooper KE (1970) Studies of the human central warm receptor. In: Hardy JD, Gagge AP, Stolwijk JAJ (eds) Physiological and behavioural temperature regulation. Thomas, Springfield, IL, pp 224–230

Davison C, Selby NE (1935) Hypothermia in cases of hypothalamic lesion. Arch Neurol Psychiatry 33:570–591

Fox RH, Davis TW, Marsh FP, Ulrich H (1970) Hypothermia in a young man with an anterior hypothalamic lesion. Lancet 2:185–188

Globus JH, Kuhlenbeck H (1942) Tumors of the striato-thalamic and related regions: their probable source of origin and more common forms. Arch Pathol 34:674–734

Hessemer V, Brück K (1985) Influence of menstrual cycle on shivering, skin blood flow and sweating responses measured at night. J Appl Physiol 59:1902–1910

Hessemer V, Zeh A, Brück K (1986) Effects of passive heat adaptation and moderate sweatless conditioning on responses to cold and heat. Eur J Appl Physiol 55:281–289

Jänisch W, Schreiber D, Güthert H (1988) Neuropathologie. Tumoren des Nervensystems. Fischer, Stuttgart

Lausberg G (1972) Zentrale Störungen der Temperaturregulation. Acta Neurochir Suppl 19

Lipton JM, Rosenstein J, Sklar FH (1981) Thermoregulatory disorders after removal of a craniopharyngeoma from the third cerebral ventricle. Brain Res Bull 7:369–373

Mathew L, Prukayastha SS, Singh R, Sen Gupta I (1986) Influence of aging in the thermoregulatory efficiency of man. Int J Biometer 30:137–145

Nadel ER, Bullarek RW, Stolwijk JAJ (1971) Importance of skin temperature in the regulation of sweating. J Appl Physiol 31:80–87

Ramanathan NL (1964) A new weighting system for mean surface temperature of the human body. J Appl Physiol 19:531–533

Strauss J, Globus HH (1931) Tumor of the brain with disturbance in temperature regulation. Arch Neurol Psychiatry 25:506

Zeh A (1985) Verhalten von Schwitz-, Vasodilatations- und Zitterschwelle im Verlauf der Hitzeadaption des Menschen. Inaugural-Dissertation am Physiologischen Institut des Klinikums der Justus-Liebig-Universität Gießen, 1985

Operations of Skull Base Processes: Value of Intraoperative Monitoring

M. Lorenz and M. R. Gaab, Hannover/FRG

Introduction

The preservation of cerebral structures and the functionality of neuronal pathways is an important aim in neurosurgical practice. Nowadays, computer-assisted investigation methods promise continuous information on cerebral functions during operations: in surgery of the main arterial vessels of the brain, somatosensory evoked potentials (SEP) and EEG evaluations can be used for intraoperative monitoring. In parasellar surgery, visual evoked potentials (VEP) can reveal changes of the function of the visual pathway.

Patients and Methods

VEP and SEP were recorded in a standard technique [1, 7, 8, 9, 15] using a Nicolet CA 1000 averager. As electrodes we used metallic wound closure clips providing a continuous impedance. A significant reduction of artifacts was achieved by using disposable plastic Codman pins for sharp fixation in the Mayfield clamp. VEP monitoring was carried out in 19 patients with parasellar tumors, including pituitary adenomas, tuberculum sellae meningeomas, clival chordomas, and an intrasellar metastasis. During transient or permanent occlusion in surgery of aneurysms or for carotid endarterectomy 33 patients were examined by continuous SEP and EEG monitoring (9 aneurysms, 24 carotid operations). This was supported by continuous transcranial Doppler sonography of the middle cerebral artery. For anesthesia fentanyl, N_2O, and Trapanal were used, which are thought to have only little influence on SEP [7].

Results

Visual Evoked Potentials. Of 19 patients only 15 showed reproducible responses at the beginning of operation. The remaining four showed poor VEP preoperatively and had to be excluded from intraoperative monitoring. Intraoperatively, four patients showed no changes in evoked responses: postoperatively, there was no deficit in these patients. In 11 patients abnormal waves (increase in latencies, decrease in amplitudes, or changes in waveforms) occurred during operation. However, in five of these records alterations were

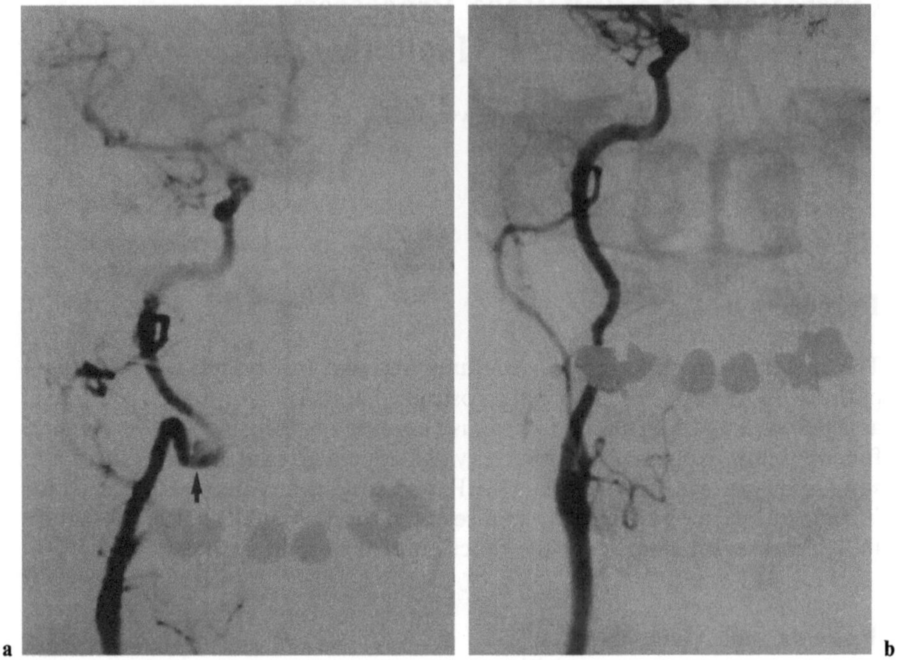

Fig. 1. **a** Pre- and postoperative angiogramms of an internal carotid aneurysm (*arrow*). **b** Excision of the carotid loop with the aneurysm and perfect reconstruction with end-to-end anastomosis

seen at times when no manipulation on the optic system was performed (trepanation, opening of dura, or without a clear reason). Postoperatively, three patients developed impairment of visual function. Only one of them had significantly abnormal VEP at the end of operation; the remaining two recovered to a status within 2 standard deviations and could not be considered as pathologic. Sometimes it took a longer time for improvement in changed waves.

Somatosensory Evoked Potentials. During temporary or permanent clipping of aneurysms marked changes occurred in three of nine cases (Figs. 1, 2). In one of these only a slight reversible prolongation of CCT was seen after transient occlusion of the middle cerebral artery distal to the thalamoperforating arteries. However, no alterations in the amplitudes occurred. The other two patients had traumatic aneurysms of the cranial base. One showed a deterioration during bleeding, and the second had a permanent change of waveform of the middle latency responses after occlusion of the ICA. Only this patient showed a deficit postoperatively, developing a psychotic disorder. We could not decide whether this could be expected after the change in the SEP. In

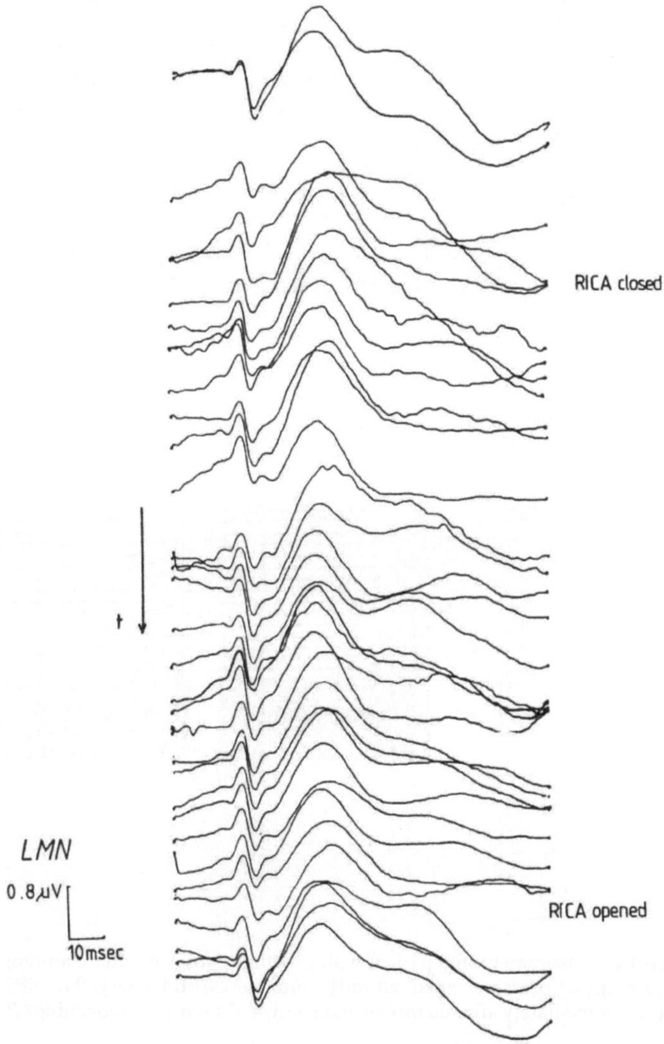

Fig. 2. Intraoperative monitoring of somatosensory evoked potentials after stimulation of the contralateral median nerve. No pathological alterations occurred. *RICA*, Right internal carotid artery

CT scan, a diffuse microembolism was presumed. There were no other postoperative deficits.

In 9 of 24 carotid endarterectomies, SEPs were abnormal during clamping (Fig. 3). All but one recovered immediately after opening the flow or inserting an intraluminary shunt (four cases). The patient with delayed recovery still showed significantly decreased amplitudes and a CCT prolongation at the end

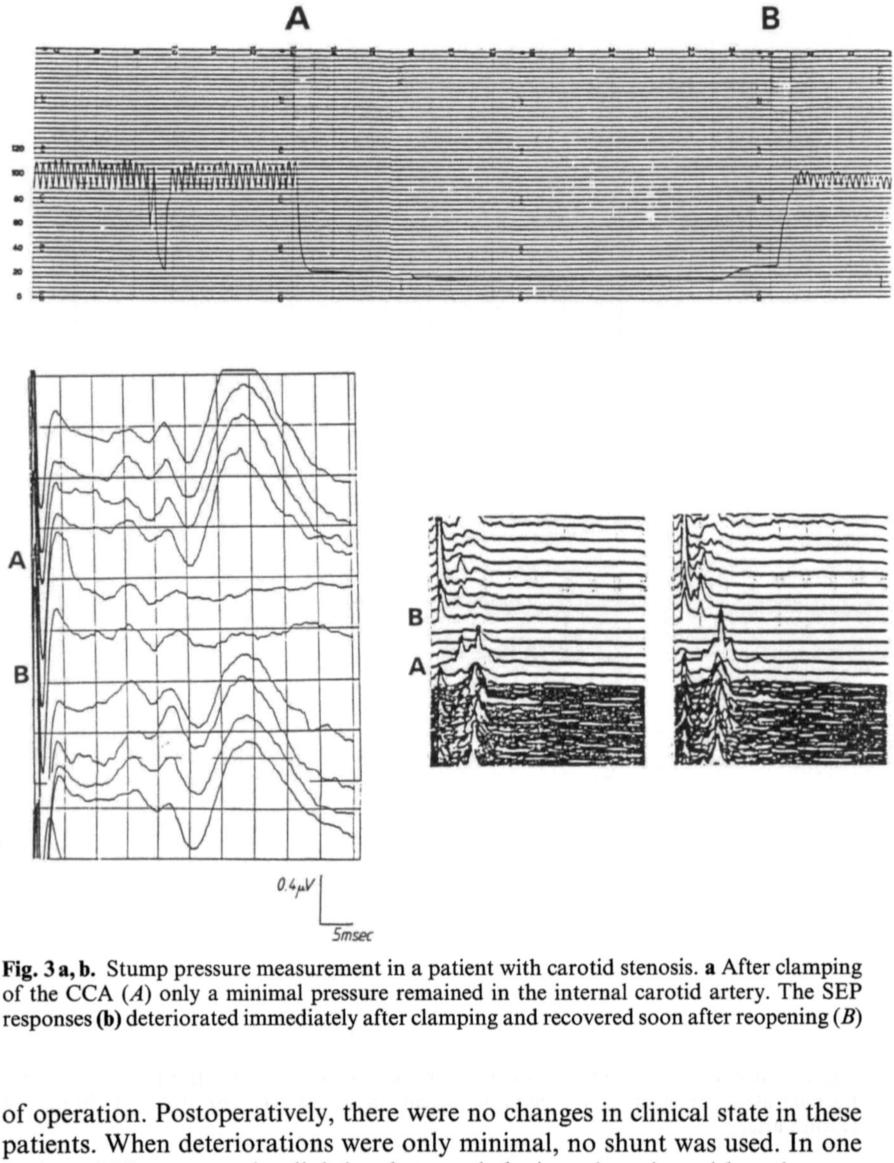

b

Fig. 3 a, b. Stump pressure measurement in a patient with carotid stenosis. **a** After clamping of the CCA (*A*) only a minimal pressure remained in the internal carotid artery. The SEP responses **(b)** deteriorated immediately after clamping and recovered soon after reopening (*B*)

of operation. Postoperatively, there were no changes in clinical state in these patients. When deteriorations were only minimal, no shunt was used. In one patient SEPs were only slightly abnormal during clamping although transcranial Doppler sonography detected a marked reduction of flow in the MCA.

Discussion

Visual Evoked Potentials. After some encouraging reports on VEP monitoring during parasellar surgery [4, 18], the opinion on the reliability of this method in

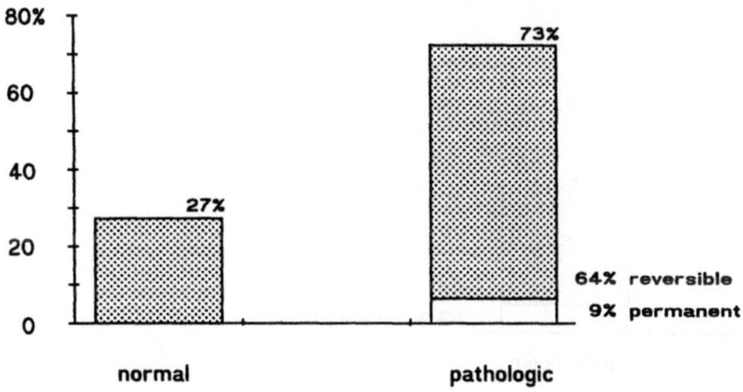

Fig. 4. Intraoperative visual evoked potential monitoring. Although 64% showed a recovery of deteriorated responses, 27% of these patients had an impairment of visual function postoperatively. Only in patients without intraoperative changes did the prediction of a good outcome seem possible

predicting postoperative results has become more critical [1, 3, 5, 6, 11, 16]. VEPs do not seem as reliable as SEPs. In some cases changes in VEPs occurred in moments when no manipulation of the optic nerve or the chiasm was performed [3]. Only in records without changes can a good prognosis be assessed [15]. In other cases a false-positive response may rather cause confusion (Fig. 4).

Somatosensory Evoked Potentials. Operations on the vascular supply of the brain often require temporary occlusion of vessels, with the risk of cerebral hypoperfusion and infarction. This risk cannot be recognized by preoperative measurements. Because of probable complications [8, 10, 13] an intraluminary shunt should be used only in cases with markedly pathologic electrophysiological signals [13]. EEG and SEP are good indicators for the cerebral blood supply [2, 7, 8–10, 12, 14–17]. They can give reliable information for predicting the postoperative neurological state [7, 8, 9]. Branston et al. found a close correlation between SEP and cerebral blood flow (CBF) [2]. Below a CBF of 30 ml 100 g^{-1} min^{-1} the EEG slows down. After an increase in CCT, amplitude alterations of SEP follow below a CBF of 18 down to 12 ml 100 g^{-1} min^{-1}. Below 12 ml 100 g^{-1} min^{-1} there is electrical silence. The critical threshold for irreversible cell damage is 8 ml 100 g^{-1} min^{-1} [14]. In our experience, SEPs give more information in critical situations of CBF, whereas the EEG is more sensitive for a beginning hypoxia [12]. If the neurophysiological parameters remain stable for more than 5 min after clamping, a sufficient collateralization can be assumed, and the time of clamping becomes less important [7–10, 14] (Fig. 5). The development of neurological deficiencies is not probable in these cases provided that no embolic complications or angiospasms occur [7, 8]. If SEPs deteriorate on one side, a control of the other can exclude metabolic or other systemic factors [7, 8, 10, 15, 17]. A reversible deterioration does not result in a neurological deficit. SEP monitoring after

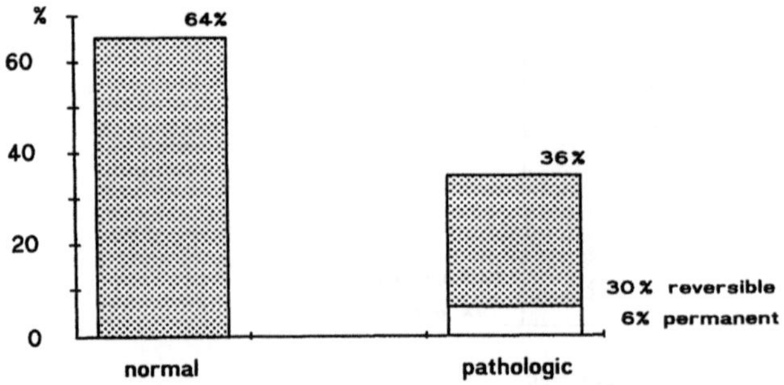

Fig. 5. Intraoperative somatosensory evoked potential monitoring. All patients with normal records or reversible changes had an unchanged neurological state postoperatively. Of two patients with significantly worse responses at the end of the operation, one did well, and the other developed a psychotic disorder due to diffuse microembolism

stimulation of the median nerves is therefore a good method for assessing the functionality of the somatosensory pathway after occlusion of the ICA and MCA. They seem to fail, however, in monitoring of the anterior circulation [7, 9, 15, 17], and in detecting local ischemic situations besides the somatosensory pathway [17]. SEPs of the lower extremities for monitoring the ACAs have their own difficulties [16].

With the use of electrophysiological neuromonitoring operations on the vascular supply of the brain will become safer. The "lethal" flow threshold for stroke is below the level that completely suppresses SEPs [12]. Some patients did well after having completely depressed signals for several minutes intraoperatively [16, 17]. Thus there must be further investigations to define the minimal tolerable changes.

References

1. Allen A, Starr A, Nudleman K (1981) Assessment of sensory function in the operating room utilizing cerebral evoked potentials: a study of fifty-six surgically anesthezised patients. Clin Neurosurg 28:457–481
2. Branston NM, Symon L, Crockard HA, Pasztor E (1974) Relationship between the cortical evoked potential and local cortical blood flow following acute middle cerebral artery occlusion in the baboon. Exp Neurol 45:195–208
3. Cedzich C, Schramm J, Fahlbusch R (1987) Are flash-evoked visual potentials useful for intraoperative monitoring of visual pathway function? Neurosurgery 21:709–715
4. Feinsod M, Selhorst JB, Hoyt WF, Wilson CB (1976) Monitoring optic nerve function during craniotomy. J Neurosurg 44:29–31
5. Goodrich JT (1987) Electrophysiologic measurements: intraoperative evoked potential monitoring. Anesth Clin North Am 5:477–489
6. Grundy BL (1983) Intraoperative monitoring of sensory-evoked potentials. Anesthesiology 58:72–87
7. Kidooka M, Nakasu Y, Watanabe K, Matsuda M, Handa J (1987) Monitoring of somatosensory-evoked potentials during aneurysm surgery. Surg Neurol 27:79–76

8. Markand ON, Dilley RS, Moorthy SS, Warren C (1984) Monitoring of somatosensory evoked responses during carotid endarterectomy. Arch Neurol 41:375–378
9. Momma F, Wang AD, Symon L (1987) Effects of temporary arterial occlusion on somatosensory evoked responses in aneurysm surgery. Surg Neurol 27:343–352
10. Moorthy SS, Markand ON, Dilley RS, McCammon RL, Warren CH (1982) Somatosensory-evoked responses during carotid endarterectomy. Anesth Analg 61:879–883
11. Nau HE, Wiedemayer H (1987) Monitoring mittels evozierter Potentiale bei intrakraniellen Eingriffen aus neurochirurgischer Sicht. 32. Jahrestagung der Deutschen EEG-Gesellschaft, Ludwigshafen
12. Ropper AH (1986) Evoked potentials in cerebral ischemia. Stroke 17:3–5
13. Spetzler RF, Martin N, Hadley MN, Thompson RA, Wilkinson E, Raudzens PA (1986) Microsurgical endarterectomy under barbiturate protection: a prospectice study. J Neurosurg 65:63–73
14. Sundt TM Jr, Sharbrough FW, Piepgras DG, Kearns TP, Messick JM Jr, O'Fallon WM (1981) Correlation of cerebral blood flow and electroencephalographic changes during carotid endarterectomy. Mayo Clin Proc 56:533–543
15. Symon L, Wang AD, Costa E Silva IE, Gentili F (1984) Perioperative use of somatosensory evoked responses in aneurysm surgery. J Neusorg 60:269–275
16. Symon L, Momma F, Schwerdtfeger K, Bentivoglio P, Costa E Silva IE, Wang A (1986) Evoked potential monitoring in neurosurgical practice. Symon L, Brihaye J, Guidetti B, Loew F, Miller JD, Nornes H, Pasztor E, Pertuiset B, Yasargil MG (eds) Advances and technical standards, vol 14. Springer, Wien New York, pp 25–70
17. Symon L (1985) Thresholds of ischaemia applied to aneurysm surgery. Acta Neurochir 77:1–7
18. Wilson WB, Kirsch WM, Neville H, Stears J, Feinsod M, Lehmann RAW (1976) Monitoring of visual function during parasellar surgery. Surg Neurol 5:323–329

Surgery of Extracerebral Tumors of the Frontal and Medial Skull Base

D. RIEDIGER, Stuttgart/FRG, M. EHRENFELD, Tübingen/FRG, and
W. HASSLER, Duisburg/FRG

Introduction

Of the seven approaches that allow access to the frontal and medial skull base, the transnasal-transethmoidal has gained worldwide acceptance in surgery of intrasellar tumors of the pituitary gland. It is commonly used by neurosurgeons, as is the transfrontal-transbasal approach. The transethmoid access is mainly used by ENT surgeons, whereas both the transoral and the unilateral transmaxillary approach are often used by maxillofacial and ENT surgeons. The transcervical approach provides access only to the lower half of the clivus (Derome 1983). In the treatment of tumors invading the frontal medial skull base from the dorsal wall of the frontal sinus, access to the skull base is provided by the tumor itself.

Case Reports

Our maxillofacial and neurosurgical experience with the transmaxillary access to the frontal and medial skull base is based on 19 patients. In one additional case the tumor originated at the skin of the forehead and invaded the skull base, so that the tumor itself determined the approach to the skull base.

Case Report 1. This case illustrates the advantages and disadvantages of the transmaxillary approach. A 36-year-old patient suffered from recurrence of a hemangioperiocytoma which had been primarily irradiated elsewhere. When we first saw the patient, the tumor extended from the right maxillary sinus via the ethmoidal cells up to the clivus. The highly vascularized tumor was embolized preoperatively to minimize the bleeding (Figs. 1, 2). The anterior wall of the maxillary sinus was exposed using a Dieffenbach incision. The underlying bone was raised atraumatically and repositioned at the end of the operation (Lindorf 1974). This procedure allows good access to the tumor and provides a good esthetic result. The operation was performed without considerable loss of blood, thus stressing the advantage of preoperative embolization. Postoperative computed tomography showed an almost complete resection of the tumor. Nine months after the operation no further tumor recurrence occurred. The transmaxillary approach can be used for all tumors located in the maxilla and basal part of the orbit, pterygopalatine fossa, upper

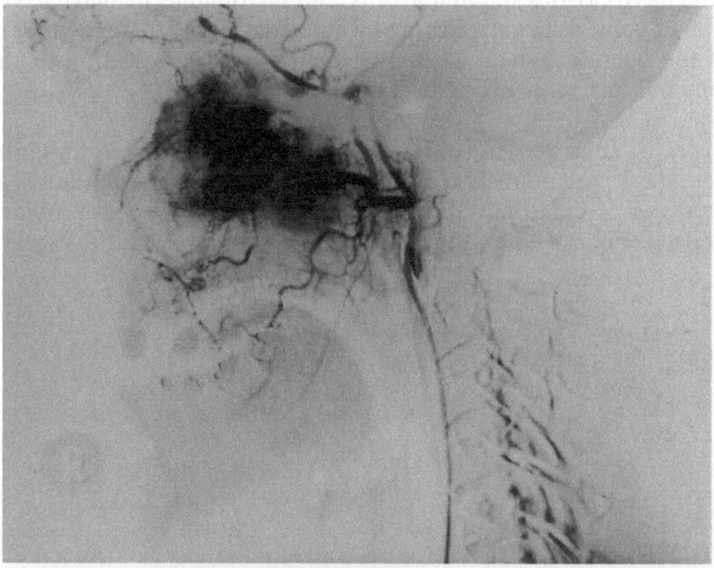

1

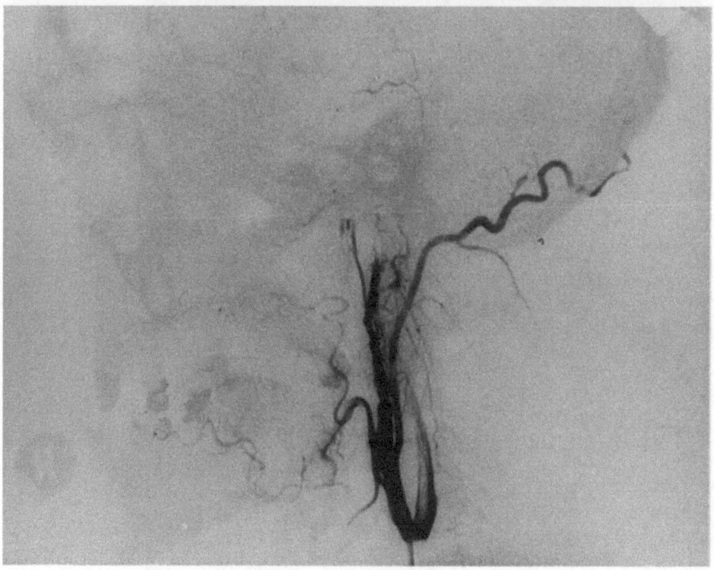

2

Fig. 1. Preoperative angiography showing a highly vascularized hemangioperiocytoma in a 36-year-old patient prior to embolization

Fig. 2. Angiography directly after embolization reveals a complete interruption of blood supply to the tumor

pharyngeal and nasal cavities, ethmoidal cells, and surrounding medial parts of the optic canal. Possible disadvantages are the limited overview on structures in the depth of the operating field, especially in cases with destruction of the vomer or bony confinements of the orbit.

Case Report 2. In a 55-year-old woman suffering from Gorlin-Goltz syndrome a basal cell carcinoma had invaded the medial and frontal skull base from the dorsal wall of the frontal sinus: access to the skull base was provided by the tumor itself (Fig. 3). After tumor resection and duraplasty the defect of

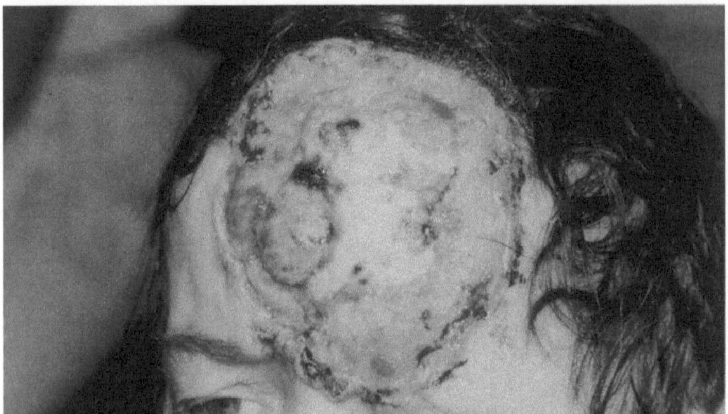
3

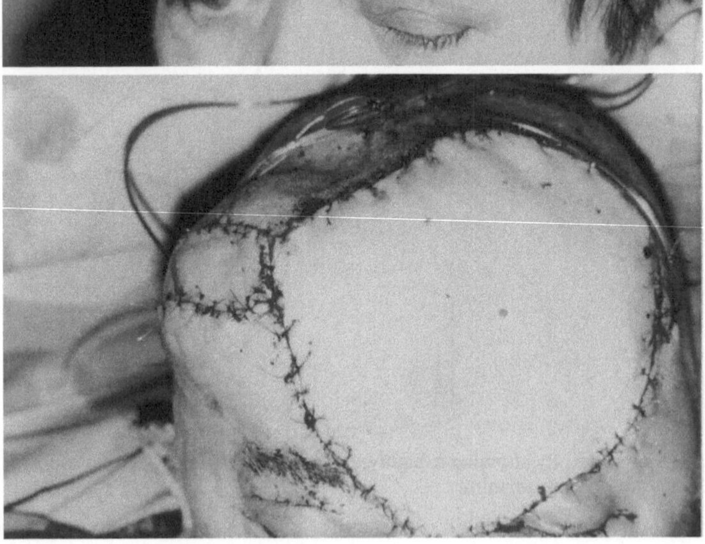
4

Fig. 3. A 55-year-old patient with a large basal cell carcinoma invading the anterior and medial skull base through the dorsal wall of the frontal sinus

Fig. 4. Directly after tumor resection, duraplasty, and coverage of the defect with a free revascularized latissimus dorsi graft

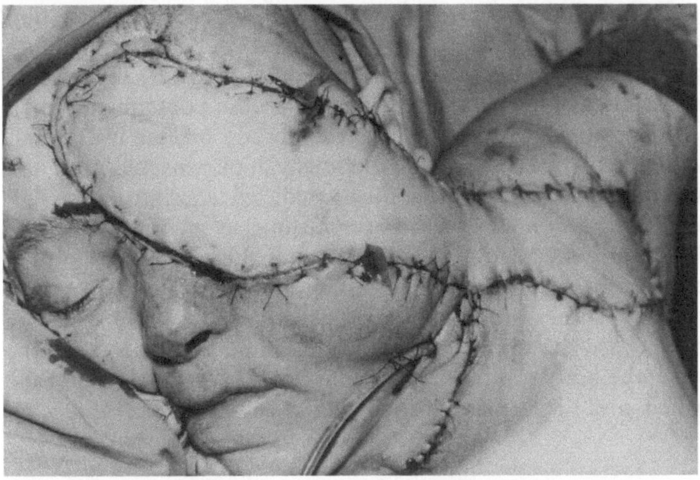

Fig. 5. The patient directly after the transfer of a long pedicled latissimus dorsi flap. Along the neck a bridge of flap skin is left in its place, a method which we find safer than transferring the flap through a subcutaneous tunnel

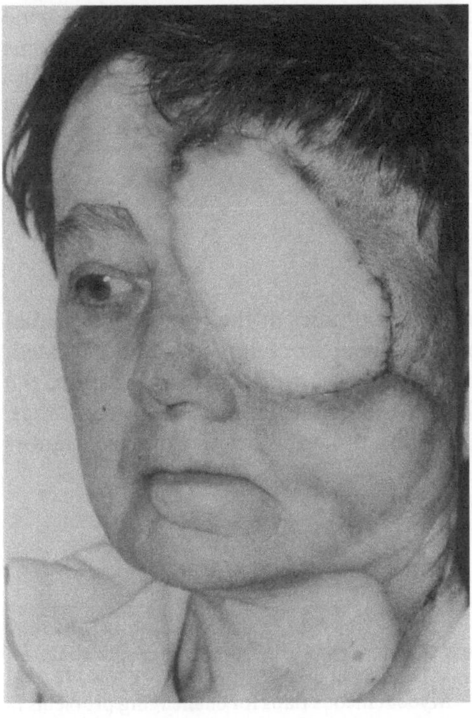

Fig. 6. After resection of the left eye and periorbit, the ethmoidal cells, and parts of the skull base including a second duraplasty and transfer of a pedicled latissimus dorsi graft. Superfluous flap skin has already been reduced

skin and bone was covered by a free revascularized latissimus dorsi flap (Watson et al. 1979; Riediger and Schmelzle 1986; Fig. 4). A partial loss of flap skin occurred, but this did not affect the incorporation of the graft. Three years later we were confronted with a relapse, which was the consequence of a compromise with the patient who refused the exenteratio orbitae in the first operation. Now removal of the eye was obviously indispensable, and the patient consented. This time we planned to raise a pedicled latissimus dorsi flap (Quillen et al. 1978; Krishna and Green 1980; Gardiner et al. 1983; Riediger and Schmelzle 1986). For coverage of the orbital region a long pedicled graft is required. In these cases it is recommended to peritomize the flap. Three days later it can be judged whether the distant part of the flap is well nourished or not. Because in this case the flap showed no signs of stasis or necrosis, the resection of the eye, the bony orbit, the ethmoidal cells and again a part of the dura was performed, and the latissimus dorsi flap was transferred (Figs. 5, 6).

Discussion

The interdisciplinary surgical management of tumors affecting the frontal medial skull base by maxillofacial surgeons and neurosurgeons provides a wide range of surgical approaches, from which the one can be chosen which seems best considering individual aspects. The possibility of simultaneous reconstruction using pedicled or free revascularized grafts enables us to perform a radical surgery and facilitates an esthetic rehabilitation.

Summary

Joint procedures in surgery of extracerebral tumors of the frontal and medial skull base by maxillo-facial surgeons and neurosurgeons are demonstrated. Two case reports out of the authors' experiences with 19 cases illustrate the advantages and disadvantages of the transmaxillary approach. Beyond that, the possibilities of immediate reconstruction of patients with extensive tumors by musculocutaneous grafts are shown.

References

Derome PJ (1983) The transbasal approach to tumors invading the base of the skull. In: Seeger W (ed) Microsurgery of the cranial base. Springer, Wien New York, pp 357–379

Gardiner LJ, Ariyan S, Pillsbury HC (1983) Myocutaneous flaps for challenging problems in head and neck reconstruction. Arch Otolaryngol 109:396–399

Krishna BV, Green MF (1980) Extended latissimus dorsi myocutaneous flap in reconstruction of the neck. Br J Plast Surg 33:233

Lindorf HH (1974) Knochendeckelverschluß nach oraler Kieferhöhleneröffnung. Dtsch Zahnärztl Z 29:587
Quillen CG, Shearing JG, Georgiade NG (1978) Use of the latissimus dorsi myocutaneous island flap for reconstruction in the head and neck area. Plast Reconstr Surg 62:113–117
Riediger D, Schmelzle R (1986) Modifizierte Anwendung des Latissimus dorsi-Lappens zur Defektdeckung im Mund-Kiefer-Gesichtsbereich. Dtsch Z Mund Kiefer Gesichtschir 10:364–374
Watson JS, Craig RDP, Orton CE (1979) The free latissimus dorsi myocutaneous flap. Plast Reconstr Surg 64:299

Craniofacial Resection for Anterior Skull Base Tumors

K. D. Post and A. Blitzer, New York/USA

Introduction

Cancers of the paranasal sinuses generally have a poor prognosis. They have an insidious onset, follow a protracted clinical course, and are poorly controlled by most forms of therapy. Their diagnosis is often delayed, and patient survival is compromised by direct extension of the tumor into the cranial cavity and orbit, as well as such surgically inaccessible sites as the retropharyngeal nodes [5, 3]. All of the paranasal sinuses, with the exception of the maxillary sinus, have a common wall with the cranial cavity. Their proximity to the cranial cavity has hindered the kind of en bloc tumor resection performed in the inferiorly located maxillary sinus. Consequently, conservative surgery in these sinuses is fraught with local recurrence at the cribriform plate and base of the skull. The exceedingly poor survival rate, 8% overall at 5 years, is caused primarily by uncontrolled local disease, with only 10% dying as a result of metastases [3, 10].

The relative safety and propriety of the combined craniofacial resection was well established by the work of Ketcham et al. [4]. In 1963, they evaluated the results obtained in a study of 30 patients with advanced paranasal sinus, nasal, and orbital cancers treated at the National Institutes of Health, most of whom were failures of other treatment modalities. After careful evaluation, 19 of these patients were found to be candidates for combined resection. With the exception of one perioperative mortality related to infection, the patients did well. In 1968, Ketcham et al. reviewed their 10-year experience, reporting a cure rate of approximately 30% in an otherwise unsalvageable group of patients. By 1973, the cure rate rose to a 56% determinate 5-year survival [12].

Development of reconstructive surgery for craniofacial malformations added to the advancement of the craniofacial resection technique [11, 2]. New modifications were made as surgeons embraced Ketcham's work. In 1975, Westbury and associates reported on using a frontal sinus osteoplastic flap as an approach for combined resection [13]. Sisson et al. expanded on Ketcham's work via a frontal craniotomy and described methods of skull base reconstruction employing bone and cartilage [9]. Shah and Galicich strongly recommended the use of a bifrontal craniotomy because it provides additional exposure, permits better assessment of tumor resectability, and facilitates dural resection and repair [7]. Schramm et al. in 1979 expanded the role of the frontal burr hole and described the use of a frontotemporal flap for cases with greater involvement of the skull base [6].

Indications

The combined craniofacial resection is indicated in patients who are good surgical risks and who have malignant tumors of the superior nasal cavity, ethmoid sinus, frontal sinus and orbit. This combined technique has also been employed for benign, but locally aggressive, tumors of this region, such as esthesioneuroblastomas, [8, 1] meningiomas, fibro-osseous lesions, chordomas and osteomas. Juvenile angiofibromas with intracranial extension also can be mangaged via this approach [6]. Patients with systemic diseases who are in poor condition or who have metastases are not candidates for craniofacial resection. Demonstration by computed tomography (CT) of involvement of the nasopharynx, superior or posterior, sphenoid or sphenoid ridge, invasion at the optic chiasm, prevertebral space extension or significant intracranial or intracerebral extension has contraindicated these procedures [4, 12]. Usually, combined craniofacial resection has been reserved for tumors which appear resectable. If there is extensive intracranial intradural extension, aggressive surgery has been excluded. We suggest that palliation of disease should be an important consideration. If intractible pain or a large fungating mass is present, surgery may be performed to alleviate problems and allow a better quality of survival.

Sixteen extracranial anterior skull base tumors have been operated upon using a combined craniofacial resection. These include seven ethmoid carcinomas, three esthesioneuroblastomas, three fibro-osseous tumors, and one each of juvenile angiofibroma, inverting papilloma, and mucocele. Of the seven patients with ethmoid carcinoma, at least two were thought to have disease too extensive for "cure." An attempt was made to resect these tumors, but this could not be achieved. Nevertheless, the outcome was extremely beneficial to both patients, each of whom experienced complete resolution of pain, and a short complication-free postoperative course.

Preoperative Evaluation

Radiographic examination should include standard sinus X-rays, including a submental vertex projection, to detect erosion of the lesser wing of the sphenoid and middle cranial fossa extension. Clouding of the sphenoid should not contraindicate surgery, since this condition may be inflammatory in nature. The limitations of resection include involvement of the optic chiasm and cavernous sinus superiorly, and the carotid canal inferiorly. Recently, we have considered resecting a tumor involving the carotid artery and cavernous sinus, and a blind eye. A bypass procedure for the carotid artery could be performed, and the eye and cavernous sinus resected.

Tomography of the skull base is also useful to evaluate erosion of the sphenoid bone. Angiography may demonstrate invasion or compression or displacement of the carotid artery or jugular venous system. It also provides information regarding the vascularity of the tumor (tumor blush) and demonstrates major feeding vessels. Embolization of feeding arteries may be considered in order to make subsequent surgery easier and safer. The CT scan and magnetic resonance imaging are excellent for evaluation of the extent of the lesion. They show dural or intracranial involvement, displacement or invasion of orbital structures, and cavernous sinus encroachment, as well as bony integrity. Axial, sagittal, and coronal projections are needed for surgical decisions.

Technique

Although several workers recommend a preliminary tracheostomy [4], this is rarely necessary, as airway compromise is uncommon in midface cancer resection. The surgical approach has been similar: a bifrontal intracranial neurosurgical exposure combined with a transfacial exposure by ENT. A bicoronal scalp flap with a bifrontal bone flap has been used in most of our cases. A unilateral exposure is indicated for asymmetric benign disease, orbital disease, or other lateral processes. The pericranium is elevated bifrontally in all cases leaving the frontal base attachment intact with the blood supply (Fig. 1). Wide intracranial exposure has been used along with lumbar spinal drainage to minimize frontal lobe retraction. This allows exposure of the skull base and dura as far posteriorly as the tuberculum sella and sphenoid wings. Intradural exposure is obtained if the basal dura is invaded by tumor. The wide basal exposure offers the following advantages:

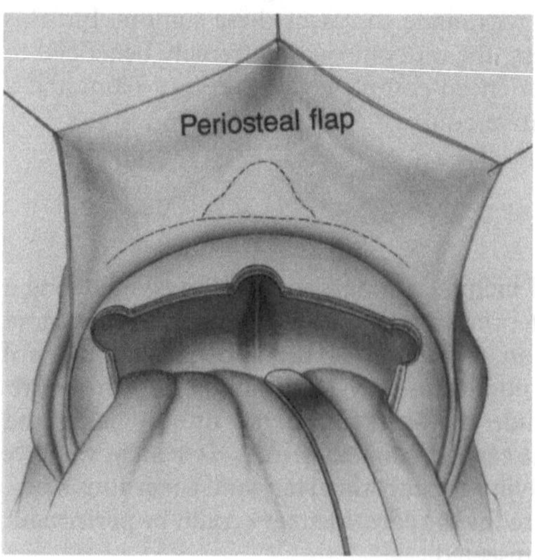

Fig. 1. Schematic drawing demonstrating the periosteal flap based along the frontal ridge leaving the blood supply intact

- Wide exposure
- Accurate evaluation of intracranial extension
- Protection of brain
- Intra- and extradural access
- Ease of dural replacement
- Security of dural closure
- Facilitates en block resection
- Adaptable
- Cosmetically acceptable

Skull base cuts can be easily varied as dictated by the pathologic anatomy. Suturing of dural patch grafts if dura is involved is facilitated. Transcranial facial dissection also facilitates wide exposures and base resection. Once the bone flap is removed, cerebral decompression is achieved by removing 20–60 ml CSF through the spinal drainage system, as well as by administering mannitol and/or prophylactic antibiotics, to allow blunt dissection of the dura. In the region of the cribriform plate, the olfactory nerves with their dural sleeves must be cut. To prevent a CSF leak, each small hole is closed immediately with a 4-0 silk suture.

If the tumor is resectable, the area of the skull base removal is outlined with a high-speed drill and osteotome. This process should include the ipsilateral ethmoid labyrinth, superior and lateral walls of the sphenoid, anterior cranial fossa, and the cribriform plate. The resection can be extended to include the orbital roof, opposite ethmoid or the frontal sinus. After the facial resection is completed, the brain is retracted again, and working from above and below the entire specimen is delivered through the facial opening. After the resection has

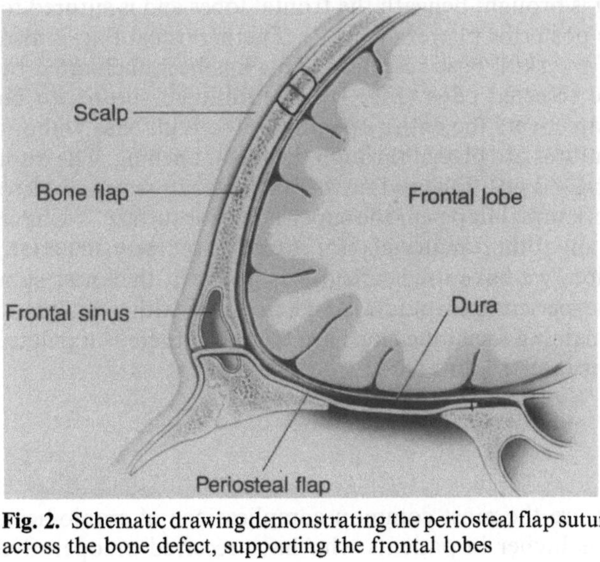

Fig. 2. Schematic drawing demonstrating the periosteal flap sutured beneath the frontal dura, across the bone defect, supporting the frontal lobes

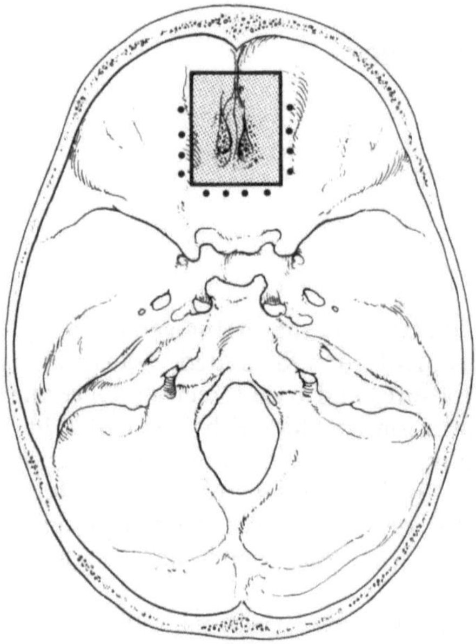

Fig. 3. Schematic drawing demonstrating the hole drilled along the edge of the bone defect to facilitate periosteal sutures

been completed and the specimen removed, repair of the anterior cranial fossa is begun, utilizing the pericranial flap raised at the beginning of the procedure. The pericranium is draped over the defect and the remaining floor of the anterior cranial fossa. It is sutured to the dura inferiorly to keep it in position and allow it to act as a sling for the frontal lobes. The technique of dural patch and periosteal sling suturing is most important to prevent CSF leakage. A frontal periosteum sling is brought beneath the frontal lobes and is sutured to the basal dura over the posterior planum (Fig. 2). The periosteal flap is also sutured to the edges of the skull base resection. This has been facilitated by drilling holes along the resected edge (Fig. 3) for additional sutures to be certain the periosteal flap covers the entire opening in the skull base without slipping forward. The sutures are placed through the facial opening. The dura is then returned to its original position, and further tacking sutures are used to attach the dura to the pericranial flap near the superoanterior surface. We have generally not needed any other materials for skull base reconstruction, although on one occasion, we have used screen mesh. A split thickness skin graft is placed over the pericranium and is held in place with a bolus of xeroform gauze; the remaining facial incisions are closed. If there is a palatal defect, a surgical stent is inserted and wired into position.

Complications

Infection and CSF leak are the most common complications of craniofacial surgery. The incidence is higher in patients who have received preoperative radiotherapy. The cases in which the dura is resected have the highest possibility of leak. The best treatment is prevention by meticulous intraoperative dural closure. Using this technique, we have had no CSF leaks or meningitis. The only complication has been one delayed bone flap infection. The cosmetic results have been excellent with only the use of an eye patch if orbital exenteration is necessary. Even if cure is not achieved, the craniofacial resection has yielded excellent pain relief. We therefore advise this procedure in most anterior skull base tumors as the benefits are significant, especially in view of the extremely low complication rate.

References

1. Chapman P, Carter RL, Clifford P (1981) The diagnosis and surgical management of olfactory neuroblastoma: The role of craniofacial resection. J Laryngol Otol 95:785
2. Derome PF (1971) Craniofacial surgery. In: Schmidek HP, Sweet WH (eds) Current techniques in operative neurosurgery. Grune & Stratton New York, p. 233
3. Donald PJ (1981) Recent advances in paranasal sinus surgery. Head Neck Surg 4:146
4. Ketcham AS, Wilkins RH, Van Buren JM, Smith RR (1963) A combined intracranial facial approach to the paranasal sinuses. Am J Surg 106:698–703
5. Miller HS, Petty PG, Wilson WF (1973) A combined intracranial and facial approach for excision and repair of cancer of the ethmoid sinuses. Aust NZ J Surg 43:179
6. Schramm VL, Myers EN, Maroon JC (1979) Anterior skull base surgery for benign and malignant disease. Laryngoscope 89:1077
7. Shah JP, Galicich JH (1977) Craniofacial resection for malignant tumors of ethmoid and anterior skull base. Arch Otol 103:514
8. Shah JP, Galicich JH (1979) Esthesioneuroblastoma – treatment by combined craniofacial resection. NY State J Med 79:84
9. Sisson GA, Bytell DE, Becker SP (1976) Carcinoma of the paranasal sinuses and craniofacial resection. J Laryngol Otol 90:59
10. Terz JJ, Young HF, Lawrence W (1980) Combined craniofacial resection for locally advanced carcinoma of the head and neck. II. Carcinoma of the paranasal sinuses. Am J Surg 140:618
11. Tessier P (1971) The definitive plastic surgical treatment of the severe facial deformities of the craniofacial dysostoses: Crouzon's and Apert's diseases. Plast Reconstr Surg 48:419
12. Van Buren JM, Ommaya AK, Ketcham AS (1968) Ten years experience with radical combined craniofacial resection of the malignant tumors of the paranasal sinuses. J Neurosurg 28:341
13. Westbury G, Wilson JSP, Richardson A (1975) Combined craniofacial resection for malignant disease. Am J Surg 130:463–469

Trigeminal Schwannoma

K. D. Post and P. McCormick, New York/USA

Schwannomas arising from the intracranial portion of the trigeminal nerve are rare. Jefferson classified these tumors into three types: tumors mainly in the middle fossa, predominantly posterior fossa tumors, and tumors with significant components in both middle and posterior fossae.

Patient Population

Fourteen patients with histologically verified schwannomas arising from the intracranial portion of the trigeminal nerve underwent operative treatment at the Neurological Institute of New York between January 1970 and December 1986. There were nine women and five men, ranging in age from 20 to 60 years (mean 40 years). None of the patients in this series had a family history or physical stigmata of neurofibromatosis. These patients accounted for 0.36% of the 3866 patients with intracranial tumor and 5.5% of 254 patients with intracranial schwannomas who received operative treatment during the same period.

Clinical Features

A summary of the presenting symptoms is shown in Tables 1 and 2. The duration of symptoms preceding diagnosis ranged from 6 days to 10 years (mean 33 months). Subjective complaints of trigeminal nerve dysfunction were the initial symptoms in eight patients. By the time of diagnosis, however, ten (71%) patients had symptoms referable to the trigeminal nerve. These complaints were sensory in nature. Only one patient complained of weakness in the muscles of mastication. Facial pain was a prominent complaint upon admission in eight patients. The pain was often initially localized to one division of the trigeminal nerve and later spread to involve other divisions. The pain was usually described as "burning" in nature and was frequently associated with subjective complaints of numbness in the same distribution. Although brief exacerbations of facial pain were common, only two patients presented with a history of typical trigeminal neuralgia, characterized by trigger points, prolonged pain-free periods, and the absence of associated signs of trigeminal nerve dysfunction. In one of these patients, the subsequent

Table 1. Clinical summary of 14 patients with trigeminal schwannoma

Case no.	Age[a], Sex	Tumor origin	First symptom	Duration of symptoms
1	49, M	V2	Pain, ear	3 months
2	45, F	Ganglion	Diplopia	4 years
3	57, F	Ganglion	Diplopia	10 months
4	28, F	Ganglion	Pain, eye; headache	1 year
5	60, F	Ganglion	Pain, V3	7.5 years
6	43, F	Ganglion	Paresthesias, V1–V3	3 years
7	31, F	Ganglion	Headache	10 years
8	20, F	Ganglion	Seizure	6 days
9	24, F	Ganglion	Pain and hypesthesia, V1	1 year
10	52, F	Root	Tic, V3	3.5 years
11	60, M	Root	Tic, V2	7 years
12	23, M	Root	Decreased hearing	2 months
13	33, M	Root	Hypesthesia, V1–V3	1 year
14	42, M	Root	Hypesthesia, V1–V3	2 years

Tic, trigeminal neuralgia; V1, ophthalmic division of trigeminal nerve; V2, maxillary division of trigeminal nerve; V3, mandibular division of trigeminal nerve.
[a] Age (years) at time of diagnosis.

Table 2. Other symptoms in patients with trigeminal schwannoma

Symptom	Trigeminal nerve deficits			Additional deficits	
	Sensory	Corneal reflex	motor	Cranial nerve	Other
Pain, V2; decreased hearing	–	–	–	–	Conductive hearing loss
–	–	–	–	VI	–
Pain, V1–V3; hypesthesia, V1–V3; ptosis; diplopia	+	+	–	III, VI	–
Hypesthesia, V3	+	+	+	–	Keratitis
Hypesthesia, V2; seizure	+	+	+	–	–
Pain, V3; hypesthesia, V1–V3; weakness, jaw	+	+	+	–	Proptosis
–	+	+	–	–	–
–	+	+	–	VII, VIII	Cerebellar pyramidal
–	+	+	–	–	–
Hypesthesia, V1–V3; abnormal gait	+	+	–	–	Cerebellar
–	–	–	–	–	–
–	+	+	–	VIII	Cerebellar
Diplopia; decreased hearing; abnormal gait	+	+	+	VI, VII VIII	Cerebellar pyramidal
Decreased hearing; abnormal gait	+	+	–	VII, VIII	Cerebellar

V1, Ophthalmic division of trigeminal nerve; V2, maxillary division of trigeminal nerve; V3, mandibular division of trigeminal nerve.

Table 3. Neurological deficits on admission examination

Neurological deficit	*n*
Trigeminal nerve	
Absent or decreased corneal reflex	11
Sensory changes	11
V1–V3 hypesthesia	7
V1 hypesthesia	2
V1–V2 hypesthesia	1
V3 hypesthesia	1
Motor	4
Other cranial nerves	
III	1
VI	3
VII	3
VIII	4
Cerebellar signs	5
Pyramidal signs	2
Exophthalmos	1
Neurotropic keratitis	1

V1, Ophthalmic division of trigeminal nerve; V2, maxillary division of trigeminal nerve; V3, mandibular division of trigeminal nerve.

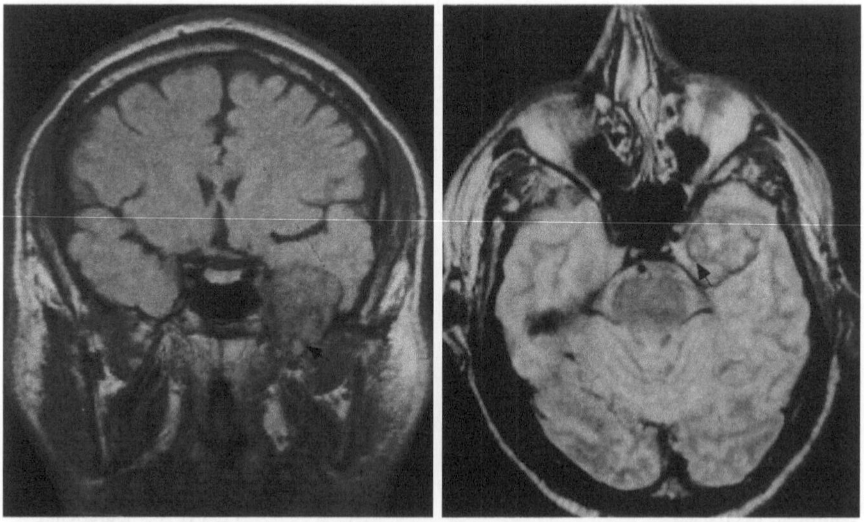

a b

Fig. 1. a T1-weighted coronal magnetic resonance image demonstrates extraaxial mass eroding through floor of midle fossa into pterygoid fossa (*arrow*). The tumor is inhomogeneous in signal intensity and somewhat hypointense with respect to surrounding brain. **b** T1-weighted axial magnetic resonance image demonstrates relationship of the tumor to the cavernous sinus. Note the integrity of the dura which clearly demarcates the tumor from the cavernous sinus (*arrow*)

appearance of facial numbness and gait difficulties raised the suspicion of tumor. Two patients complained of facial numbness without associated facial pain. Dysfunction of other cranial nerves, manifested by complaints of decreased hearing or diplopia, was elicited during admission history in five patients. Gait abnormalities were noted by three patients. Two patients (cases 7 and 8) had no history consistent with cranial nerve, pyramidal tract, or cerebellar dysfunction.

Individual neurological deficits found on admission examination are listed in Tables 1 and 2. A summary of these deficits is shown in Table 3. Diminished facial sensation and a diminished or absent corneal reflex was present in 11 (79%) patients. The hypesthesia was never complete and usually involved all three divisions of the trigeminal nerve. Four patients demonstrated weakness in the muscles of mastication. One patient had marked atrophy of the masseter and temporalis muscles. Deficits in the cranial nerves controlling the extra-ocular muscles were present in three patients. Varying degrees of sensorineural hearing loss were demonstrated in four patients, and a conductive hearing loss was present in one patient. A peripheral seventh nerve facial paresis was noted in three patients. Cerebellar signs, manifested by limb or gait ataxia, were present in five patients. Two patients demonstrated hyperactive deep-tendon reflexes with extensor plantar responses. One patient had a normal admission examination.

Tumor Characteristics

The tumors were on the left side in eight patients and on the right side in six; they were thought to originate in the gasserian ganglion in eight patients and from the trigeminal root in five. A large tumor component extended into the pterygoid fossa in one patient (Fig. 1). The size of the tumors ranged from 1.5 to 8 cm in diameter (mean 5 cm).

A literature review provided an additional 106 cases to our 14. A clinical analysis of 120 cases of trigeminal schwannoma is summarized in Tables 4 and 5 (McCormick et al. 1988). The location of the tumors, confirmed during surgery or at autopsy was: totally or predominantly middle fossa, 61 cases; totally or predominantly posterior fossa, 38 cases; and dumbbell tumors with significant middle and posterior fossae components, 21 cases. There was no association between age or sex of the patient and tumor location. Facial pain was a predominant complaint on admission history in 45 (36%) patients; it was more common with tumors arising from the trigeminal root (28%).

Surgical Treatment

Twenty operative approaches to tumor were performed on 14 patients. Tumors located entirely or predominantly within the middle fossa were approached through a standard subtemporal intradural route. Cerebrospinal fluid

Table 4. Initial symptom in 114 patients with trigeminal schwannoma

Symptom	*n*	%
Trigeminal nerve dysfunction	67	59
Numbness	34	30
Pain	26	23
Paresthesias	7	6
Headache	18	16
Diplopia	10	9
Hearing loss/tinnitus	7	6
Visual loss	5	4.5
Ear pain	4	3.5
Other[a]	5	4.5
Total	116[b]	

[a] Includes subarachnoid hemorrhage, vertigo, seizure, exophthalmos, gait difficulty (one patient each).
[b] Three patients presented with more than one initial complaint; one patient was asymptomatic.

Table 5. Abnormal findings on admission examination in 120 patients with trigeminal schwannoma[a]

Neurological abnormality	*n*	%
Trigeminal nerve		
Decreased sensation	89	74
Diminished or absent corneal reflex	86	72
Pain	45	36
Motor weakness	50	42
Other cranial nerve deficits		
II	12	10
III	18	15
IV	8	7
VI	42	35
VII	27	23
VIII	37	31
IX, X	9	8
XI	2	1
XII	4	3
Cerebellar signs	28	23
Long tract signs	19	16
Exophthalmos	20	17
Papilledema	12	11

[a] Only 26 patients (22%) had abnormal findings limited to the trigeminal nerve.

drainage through a lumbar spinal catheter and mannitol were employed to assist in temporal lobe elevation in most cases. A suboccipital approach was utilized in patients with tumor located predominantly in the posterior fossa. One patient underwent a combined transtentorial approach, and in one patient a staged subtemporal and suboccipital procedure was performed. Both of these patients had dumbbell-shaped tumors with significant middle and posterior fossae tumor components.

The initial surgery resulted in gross total removal in six patients, near total removal in four, and partial removal in four. The factor limiting the extent of resection was tumor adherence to the lateral wall of the cavernous sinus or the brainstem. Partial or total removal of portions of the trigeminal nerve complex was required in eight patients because of involvement with tumor. All four patients who had only partial tumor removal presented with symptomatic recurrences at 52, 13, 24, and 17 months (mean 26.5), respectively. A second operation resulted in near total removal in these patients. Three of these patients are without clinical evidence of recurrence at 12, 28, 72 months following the second operation. One patient presented with symptomatic recurrence 21 years after the second resection. A third operation resulted in near total removal. She is currently without clinical recurrence 7 years following the third operation. Of the four patients whose initial operation resulted in near total removal, two are clinically well at 40 and 24 months, respectively. One patient died 3 weeks postoperatively, the fourth from massive tumor recurrence 4 years postoperatively. All six patients whose initial operations were curative are clinically well without computed tomographic or clinical evidence of tumor recurrence at follow-up ranging from 4 to 177 months (mean 51.5; median 27.5)

Postoperative Course

In the immediately postoperative period, the appearance of additional cranial nerve deficits was common. Abnormalities of trigeminal nerve function, either new onset or worsening of existing deficits, were seen in nine patients. These deficits tended to be permanent. Other cranial nerve deficits were transient and resolved between 3 weeks and 4 months postoperatively. Improvement of preoperative deficits was also common. Diplopia, which was present in three patients preoperatively, resolved after operation in two. Hearing returned to a functional level in three of four patients and facial weakness improved in two of three. Cerebellar and pyramidal tract signs resolved in all patients. Facial pain, present in eight patients preoperatively, continued to be a problem in only one patient.

Long-Term Follow-up

Aside from one postoperative death and one patient who died from tumor recurrence 4 years after operation, all patients are clinically well and without clinical evidence of tumor recurrence at follow-up ranging from 4 to 177 months (47 most; median 31). One patient, who was well 6 years after operation, has been lost to follow-up. Recent computed tomography demonstrated no tumor recurrence in seven patients, minimal residual tumor in two, and slight tumor progression on serial computed tomographic scans in two.

Some degree of trigeminal sensory loss remains in nine patients. Weakness in the muscles of mastication is evident in five patients. All of these patients had resection of some part of the trigeminal root, ganglion, or peripheral divisions during operation. Two patients have required tarsorrhaphy for corneal ulceration secondary to neurotropic keratitis. One patient underwent a lateral rectus recession procedure for a persistent abducens nerve palsy.

References

Jefferson G (1955) The trigeminal neurinoma with some remarks on malignant invasion of the gasserian ganglion. Clin Neurosurg 1:11–54

McCormick PC, Bello JA, Post KD (1988). Trigeminal schwannoma: surgical series of 14 patients and a review of the literature. J Neurosurg 69:850–860

Parasellar Chondrosarcoma: Case Report and Literature Review

A. DAHER and M. SINDOU, Lyon/France

Cartilaginous tumors represent 0.16% of all intracranial tumors. Of these, 14% are chondrosarcomas. Most of them arise from the base of the skull and particularly from sphenotemporal and sphenooccipital basal synchondroses [1, 2, 5, 7, 8, 9, 12, 16, 18, 21, 22]; other localizations have been reported in the literature concerning the convexity [2], the falx [6], and the choroid plexus [20] from ectopic cartilages. A literature review enabled us to find 26 cases of parasellar chondrosarcomas; these represented 61% of chrondrosarcomas of the base (41 cases) and 35.6% of all the intracranial chondrosarcomas (73 cases). From this study, the clinical and radiological characteristics and the surgical problems of parasellar chondrosarcomas are discussed here.

Case Report

The clinical history of a 40-year-old woman started in 1979, when she had a diplopia in relation with paralysis of the left sixth nerve. No cause was found; therefore the diplopia was corrected by an ophthalmological operation. In April 1986, the patient was affected by a reduction of her visual acuity and a left ptosis.

On admission, the patient was alert and well oriented. The neurological examination stressed the fact that she had a complete left ophthalmoplegia. The visual acuity was 4/10 on the left side and 9/10 on the right. Skull X-rays and tomograms showed laterosellar calcifications on the left side and a widening of the left sphenoidal fissure with an erosion of its external edge (Fig. 1). Computed tomography (CT) revealed a left expansive parasellar lesion which was heterogeneous with many calcifications and without enhancement (Fig. 2). The left carotid angiogram confirmed the presence of an avascular lesion compressing the intracavernous carotid artery and displacing C4 downward and C3 forward (Fig. 3).

A left pteriono-subtemporal approach was performed, with resection of the anterior clinold process and opening of the external and upper walls of the optic canal. The lateral wall of the cavernous sinus was distended by the tumor. We noted that the left optic nerve was compressed between left CA_1 and the supraclinoid carotid artery (pushed away by the tumor; Fig. 4). The lateral wall of the cavernous sinus was incised between V1 and V2. The tumor was entirely extradural and avascular. A complete removal was obtained and the

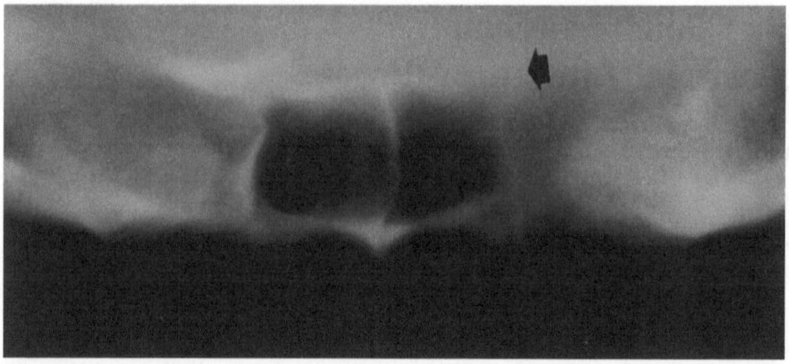

Fig. 1. X-ray tomography showing widening of the left sphenoidal fissure and left parasellar calcifications (*arrow*)

venous bleedings were easily controlled by packing with Surgicel. The histopathological examination confirmed the diagnosis of low-grade myxoid chondrosarcoma. The postoperative course was uneventful, and the post-operative CT scan confirmed the total removal of the tumor (Fig. 5). No complimentary radiotherapy was carried out. Three months later the patient presented a partial recovery of the third nerve and an improvement of her visual acuity (6/10).

Literature Review

Findings reported in the literature and concerning parasellar chondrosarcomas can be summarized as follows.

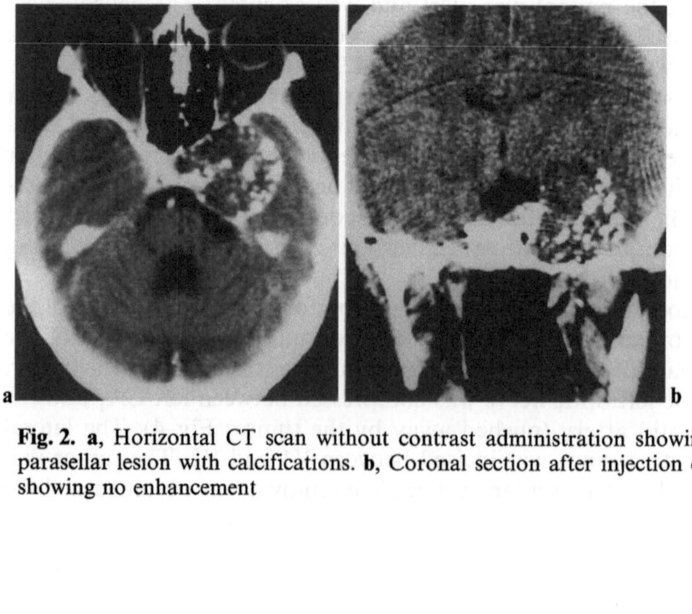

Fig. 2. a, Horizontal CT scan without contrast administration showing a heterogeneous parasellar lesion with calcifications. **b,** Coronal section after injection of contrast medium showing no enhancement

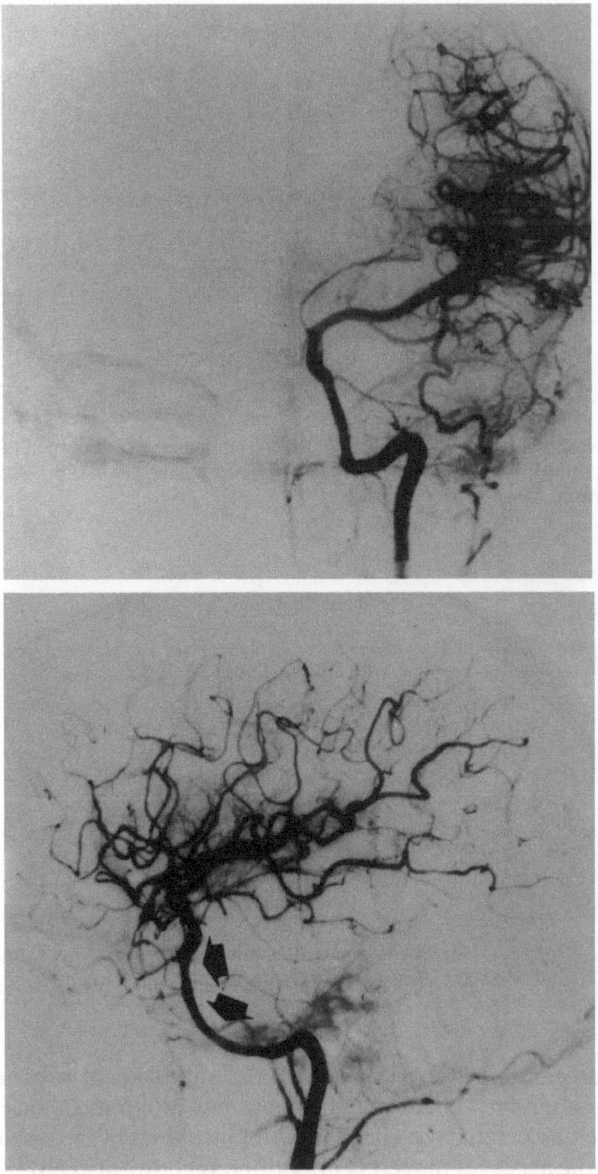

a

b

Fig. 3. **a**, Frontal view of the left carotid angiography showing an avascular lesion in the parasellar area. **b**, Lateral view shows marked displacement of C4 and C3 (*arrows*)

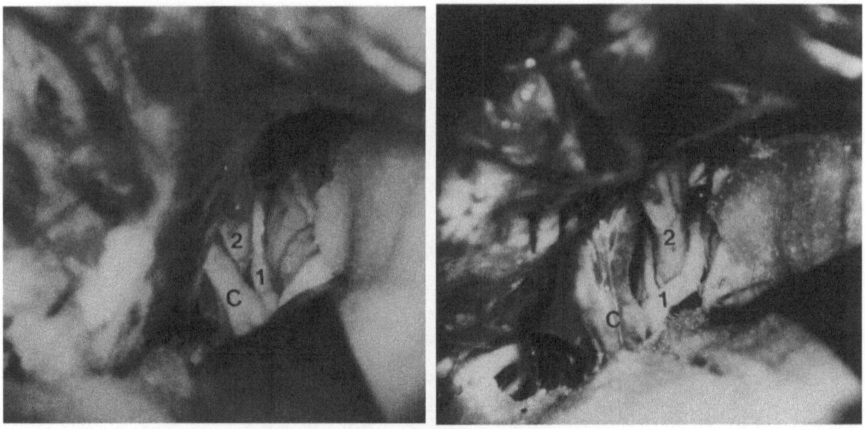

a b

Fig. 4. **a**, Intraoperative view showing the compression of the left optic nerve (*2*) between the left CA_1 (*1*) and the supraclinoid portion of the ICA (*c*) displaced by the tumor. **b**, After removal of the tumor, the optic nerve is decompressed

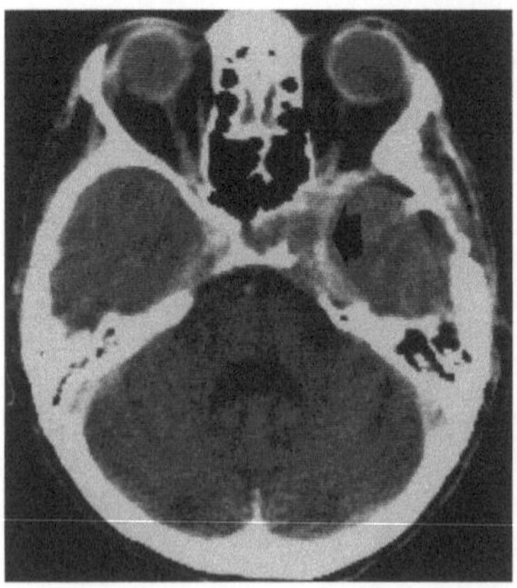

Fig. 5. Postoperative axial CT scan demonstrating the excision of the tumor

1. Parasellar chondrosarcomas have a clinical and radiological presentation that is quite characteristic. The revealing symptoms are in most cases oculomotor palsies, often associated with headaches and troubles of the visual acuity, and more rarely the nerves of the cerebellopontine angle are affected. X-rays and tomograms often show calcifications and bone destruction of the sellar area. CT scans without contrast enhancement show heterogenous lesions, and angiography reveals expansive avascular processes pushing away the carotid artery (11 cases out of 14 detailed).

2. In 13 cases in the literature where the authors mentioned the surgical removal of tumors 10 had a biopsy or a partial removal [1, 2, 5, 7, 9, 12, 16, 18, 21] and only 3 a complete removal [9, 12, 22]. The surgical approach was

subtemporal five times [1, 8, 21], subfrontal three times [8, 18, 21], transsphenoidal four times [5, 8, 9], and orbitomaxillary once [7].

3. In the 11 cases in which a radiotherapy was carried out it was done once preoperatively [2] and ten times postoperatively [7, 8, 16, 18, 21].

4. The most important problem in these tumors is that of prognosis. The frequency of local long-term recurrences is hard to quantify; the authors rarely mention the long-term follow-up (12 months – 5 years). The risk of recurrence requires regular neuroradiological supervision even in the absence of neurological signs.

5. The probability of extracranial metastasis through the blood tract was mentioned in only two publications [8, 14].

Discussion

These tumors, although located in the cavernous sinus, are relatively accessible to surgical treatment because they do not adhere to the vascular-nervous structures. They have a friable constitution and a weak vascularization. Their removal is relatively easy due to the fact that the cavernous sinus is filled in by the tumor; therefore venous hemorrhage is well controlled. If necessary, the carotid control could be done either at the cervical level or in the Glascock space. In fact, the most difficult problem is to obtain a complete removal of the tumor, which is neither radiosensitive nor chemosensitive. Indeed, if the tumor is generally well separated from the vascular-nervous structures, it is often not very limited from the bone tissue from which it arises.

The approach used in the majority of cases published was the subtemporal. The extradural subtemporal approach, very satisfactory theoretically, could have been used for our patient due to the fact that the intradural extension was absent, but lack of visual acuity required intradural access to explore and decompress the optic nerve. This approach has the disadvantage that it is necessary to know that here is no intradural extension preoperatively, which is difficult to ascertain in practice.

The intradural subtemporal approach provides good access to the whole floor of the middle fossa and sphenotemporal sutures, which are the starting points of the tumor. Moreover, it provides good control of the distal internal carotid artery in its juxtaclinoid segment as well as proximal control in the Glascock space. Its risk is the excessive retraction of the temporal lobe. This is why the pterionosubtemporal approach with an opening of the Sylvian valley seemed to be preferable. The mobilization of orbito-zygomatico-malar block as described by Lesoin [15], which we have used in several subsequent cases of tumors of the cavernous sinus, provides a much better access to the lateral region of the skull base with a minimum retraction of the temporal lobe.

References

1. Adegbite A et al. (1985) Primary intracranial chondrosarcoma: a report of two cases. Neurosurgery 17:490–494
2. Bahr A, Gayler B (1977) Cranial chondrosarcomas. Report of four cases and review of the literature. Radiology 124:151–156
3. Byers PD (1981) A study of histological features distinguishing chordoma from chondrosarcoma. Br J Cancer 43:229–232
4. Cianfriglia F et al. (1978) Intracranial malignant cartilaginous tumours. Report of two cases and review of the literature. Acta Neurochir 45:163–175
5. Currie J et al. (1983) Chronic isolated abducens paresis from tumors at the base of the brain. Arch Neurol 40:226–229
6. Cybulski G et al. (1985) Falcine chondrosarcoma: case report and literature review. Neurosurgery 16:412–415
7. Gay I et al. (1981) Chondrosarcoma at the skull base. Ann Otol 90:53–55
8. Grossman R, Davis K (1981) Cranial computed tomography of chondrosarcoma of the base of the skull. Radiology 141:403–408
9. Hardy J et al. (1966) Volumineux chondrosarcome calcifié de la région sellaire. Exérèse trans-sphénoïdale sous contrôle radioscopique télévisé. Neurochirurgie 12:491–502
10. Hassounah M et al. (1985) Primary cranial and intracranial chondrosarcoma. A survey. Acta Neurochir 78:123–132
11. Heffelfinger MJ et al. (1973) Chordomas and cartilaginous tumors at the skull base. Cancer 2:410–420
12. Heros R et al. (1980) Intracranial mesenchymal chondrosarcoma. Surg Neurol 14:311–317
13. Kubota T et al. (1982) Primary intracranial mesenchymal chondrosarcoma: case report with review of the literature. Neurosurgery 10:105–110
14. Leedham PW Swash M (1972) Chondrosarcoma with subarachnoid dissemination. J Pathol 107:59–61
15. Lesoin F et al. (1986) Intérêt de la mobilisation du volet orbitozygomatico-malaire. Neurochirurgie 32:90–93
16. Minagi H Newton T (1969) Cartilaginous tumors of the base of the skull. Am J Roentgenol 105:308–313
17. Moore T et al. (1985) CT and angiography of primary extradural juxtasellar tumors. Am J Radiol 145:491–496
18. Reuter K, Weber AL (1981) Parasellar chondrosarcoma in a patient with Ollier's disease. Case report. Neuroradiology 22:151–154
19. Richardson R et al. (1981) Chondrosarcoma of the middle cranial fossa: case report. Neurosurgery 8:707–711
20. Scott RM et al. (1976) Myxochondrosarcoma of the fourth ventricle: case report. J Neurosurg 44:386–389
21. Sekhar L, Meller A (1986) Operative management of tumors involving the cavernous sinus. J Neurosurg 64:879–889
22. Waga S et al. (1981) Chondrosarcoma of the ethmoid sinus extending to the anterior fossa. Surg Neurol 16:324–328

Cholesterol Granuloma of the Skull Base: A Review

F. W. GAMACHE Jr. and T. MCLURE, New York/USA

The literature describing cholesterol granuloma is confusing. Cholesterol granuloma, cholesterol cyst, cholesteatoma, cholesteatomatous endothelioma are terms often used interchangeably in the literature for a lesion encountered at the skull base, composed of a collection of cholesterol crystals, surrounded by foreign body giant cells, hemosiderin granules, and soft tissue stroma. The term cholesteatoma has also been used to describe the pathologically distinct entities of cholesterol granuloma, epidermoid (intradural and acquired), and craniopharyngioma. Such improper use of the above terms is also encountered in the radiographic literature and has been critically reviewed by Lo et al. [14]. From a practical point of view, there is a significant difference in the tissue consistency encountered at surgery between a cholesteatoma and a cholesterol granuloma. Such tissue differences may significantly alter treatment plans. Cholesterol granuloma is not a rare tumor, and since most of the literature regarding cholesterol granuloma is found in the otolaryngology journals, a review of the subject for neurosurgeons and head and neck surgeons seems worthwhile.

The exact incidence of cholesterol granuloma is not well known. Sabin et al. reported six cholesterol granulomas among nine petrous apex cystic lesions observed over a 20-year period [25]. The Otologic Medical Group of Los Angeles estimated the incidence of cholesterol granuloma to be approximately 1 for every 30 acoustic tumors and to be certainly less common than cholesteatoma. Cholesteatomas represent less than 1% of all intracranial neoplasms but 6%–7% of all tumors of the cerebellopontine angle [24].

The exact etiology of cholesterol granuloma remains controversial. Cholesterol granuloma is believed to result from an inflammatory response of the temporal bone to obstruction of air-containing spaces. Lack of ventilation and drainage of any part of the air-containing spaces of the temporal bone seems to be the most frequent cause. Experiments by Beaumont [1], Hiraide et al. [11], Main et al. [15], and Ojala [21] support this line of reasoning. Cholesterol and iron products of blood degradation (as might occur after local hemorrhage) are believed to incite a foreign body response.

Microscopically, cholesterol granuloma is comprised of cholesterol crystals associated with foreign body giant cells in a matrix of fibrous connective tissue, round cell infiltration, macrophages, and frequently pathological blood vessels. These abnormal blood vessels frequently rupture into areas of old hemorrhage so that a mixture of recent and old hemorrhage may be seen radiographically side by side. The cyst wall is composed of fibrous connective tissue unlike the keratinizing squamous epithelium composing the wall of a cholesteatoma [6]. The cysts which develop, because of the dark material within, are frequently referred to as chocolate cysts. When the ear is involved, the patients frequently have a past history of otitis media with a dark bluish discoloration of the ear drum [13, 19]. While iron deposits and the iron chelating substance lactoferrin have both been found in such granulomas, hemorrhage has not been clearly linked with the source of the iron [26]. Metabolic differences between cholesterol granuloma and cholesteatoma regarding ratios of cholesterol to lipid-bound phosphorus and Δ-7-cholesterol, have been described by Hayashida et al. [10] and Nomura [20]. Other histologic and biochemical differences between cholesterol granuloma and cholesteatoma have been described by Sade and Teitz [27] and Ferlito [2].

Radiographically, cholesterol granuloma generally is a sharply delineated, smoothly marginated expansive lesion with a rim which enhances with intravenous contrast on computed tomography, but whose center is isodense with surrounding brain [7]. The lesion is generally extradural. The major differential diagnosis includes cholesteatoma and mucocele, both of which are sharply and smoothly marginated, ovoid, expansive, and neither of which is contrast enhancing. Unlike cholesterol granuloma, however, the centers of these common lesions approximate cerebrospinal fluid density and are hypodense relative to brain. Recently, with the advent of magnetic resonance imaging, it has become apparent that cholesterol granuloma presents with predominantly hyperintense signals on both short and long TR sequences unlike cholesteatoma which is usually hyperintense only on T2 images [28].

Manasse first described cholesterol granuloma of the middle ear in 1917 [16]. Shambaugh in 1929 described idiopathic hemotympanum as a prelude to the development of cholesterol granuloma of the middle ear [29]. Much later Plester and Steinbach [23] noted the association between cholesterol granuloma to be a "harmless clinical phenomenon." House and Brackmann [12] related massive erosion of the petrous apex in two of their patients to cholesterol granuloma. Eventually cholesterol granuloma of the cerebellopontine angle was reported as well [6, 12]. Gray et al. reported cholesterol granuloma involving the sphenoid sinus [9], and Milton and Bickerton reported nine cases of cholesterol granuloma involving the maxillary sinus [17]. Wyler et al., in reviewing the literature up to 1974, concluded cholesterol granuloma rarely produced neurological signs [30]. Sabin et al. [25], however, recently reviewed cystic lesions of the petrous apex, including cholesterol granuloma, which presented because of neurological symptoms.

Cholesterol granuloma of the petrous apex commonly grows silently eroding surrounding bone and eventually exerting pressure on cranial nerves. Lo et al. observed a growth rate of approximately 3 mm/year in one patient [14]. As a result, patients may present with hearing loss, tinnitus, vertigo, facial pain or weakness, diplopia, or other signs of cranial nerve involvement. A significant hearing loss is usually sensory-neural but may be mixed [7]. No pathognomonic ENG findings exist.

The treatment of choice for cholesterol granuloma has been to provide drainage for the cyst with total excision of the cyst wall where possible. In some cases, however, total excision of the cyst wall is not practical. Thus, a procedure which allows for marsupialization of the cyst or chronic drainage of the cyst into another cavity (such as the mastoid) has been recommended. In addition, the semiliquid consistency of cholesterol granuloma is considerably different from the caked rice consistency of an epidermoid. Such a difference in internal consistency allows for satisfactory drainage of the cholesterol granuloma by simple fenestration of the cyst capsule – unlike cholesteatoma which requires more concerted debulking of the internal contents. Unfortunately, pathways for drainage frequently become clogged with time requiring additional surgery. Some authors have suggested the use of a silastic drain tube to provide a permanent drainage pathway between the granuloma and a final drainage cavity [7]. While the idea is an attractive one, unfortunately obstruction of this silastic tube with inflammatory tissue has been noted to occur with time (Graham, personal communication).

For the patient with a cholesterol granuloma of the petrous bone in whom hearing remains intact, a translabyrinthine exploration would result in a loss of hearing. Treatment by a different approach would thus be preferable. Treatment of these lesions by various sinus approaches has been reported by several authors. A transethmoid-transsphenoidal drainage of a cystic skull base lesion was described by Montgomery [18]; however, the pathologic identification of the lesion was not made. Transsphenoidal fistulization of a petrous apex cholesteatoma has been described by Gacek [4] and Gray et al. [9]. Bilateral congenital cholesteatoma of the middle ear has been described by Peron and Schuknecht [22]. Giant bilateral cholesterol cyst of the petrous apex has been described by Graham et al. [8]. Description of bilateral cholesterol granulomas involving the petrous bone has recently been described by Gamache et al. [5]. Both lesions were treated simultaneously by a transsphenoidal-transclival approach. By avoiding a transdural approach, the possibility of aseptic meningitis (immediate or delayed) may be avoided. A subtemporal extradural approach may offer limited exposure because of cranial nerves and vascular structures traversing the skull base. While large bony defects in the sphenoid bone and clivus do not guarantee permanent drainage of the cyst, this route has been useful in several cases for the current authors. Clinically, the patients report drainage as a "funny taste" in the back of the throat once every few weeks which lasts for minutes. Should the drainage pathway plug, using local anesthesia, a large bore needle might be passed under direct vision with a nasal speculum into the sphenoid sinus. The needle could then be guided with biplanar fluoroscopy through the (surgically created) aperture in the clivus and into the granuloma. The granuloma could be drained and the drainage pathway reopened. If unsuccessful, the pathway could be easily reopened surgically.

In conclusion, cholesterol granuloma appears to be differentiated from cholesteatoma by characteristic findings on magnetic resonance imaging. Because of its semiliquid nature, cholesterol granuloma lends itself more easily to surgical drainage – for example through a transsphenoidal, transclival approach. Such an approach is cosmetically acceptable and affords easy repeat drainage with a nasal speculum and fluoroscopy.

References

1. Beaumont G (1966) The effects of exclusion of air from bones. J Laryngol Otol 80:236–249
2. Ferlito A (1973) Middle ear cholesteatoma: and cholesterol granuloma: Some histopathological and clinical considerations. Ann Otolaryngol (Paris) 90:697–710
3. Friedmann I, Graham M (1979) The ultrastructure of cholesterol granuloma of the middle ear: an electron microscope study. J Largyngol Otol 93:433–443
4. Gacek R (1980) Evaluation and management of primary petrous apex cholesteatoma. Otolaryngol Head Neck Surg 88:519–523
5. Gamache FW, McLure T, Deck M (1988) Cholesterol granuloma of the skull base: case Report. Neurosurgery 22(6):1098–1101
6. Gherini S (1986) Pathologic quiz case 2. Arch Otolaryngol Head Neck Surg 112:674–677
7. Gherini S, Brackmann D, Lo W, Solti-Bohman L (1985) Cholesterol granuloma of the petrous apex. Laryngoscope 95:659–664
8. Graham M, Kemink J, Latack J, Kartush J (1985) The giant cholesterol cyst of the petrous apex: a distinct clinical entity. Laryngoscope 95:1401–1406
9. Gray W, Salcman M, Rao K, Hafiz M (1985) Cholesterol granuloma of the petrous apex and sphenoidal sinus: a case report. Neurosurgery 17:67–69
10. Hayashida T, Iwamori M, Kitsuwa T, Nagai Y, Nomura Y (1984) Biochemical study of cholesteatoma and cholesterol granuloma – occurrence of 7 cholesterol in the tissues of cholesteatoma. ORL 46:242–247
11. Hiraide F, Inouye T, Miyokogawa N (1982) Experimental cholesterol granuloma histopathological and histochemical studies. J Laryngol Otol 96:491–501
12. House J, Brackmann D (1982) Cholesterol granuloma of the cerebellopontine angle. Arch Otolaryngol 108:504–507
13. Linthicum F Jr (1971) Cholesterol granuloma (iatrogenic). Further evidence of etiology, a case report. Ann Otol Rhinol Laryngol 80:207–209
14. Lo W, Solti-Bohman L, Brackmann D, Gruskin P (1984) Cholesterol granuloma of the petrous apex: CT diagnosis. Radiology 153:705–711
15. Main T, Shimada T, Lin D (1970) Experimental cholesterol granuloma. Arch Otolaryngol 91:356–359
16. Manasse P (1917) Otitis media catarrhalis chronica: Handbuch der pathologischen Anatomie des menschlichen Ohres. In: Manasse P, Korner H (Hrsg) Die Ohrenheilkunde der Gegenwart und ihre Grenzgebiete, Bd IX. Bergmann, Wiesbaden, S 46–53
17. Milton C, Bickerton R (1986) A review of maxillary sinus cholesterol granuloma. Br J Oral Maxillofac Surg 24:293–299
18. Montgomery W (1977) Cystic lesions of the petrous apex: transsphenoid approach. Ann Otol 86:429–435
19. Nager G, Vanderveen T (1976) Cholesterol granuloma involving the temporal bone. Ann Otol 85:204–209
20. Nomura Y (1984) Cholesterol metabolism in cholesteatoma and cholesterol granuloma. Ann Otorhinol Laryngol [Suppl] 112:129–132
21. Ojala L (1957) Pneumatization of the bone and environmental factors: experimental studies on chick humerus. Acta Otolaryngol [Suppl] 133:1–28
22. Peron D, Schuknecht H (1975) Congenital cholesteatomata with other anomalies. Arch Otolaryngol 101:498–505
23. Plester D, Steinbach E (1982) Cholesterol granuloma. Otolaryngol Clin North Am 15:665–672

24. Rosenberg R, Hammerschlag P, Cohen N, Bergeron R, Reede D (1986) Cholesteatoma vs. cholesterol granuloma of the petrous apex. Otolaryngol Head Neck Surg 94:322–327
25. Sabin H, Bordi L, Symon L (1987) Epidermoid cysts and cholesterol granuloma centered on the posterior fossa. Neurosurgery 21:798–805
26. Sade J, Halevy A, Klajman A, Mualem T (1980) Cholesterol granuloma. Acta Otolaryngol 89:233–239
27. Sade K, Teitz A (1983) Cholesterol and cholesteatoma. Acta Otolaryngol 95:547–553
28. Saint Louis L, Touri S, Deck M (1987) MRI of petrous apex cholesterol cysts. Abstract 25th Annual Meeting of Amer Soc of Neuroradiology, New York, May 1987, p. 107
29. Shambaugh G (1929) The blue drum membrane. Arch Otolaryngol 10:238–240
30. Wyler A, Leech R, Reynolds A, Ojemann G, Mead C (1974) Cholesterol granuloma of the petrous apex. J Neurosurg 41:765–768

The Transpalatal-Transpharyngeal Approach to Chordomas of the Ventral Sphenoclival Regions

C. ZINI, A. GANDOLFI, and F. PIAZZA, Parma/Italy

Introduction

It is well recognized that chordomas originate from intraosseous notochordal remnants in the cranial and sacrococcygeal regions [20]. During embryonic developments the mesodermal tissue envelops the notochord and neural tube, laying down cartilage in the midsagittal plane of the skull base. At the cranial end of neuraxis, notochord is brought into close proximity to the surface of the cartilage dorsally to the hypophyseal fossa and all along the clivus of Blumenbach [7, 8, 25–27]. Normally the captured cells of the notochord degenerate and disappear, but in rare instances they remain as rests of primitive notochordal cells. It is believed that chordomas develop from these enclosed vestiges [14]. They are found on the clivus, parasellar region, nasopharynx odontoid process, vertebral and sacrococcygeal region.

Chordomas are slowly growing tumors, usually encapsulated, and varying greatly in size at the time of diagnosis. They are definitely rare pathological entities, representing between 0.1 %–0.2 % of intracranial tumors [14]. Midline lesions at the base of the skull have been considered in the past to be either inoperable or, in any case, difficult, sometimes discouraging, and hard-to-manage surgical problems [4, 21]. The evolution and recent progress in the feasibility of approaches to the skull base and the development of microsurgery applied to the exploration of these regions have made the approach and treatment of these lesions an affordable and feasible surgical task [4]. The operative techniques used today in dealing with tumors of the sphenoclival and ventral clivooccipital plates (and also upper cervical spine) generally employ the so-called transbasal approaches [14, 4, 22, 19, 23, 15, 12, 24, 16, 6, 11, 17], including the transethmoidal and transsphenoidal approaches to the sella and parasellar region [22, 19, 23, 15, 12]; the transoral-transpalatal approach to middle and lower clivus [16]; and the transcervical-transclival to lower clivus and upper spinal column [24]. To emphasize the optimal results obtained in treating chordomas of the ventral clivus using the transbasal approach, we refer to a case operated upon in our Clinic using the transoral-transpalatal approach. Technical details and indications of this technique of particular interest to the base of the skull surgeon are presented.

Case Report and Operative Technique

A 51-year-old man was referred to our Clinic with a 10-month history of progressive respiratory distress from obstruction of both nasal fossae. There had been no episodes of bleeding but an infrequent yellowish, mucous rhinorrhea. Additional complaints were: anosmia, hypogeusia, halitosis, rhinolalia, and mild hypoacusia. Physical and instrumental examinations revealed the presence of a lobulated mass occupying the posterior superior epipharyngeal compartments and obliterating the choanae. At anterior rhinoscopy both the nasal fossae were occupied by polypoid vegetations, grayish and translucent in appearance, apparently coming off and down from posterior ethmoid cells. Direct rhinopharyngoscopy was performed with the intention of removing pieces of tumor and reaching a preoperative histological diagnosis.

The tumor was revealed to be a chordoma. High-resolution computed tomography of the base of the skull showed wide destruction at the level of sphenoid body, upper and middle regions of the clivus, and posterior ethmoid without apparent intracranial extension. The base of left pterygoid process was also eroded and washed out. The sphenoid sinus appeared invaded by tumor. The wall of internal carotid artery in petrous bone was intact. Due to the anterior inferior and caudal extension of tumor, respectively, into the ethmoid and epipharyngeal compartments, a transoral-transpalatal approach to the base of the spenoid bone and ventral clivus was carried out. Removal of hard palate, mainly on left side, after incision of mucosa and division of soft palate allowed complete control of the bulging mass at the level of posterior choanal openings.

The anteroinferior pole of the tumor in the nasal fossae could be isolated, together with lateral extension coming into contact with tubal ostium and base of pterygoid process. The surface of the lobulated tumor mass was covered by intact mucosa. Intracapsular tumor removal using suction was carried out. Prolonged debulking of the tumor mass finally disclosed the osseous plate of ventral clivus. The floor of sphenoid sinus was eroded, but the septum was maintained. Using small curettes, tumor digitations were displaced from residual posterior ethmoidal cells. The eroded clival surface was levigated with diamond burr, and the tumor was followed to capture its extreme lateral extensions in the paraclival regions. Prevertebral, pharyngeal soft tissues were incised, and the tumor was squeezed out from their cut edges. The tip of petrous bone at the level of the horizontal precavernous carotid canal was intact. Suturing of mucosal planes and careful packing of empty spaces left after tumor removal were performed. The patient did well postoperatively. He was given a full course of radiation therapy in the regions of previous formal operation to ensure complete eradication of possible microscopic tumor remnants left behind.

Discussion

Chordomas are primarily extradural tumors showing a tendency over time to destroy and invade the basal dura, eventually occupying the intradural and intra-arachnoid compartments of the cranial cavity in their later stages. The direction of growth is variable [14]. We may differentiate sellar-parasellar and clival groups, showing either an extra- or intracranial main extension. Those following an anterior-inferior direction of growth become primarily nasopharyngeal masses and fall into the fields of both otologists and neurosurgeons [4]. Depending on localization, direction of infiltration, size and speed of growth, manifold symptoms are to be expected from chordomas at the skull base. Intracranial forms usually determine signs and symptoms of cranial nerves involvement, either in a single or in a combined fashion, often alternating from side to side. Headache is a common symptom in tumors with transdural extension [1, 3, 18, 5, 9, 10, 13]. Those with a primarily rhinopharyngeal spread and growth present clinically with symptoms related to respiratory difficulty, frank dyspnea, a sense of tension and fullness back in the nose, and sometimes epistaxis and infections. The derangement of olfactory function is also a common complaint. The transoral-transpalatal approach, first described by Mullan et al. [16] offers the possibility of eradicating the latter variant of chordomas, using microscope and high-magnification surgery. These can be removed without damaging the ventrobasal dura covering the inner surface of clivus and arachnoid layers and brain stem basal vessels, together with carotid arteries at their entry into cavernous sinus. Carotid arteries are brought well under control using the transpalatal approach if sufficiently widened toward lateral skull base. The posterior ethmoidal compartment and ventral clivus are also adequately exposed and can be explored throughout. By removing the posterior third of hard palate, the direction of sight of the operating surgeon can safely control the body of sphenoid bone, floor and cavity of sphenoid sinus, posterior ethmoidal compartment, and choanal openings with interchoanal septum. The region delimiting Rosenmüller recess, base of pterygoid process, and tip of pyramids is fully in view. With this approach radical tumor removal without recurrence over the long term is possible, particularly in ventral extradural tumors. The approach also offers the possibility of operating and excising infiltrated areas of basal dura and adequately correcting the consequent defect in CSF pathway. Possible infections brought into the cranial cavity from a nonsterile field can be prevented by adequate perioperative antibiotic management of the patient. Nowadays, high-resolution computed tomography and magnetic resonance imaging can precisely define extension and peripheral limits of skull base tumors, thus allowing the skull base surgeon to choose the most adequate microsurgical approach among the available basal craniotomies for controlling this region.

The transoral transpalatal approach to ventral sphenoclival area is a safe and effective surgical technique to eradicate chordomas having their main extension in the epipharyngeal compartment.

Summary

A case of sphenoclival chordoma, with extension toward the epipharyngeal compartment, is described. Origin, direction of spread, clinical presentation of ventrally placed chordomas with respect to clival plate are presented. Surgical treatment of this lesion using the transpalatal-transpharyngeal approach to midline exophytic chordomas are presented. Advantages and technical details of this approach are stressed.

References

1. Bailey P, Bagdasa B (1929) Intracranial chordoblastoma. Am J Pathol 5:439–449
2. Congdon CC (1952) Benign and malignant chordomas. A clinico-anatomical study of twenty-two cases. Am J Pathol 28:793–822
3. Dandy WE (1938) Hirnchirurgie. Barth, Leipzig
4. Decker R, Malis LI (1970) Surgical approaches to midline lesions at the base of the skull: a review. J Mount Sinai Hospital 37:84–102
5. Epple S, Ruckensteiner E (1946) Die Röntgendiagnose des Clivuschordoms. Schweiz Med Wochenschr 76:764–766
6. Falconer MA, Bailey IC, Duchen LW (1968) Surgical treatment of chordoma and chondroma of the skull base. J Neurosurg 29:261–275
7. Gandolfi A (1977) Il condroma della base cranica: discussione a proposito di una caso anatomo clinico. Acta Neurol 32:833–839
8. Gandolfi A (1979) Vertebral chordoma: intraoperative cytological diagnosis. Acta Neurol 34:156–160
9. Givner I (1945) Ophthalmologic features of intracranial chordoma and allied tumors of the clivus. Arch Ophthalmol (Chicago) 33:397–403
10. Godtfredsen E (1943) Eye and nerve symptoms in connection with cranial chordomas. Acta Ophtalmol (Copenh) 21:224–236
11. Guthkelch AN, Williams RG (1970) Recurrent chordoma of the clivus. Acta Neurol Scand 46:622–630
12. Hardy J, Wigser SM (1965) Transsphenoidal surgery of pituitary fossa tumors with televised radiofluorescopic control. J Neurosurg 23:612–619
13. Kleinsasser O, Friedmann G (1958) Die Chordome der Schädelbasis. Ein Beitrag zur Klinik, Pathologie und zur Differentialdiagnose am Röntgenbild. Dtsch Z Nervenheilkd 177:263–285
14. Krayenbuhl H, Yasargil MG (1975) Cranial Chordomas. Progr Neurol Surg 6:380–434
15. Mouren P, Poinso Y, Sebahoun G, Pellissier JF (1970) Les chordomes de la base du crane. Semin Hop (Paris) 46:501–510
16. Mullan S, Naunton R, Hekmat-Panah J, Vailati G (1966) The use of an anterior approach to ventrally placed tumors in the foramen. J Neurosurg 24:536–534
17. Nissen R, Klinger M (1964) Clivustumoren. Operative Möglichkeiten. Schweitz Arch Neurol Neurochir Psychiat 94:122
18. Psenner L (1952) Beitrag zur Klinik und zur Röntgendiagnostik des Chordoms der Schädelbasis. Fortschr Röntgenstr 77:425–433
19. Rougerie J, Guiot G, Bouche J, Trigo JC (1967) Les voies d'abord des chordomes du clivus. Neurochirurgie 13:559–570
20. Rubinstein L (1972) Tumors of the central nervous system. In: Atlas of tumor pathology, 2nd series, fascicle 6, Armed Forces Institute of Pathology, Washington, D.C., pp 315–318
21. Schisano G, Tovi D (1969) Clivus chordomas. Neurochirurgia 5:99–120

22. Spiess G (1911) Tumor der Hypophysengegend auf endonasalem Wege erfolgreich operiert. Munch Med Wochenschr 58:2503–2503
23. Steimle R, Coudry H, Pageaut G, Mallie M, Comte D, Jaquet G (1969) Chordome sellaire. Voie d'abord transphenoidale. Acta Neurochir 20:237–247
24. Stevenson GC, Stoney RJ, Perkins RK, Adams JE (1966) A transcervical transclival approach to the ventral surface of the brainstem for removal of a clivus chordoma. J Neurosurg 24:544–551
25. Tondury G (1964) Anatomie und Entwicklungsgeschichte des Schädel-Wirbelsäulen-überganges. Deutscher Röntgenkongreß 1963, Thieme, Stuttgart S 45–53
26. Williams LW (1908) The later development of the notochord in mammals. Am J Anat 8:251–282
27. Wood EH, Himadi GM (1950) Chordomas. A roentgenologic study of sixteen cases previously unreported. Radiology 54:706–716

Transnaso-Sphenoidal Approach to Skull Base Lesions Other than Pituitary Adenomas

F. Calbucci, F. Tognetti, and G. Frank, Bologna/Italy

Since its technical refinement and popularization by Hardy, the transnaso-sphenoidal (TNS) approach in microsurgery has increasingly been used for entering the sella turcica. Besides the removal of pituitary adenomas, which is the chief indication, this route allows the treatment of other intrasellar lesions as well as of some tumors of the sphenoid sinus and the clivus. In our Department the ratio of pituitary adenomas to other lesions treated by the TNS approach is 5:1. We present 36 cases of skull base lesions other than pituitary adenomas approached transsphenoidally during the last 9 years.

Case Material

Skull Base Cancer (Fig. 1). Five patients (two women and three men) were treated; the mean age was 53 years. Symptoms were variably combined, facial pain being the most frequent complaint, in three cases; complete ophthalmoplegia, third nerve and sixth nerve palsy came next in frequency with two instances each. In three patients the lesion was primary, while two cases were metastases from breast cancer. The postoperative course was consistent with amelioration of the oculomotor and the abducens nerve in three cases. The condition in two patients was not improved by surgery.

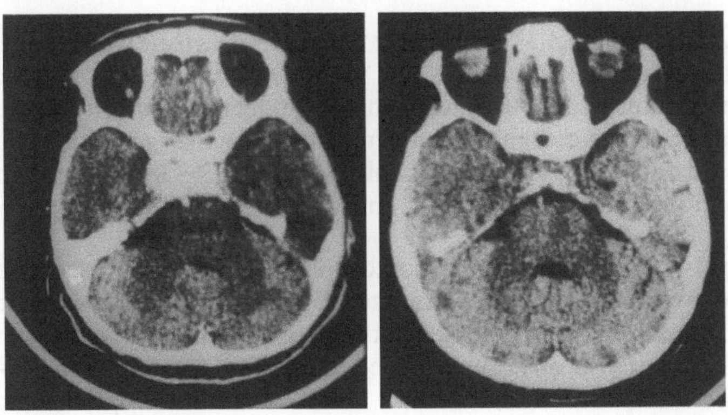

Fig. 1. Skull base cancer. Pre- (*left*) and postoperative (*right*) CT scan

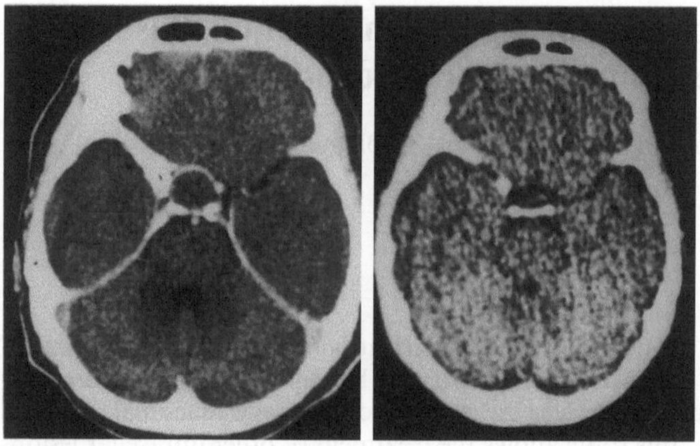

Fig. 2. Cystic craniopharyngioma. Pre- (*left*) and postoperative (*right*) CT scan

Empty Sella. Seven patients were operated on (six women and one man), with a mean age of 46 years. Spontaneous rhinorrhea was the chief complaint and was observed in five cases. Impairment of the chiasmatic function was present in the two other patients. At operation the sella was plugged with fat and muscle. Results were discouraging: only two patients with CSF leak could be cured, while the remaining three demanded spinal fluid diversion through a spinoperitoneal shunt. Visual function was unchanged.

Craniopharyngioma (Fig. 2). Seven patients (three women and four men) were in this group; age averaged 43 years. All patients suffered a chiasmatic syndrome of some degree; three patients had secondary amenorrhea, while two had panhypopituitarism. Visual deficits ameliorated constantly after surgery, while endocrine disturbances were left unchanged and needed hormone replacement. We had four cases of postoperative, permanent diabetes insipidus. The tumor was reoperated transcranially in two circumstances.

Chordoma/Chondroma (Fig. 3). There were five such cases (one woman and four men); the mean age was 49 years. The main symptom was headache, occurring in three instances. Epistaxis, chiasmatic syndrome, and third nerve palsy were present in one case each. Patients improved after surgery, with the exception of those two having visual and oculomotor deficit, respectively. In two cases a second operation, transcranial and transoral, was performed.

Mucocele (Fig. 4). Four patients harbored a sphenoethmoidal mucocele (two women and two men), with a mean age of 48 years. Symptoms were variable, with facial pain being represented in three cases, visual field deficits in two, and third nerve palsy in one. Disturbances had all improved postoperatively, except for one case of monocular blindness due to optic nerve compression.

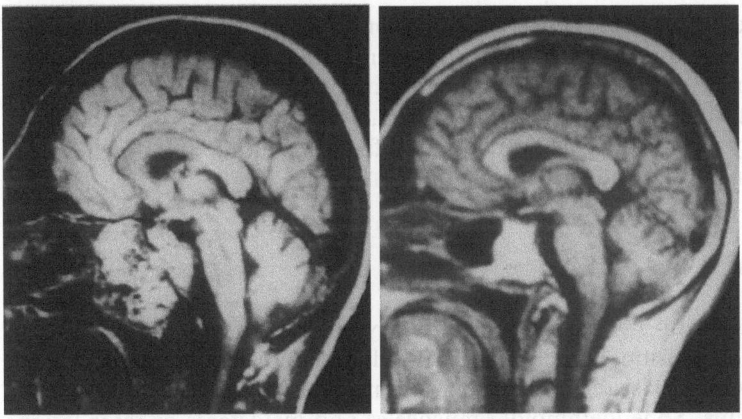

Fig. 3. Chordoma of the clivus growing anteriorly toward the nasal fossa and sphenoid sinus. Pre- (*left*) and postoperative (*right*) MRI

Meningioma. Two patients with a clivus meningioma and another with tuberculum sellae meningioma were approached transsphenoidally. Two were women and one a man, with age averaging 53 years. Although this tumor is not an indication for the TNS approach, this was carried out in order to have some material for diagnosis (two cases with intracranial hypertension) and to debulk the tumor (one patient with rapid visual loss). This latter case was the only one which improved postoperatively. All three patients were subsequently approached transcranially.

Miscellaneous Lesions Within the Pituitary Gland. Five further patients harbored unusual intrapituitary lesions. Three were females with a metastasis from breast cancer; mean age was 62 years. Symptoms were diabetes insipidus in one case, chiasmatic syndrome in two cases, and third nerve palsy in two

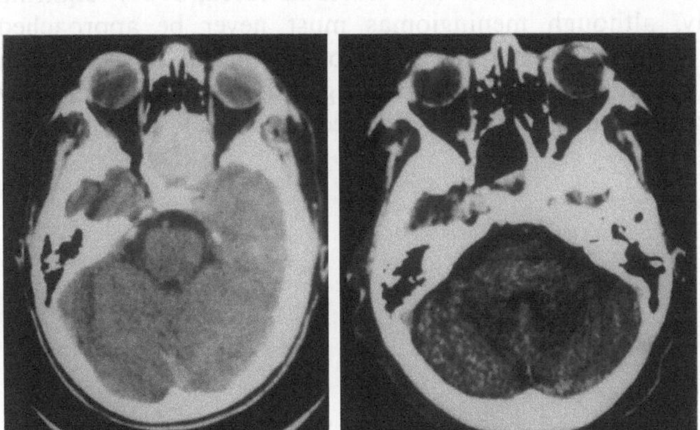

Fig. 4. Sphenoethmoidal mucocele. Pre- (*left*) and postoperative (*right*) CT scan

others. The chiasmatic deficit was the only one to show postoperative amelioration. A fourth patient was a 12-year-old boy with diabetes insipidus; an eosinophilic granuloma was extirpated, with no effects on the symptom. The last patient was a woman aged 20 years, with complete pituitary gland insufficiency. A mucoid cyst was delivered from within the gland, with some improvement in the endocrine function.

Discussion

The indication for TNS surgery in pituitary adenomas with limited, if any, suprasellar expansion is well standardized and indisputable. In contrast, controversy may exist regarding other lesions of both the sellar area and the sphenoid sinus. Sphenoethmoidal mucoceles and metastases or rarer lesions within the pituitary parenchyma can be safely and totally extirpated by this approach, with minimal morbidity and with sparing of the pituitary function. Debate, however, may arise concerning the empty sella, particularly in light of the scarce results observed: only two patients of seven benefited from operation. The only possible indication is documented rhinorrhea of sphenoidal origin, although a CSF diversion is likely to be used later in a number of cases due to persistence of the CSF leak. Other symptoms such as visual failure do not indicate operation.

In skull base cancer, whether primary or metastatic, the TNS route usually permits only partial removal, and symptoms do not always improve. The approach is, however, necessary for a diagnostic biopsy. Also in chordomas and chondromas excision is more frequently incomplete; in such cases a second surgical procedure through a different route is the rule. Intra- and suprasellar "cystic" craniopharyngiomas are a good indication for TNS surgery, while operation may not be definitive in "solid" cases. Side effects (e.g., diabetes insipidus) influence results heavily, with more than 50% incidence in our series. We also had a high rate of recurrence (two cases in seven, 29%) requiring craniotomy. Finally, although meningiomas must never be approached transsphenoidally, we found this route occasionally useful in cases with doubtful preoperative diagnosis or when debulking of the tumor on emergency basis (presence of rapid visual loss) was requested (one case).

Transmaxillar-Transnasal Approach: A Microsurgical Anatomical Model

A. RABADÁN, D. BENEDETTI, H. CONESA, J. DERKRIKORIAN, and M. PEIRANO, Buenos Aires/Argentina

Introduction

Many different procedures have been described for exposing the anterior clival region, but the transoral [3, 20, 22] and transcervical [11, 25, 32] are the most frequently used. Other approaches, such as the transseptal-transsphenoidal [14, 24], transantral [18], transnasal [26], bilateral Le Fort I maxillotomy [4, 19], transbasal [9] and transpalatal [12, 31] are also valid options. Modifications of the Caldwell-Luc approach, resecting bone from the anterior maxillary antrum, its frontal process and the lateral nasal wall, to remove tumors of the nasal cavity were first published by Denker and Khaler [8, 28]. Then Loyo et al. modified this approach for pathology of the cavernous sinus [16]. Casson et al. reported "the mid-facial degloving approach," which was then used by ENT surgeons to manage tumors of the nasal cavity, paranasal sinuses, and nasopharynx [1, 2, 5, 6, 23]. These approaches were used to reach the clivus.

However, it may be necessary to have a wider access to reach midline skull base tumors (extracranial, intracranial, or combined) or intracranial lesions, such as aneurysms arising from the middle third portion of the basilar artery, or small lesions of the anterior surface of the brain stem. For this purpose, we have developed a microsurgical model, transmaxillar-transnasal in both dry skull and cadaver.

Methods

The surgical technique has four stages: the antromaxillary, nasal, sphenoidal, and clival. *The antromaxillary stage* consists of:

- Superior sublabial incision.
- Right unilateral soft tissue dissection up to the infraorbitary foramen.
- Maximal antrotomy, preserving the neurovascular structures coming from the infraorbitary foramen, at the top of the field (Fig. 1).
- Removal of the medial wall of the antrum, the palatine vertical lamina, and the medial aspect of the posterior wall of the antrum, considering the pterygoid process as the posterior limit (Fig. 2).
- Ligature of the sphenopalatine artery.

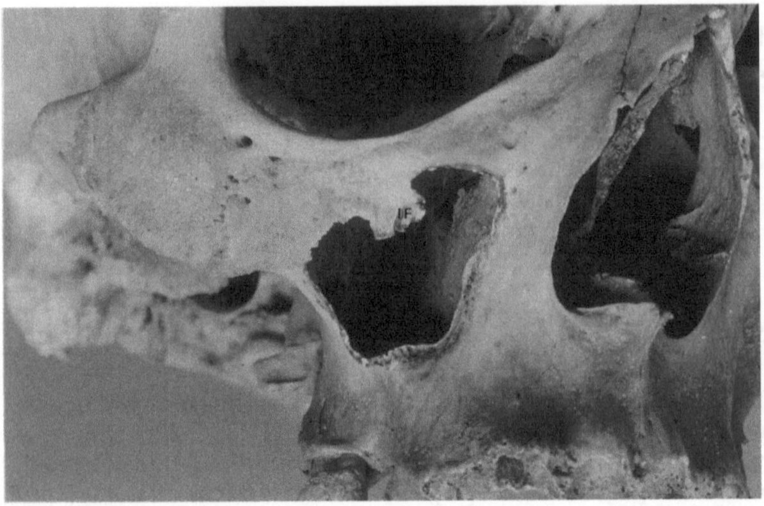

Fig. 1. Dry skull. Maximal antrotomy up to the infraorbitary foramen. *IF*, Infraorbitary foramen

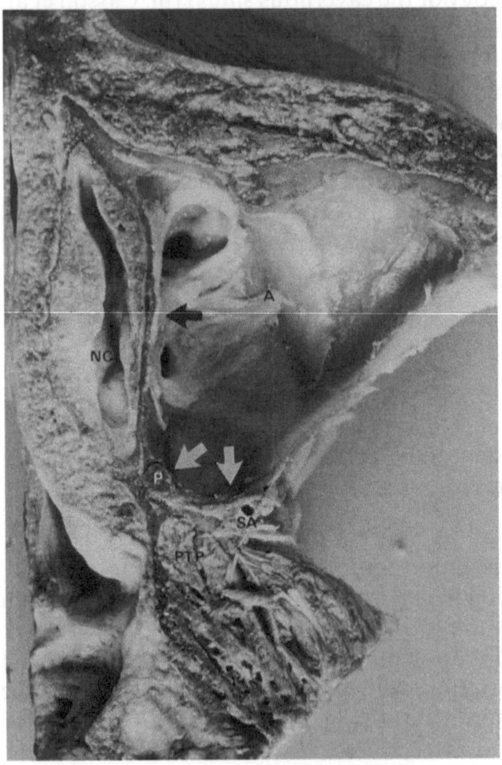

Fig. 2. Anatomical preparation (horizontal section). *Arrows* point to the area to be removed: medial wall of the antrum, the palatine vertical lamina, and the medial aspect of the posterior wall of the antrum. *A*, Antrum; *NC*, nasal cavity; *P*, palatine vertical laminae; *PTP*, pterygoid process; *SA*, sphenopalatine artery

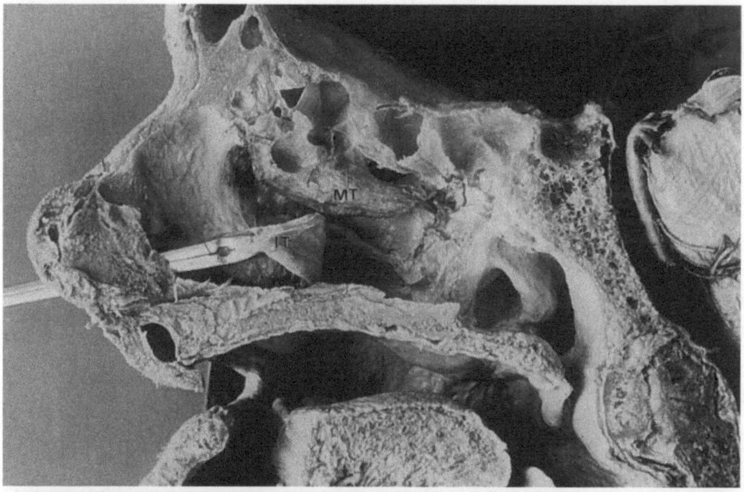

3

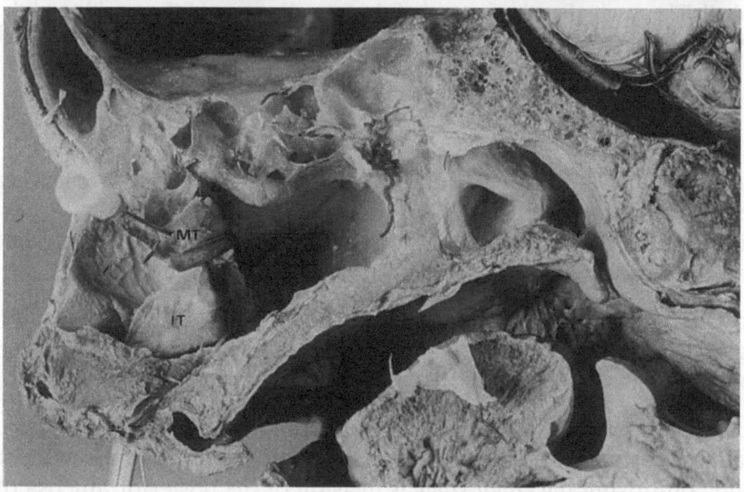

4

Fig. 3. Anatomical preparation (view from the left side). Disinsertion of the posterior end of the inferior turbinate. *IT*, Inferior turbinate; *MT*, middle turbinate

Fig. 4. Anatomical preparation (view from the left side). Disinsertion of the posterior and middle sectors of middle and inferior turbinates, demonstrating the wide communication between the antrum and the nasal cavity. *MT*, Middle turbinate; *IT*, inferior turbinate

The nasal stage includes two parts, the external and the internal. The former consists of:

- Longitudinal incision of the nasal external mucosa at the upper part of the antrum, so as to enter the nasal cavity between the middle and the inferior turbinates.
- Disinsertion of the posterior and middle sectors of both turbinates (Fig. 3).

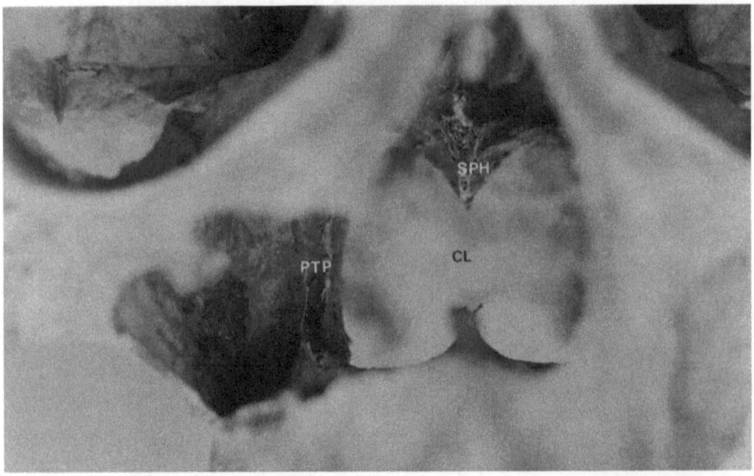

Fig. 5. Dry skull. The result after the removal of the frontal process of the maxilla and the nasal septum, showing the pterygoid process (*PTP*), the rostrum of the sphenoid (*SPH*), and the clivus (*CL*)

- They are then pulled transnasally by their posterior end, remaining attached to the mucosa only by their anterior portion (Fig. 4).
- Osteotomy of the frontal process of the maxilla as inferiorly as possible, but preserving the canine root.

For the internal part, after dissection of the septal mucosa, the cartilaginous septum is disinserted and displaced to the opposite side, and the septal bone is removed (Fig. 5).

The sphenoidal stage consists of opening the sphenoid sinuses; removal of the anterior wall, the rostrum, and the floor; and complete resection of the mucosa (Fig. 6). And comprising *the clival stage* is incision of the posterior mucosa of the cavum, exposing the clivus; drilling of the clivus until the dura is reached.

At the end of the procedure, an osteosynthesis of the frontal process of the maxilla is performed (Fig. 7).

Fig. 6. Dry skull. The final result of the sphenoidal stage, demonstrating the access to the entire clivus (*CL*)

Fig. 7. At the end of the procedure, osteosynthesis of the frontal process of the maxilla is performed

Fig. 8. CT scan indicating the three anatomical coronal sections. *First section*, at the level of the cavum; *second section*, at the level of the anterior clinoid process; *third section*, at the retroclival level

6

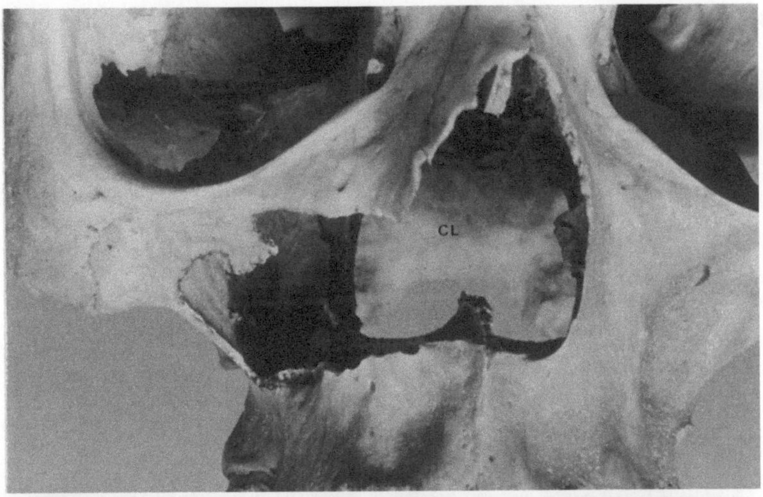

7

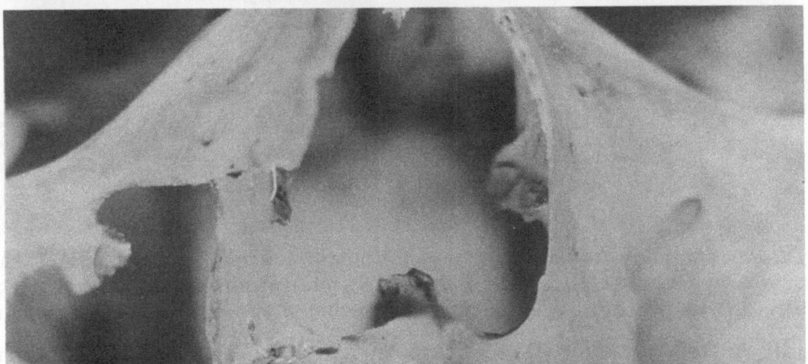

8

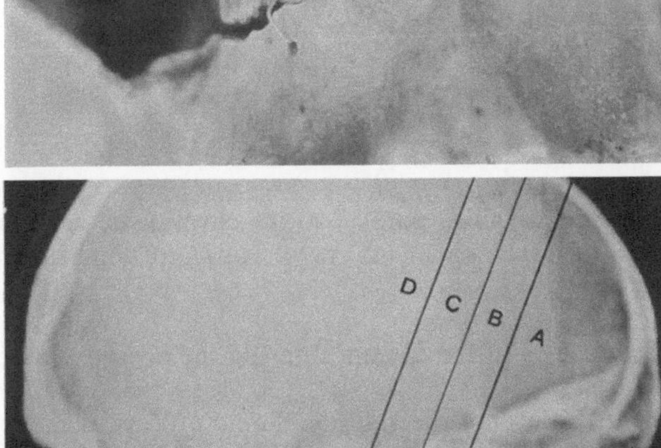

Figs. 6–8

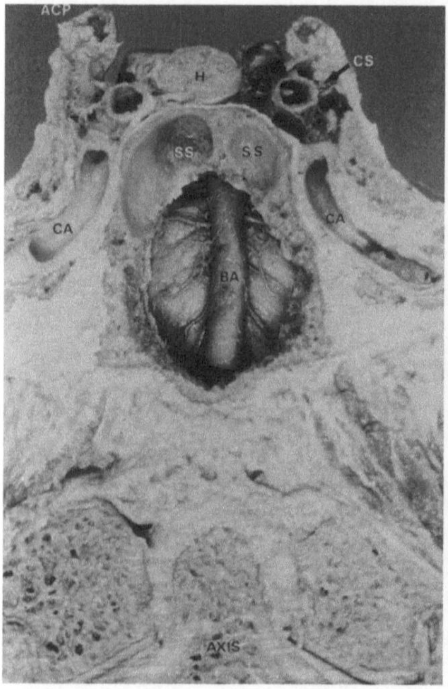

9

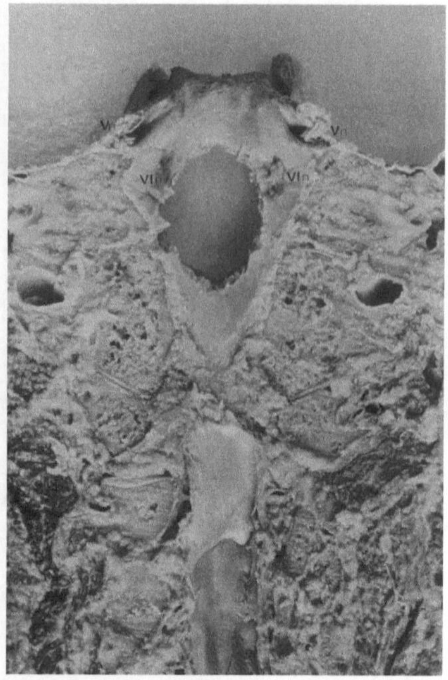
10

Fig. 9. Anatomical section at the level of the anterior clinoid process (*ACP*). It shows the sellar area, hypophysis (*H*), cavernous sinus (*CS*), intrapetrous carotid artery (*CA*), sphenoid sinuses (*SS*), surgical opening at the clivus, brain stem (*BS*), basilar artery (*BA*), atlas and axis

Fig. 10. Anatomical section at a retroclival level, demonstrating the surgical opening at the clivus, the dura, the fifth and sixth nerves

Anatomical Results

Three anatomical coronal sections parallel to the clival plane were made to verify the precision of each stage and preservation of the surrounding structures (Fig. 8):

- First section, at the level of the cavum, showing the operative access and sphenopalatine artery.
- Second section, at the level of the anterior clinoid process, demonstrating the sellar area, the hypophysis, the cavernous sinus, the intrapetrous carotid artery, the sphenoid sinuses, the brain stem, basilar artery, and the atlas and axis (Fig. 9).
- Third section, at the retroclival level, showing the surgical opening at the clivus, the dura, and the fifth and sixth nerves (Fig. 10).

Discussion

The number of surgical techniques described to reach the clivus is notoriously high. From an anatomical point of view, a direct approach, such as the anterior routes, is more attractive than the others. All the anterior approaches have some disadvantages. The transoral and the transcervical ones expose well only the middle and the lower clivus [7, 9, 10, 11, 15, 18]. The transseptal-transsphenoidal route does not provide enough room due to the frontal process of the maxilla and the lateral nasal walls [9, 10, 18]. Something similar happens with the transantral approach, but the clivus is also reached in an oblique way [17, 1]. The transbasal approach, described for tumors involving the ethmoid and sphenoid in the midline of the anterior and middle cranial fossa, is rarely sufficient when the tumor grows posteriorly into the upper half of the clivus, below the sella [9, 10]. Although the bilateral Le Fort I maxillotomy could offer a better access, severe complications have been described [13, 21, 27, 29, 30]. The tránsnasal and transpalatal approaches are not adequate as a single procedure [7, 17]. The crusting or the atrophic rhinitis, a consequence of the extensive resection of mucosa and turbinates, is the principal disadvantage of the enlarged Caldwell-Luc approaches [6, 23].

Therefore, we have tried to develop an approach that provides a wider access to the midline of the skull base. The wider access is achieved mainly by the osteotomy of the frontal process of the maxilla, which transforms the nasal cavity and the antrum in a single cavity preserving the functional anatomy of the nose. The cosmetic effect is achieved by the replacement of the frontal process at the end of the procedure. The technique includes the well-known good cosmetic result of the sublabial approaches and prevents vascular and neural damages in the same way that other anterior approaches do. Finally, this transmaxillar-transnasal technique could be additionally used in combination with other approaches in extensive tumors.

Conclusions

We can summarize the advantages of our approach as follows:

- Good cosmetic results
- Preservation of nasal physiological anatomy
- Wide access
- No vascular or neural impairment
- Possibility of combination with other approaches

References

1. Allen GW, Siegel GJ (1981) The sublabial approach for extensive nasal and sinus resection. Laryngoscope 91:1635–1639
2. Anand VK, Conley JJ (1983) Sublabial surgical approach to the nasal cavity and paranasal sinuses. Laryngoscope 93:1483–1484
3. Apuzzo MLJ, Weiss MH, Heiden JS (1978) Transoral exposure of the atlantoaxial region. Neurosurgery 3:201–207
4. Archer DJ, Young S, Uttley D (1987) Basilar aneurysms: a new transclival approach via maxillotomy. J Neurosurg 67:54–58
5. Casson PR, Bonnano PC, Converse JM (1974) The midfacial degloving procedure. Plast Reconstr Surg 53:102–106
6. Conley J, Price JC (1979) Sublabial approach to the nasal and nasopharyngeal cavities. Am J Surg 138:615–618
7. Decker RE, Malis LI (1970) Surgical approaches to midline lesions at the base of the skull. Mount Sinai J Med 37:84–102
8. Denker A, Khaler O (1926) Handbuch der Hals Nasen Ohren Heilkunde. Springer, Berlin
9. Derome PJ (1985) Surgical management of tumors invading the skull base. Can J Neurol Sci 12:345–347
10. Derome PJ, Visot A, Monteil JP, Maestro JL (1987) Management of chronical chordomas. In: Sekhar LN, Schramm VL Jr. (eds) Tumors of the cranial base: diagnosis and treatment. Futura Mount Kisco New York, pp 607–622
11. Fox JL (1967) Obliteration of midline vertebral artery aneurysm via basilar craniectomy. J Neurosurg 26:406–412
12. Kennedy DW, Papel ID, Holliday M (1986) Transpalatal approach to the skull base. Ear Nose Throat J 65:127–133
13. Lanigan DT, West RA (1984) Management of postoperative hemorrhage following the Le Fort I maxillary osteotomy. J Oral Maxillofac Surg 42:367–375
14. Laws ER Jr (1984) Transsphenoidal surgery for tumors of the clivus. Head Neck Surg 92:100–101
15. Lesoin F, Jomin M, Pellerin P, Pruvo JP, Carini S, Servato R, Rousseaux M (1986) Transclival transcervical approach to the upper cervical spine and clivus. Acta Neurochir 80:100–104
16. Loyo M, Sanchez Marle JF, Del Valle R, Mares M (1987) The Transnasal Approach to the Cavernous Sinus. In: Dolenc VV (ed) The cavernous sinus. Springer, Wien New York, pp 397–403
17. Mann WJ, Gilsbach J, Seeger W, Flöel H (1985) Use of a malar bone graft to augment skull-base access. Arch Otolaryngol 111:30–33
18. Miller RH, Woodson GE, Neely JG, Murphy EC (1982) A surgical approach to chordomas at the base of the skull. Otolaryngol Head Neck Surg 90:251–255
19. Moloney F, Worthington P (1981) The origin of the Le Fort I maxillary osteotomy: Cheever's operation. J Oral Surg 39:731–734
20. Mullan S, Nauton R, Hekmat-Panah J, Vailati G (1966) The use of an anterior approach to ventrally placed tumors in the foramen magnum and vertebral column. J Neurosurg 24:536–543
21. Parner EI, Becker ML (1972) Necrosis of the anterior maxilla following osteotomy. Oral Surg 33:326–330
22. Pasztor E, Vajda J, Piffkó P, Horvath M, Gador I (1984) Transoral surgery for craniocervical space-occupying processes. J Neurosurg 60:276–281
23. Price JC (1986) The midfacial degloving approach to the central skull-base. Ear Nose Throat J 65:174–180
24. Rougerie J, Guiot G, Bouche H, Trigo JC (1967) Les vois d'abord des chordomes du clivus. Neurochirurgie 13:559–570

25. Stevenson GC, Stoney RJ, Perkins RK, Adams JE (1966) A transcervical transclival approach to the ventral surface of the brain stem for removal of a clivus chordoma. J Neurosurg 24:544–551
26. Stewart MJ, Morin JE (1926) Chordoma: a review, with report of a new sacrococcygeal case. J Pathol Bacteriol 29:41–60
27. Stringer DE, Dolwich MF, Steed DL (1979) Subcutaneous emphysema after Le Fort I osteotomy: report of two cases. J Oral Surg 37:115–116
28. Tapia Acuña R (1981) Juvenile nasopharyngeal fibroma. Ann Otol 90:420–422
29. Watts PG (1984) Unilateral abducens nerve palsy: a rare complication following a Le Fort I maxillary osteotomy. Br J Oral Maxillofac Surg 22:212–215
30. Westwood RM, Tilson HB (1975) Complications associated with maxillary osteotomies. J Oral Surg 33:104–115
31. Wilson CP (1951) The approach to the nasopharynx. Proc R Soc Med 44:353–358
32. Wissinger JP, Danoff D, Wisiol ES, French LA (1967) Repair of an aneurysm of the basilar artery by a transclival approach. J Neurosurg 26:417–419

Transsylvian Pretemporal Approach to the Infundibular Region

J. C. Chiaradio, R. Schreck, V. Chavez, and M. C. Lozano,
Buenos Aires/Argentina

Most craniopharyngiomas are seated in the suprasellar region and reach the third ventricle behind the optic chiasm. The retrochiasmatic, or postero-superior or superior portion of the tumor cannot be resected under direct vision when the subfrontal or pterional approach is used. New approaches have been described after the advent and improvement of microsurgical techniques. We became particularly interested in the transsylvian approach to the infundibular region and mammilar bodies which permits direct visualization of these structures; this we discuss in the present paper. It is based on our experience with a series of eight adult craniopharyngiomas and two tuberculum sellae meningiomas, one of them giant.

Surgical Technique

The operation is usually performed on the right side, except when the tumor has a net extension to the left. The patient is placed on his left side with the head falling slightly from the horizontal plane. A lumbar puncture is done, and a continuous drain inserted. The skin incision begins at the zygomatic arch near the tragus and is directed upward and forward in the form of curve. Craniotomy is done with a 38- or 44-mm Scovile trephine, so that the sylvian valley and the temporal tip can be exposed simultaneously. The duramater is opened with a Y incision, and the carotid region is approached following the lesser wing of the sphenoid. Once this has been achieved, the dissection of the sylvian valley is begun, from the carotid artery outward, paying special attention to freeing the temporal lobe from the posteroinferior border of the sylvian artery. Sometimes, the ample opening of the sylvian valley is impeded by a perforating vessel, the anterior temporal artery.

The freeing of the temporal tip from its venous attachment is performed afterwards. The veins can be handled by subpial dissection, but in about 30% of cases they must be coagulated and sectioned. In this way the temporal tip is freed, and then with the autostatic spatulas already in position the arachnoidal attachments between the anteromedial border of the temporal lobe and the free edge of the tentorium and oculomotor nerve are dissected. Continuous aspiration of cerebrospinal fluid helps retraction with only slight pressure on the brain.

The carotid artery dissection is done by both edges till bifurcation. In this way we can visualize the external aspect of the supraclinoid portion of the carotid artery and the posterior communicating artery with its perforating arteries which, wrapped in a thin arachnoidal sheath, lie obliquely to the optic chiasm and optic tract. The perforators can be freed in such a manner that dissecting maneuvers on the tumor can be performed through them.

The exposed area is thus divided into four triangles: (a) retrocarotid-supraposterior communicating artery space, (b) the retroposterior communicating space, (c) the interoptic-carotid artery space, and (d) the space limited by A1 segment, optic chiasm, and carotid artery bifurcation. Any or all of these can be used depending on the displacement which the tumor has produced on the different structures, but the upward and forward displacement of the internal carotid artery allows the dissection in the retrocarotid-supraposterior communicating artery space, with high magnification, through the perforating vessels. This almost surely allows the visualization of the inferior aspect of the optic chiasm and the mammilar bodies region.

Discussion

The retrochiasmatic-mammilar region is situated in a higher level than the optic nerves, chiasm, and optic tract, and even higher when they have been dislodged by a slowly growing tumor that does not always elongate the chiasm proportionately. The tumors located in this region are thus difficult to approach in their superior aspect, which contacts the tuber cinereum and the mammilary bodies. Classically, this portion of the tumor has been left in place, giving rise to relapses, or dissected by blind traction maneuvers which frequently cause the rupture of hypothalamic perforators and serious damage in the area. This is what happens with the subfrontal and classical pterional approaches.

The transsylvian approach by mobilization of the temporal lobe gains 2.5–3.0 cm of space from the tip, basal to the chiasm and by this means permits an ascending way to the lesion, since the concavity of the anterior fossa allows only a descending approach angle when this route is used to operate in this area and hence is only apt to resect small tumors, those which can be easily aspirated or those which have smooth attachments with the infundibulum and can be tractioned.

The approaches through the lamina terminalis have been carefully avoided by us, as there are cerebral areas whose function is not well understood yet and, when harmed, can give rise to minor symptoms. We believe that surgery must tend to anatomic and functional preservation, this being altered only when there are no other options.

Hakuba (1986), although without specifically mentioning it in his paper, describes an ascending approach (caudocephalic) which allows him to divide the hypothalamus from the tumor capsule, under direct vision. The approach we propose allows the same thing, that is, the direct surgical visualization by an

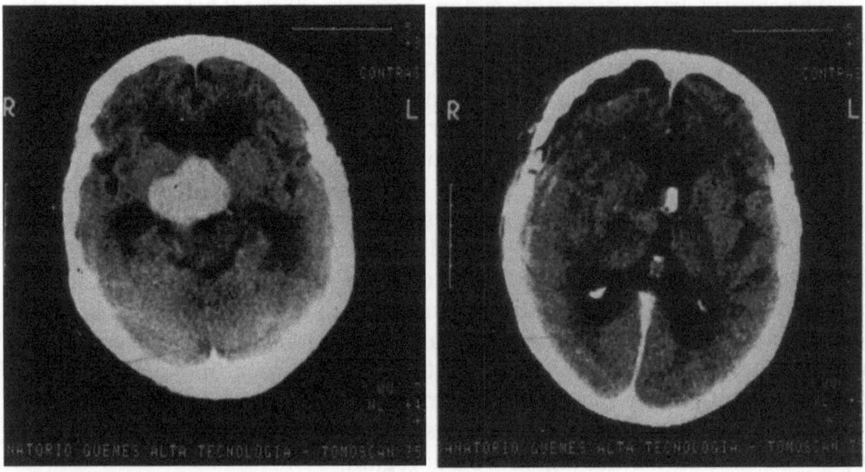

a b

Fig. 1 a, b. This approach was extremely useful in retrochiasmatic tumor even in those located deeply in the infundibular region

ascending route to the lesion, without harming the oculomotor nerves, conservation of the temporal lobe, and by a shorter, direct, and safer way than the transpetrous-transtentorial route. The approach proposed here permits the dissection of tumors seated high in the region (Fig. 1 a, b), even those which have a high percentage of their volume located intraventricularly. The dissection of the tumor through perforating vessels previously freed from their arachnoidal attachments has been found to be a safe procedure on condition that high magnification and moderate aspiration is used.

Conclusion

Three different types of approaches to the mammilo-infundibular region can be considered: (a) descendent, with access to the region downwards: subfrontal, pterional; (b) interhemispheric: translamina terminalis; and (c) ascendent, with access to the region upwards: transsylvian pretemporal, transpetrous-transtentorial. We believe that the latter allows the resection of retrochiasmatic tumors in a more anatomic fashion and with less vascular damage to the delicate surrounding structures.

References

Hakuba A (1986) The transpetrosal-transtentorial approach and its application in the therapy of retrochiasmatic craniopharyngiomas. In: Samii M (ed) Surgery in and around the brain stem and the third ventricle. Springer, Berlin Heidelberg New York Tokyo, pp 396–404

Symon L, Sprich W (1985) Radical excision of craniopharyngioma. Results in 20 patients. J Neurosurg 62:174–181

Meningioma and Parasellar Pituitary Adenoma Affecting the Cavernous Sinus: Radical Tumor Extirpation?

K. von WILD, Münster/FRG and N. ESKINJA, Opatija/Yugoslavia

Introduction

Thanks to the introduction of modern diagnostic procedures (cranial computed tomography, CCT; magnetic resonance imaging, MRI; digital subtraction angiography, DSA) careful anesthesiology, and the development of microsurgical techniques with growing experience in the management of mass lesions within the tentorial notch, occupying and vascular lesions of the cavernous sinus have been attacked more frequently in recent years by the direct approach as pioneered by Parkinson 1965. His outstanding work has influenced some excellent anatomical studies of the topography of the cavernous sinus. We were therefore encouraged to treat 28 tumors invading the cavernous sinus more aggressively by direct microsurgical approach from different sides, some of them via combined approaches in two stages and in cases of tumor recurrence (Table 1). As in the past, the possibility of bleeding from the venous structures and the potential risk of injury to the intracavernous carotid artery or cranial nerves III–VI kept the neurosurgeon from direct approach to intracavernous tumors. We discuss our results in the light of microsurgical possibilities and limitations in cases of meningiomas and parasellar adenomas.

Anatomy

Although several fundamental studies of the anatomy of the parasellar space, known since Winslow (1732) as the cavernous sinus, have been performed during the past two decades [19, 21, 23, 24, 32, 35, 46, 47, 49], a definite topographical description of the region still seems necessary [47]. As for the operative treatment of lesions involving the cavernous sinus, it is generally accepted that the cavernous sinus is not a trabeculated venous canal but a network of extradural veins, enclosed by leaves of dura and periosteum, located on either side of the sella turcica with intrasellar venous connections across the midline area [21, 49]. The periosteal layer forms the floor and most of the medial wall. The lateral wall is formed of two layers. A smooth superficial dural layer is attached to the deep one, which is formed by the dural sleeves or sheaths accompanying the corresponding cranial nerves III, IV, and V1 and V2. The deep layer is thin or even missing in the area of Parkinson's triangle [32, 33,

Table 1. Tumors invading the cavernous sinus

Tumor type	*n*	Extent of tumor	Cavernous sinus operation
Meningiomas	1	Intracavernous, middle fossa	Pterional-subtemporal lateral (subtotal)
	2	Petrous apex, tentorial notch	Two-stage suboccipital-subtemporal, dorsal (total)
	1	Petrous apex, tentorial edge	Subtemporal, lateral (total)
Recurrent meningiomas	1	Intracavernous, intrasellar, tentorial edge	Dolenc approach frontosubtemporal, anterior lateral patent ICA (subtotal)
	1	Medial sphenoid wing, middle fossa	Pterional, anterior-lateral, patent ICA (subtotal)
	1	Petrous apex, tentorial edge	Pterional-subtemporal dorsal lateral (total)
	2	Diaphragma sellae, tentorial edge, petroclival area	Pterional-subtemporal dorsal-medial (one total), dorsal-lateral (one subtotal)
Pituitary adenomas	5	Intra-, para-, and suprasellar	Bilateral subfrontal, medial (one case total) Subfrontal, opticocarotid, medial (two cases total) Pterional, lateral (two cases subtotal)
	5	Infra-, para-, supra-, and infrasellar	Transethmoidal-transsphenoidal, inferior-medial, occlusion of ICA (two subtotal), transnasal-transsphenoidal inferior-medial, (three total)
	2	Intraparasellar	Transsphenoidal (one total)
Recurrent adenomas	1	Intra-, supra-, para-, and infrasellar	Two-stage transethmoidal-transsphenoidal and pterional with inferior-medial superior opticocarotid, and lateral approach (total)
	1	Intra-, para-, and suprasellar	Subfrontal, opticocarotid medial (subtotal),
	2		Pterional, lateral (one total, one subtotal)
Fibrous dysplasia	1	Sphenoid, temporal, petrous bones, and zygomatic	Orbitozygomatico-subtemporal, anterior (subtotal)
Malignant meningioma	1	Medial sphenoid wing, infratemporal, and middle fossae	Two-stage embolization and orbitopterional, anterior and lateral (subtotal)
Carcinoma metastasis	1	Sphenoid wing, middle and infratemporal fossae	Orbitozygomatico-subtemporal, inferior lateral anterior (subtotal)

34]. The cavernous sinus space contains the internal carotid artery (ICA) with its important branches, described elsewhere in detail including anatomical variants [2, 12, 21, 23], sympathic nerves, and the abducens nerve, surrounded by an interrupted arachnoid network and venous structures. The intracavernous course and the indentations of the ICA into the sphenoid, where the bone occasionally may be absorbed, vary in shape depending on the length and breadth of the sphenoid sinus and the age of the patients [12, 15, 16, 24, 37, 38].

Patients, Methods, and Results

Between March 1983 and April 1988, 25 patients underwent direct surgical approach to the cavernous sinus because of 28 tumors. There were 9 meningiomas, 1 malignant meningioma, 16 pituitary adenomas, 1 carcinoma metastasis, and 1 fibrous dysplasia. There was only one patient (Figs. 1, 2) in whom the tumor, a typical fibroblastic meningioma, has primarily originated inside the cavernous sinus. In general, the tumor invasion into the cavernous sinus depended on the site of its origin at the base, e.g., sellar and parasellar region, sphenoid ridge, petrous apex, or tentorial edge. The operative approach was chosen to avoid brain damage and injury to the neurovascular structures.

As the operative approaches available to cavernous sinus may be divided into primarily intradural and extradural approaches, we used the intradural one in all but one, in whom Prof V. Dolenc of Ljubljana and the senior author

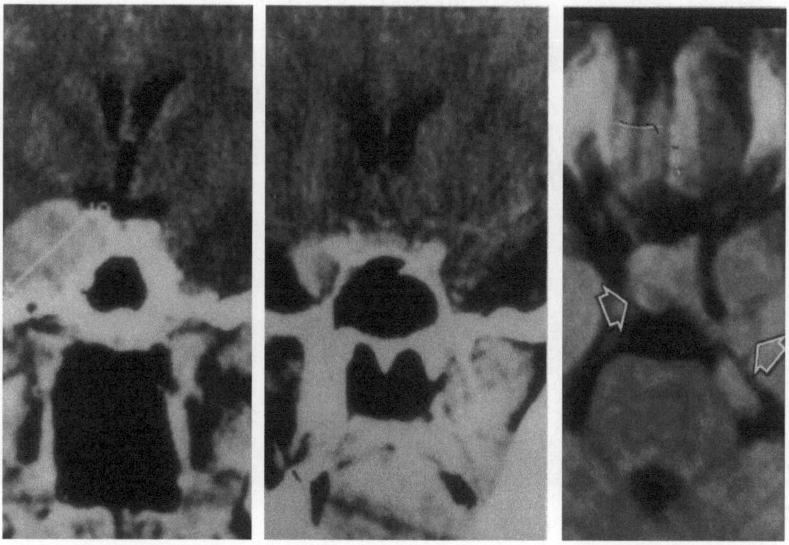

Fig. 1. Meningioma of cavernous sinus before and after right subtemporal subtotal removal 1985 (CCT) and tumor progression into the sella after 3 years despite postoperative radiosurgery (MRI)

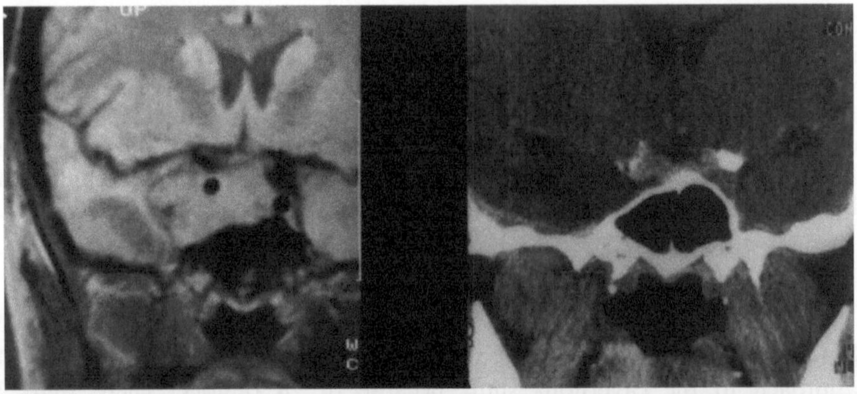

2

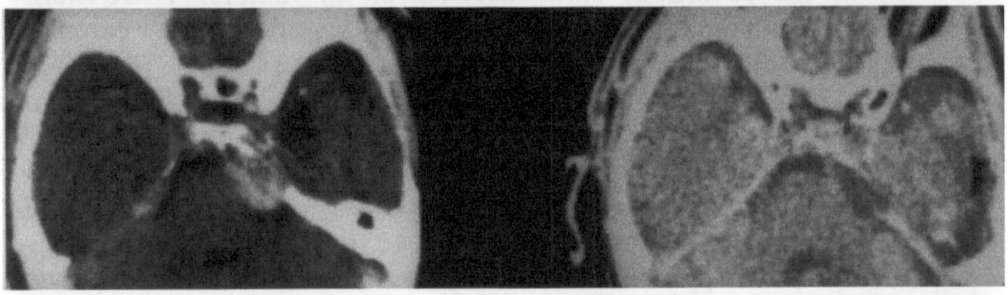

3

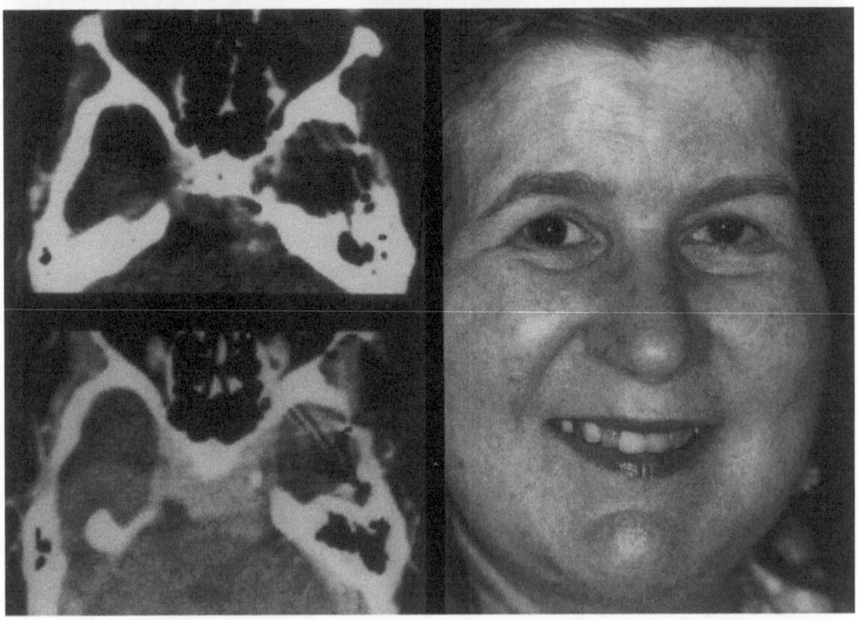

4

Figs. 2–4

performed an extradural-intradural direct approach to the cavernous sinus (case no. 1). In four out of six cases the subtemporal approach to the petrous apex was followed by partial resection of the tentorium with preservation of cranial nerve IV. Two recurrent parasellar meningiomas had invaded the cavernous sinus along the sheaths of the third and fourth nerves (Figs. 3, 4).

Pituitary adenomas invading the cavernous sinus were approached via a transnasal, transsphenoidal, transsellar route together with the ENT surgeon in one or two stages in respect to their size and invasion of the skull base. In case of extensive supra- and parasellar tumor growth the adenoma was operated on via an intracranial subfrontal or pterional approach or using a combined transnasal intradural approach (case no. 6).

The opening of cavernous sinus followed microsurgical anatomy:

Inferior medial, seven adenomas, one ICA injury with secondary occlusion due to a false aneurysm after 10 days because of previous chronic inflammation and scar tissue of the parasellar paravascular structures
Anterior lateral after resection of the orbital roof and anterior clinoid process with control of extradural and subarachnoid ICA, two cases, one ICA injury (ICA patent after temporary clipping)
Anterior medial via the medial triangle of Hakuba (1982), subchiasmatic transsellar or opticocarotid following along adenoma tissue invasion
Lateral via Parkinson's triangular space (1965)
Lateral inferior when nerves V2 and V3 were displaced in case of metastatic carcinoma, malignant meningioma, fibrous dysplasia
Dorsal superior via oculomotor trigone (Ono et al. 1984) in case of giant adenoma and meningioma of the diaphragma

Total removal was accomplished in 13 out of 26 tumors invading the cavernous sinus. There were four meningiomas and nine pituitary adenomas, as demonstrated in postoperative CT and/or MRI follow-up examinations as well as by hormone measurements (Figs. 2, 3, 6).

Only subtotal removal could be achieved in five out of ten meningiomas because of tumor tissue surrounding the intracavernous ICA, with one ICA injury and temporay clipping during subtemporal approach, and to preserve nerve formations of III–VI (Figs. 1, 4, 5).

Fig. 2. Coronar CCT (*right*) after contrast medium application 3 weeks postoperatively, demonstrating total removal of progressive meningioma of cavernous sinus (MRI, *left*) via Dolenc approach March 1988

Fig. 3. Total removal of a recurrent meningioma of the petrous apex invading right cavernous sinus with preservation of the cranial nerves as demonstrated in axial CCT sections after contrast medium application

Fig. 4. Axial CCT sections after contrast medium application demonstrating a meningioma of the diaphragma sellae invading the right cavernous sinus. Total removal followed by complete recovery of both III nerves

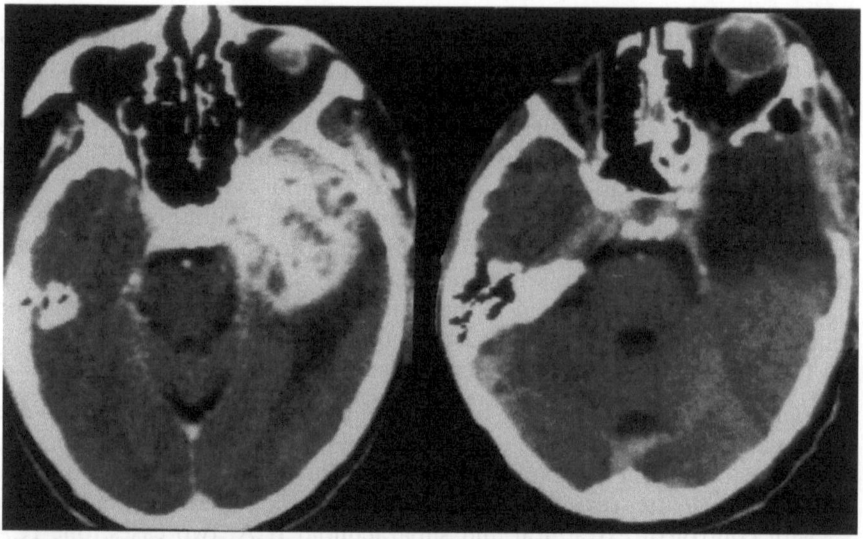

5

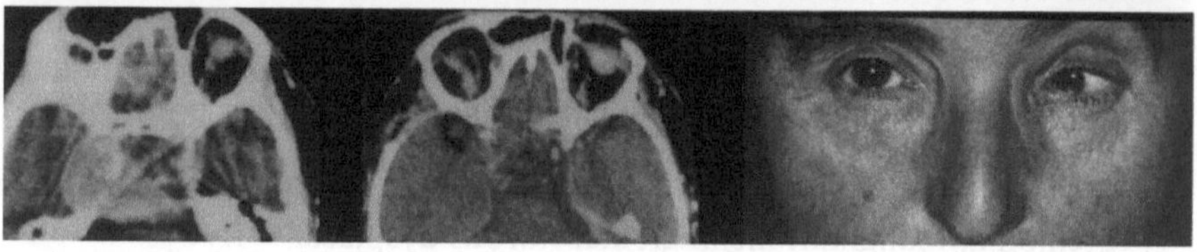

6

Fig. 5. Axial CCT sections with contrast medium application demonstrating a malignant right temporal meningioma invading the cavernous sinus before and after radical removal

Fig. 6. Giant pituitary adenoma invading left cavernous sinus before and after total removal with 3-year follow-up

Pituitary adenoma resection was subtotal in 7 out of 16 cases. There was one prolactinoma with invasion of the skull base, in which intraoperative injury of ICA occurred. In two hormone-secreting adenomas some tissue had to be left inside the cavernous sinus because of transsphenoidal limited space and view, although growth hormone (GH) levels were brought down to below 10 ng/ml in one GH-secreting tumor and in the prolactinoma serum prolactin levels were in the normal range under dopamin agonist medication. In four giant adenomas resections were incomplete because of limited space and view within the cavernous sinus via lateral subtemporal and anterior subfrontal approaches.

Cranial nerve disturbances due to menigiomas were present as follows: isolated sixth nerve palsy in the intracavernous meningioma (Fig. 1), neuralgic facial pain and numbness V1 in one tumor; V2, V3 in two petrous apex menigiomas; and V1–V3 in the malignant meningioma, the latter combined with a complete ophthalmoplegia.

Adenomas extension into the tentorial notch caused nerve III paresis in six patients, two of them were combined with additional involvement of nerves IV and VI. Optic nerve compression was evident in all suprasellar adenomas and in two medial sphenoid wings and in one diaphragma sellae meningioma. There was a complete visual loss preoperatively in one recurrent meningioma of the sphenoid wing for years.

The following cranial nerves lesions have been demonstrated by postoperative follow-up findings. In complete removal of meningioma: V1, same as before in one; V2, V3 with additional V1 numbness in one and IV palsy in another and incomplete III paresis with permanent IV in one recurrent tumor. In subtotal removal of malignant meningioma: additional V3 palsy and partial V2 after lateral subtemporal approach in combination with preexisting total ophthalmoplegia, although III, IV, V, and V1, V2 have been preserved showing some recovery over months. In recurrent diaphragma sellae meningioma: there was an incomplete III on both sides with good recovery and with a permanent IV left; however, she still complains of diplopia (Fig. 4). In four giant adenomas, partial III paresis remained unchanged in three patients and recovered in one. In those cases with intraoperative partial III, IV, V1, V2, and VI injuries, III and IV showed partial recovery after total removal; however, VI remained permanent in one (Fig. 6).

There was no postoperative death in 26 procedures. One patient died of myocardial infarction and pulmonary complications 5 months after transsphenoidal operation with secondary ligation of the injured ICA in the neck, although angiography had shown patent intracranial arteries in addition to multiple arteriosclerotic plaques.

Postoperative extracranial irradiation has not yet been a routine procedure in total or subtotal removal of meningiomas or pituitary adenomas except in the malignant tumor. Radiosurgery has been carried out by Prof. L. Steiner of Stockholm in one rare case of primarily intracavernous meningioma; however, this tumor showed further tumor progression 1 year after subtotal resection (case no. 1).

Case Reports

Case 1: Intracavernous Meningioma (Figs. 1, 2). A 28-year-old woman was admitted with a 5-year history of headaches and intermittent diplopia that had increased periodically during pregnancy and menstruation. Partial paresis of IV with numbness of V1, V2 was observed. CCT and angiography demonstrated a typical meningioma of the right cavernous sinus extending into the middle fossa and into the tentorial notch. As the potential risk of III, IV and ICA injury during tumor resection within the cavernous sinus had to be avoided by all means, on 25 March 1985 the meningioma was subtotally removed by subtemporal approach with opening of the cavernous sinus in the anterior-lateral part with injury of the ophthalmic and partial V2 and V3 as they were surrounded by the tumor (Fig. 1). In May 1986, radiosurgery was performed by Prof. L. Steiner. However, 1-year MRI demonstrated tumor progression along the floor of the sella turcica across the midline with intra- and suprasellar tumor extension causing visual disturbances and amenorrhea due to compression of the optic chiasma and pituitary gland (Fig. 2). On 18 March 1988 Prof. V. Dolenc and the senior author approached the tumor extra-intradurally with total removal of the intracavernous and intrasellar tumor mass (Fig. 2). Intracavernous ICA was injured but kept patent after temporary clipping; nerves III and IV have been decompressed and showed partial recovery within 3 months. The postoperative course was uneventful with discharge of the patient after 2 weeks.

Case 2: Petrous Clival Meningioma (Fig. 3). A 57-year-old man was admitted because of neuralgic facial pain, V1, V2, and grand mal epilepsy. CCT revealed a tumor involving the petroclival area and cavernous sinus. Total removal of the meningioma was performed in two stages: on 22 March 1983, via a retrosigmoid approach, resection of the clival part within the posterior fossa, and decompression of nerve V, and on 3 May 1983, via a subtemporal approach, with partial resection of the tentorium with total removal of the residual tumor at the petrous apex and tentorial edge. However, there was a tumor recurrence after 2 years with invasion of the cavernous sinus. On 24 April 1985, total removal of the tumor using a pterional-subtemporal approach to the clival area and cavernous sinus with opening of Parkinson's triangle was performed. Cranial nerves III and IV were dissected free, but IV could not be preserved as it was embedded within the meningioma. Two years later, we saw him with permanent IV partial III paresis without recurrency (Fig. 3).

Case 3: Meningioma of the Diaphragma (Fig. 4). A 50-year-old woman admitted because of headache and diplopia, with III, IV, and VI paresis left 5.5 years after partial removal of a diaphragma sellae–meningioma via a right subtemporal approach in another department. CCT and angiography demonstrated the tumor mass lesion of the sellar and parasellar region with extension into both cavernous sinus and along the clivus. She was operated on in two stages because of expected periodical ophthalmoplegia following tumor removal. On 21 September 1984, tumor resection was carried out via left pterional approach with opening of cavernous sinus in the oculomotor triangle of Ono with preservation of III, VI, and pituitary function but injury of IV, and on 26 May 1986 total removal of the right supra- and parasellar tumor and the portion within the cavernous sinus following the sheaths of III and IV with preservation of all cranial nerves and subsequently full recovery of III. However, she suffers from diplopia and anterior pituitary insufficiency and needs cortisol medication.

Case 4: Temporal Malignant Meningioma (Fig. 5). A 53-year-old man had at admission suffered complete disablement for 1 year, caused by left hemiplegia and right ophthalmoplegia and severe headache due to malignant angioblastic meningioma of the temporal skull base with invasion of the cavernous sinus. He was transferred from our X-ray department where he had been sent for radiotherapy some months after tumor biopsy in another clinic. On 3 March and 10 March 1988, we performed a two-stage operative approach and removed the tumor nearly completely from the temporal and subtemporal region after embolization of the tumor vessels by Dr. Fischedick. Removal of the tumor within the cavernous sinus by orbitopterional frontolateral opening with preservation of cranial nerves III, IV, VI, and partial V2 and V3. The postoperative course was uneventful. Partial recovery of hemiplegia and cranial nerves during postoperative radiotherapy was noted.

Case 5: Invasive Pituitary Adenoma (Fig. 6). A 43-year-old woman on CCT and angiography demonstrated a giant adenoma invading the base, and left supra- and parasellar region with cavernous sinus, causing right hemiparesis, paresis of II, III, IV, V1 and VI nerves with impaired visual acuity and visual field defects 2 years after left frontotemporal approach with partial removal of the suprasellar tumor and radiotherapy in another department. Postoperative CSF leakage and meningitis with ENT-surgical occlusion of the fistula. The giant adenoma was completely resected in two stages via a combined extradural ENT-neurosurgical transethmoidal-transsphenoidal approach on 9 July 1984 together with Prof. Dr. Draf of Fulda, and on 30 January 1985 via an intradural pterional approach. The cavernous sinus was entered, medial, transsellar opticocarotid, and dorsalsuperior along tumor extension. Scar tissue around III–VI caused partial III, IV, and V2, V3 (Fig. 1) palsy and permanent V1 and VI deficits. The postoperative course and follow-up was uneventful with normal capacity. She refused operative correction of the musculus rectus lateralis and reconstruction of V1 with graft as described by Samii (1972).

Discussion

Fundamental studies of the anatomy of the parasellar space known as the cavernous sinus, have been performed [19, 21, 23, 24, 32, 33, 35, 38, 46, 47, 49]. The anatomy and topography of the cavernous sinus and its neurovascular structures including anatomical variants are described elsewhere in detail [12, 15, 19, 23, 24, 35].

Tumors originating outside the cavernous sinus with secondary invasion of the cavernous sinus most commonly arise from parasellar structures and the hypophyseal region. Tumors originating primarily inside the cavernous sinus are rare. Because of the arachnoid network inside the cavernous sinus and the intracavernous intrasellar venous connection [21, 49], tumors may cross from one side to the other. They may follow the dural sheaths along cranial nerves III to VI into the cavernous sinus [24, 48]. Therefore, total removal of tumors invading the cavernous sinus could hardly be accomplished. Radical excision within this space has its limitations. When ICA and nerve III are to be preserved, the neurosurgeon must focus on tumor mass reduction without damaging the functionally important neurovascular structures in this area. However, based on recent anatomical studies that have given a better insight into the cavernous sinus topography and microsurgical experience with different approaches, a number of tumors can now be completely excised without additional severe injury (Figs. 1–6) [1, 3, 6, 10, 17, 18, 22, 25, 26, 28, 30, 39, 40]. This was possible in as many as two-thirds of 63 cases in Dolenc's series (1987) via a direct extra-intradural, anterior lateral approach. Sekhar and Moller (1986) reported on total removal in six out of seven tumors involving cavernous sinus by a lateral superior or inferior approach, with temporary clipping and suture of ICA in one patient. Lesoin et al. [27, 29] preferred to approach intracavernous tumors via the pterional route together with an orbitozygomaticomolar flap and achieved extensive but incomplete resection in 18 out of 21 cases. Kawase et al. (1987) advocate the transzygomatic approach for orbit-cranial boundary tumors with five total removals in seven cases. Moreover, these authors used a transpetrosal approach in 12 patients and removed two clival meningiomas invading the cavernous sinus totally via this route [22, 42, 43]. Cioffi et al. (1987) reported on 11 subtotal and one total removal of the cavernous sinus meningiomas with postoperative radiotherapy for partially resected tumors. In large lateral and posterior cranial base neoplasms invading the cavernous sinus, Sekhar et al. [41] described a subtemporal-preauricular infratemporal fossa approach that may be combined with an extradural transethmoidal or an intradural frontotemporal or retromastoid approach. Guiot and Derome [16], as Jefferson [20] and Sekhar et al. [41], thought a certain parasellar tumor invasion to be inaccessible via an inferior medial approach [18] because of inadequate exposure, poor proximal and distal control of ICA and the danger of CSF leakage. From our experience we can say that in soft adenomas tumor tissue can easily be sucked out of the cavernous sinus, whereas scar tissue secondary to inflammation of parasellar sinus contains a high risk of neurovascular lesions especially when the medial wall of the cavernous sinus has been destroyed. This happened in one of our cases with bony erosion of the lateral wall of the sphenoid sinus. Moreover, in some cases the bony wall may be missing, with ICA bulging into the sellar space [24, 37].

Intracavernous meningiomas are usually very hard to remove due to their tissue type and because they are mixed with the trabeculated intracavernous network so that an ultrasonic surgical aspirator may be helpful, but dangerous, next to the carotid wall. The abducens nerve, running freely in the cavernous sinus, was found to be most vulnerable, whereas the ophthalmic, more often than IV or V branches may be sacrificed when they are surrounded by scar and tumor tissue or in order to get better access for tumor removal. Intraoperative extracranial monitoring of cranial nerves II–VIII has proven helpful in locating during microdissection [40, 41], as has been described for the ICA with the aid of Doppler sonography [14]. We have started neuromonitoring, but so far our experiences are still too limited as to draw our conclusions regarding preservation of cranial nerves function.

Table 2. Operative management of tumor involving the cavernous sinus

Inferior medial approach	Transsphenoidal route to the sella; limited space, potential danger of ICA injury and severe hemorrhage
Medial anterior approach "Medial triangle" (Hakuba 1982)	Subfrontal, frontopterional; opticocarotid; excellent overview after resection of anterior clinoid process and tuberculum sellae
Anterior lateral approach "Epi-subdural direct" (Dolenc 1983)	Frontotemporal, pterional, resection of the orbital roof and anterior clinoid process; excellent control of intracavernous ICA
"Transzygomatic" (Kawase 1986)	Orbit-cranial boundary tumors, wide access
"Orbitozygomaticomolar" (Pellerin 1984)	Frontotemporosphenoid tumors, e.g., meningiomas, less traumatic, wider access to the middle fossa
Lateral approach "Triangular space" (Parkinson 1965)	Temporal, subtemporal, pterional; preservation of ICA and cranial nerves III–VI, limited anterior space
Inferior lateral approach "Subtemporal-preauricular infratemporal fossa approach" (Sekhar 1986)	Possible only when mandibular branch of V is divided because of tumor invasion or intraoperative control of ICA from the petrous apex
Dorsal superior approach "Oculomotor trigone" (Ono et al. 1984)	Pterional, subtemporal, subfrontal, ipsi- and contralateral; limited space and overview

In general, nerve III recovers after decompression within 3–6 months postoperatively. Nerves IV and V1 should be reconstructed with sural grafts if they are sacrificed [39 a]. Inside the cavernous sinus, venous bleeding was rare because of tumor infiltration and could be controlled easily by packing with surgicel or gelfoam. Tumors surrounding the intracavernous ICA erode the arterial wall, so that the slightest operative trauma can bring about rupture of the artery. We therefore advocate leaving a part of tumor around the wall in these cases. Severe bleeding from ICA injury (case no. 1) can be prevented by previous control of the petrous ICA for a length of about 2 cm and distally at the supraclinoid subarachnoidal segment, distal to the ophthalmic artery, with temporary clipping [7, 8, 10]. Previous inflammation of paranasal sinus and parasellar dural structures is a dangerous handicap. Moreover, in preserving important neurovascular structures that are involved by the tumor, the question of radical tumor excision becomes a question of life quality.

Postoperative radiation therapy for incompletely resected meningiomas and invasive pituitary adenomas is recommended today [4, 5, 11, 13, 16, 20, 25, 36, 42, 44, 45, 50], but the number of cases and the follow-up period in meningiomas is still too limited to draw final conclusions. The most important limiting factor of radiation therapy in this region remains the vulnerability of the surrounding brain, optic pathways, and pituitary gland, even in radiosurgery [45].

Up to now intracranial intracavernous tumor surgery (Table 2) is limited in preservation of cranial nerve functions and ICA and because of its duration it should be restricted to tumors in younger patients in optimal general condition.

References

1. Beck DW, Menezes AH (1987) Lesions in Meckel's cave: variable presentation and pathology. J Neurosurg 67:684–689
2. Brassier G, Lasjaunias P, Guegan Y, Pecker J (1987) Microsurgical anatomy of collateral branches of the intracavernous internal carotid artery. In: Dolenc VV (ed) The cavernous sinus. Springer, Wien New York, pp 81–103
3. Butti G, Gaetani P, Giordana MT, Paoletti P (1983) Meningiomas of Meckel's cave. Surg Neurol 20:305–309
4. Carella RJ, Ransohoff J, Newall J (1982) Role of radiation therapy in the management of meningioma. Neurosurgery 10:332–339
5. Cioffi FA, Bernini FP, Punzo A, Natale M, Muras I (1987) Cavernous sinus meningiomas. Neurochirurgia 30:40–47
6. Dolenc VV (1979) Microsurgical removal of large sphenoidal bone menigiomas. Acta Neurochir [Suppl] 28:393–401
7. Dolenc VV (1983) Direct microsurgical repair of intracavernous vascular lesions. J Neurosurg 58:824–831
8. Dolenc VV (1985) A combined epi- and subdural direct approach to carotidophthalmic artery aneurysms. J Neurosurg 62:667–672
9. Dolenc VV, Skrap M, Sustersic J, Skrbec M, Morina A (1987) A transcavernous-transsellar approach to the basilar tip aneurysms. Br J Neurosurg 1:251–259
10. Dolenc VV, Kregar T, Ferluga M, Fettich M, Morina A (1987) Treatment of tumors invading the cavernous sinus. In: Dolenc VV (ed) The cavernous sinus. Springer, Wien New York, pp 377–391

11. Effenterre R van, Bataini JP, Cabanis EA, Iba-Zizen MT (1979) High energy radiotherapy in the treatment of meningiomas of the cavernous sinus. Acta Neurochir 28:464–467
12. Fujii K, Chambers SM, Rhoton AL (1979) Neurovascular relationships of the sphenoid sinus. J Neurosurg 50:31–39
13. Fukui M, Kitamura K, Nakagaki H, Yamakawa Y, Kinoshita K, Hayabuchi N, Jingu K, Numaguchi Y, Matsuura K, Watanabe K (1980) Irradiated meningiomas; a clinical evaluation. Acta Neurochir 54:33–43
14. Gilsbach JM (1983) Intraoperative Doppler sonography in neurosurgery. Springer, Wien New York
15. Grisoli F, Vincentelli F, Henry JF, Thomassin JM (1982) Anatomical bases for the transphenoidal approach to the pituitary gland. Anat Clin 3:207–220
16. Guiot G, Derome P (1976) Surgical problems of pituitary adenomas. In: Krayenbühl H et al. (ed) Advances and technical standards in neurosurgery, vol 3. Springer, Wien New York, pp 1–33
17. Hakuba A, Matsuoka Y, Suzuki T, Komiyama M, Jin TB, Inoue Y (1987) Direct approaches to vascular lesions in the cavernous sinus via the medial triangle. In: Dolenc VV (ed) The cavernous sinus. Springer, Wien New York, pp 272–284
18. Hardy J, Wigser SM (1965) Transsphenoidal surgery of pituitary fossa tumors with televised radiofluoroscopic control. J Neurosurg 23:612–620
19. Harris FS, Rhoton AL (1976) Anatomy of the cavernous sinus. A microsurgical study. J Neurosurg 45:169–180
20. Jefferson G (1955) The invasive adenomas of the anterior pituitary. Liverpool, University Press
21. Kaplan HA, Browder J, Krieger AJ (1976) Intercavernous connections of the cavernous sinuses. J Neurosurg 45:166–168
22. Kawase T, Toya S, Shiobara R, Kimura C, Nakajima H (1987) Skull base approaches for meningiomas invading the cavernous sinus. In: Dolenc VV (ed) The cavernous sinus. Springer, Wien New York, pp 346–354
23. Knosp E, Müller G, Perneczky A (1987) The blood supply of the cranial nerves in the lateral wall of the cavernous sinus. In: Dolenc VV (ed) The cavernous sinus. Springer, Wien New York, pp 67–80
24. Lang J (1985) Hypophyseal region – anatomy of the operative approaches. Neurosurg Rev 8:93–124
25. Lesoin F, Jomin M, Bouchez B, Duret MH, Clarisse J, Arnott G, Pellerin P, Francois P (1985) Management of cavernous sinus meningiomas. Neurochirurgia 28:195–198
26. Lesoin F, Jomin M (1987) Direct microsurgical approach to intracavernous tumors. Surg Neurol 28:17–22
27. Lesoin F, Autricque A, Vilette L, Franz K, Jomin M, Pellerin P (1987) The antero-external approach to the internal carotid artery at the base of the skull and intrapetrously. In: Dolenc VV (ed) The cavernous sinus. Springer, Wien New York, pp 311–319
28. Lesoin F, Pellerin P, Autricque A, Clarisse J, Jomin M (1987) The direct microsurgical approach to intracavernous tumors. In: Dolenc VV (ed) The cavernous sinus. Springer, Wien New York, pp 323–331
29. Lesoin F, Franz K, Pellerin P, Villette L, Autricque A, Jomin M (1987) An orbito-zygomaticomalar bone flap approach: a technical note. In: Dolenc VV (ed) The cavernous sinus. Springer, Wien New York, pp 341–345
30. MacKay A, Hosobuchi Y (1978) Treatment of intracavernous extensions of pituitary adenomas. Surg Neurol 10:377–383
31. Ono M, Ono M, Rhoton AL, Barry M (1984) Microsurgical anatomy of the region of the tentorial incisura. J Neurosurg 60:365–399
32. Parkinson D (1965) A surgical approach to the cavernous portion of the carotid artery. J Neurosurg 23:474–483
33. Parkinson D (1973) Carotid cavernous fistula: direct repair with preservation of the carotid artery. J Neurosurg 38:99–106

34. Parkinson D (1987) Carotid cavernous fistula. History and anatomy. In: Dolenc VV (ed) The cavernous sinus. Springer, Wien New York, pp 3–29
35. Paullus WS, Pait TG, Rhoton AL (1977) Microsurgical exposure of the petrous portion of the carotid artery. J Neurosurg 47:713–726
36. Petty AM, Kun LE, Meyer GA (1985) Radiation therapy for incompletely resected meningiomas. J Neurosurg 62:502–507
37. Renn WH, Rhoton AL (1975) Microsurgical anatomy of the sellar region. J Neurosurg 43:288–298
38. Rhoton AL, Hardy DG, Chambers SM (1979) Microsurgical anatomy and dissection of the sphenoid bone, cavernous sinus and sellar region. Surg Neurol 12:63–104
39. Samii M, Wild K von (1981) Operative treatment of lesions in the region of the tentorial notch. Neurosurg Rev 4:3–10
39a. Samii M (1972) Autologe Nerventransplantation im Trigeminusbereich. Med Mitt (Melsungen) 46:189–194
40. Sekhar LN, Moller AR (1986) Operative management of tumors involving the cavernous sinus. J Neurosurg 64:879–889
41. Sekhar LN, Schramm VL, Jones NF (1987) Subtemporal-preauricular infratemporal fossa approach to large lateral and posterior cranial base neoplasms. J Neurosurg 67:488–499
42. Shibata S, Mori K (1987) Effect of radiation therapy on extracerebral cavernous hemangioma in the middle fossa. J Neurosurg 67:919–922
43. Shiobara R, Ohira T, Kanzaki J, Toya S (1988) A modified extended middle cranial fossa approach for acoustic nerve tumors. J Neurosurg 68:358–365
44. Symon L, Jakubowski J (1979) Clinical features, technical problems and results of treatment of anterior parasellar meningiomas. Acta Neurochir [Suppl] 28:367–370
45. Steiner L (1982) Radiosurgery in intracranial tumor and arteriovenous malformations in children. In: Voth D, Gutjahr F, Langmaid C (eds) Tumors of the central nervous system in infancy and childhood. Springer, Berlin Heidelberg New York, pp 315–324
46. Taptas JN (1982) The so-called cavernous sinus: a review of the controversy and its implications for neurosurgeons. Neurosurgery 11:712–717
47. Taptas JN (1987) Must we still call cavernous sinus the parasellar vascular and nervous crossroads? The necessity of a definite topographical description of the region. In: Dolenc VV (ed) The cavernous sinus. Springer, Wien New York, pp 30–40
48. Tsuha M, Aoki H, Okamura T (1987) Roentgenological investigation of cavernous sinus structure with special reference to paracavernous cranial nerves. Neuroradiology 29:462–467
49. Umansky F, Nathan H (1982) The lateral wall of the cavernous sinus. J Neurosurg 56:228–234
50. Yamashita J, Handa H, Iwaki K, Abe M (1980) Recurrence of intracranial meningiomas, with special reference to radiotherapy. Surg Neurol 14:33–40

The Microanatomical Basis for Cavernous Sinus Surgery

B. SADASIVAN, M. DUJOVNY, S. MA, L. J. ZAMORANO, H. G. MIRCHANDANI, S. C. OHAEGBULAM, and J. I. AUSMAN, Detroit/USA and Enugu/NIGERIA

The cavernous sinus (CS) is the venous space on either side of the sella turcica. It extends from the superior orbital fissure to the apex of the petrous temporal bone. There have been several reports on direct surgery within the CS. Parkinson (1965) introduced direct surgery within the CS through a transdural opening in the dural wall. We have used a superior approach through the roof of the CS for direct surgery on intracavernous aneurysms. This study evaluates the microanatomical features of the CS that form the basis for surgery through the roof or the lateral wall of the CS.

Materials and Methods

Twenty-five tissue blocks containing 50 CS were removed from the cranial base of adult cadavers with the help of an electric saw. Each block contained the sphenoid bone, part of the petrous temporal bone, the anterior and posterior clinoids, pituitary gland, both CS, carotid artery and cranial nerves II, III, IV, V, and VI. The petrous carotid artery was cannulated and injected with colored silicon rubber (Microfil, Canton Biomedical Products, Boulder, Colorado). The microdissection was performed using standard microsurgical instruments, a high-speed air drill and the Zeiss OPMI operating microscope. A Pentax K-1000 camera was used for documentation. Simulated surgery with microdissection was performed to evaluate the superior and the lateral approaches (Fig. 1).

Results

The Lateral Approach. The lateral wall of the CS consists of two layers. The outer layer is firm and fibrous. The cranial nerves were embedded in the less fibrous and sometimes incomplete inner layer. We incised the outer layer parallel to and above the oculomotor nerve and separated it from the inner layer. The outer layer was excised from the petrous ridge to the lesser wing of the sphenoid such that the oculomotor, trochlear, and ophthalmic nerves were exposed. We identified four patterns in the arrangement of the cranial nerves based on how it affects surgical access through the lateral wall (Fig. 2). In type A (65% of the specimens), the trochlear nerve runs close to the oculomotor

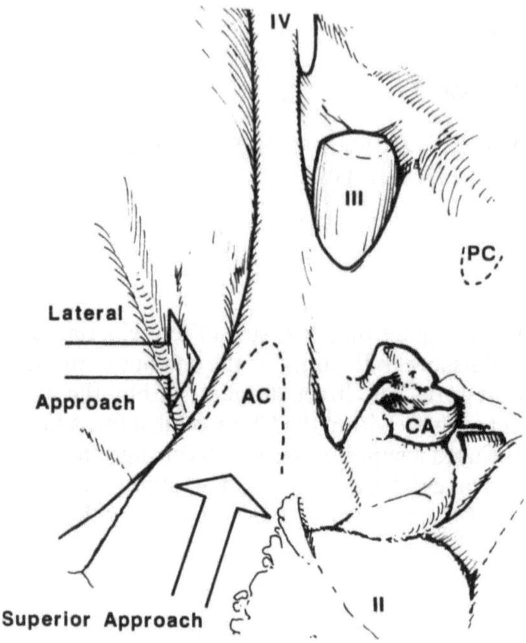

Fig. 1. The superior and lateral approaches to the cavernous sinus *AC*, anterior clinoid; *PC*, posterior clinoid; *CA*, carotid artery; *III*, optic nerve; *IV*, trochlear nerve

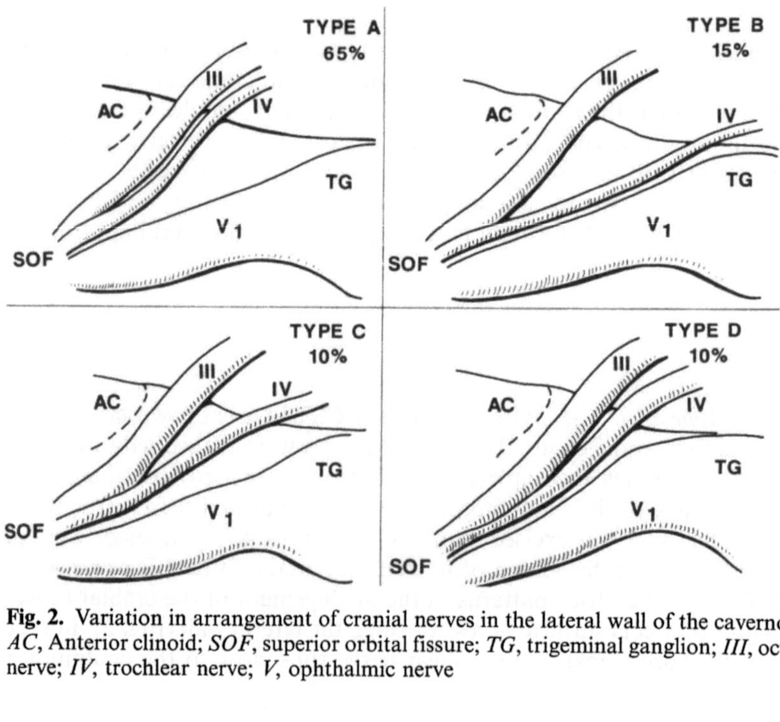

Fig. 2. Variation in arrangement of cranial nerves in the lateral wall of the cavernous sinus. *AC*, Anterior clinoid; *SOF*, superior orbital fissure; *TG*, trigeminal ganglion; *III*, oculomotor nerve; *IV*, trochlear nerve; *V*, ophthalmic nerve

nerve. The CS is best entered in the triangular space between the trochlear and ophthalmic nerves, as described by Parkinson. In type B (15% of the specimens), the trochlear nerve runs close to the ophthalmic nerves. The triangular space between the oculomotor and trochlear nerves can be incised to enter the CS. In type C (10% of the specimens), the trochlear nerve runs midway between the oculomotor and ophthalmic nerves. There is a triangular space both above and below the trochlear nerve. Either or both spaces could be incised to enter the CS. An incision into the superior triangle gives better access to the posterior superior CS and an incision in the inferior triangle gives better access to the posterior inferior. In type D (10% of the specimens), the oculomotor, trochlear, and ophthalmic nerves run close together. To incise the lateral wall and expose the CS, the trochlear nerve must be retracted either superiorly or inferiorly.

The average size of the triangular space through which the CS is entered (types A, B and C) is 17.4 mm^2 (range, 3–41 mm^2). With nerve retraction the area enlarges, but retraction may stretch and injure the cranial nerves.

The tentorial branch from the meningohypophyseal trunk crosses the roof of the CS in the posterosuperior corner of the CS and provides the blood supply to the oculomotor nerve. The inferior CS artery runs on the floor of the CS and supplies the abducens and ophthalmic nerves. The area along the trochlear nerve is the watershed area, making it the optimum place to incise the lateral wall without compromising the vascular supply to the cranial nerves. Table 1 shows the average length of the cranial nerves in the CS and the segment with the CS that is surgically accessible in the lateral approach. The entire length of the trochlear and ophthalmic nerves within the CS is accessible. As the cranial nerves approach the superior orbital fissure, the trochlear nerve moves to a position superolateral to the oculomotor nerve and the ophthalmic nerve moves to a position lateral to the oculomotor one. Hence, the anterior 5 mm of the oculomotor nerve in the CS is not accessible. After the lateral wall is incised, the abducens nerve can be identified, and the segment from Dorello's canal to the inferior CS artery can be exposed. Beyond that, the abducens nerve crosses the CS to enter the lateral wall and lies medial to the ophthalmic nerve and

Table 1. Surgical access to CS structures in the lateral approach

Anatomical structure	Average length in CS	Surgically accessible segment
III cranial nerve	16 mm	Entire length excect anterior 5 mm
IV cranial nerve	24 mm	Entire length
V cranial nerve	27 mm	Entire length
VI cranial nerve	32 mm	From Dorello's canal to origin of inferior cavernous sinus artery
Carotid artery		CCA1 and CCA2 segment up to origin of inferior cavernous sinus artery

inferolateral to the oculomotor nerve. This segment of the abducens nerve is not accessible.

The lateral approach gives good exposure of the proximal half of cavernous carotid artery. This includes the ascending segments, the origin of the meningohypophyseal trunk, and the horizontal segment up to about the origin of the inferior CS artery.

The Superior Approach. There is a triangular space between the anterior clinoid, the posterior clinoid, and the entrance of the oculomotor nerve in the roof of the CS. The CS can be entered by incising this space. The average size of this space is 13.5 mm^2 (range, 2.5–27 mm^2) and provides a limited access to the proximal cavernous carotid artery near the origin of the meningohypophyseal trunk.

To have good access to the CS through the superior approach, the anterior clinoid must be excised. We excised the dura over the anterior clinoid and medial sphenoid ridge. Then using a high-speed air drill the anterior clinoid, the medial sphenoid ridge as well as the roof and the lateral wall of the optic canal were excised. This allows for mobilization of the optic nerve medially if necessary. The average area of the quadrilateral roof of the CS exposed is 60 mm^2 (range, 28–84 mm^2).

On the inferior aspect of the anterior clinoid there is often a small vessel from the carotid artery that supplies the anterior clinoid (anterior clinoid artery).

We opened the roof of the cavernous sinus by making an incision parallel and medial to the oculomotor nerve. Medial to the entrance of the oculomotor nerve, the tentorial branch of the meningohypophyseal trunk is found crossing beneath the roof to supply the oculomotor nerve. The superior approach gives good access to the entire cavernous carotid artery as well as the meningohypophyseal trunk, the inferior CS artery, and McConnell's capsular artery. The whole length of the cavernous segment of the abducens nerve except for the distal 4 mm can be exposed through this approach. The superior aspect of the oculomotor nerve is seen, but the rest of the cranial nerves in the lateral wall are either poorly visualized or not visualized at all.

Discussion

Parkinson (1965) introduced the direct transdural approach through the lateral wall of the CS for surgical treatment of caroticocavernous fistula. This method has been used in several reports on direct CS surgery. His landmark was the entrance of the third and fourth cranial nerves. The incision was placed 4 mm beneath the point of disappearance of the third nerve and extended anteriorly approximately 2 cm parallel to the slope of the third and fourth nerves. Our study shows that because of the variation of the cranial nerve arrangement in the lateral wall, in 25% of cases (types B and D), the cranial nerves may be injured by such an incision.

To prevent injury to the cranial nerves in the transdural lateral approach to the CS, it is necessary to dissect the outer dural layer and visualize the course of the cranial nerves before the inner layer is incised to enter the cavernous sinus.

The lateral approach provides easy exposure of the oculomotor, trochlear, ophthalmic, and abducens nerves. However, it provides only a limited access to the cavernous carotid artery. The capsular and the ophthalmic arteries cannot be visualized. Retraction of the nerves to expose the carotid artery puts the nerves, especially the trochlear nerve, at risk from surgical trauma. This approach is good for surgery of the cranial nerves and for tumors that invade the lateral wall of the CS.

The superior approach to the CS through the roof provides a better exposure of the entire cavernous carotid artery and all its branches. Even the most distal segment of the cavernous carotid artery and the ophthalmic artery can be visualized by mobilizing the optic nerve. Although the abducens nerve can be exposed through this approach, there is very limited exposure of the other cranial nerves which puts them at less risk from surgical trauma. Although oculomotor palsy has been reported in this approach, this was transient.

Matsuoka et al. (1986) have reported using the superior approach in four cases to operate directly on cavernous aneurysms. We have also used this approach in nine cases to operate directly on cavernous aneurysms. We feel the superior approach is the better approach for direct surgery on vascular structures within the cavernous sinus.

References

Matsuoka Y, Hakuba A, Kshi H, Nishimura S (1986) Direct surgical treatment of intracavernous internal carotid artery aneurysms: report of four cases. Surg Neurol 26:360–364

Parkinson D (1965) A surgical approach to the cavernous portion of the carotid artery. Anatomical studies and case report. J Neurosurg 23:474–483

The Cavernous Sinus Syndrome: An Anatomical and Clinical Study

J. J. van OVERBEEKE, Utrecht/The Netherlands

Introduction

Because of the close relationships among the cranial nerves in the vicinity of the cavernous sinus, lesions in this region may cause various multiple cranial nerve palsies. Many clinical descriptions of these nerve palsies have been given, which has resulted in a confusing number of syndromes having in common partial or complete ophthalmoplegia, perhaps combined with a partially or completely affected trigeminal nerve. These syndromes have generally been named after the first author to describe them (Jacod 1921; Raeder 1924; Jefferson 1953; Lakke 1962; Roger et al. 1969; Bajandas 1977). The syndromes listed in Table 1 have a complete ophthalmoplegia in common. These symptoms are caused by a pathological involvement of the anterior site near the apex of the cavernous sinus. A trigeminal nerve involvement is caused by pathological involvement of the posterior site of the cavernous sinus (Table 2).

Raeder early (1924) drew attention to autonomic ocular disturbances caused by lesions within the cavernous sinus between the trigeminal nerve and the vertical portion of the intracavernous internal carotid artery. In 1938

Table 1. Syndromes related to pathological involvement of the cavernous sinus

	Affected nerves	Site	Etiology
Rollet syndrome (1865)	III, IV, VI, V1	Superior orbital fissure	Syphilitic periostitis
Rochon-Duvigneaud syndrome (1896)	III, IV, VI, V1	Superior orbital fissure	Syphilitic periostitis
Foix syndrome (1922)	III, IV, VI, V1	Cavernous sinus	Pituitary adenoma
Jacod syndrome (1921)	II, III, IV, VI, V1, V2, V3	Sphenoid sinus	Nasopharynx tumors
Dandy syndrome (1944)	III, IV, VI, V1	Anterior part of the cavernous sinus	Aneurysm of the internal carotid artery
Tolosa-Hunt syndrome (1966)	III, IV, VI, V1 (variable)	Cavernous sinus	Periarteritic lesions of the internal carotid artery

Table 2. Syndromes related to pathological involvement of the cavernous sinus

	Affected nerves	Site	Etiology
Jefferson syndrome (1938), cavernous sinus syndrome (CSS)			
Anterior CSS	V1, III, IV, VI (variable)	Anterior part of the cavernous sinus	Aneurysm of the ICA
Middle CSS	V1, V2, III, IV, VI (variable)	Middle part of the cavernous sinus	Tumors, traumatic lesions
Posterior CSS	V1, V2, V3, III, IV, VI (variable)	Posterior part of the cavernous sinus	Caroticocavernous fistula
Raeder syndrome (1924)	V1, V2, V3, VI (variable), sympathetic fibers	"Paratrigeminal" region	Tumor, fractures skull base, aneurysms
Gradenigo syndrome (1904)	V1, V2, V3, VI	Petrous apex	Otitis media, tumor, fracture petrous bone

Jefferson simplified the classification of the syndromes and introduced an anatomical-topographical classification (Table 3). The lateral wall of the cavernous sinus was divided into anterior, middle, and posterior parts. The main feature was trigeminal nerve involvement. On the anterior part the ophthalmic nerve was considered to be involved, on the middle part the ophthalmic and maxillary nerves, and on the posterior part the complete trigeminal nerve. The clinical pictures speak for themselves and are divided into anterior, middle, and posterior cavernous sinus syndromes. Many of the syndromes mentioned in Table 1 can be grouped in the anterior cavernous sinus syndrome. However, autonomic disturbances that may occur because of pathological involvement of the cavernous sinus were not described. Also disturbances of the ocular movements were only sparsely described.

Table 3. Jefferson syndrome (cavernous sinus syndrome; 1938)

Anatomical localization	Affected nerves	Syndromes
Anterior part of the cavernous sinus	V1, III, IV, VI (variable)	Anterior cavernous sinus syndrome (synonyms: Rollet syndrome, Rochon-Duvigneaud syndrome, Foix syndrome, Jacod syndrome, Tolosa-Hunt syndrome)
Middle part of the cavernous sinus	V1, V2, III, IV, VI (variable)	Middle cavernous sinus syndrome
Posterior part of the cavernous sinus	V1, V2, V3, III, IV, VI (variable)	Posterior cavernous sinus syndrome

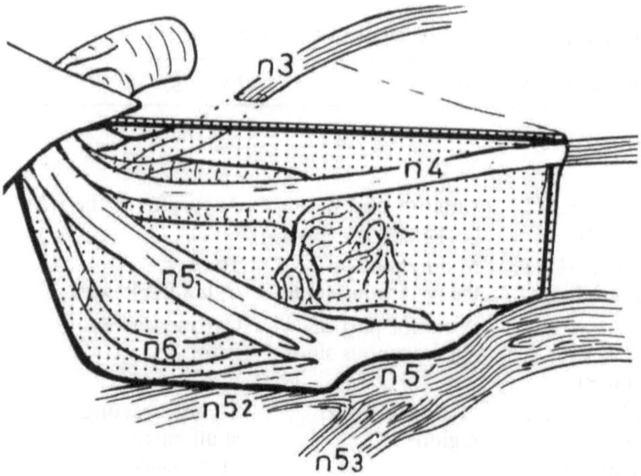

Fig. 1. Schematic drawing of the lateral wall of the cavernous sinus. *Dotted area* the lateral wall. The trochlear and ophthalmic nerves run within the lateral wall; the oculomotor nerve joins the lateral wall in the anterior part; the abducens nerve joins the lateral wall near its entrance into the superior orbital fissure

Therefore, based on an anatomical study of 26 adult cavernous sinuses, another anatomical-topographical classification of these syndromes is presented. Here the following anatomical features are important. Near the superior orbital fissure the ophthalmic, trochlear, oculomotor, and abducens nerves lie close together intersected by numerous thick irregular fibrous septa from the dura of the lateral wall. Here these nerves form the anterior lateral wall of the

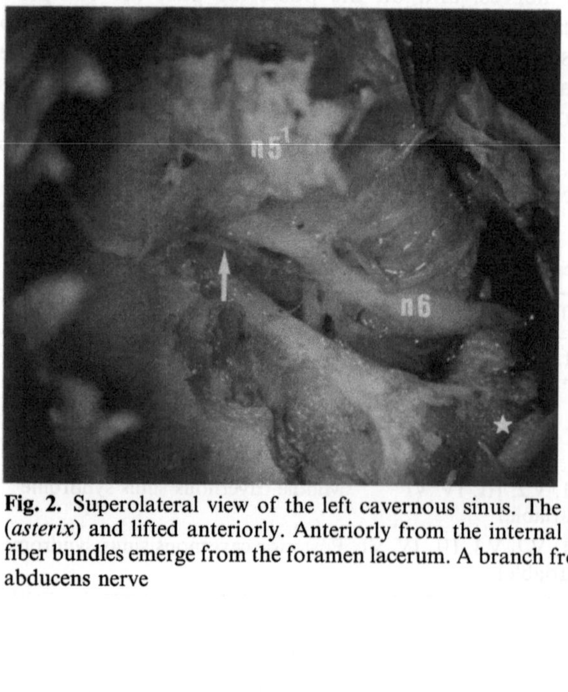

Fig. 2. Superolateral view of the left cavernous sinus. The ophthalmic nerve has been cut (*asterix*) and lifted anteriorly. Anteriorly from the internal carotid artery two sympathetic fiber bundles emerge from the foramen lacerum. A branch from the anterior bundle joins the abducens nerve

Table 4. Anatomical-topographical classification of syndromes related to pathological involvement of the cavernous sinus

Syndrome	Site	Affected nerves	Clinical symptoms
Superior orbital fissure syndrome	Superior orbital fissure, anterior part of the lateral wall of the cavernous sinus	III, IV, VI, V1 sympathetic and parasympathetic fibers	Total ophthalmoplegia, altered sensation V1, moderate autonomic, pupillary and ocular symptoms, exophthalmus
Syndrome of the lateral wall of the cavernous sinus	Middle and posterior parts of the lateral wall of the cavernous sinus	III (variable), IV, V1, V2, V3 (variable), parasympathetic fibers	Trochlear nerve paresis, oculomotor nerve paresis (variable), altered sensation V
Cavernous sinus	Cavernous sinus	V1, V2, V3 (variable), sympathetic fibers	Altered sensation V, partial Horner syndrome, exophthalmus

cavernous sinus (Fig. 1). The oculomotor nerve enters the lateral wall of the cavernous sinus in the anterior part after piercing the roof (Fig. 1; Rhoton 1977; Lang and Reiter 1984). Then it courses between the two dural leaves of the anterior half of the lateral sinus wall (Umanski and Nathan 1982). The abducens nerve is within the sinus, adhering to the internal carotid artery and joining the other nerves in the lateral wall near the apex of the cavernous sinus (Fig. 1). The trochlear and ophthalmic nerves are both closely related to the lateral wall of the cavernous sinus (Umanski and Nathan 1982; Lang and Reiter 1984). Sympathetic fibers run in bundles through the foramen lacerum on the surface of the internal carotid artery. A branch of sympathetic fibers emerging mostly from the anterior chain joins the abducens nerve within the sinus and is ultimately distributed to the other cranial nerves in the apex of the cavernous sinus (Fig. 2); (Harris and Rhoton 1976; Parkinson 1978).

According to these anatomical observations the following three syndromes are proposed (Table 4).

The Superior Orbital Fissure Syndrome

Lesions in the vicinity of apex of the cavernous sinus are considered to cause this syndrome. A total ophthalmoplegia with altered sensation within the first division of the trigeminal nerve occurs. Because the ophthalmic veins are compressed, and the ocular muscles are paralyzed, there is a marked exophthalmus. Parasympathetic and sympathetic disturbances are present because of oculomotor nerve involvement and involvement of the sympathetic parasellar connections. However, the autonomic nerve disturbances may be obscured because of the simultaneous occurrence of parasympathetic and sympathetic disturbances (Trobe 1978). The conjunctival injection as part of a sympathetic involvement is hidden by the exophthalmus and venous congestion. The ptosis as a part of parasympathetic disturbances is often not visible due to the simultaneous oculomotor nerve paresis.

The Syndrome of the Lateral Wall of the Cavernous Sinus

Lesions adjacent to the middle and posterior part of the lateral wall of the cavernous sinus may present in this syndrome. Because of the sheltered position of the abducens nerve within the cavernous sinus and the course of the oculomotor nerve halfway into the lateral sinus wall, the autonomic ocular innervation is largely spared in this syndrome. If present, the parasympathetic innervation is affected because of the oculomotor nerve involvement. The trigeminal nerve may be involved totally. In this syndrome the ophthalmoplegia is limited to a trochlear and oculomotor nerve palsy. If only the posterior part of the lateral wall is affected, there may be only a trochlear nerve palsy.

The Cavernous Sinus Syndrome

This syndrome is caused by lesions inside the cavernous sinus. The abducens nerve is involved. Lesions in the middle of the sinus affect simultaneously the accompanying sympathetic fibers, and may cause a partial Horner syndrome. This was described by Raeder in 1924. Lesions posteriorly at the site of the petrous bone affect only the abducens nerve and may spare the sympathetic fibers. This clinical picture has been described by Gradenigo in 1904. In both cases extension of the lesion laterally to the lateral wall may affect the trigeminal and trochlear nerves. In fact, the Raeder and Gradenigo syndromes enclose the cavernous sinus syndrome and are partial cavernous sinus syndromes. Nevertheless, we prefer to use the term cavernous sinus syndrome itself because this term indicates that the cavernous sinus is involved.

The importance of this anatomical-topographical simplification of the various syndromes related to pathological involvement of the cavernous sinus is that it gives a clear view of the symptomatology of the syndromes described. It enables the physician to locate the lesions by means of a simple neurological examination and facilitates the interpretation of the neuroradiodiagnostic examinations. The clinical symptoms seen in these syndromes can be transient and may change from one syndrome into another. A recognition of these clinical features may focus attention on the mode and direction of growth of the lesion which may indicate a malignancy. Because clinical signs always precede further diagnostic and clinical treatment, an early and good understanding of the clinical signs still remains very important.

References

Bajandas J (1977) The six syndromes of the sixth nerve. In: Lawton Smith (ed) Neuro ophthalmology update. Masson, Paris, pp 49–65
Harris FD, Rhoton AL (1976) Anatomy of the cavernous sinus. A microscopical study. J Neurosurg 45:169–180
Jacod M (1921) Sur la propogation intracranienne des sarcomes de la trompe d'eustache syndrome du carrefour petro-sphenoidal paralysie des 2e, 3e, 4e, 5e et 6e paires cranienne. Rev Neurol 28:33–38
Jefferson G (1953) The Bowman lecture. Concerning injuries, aneurysms and tumors involving the cavernous sinus. Trans Ophthalmol Soc UK 73:117–152
Lang L, Reiter U (1984) Ueber den Verlauf der Hirnnerven in der Seitenwand des Sinus cavernosus. Neurochirurgica 27:93–97
Lakke PWF (1962) Superior orbital fissure syndrome. Arch Neurol 7:289–300
Parkinson D et al. (1978) Sympathetic connections to the fifth and sixth cranial nerves. Anat Rec 191:221–226
Raeder JG (1924) "Paratrigeminal" paralysis of oculo-pupillary sympathetic. Brain 47:149–158
Roger J et al. (1969) Multiple cranial nerve palsies. In: Vinken, de Bruyn. Handbook of clinical neurology, vol 2, chap 5. North-Holland, Amsterdam, pp 86–106
Rhoton AL et al. (1977) Microsurgical anatomy of the sellar region and the cavernous sinus. In: Glaser JS (ed), Neuro-ophthalmology, vol X. Mosby, St. Louis, pp 75–101
Trobe JD et al. (1978) Meningiomas and aneurysms of the cavernous sinus. Arch Ophthalmol 96:457–467
Umanski F, Nathan H (1982) The lateral wall of the cavernous sinus. With special reference to the nerves related to it. J Neurosurg 56:228–234

Vascular Lesions in the Sellar Region

J. I. Ausman, F. G. Diaz, M. Dujovny, B. Sadasivan, S. C. Ohaegbulam, and L. J. Zamorano, Detroit/USA and Enugu/NIGERIA

Aneurysms in the cavernous sinus region produce mechanical changes which affect the cranial nerves contained within the confines of the venous sinus, producing the symptom complex known as the superior orbital fissure syndrome. The size of the aneurysm frequently exceeds 25 mm, and therefore these aneurysms are often giant in nature. Parkinson (1973) reported the direct approach to the sinus with the aid of cardiac standstill. Dolenc (1985) reported on the approach to aneurysms within the cavernous sinus through an extradural route. We present our experience with aneurysms in the cavernous sinus region treated through an intradural approach.

Materials and Methods

We present our experience with 32 patients with aneurysms in the cavernous sinus region from among 356 aneurysm patients treated at Henry Ford Neurosurgical Institute from October 1978 to June 1987. Our patients were divided in two groups: those with aneurysms contained entirely within the cavernous sinus ($n = 22$) and those with aneurysms arising from the carotid artery in the cavernous sinus extending into the subarachnoid space ($n = 10$). Purely intracavernous aneurysms were further divided according to their size into small (<25 mm, $n = 7$) and giant (>25 mm, $n = 15$). All patients with purely intracavernous lesions presented with a superior orbital fissure syndrome. Nine of ten patients with subarachnoid extension presented with a hemorrhage (grade III).

All patients are operated on under barbiturate anesthesia through a pterional approach; a lumbar subarachnoid drain is inserted preoperatively, and intravenous furosemide and mannitol were given to ease brain retraction. The cavernous sinus region is exposed after the Sylvian fissure was split widely, and the temporal lobe tip was freed of its venous attachments. The cavernous sinus is exposed by first dissecting cranial nerves III and IV from their subarachnoid location and following them into the cavernous sinus. Control of venous bleeding is achieved by packing the sinus with oxidized cellulose. Proximal control of the internal carotid artery is obtained in the neck through a separate incision and placing a Silverstone clamp distal to the common carotid bifurcation. The clamp may be used to obtain partial or total arterial occlusion during the dissection of the aneurysm. The intracranial internal carotid artery is then exposed by radical removal of the superior and lateral orbital walls, anterior clinoid process, and unroofing of the optic nerve canal. The cranial nerves are dissected in the confines of the sinus, and the dura over the carotid artery is resected. With distal and proximal control, the carotid may be temporarily occluded under barbiturate and lidocaine protection, and the aneurysm may then be clipped ($n = 16$). The complete dissection of the internal carotid artery and the aneurysm is not always possible because of the size of the aneurysm. In these patients, we have selected to trap the aneurysm ligating the internal carotid artery in the neck and in its supraclinoid portion distal to the aneurysm. An extracranial-intracranial (EC-IC) anastomosis is then performed to a proximal branch of the middle cerebral artery ($n = 6$). Our initial approach to these aneurysms consisted of the ligation of the internal carotid artery in the neck followed by a cortical EC-IC anastomosis ($n = 10$).

Two patients who had severe visual loss preoperatively, progressed to complete visual loss. Two patients developed a moderate hemiparesis which resolved during the first 6 weeks after surgery, two developed persistent diplopia, and two patients died; one from a massive cerebral infarction after a trapping procedure and one from an intraoperative hemorrhage. The rest have had complete resolution of their symptoms and remain well.

Discussion

Aneurysms in the region of the cavernous sinus represent a major challenge because of the intimate relationship that the internal carotid artery has to the cranial nerves contained within the sinus and because the internal carotid artery is fully surrounded by the venous structures as it courses from its intrapetrous portion to its subarachnoid location. Parkinson described that when the veins draining the cavernous sinus obtained from cadavers were selectively injected with resin, and the anatomical specimens were decomposed, he could demonstrate distinct venous channels; therefore, he believed the cavernous sinus to be series of interconnecting veins contained within the leaves of the parietal dura located around the sella turcica. Others have reported that the cavernous sinus is in fact a true sinus without any distinct venous structures containing the blood within the dural leaves.

A major problem posed by a direct approach to lesions contained within the cavernous sinus has been the copious bleeding which is frequently encountered when the sinus is opened. This motivated Parkinson to combine his direct approach to the sinus with the simultaneous use of extracorporal circulation and cardiac arrest. The safe time of complete cardiac arrest obtained by this technique is limited to 1 h; it is therefore imperative that the entire procedure conducted in the confines of the cavernous sinus be limited to 1 h. In most cases Parkinson found this to be sufficient to complete the procedure; but it is conceivable that if a problem arose, and it were necessary to work on the carotid artery for longer than 1 h, one may be unable to do so because of the intrinsic limitations imposed on the surgeon by the need for cerebral and myocardial reperfusion.

Dolenc has presented a direct attack on the internal carotid artery within the cavernous sinus without the need for cardiac arrest. Through an extradural pterional or subtemporal approach, Dolenc exposes the lesser sphenoid wing from the anterior clinoid process to its most lateral extent. He removes the lesser sphenoid wing, the orbital roof and the anterior clinoid to expose the cavernous portion of the internal carotid artery. He gains proximal arterial control by dissecting out the intrapetrous portion of the internal carotid artery lateral to the gasserian fossa. A small dural incision is made to expose the subarachnoid portion of the internal carotid artery to gain distal control. The cavernous sinus is opened and control of any venous bleeding is obtained by packing the cavernous sinus with oxidized cellulose. Control of bleeding from the internal carotid artery may be easily obtained by temporarily ligating the artery between the proximal and distal sites exposed. To enter the cavernous sinus, Dolenc prefers first to dissect cranial nerves III, IV, and V1 to mobilize them free of the internal carotid artery. With the maneuver one gains gradual control of the bleeding from the sinus, preserves the integrity of the cranial nerves, and exposes safely the internal carotid artery wall.

Through a strictly pterional approach one may expose the entire lesser wing of the sphenoid, anterior clinoid, orbital roof, and if necessary the lateral orbital apex. A transdural exposure facilitates the direct observation of the brain and cranial nerves as well as their preservation. The dura over the orbital roof, anterior clinoid and lateral orbital apex may be dissected off readily, and these structures are quickly removed with a high-speed drill. Following Dolenc's description, cranial nerves III, IV, and V1 may then be dissected from their subarachnoid portion into and through the cavernous sinus to the superior orbital fissure. Venous bleeding may be controlled by packing the cavernous sinus with oxidized cellulose or by direct coagulation of those venous channels which are visualized under the microscope. The entire dissection is thus carried out through a pure pterional approach, by a limited craniotomy, and with minimal brain retraction. Since the area of the lateral temporal fossa is not approached, one does not have to deal with any of the potential problems previously listed for this approach. To gain proximal control, one must expose the internal carotid artery in the neck through a small submandibular incision. The potential risks of this dissection are considerably less than those listed for the exposure of the intrapetrous portion of the internal carotid artery. The one distinct disadvantage to this procedure is the need for an additional cervical incision, but the benefits gained from a rapid exposure and minimal additional morbidity, in our opinion, outweigh the disadvantages.

We initially approached aneurysms of the cavernous internal carotid artery by a gradual occlusion of the internal carotid artery coupled with a cortical EC-IC anastomosis. We later proceeded to a trapping procedure with an immediate ligation of the internal carotid artery in the neck and in the subarachnoid area and an immediate high-flow anastomosis of the superficial temporal to the proximal middle cerebral artery because some patients developed transient ischemic complications while the internal carotid artery clamp was being closed. After Dolenc reported his experience with the cavernous internal carotid artery dissection, we proceeded with this approach because it attacks the problem directly and may preserve the internal carotid artery. Sadasivan et al. (1988) have studied the operative approaches to the cavernous sinus using skull base specimens in our laboratory. We have found that the best approach to the intracavernous vascular structures is through the roof of the cavernous sinus after the anterior clinoid and the medial sphenoid wing are drilled off. The direct approach to aneurysms in the intracavernous internal carotid artery is not always feasible; and in these patients with inaccessible lesions the trapping procedure with an immediate high-flow EC-IC anastomosis to the proximal middle cerebral artery should be used.

In our limited experience, we have had some ischemic complications which resulted initially from our using a small cortical bypass and gradual proximal internal carotid artery ligation. The patient who died with a massive cerebral infarction in the trapping group confirms the fact that not all patients can be satisfactorily protected even by a high-flow anastomosis. Therefore, we believe that a direct approach to the aneurysms is probably the best initial alternative in the management of these patients and the others should be considered only after the direct approach has been attempted.

References

Dolenc VV (1985) A combined epi-and subdural direct approach to carotid ophthalmic artery aneurysms. J Neurosurg 62:667–672

Parkinson D (1973) Carotid cavernous fistula: direct repair with preservation of the carotid artery. Technical note. J Neurosurg 38:99–106

Sadasivan B, Dujovny M, Ma SH, Zamorano L, Mirchandani HG, Ohaegbulam SC, Ausman JI (1988) The microanatomical basis for cavernous sinus surgery. IV International Congress, Skull Base Study Group 1988, this volume: 479–483

Surgery of Vascular Lesions in the Cavernous Sinus

A. Perneczky, E. Knosp, Mainz/FRG and A. Zohdi, Cairo/Egypt

The development of technical standards in neurosurgery has led to solution of formerly difficult problems without great risks to patients. One of these problems is the surgery of vascular lesions involving the cavernous sinus region.

Surgical Technique

From the surgical point of view the cavernous sinus can be divided into three parts (Fig. 1). The anterior part corresponds to the para-infraclinoidal portion of the internal carotid artery and involves mainly the problems of surgery of para- and infraclinoidal aneurysms (Iwabuchi et al. 1978; Milenkovic et al. 1982; Knosp et al. 1988; Perneczky et al. 1985, 1986, 1987, 1988). The middle part represents the field of the lateral sinus wall consisting of cranial nerves III, IV, and V and the underlying posterior knee of the carotid siphon with the abducens nerve (Dolenc 1983; Parkinson 1965, 1973; Perneczky et al. 1987; Sekhar et al. 1986; Umansky and Nathan 1982). The posterior part involves the region of the petrous bone tip including the gasserian ganglion and the canal of the internal carotid artery (Dolenc 1983; Perneczky and Knosp 1986; Sekhar et al. 1986).

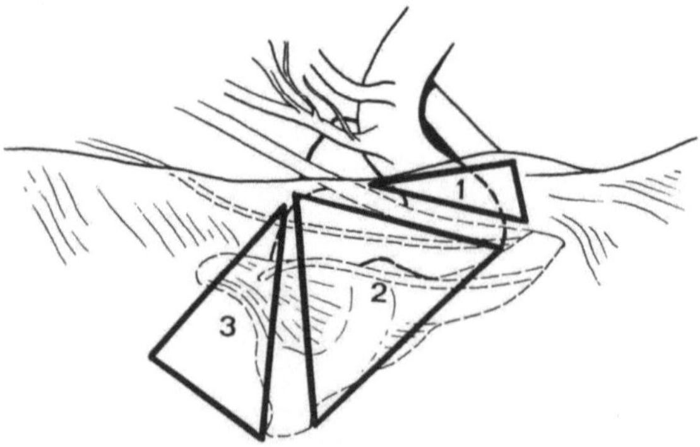

Fig. 1. Schematic presentation of the three surgical parts of the cavernous sinus. *1*, Anterior part; *2*, middle part (lateral sinus wall); *3*, posterior part (pyramid tip)

Ipsilateral Approach to the Anterior Siphon Knee (Perneczky et al. 1985, 1987). In cases with para- and infraclinoidal aneurysms, where the aneurysm sack points anterolaterally, posterolaterally, or laterally, we choose an ipsilateral pterional approach. The steps in the ipsilateral approach are:

a) unroofing of the ipsilateral optic canal for gentle mobilization of the nerve;
b) removal of the anterior clinoid process;
c) identification of the roof of the cavernous sinus;
d) dissection of the fibrous ring sharply around the carotid artery;
e) blunt dissection along the internal carotid artery up to the anterior siphon knee.

Contralateral Approach to the Anterior Siphon Knee (Milenkovic et al. 1982; Perneczky et al. 1985, 1987; Vajda et al. 1988). In cases where the aneurysm arises from the internal carotid artery infraclinoidally and medially, the contralateral approach is appropriate. This approach leads obliquely below the optic chiasm to the medial wall of the contralateral internal carotid artery. The craniotomy is performed fronto-laterobasally. The steps in the contralateral approach are:

a) opening of the chiasmatic cistern;
b) removal of the tuberculum sellae, the mucosa of the sphenoid sinus being left intact and retracted;
c) removal of the medial wall of the contralateral optic canal;
d) identification of the contralateral transversal plate (roof of the cavernous sinus) and fibrous ring;
e) sharp dissection of the fibrous ring;
f) blunt dissection along the medial and anterior wall of the carotid artery up to the anterior siphon knee.

Approach Through the Lateral Wall of the Cavernous Sinus (Dolenc 1983; Parkinson 1965, 1973; Perneczky et al. 1987; Umansky and Nathan 1982; Johnston 1979; Browder 1937). Aneurysms located in the middle and posterior parts of the cavernous sinus can be operated on by direct dissection through the lateral sinus wall, but in the majority of vascular lesions of this area the interventional neuroradiological methods are preferable to direct surgery. In the few cases in which the endovascular technique does not lead to a satisfactory treatment of the lesion the approach through the lateral sinus wall seems to be the method of choice. The steps in the transcavernous approach are:

a) Extradural removal of the lesser sphenoid wing and anterior clinoid process;
b) extradural subtemporal exposure of the intrapetrous carotid artery;
c) opening of the sylvian fissure and mobilization of the temporal pole;
d) flaplike dissection of the superficial layer of the lateral sinus wall;
e) dissection of the profound layer of the lateral wall with the divisions of the trigeminal nerve;

f) approaching of the intracavernous portion of the internal carotid artery;
g) temporary selective occlusion of the intracavernous segment of the internal carotid artery by the application of temporary clips or by temporary endovascular balloon placement;
h) the flaplike opening of the superficial wall layer enables a good reconstruction of the lateral wall of the cavernous sinus at the end of the procedure.

Clinical Cases and Results

Between 1980 and January 1988, we carried out an intraoperative dissection of one or more parts of the cavernous sinus in 70 cases. In the anterior cavernous sinus portion the vascular lesions were predominant, there were 24 aneurysm cases and six patients with tumor. In the middle and posterior parts of the cavernous sinus the tumorous processes are most frequent, in our material 2 aneurysms and 38 tumorous lesions (Tables 1, 2). Out of the 26 vascular cases, there were 21 women and 5 men. Sixteen patients had subarachnoid hemorrhage and 5 of them underwent acute surgery. In 11 cases the aneurysm had a diameter of 2.5 cm and more. Two other patients had an aneurysm size between 1.5 and 2.5 cm. Table 3 shows our case collection.

Table 1. Cavernous sinus surgery: dissected portion of the cavernous sinus (1980–1987)

	n
Anterior	
Aneurysm	24
Tumor	6
Middle (lateral wall)	
Aneurysm	2
Tumor	14
Posterior	
Tumor	2
Middle + posterior	
Tumor	22
Total	70

In none of the 26 cases were there intraoperative complications. Eighteen patients had carotid-ophthalmic aneurysms with partial extradural extension of the aneurysm neck. Six other patients had an extradural aneurysm with intradural bulging of the aneurysm dome. The surgical methods used are listed in Table 4. In 23 cases the aneurysm sack was clipped by one or more clips; in 9 cases, an additional resection of the sack was carried out. The operative results (Table 5) showed a benefit for the patient in 21 cases.

One patient died on the 5th postoperative day following epidural hemorrhage. Another patient died on the 2nd day of brain edema and vasospasm.

Table 2. Cavernous sinus surgery: Lesions

	n
Aneurysms	
Para-infraclinoidal	24
Intracavernous (giant)	2
Tumors	
Meningioma	20
Pituitary adenoma	8
Skull base tumors	
Carcinoma	3
Chondrosarcoma	3
Chondroma	1
Neurinoma	2
Cholesteatoma	2
Dermoid	2
Plasmocytoma	1
Tumors of the orbita	
Meningioma	1
Cylindroma	1
Cases	
Vascular	26
Tumors	44
Total	70

Three patients had fatal late complications. One died on the 6th day following myocardiac infarction, one on the 21st day of cerebellar hemorrhage caused by an intracerebellar AVM, and one on the 42nd day following pneumonia (Table 6).

Discussion

Aneurysms originating in the area of the anterior siphon knee map partly reach the subarachnoid space and thus cause subarachnoid hemorrhage. Some authors indicate direct surgical management of such aneurysms by a transcavernous approach (Dolenc 1983; Parkinson 1973; Perneczky et al. 1987, 1988). This approach presents considerable risks. It requires a major amount of anesthesiological attention and is inferior to the interventional neuroradiological methods. In very few cases only, in which the endovascular technique could not lead to the solution of the problem, the indication to open intracavernous surgery is given.

The anatomical investigations that we performed (Perneczky et al. 1986, 1988; Knosp et al. 1988) have shown that it is possible to expose the anterior siphon knee without opening the cavernous sinus in a way as to provide good

Table 3. Para- and infraclinoidal aneurysms treated by direct surgery ($n=26$)

No.	Name	Sex	Location	Ø	SAH	Surgery	Result	Remarks
1	S.E.	F	Right ophth.	1 cm	+	Clipping (1 clip)	Excellent	
2	M.G.	F	Right ophth.	3 cm	–	Clipping (3 clips) resection	Excellent	
3	B.M.	F	Left ophth.	2.5 cm	+a	Clipping (2 clips) resection	Excellent	
4	L.E.	F	Right ophth.	3 cm	–	Clipping (3 clips) resection suture	Excellent	Posttraumatic aneurysm
5	S.A.	M	Left ophth.	2.5 cm	–	Clipping (4 clips) resection	Excellent	Posttraumatic aneurysm
6	S.M.	F	Right ophth.	5 cm		Clipping (3 clips)	Good	+, on 6th postop. day myocard infarction
7	S.H.	F	Left ophth.	2.5 cm	+a	Clipping (2 clips) resection	Excellent	Vasospasm
8	J.M.	F	Left ophth.	1 cm	+a	Clipping (1 clip) resection	Excellent	Intraoperative rupture
9	H.H.	F	Left ophth.	1.5 cm	–	Clipping (1 clip)	Excellent	
10	M.O.	F	Right carot. Right ophth.	1 cm 3 mm	+	Clipping (1 clip) Clipping (1 clip)	Excellent	
11	W.E.	F	Comm. ant. Left ophth.	3 mm 3 mm	+	Clipping (1 clip) Clipping (1 clip)	Excellent	
12	L.R.	M	Right ophth.	1 cm	+	Clipping (1 clip)	Good	
13	F.M.	F	Right ophth.	2.5 cm	+	Clipping (2 clips) resection	On 42nd day, stage III, pulm. +	Temp. clipping of internal carotid, 4.5 min
14	S.W.	F	Left ophth.	2.5 cm	+	Clipping (2 clips)	Excellent	

15	R.J.	F	Left ophth.	5 cm	–	Clipping (3 clips)	On 2nd day edema +	Intraop. rupture temp. clip. of carotid 10 min
16	Z.M.	F	Extradural right	3.5 cm	–	Direct surg., (lat. sinus wall) Balloon occlus., suture	No change	2nd stage balloon embolization
17	Z.R.	F	Partial intradural right	5 mm	–	Clipping (1 clip)	EDH +	
18	T.R.	M	Partial intradural right	4 mm	+	Suture	Excellent	
19	J.S.	F	Partial intradural left	5 mm	+	Contra-approach, clipping (1 clip)	Excellent	+ after 3 weeks, cerebell. hemorrh. AVM
20	A.M.	F	Left ophth.	8 mm	–	Clipping (1 clip) resection	Excellent	Tuberculum sellae meningioma
21	W.C.	M	Partial intradural left carotid	8 mm	+	Clipping (1 clip)	Excellent	Two more Aneurysms, all clipped
22	I.R.	F	Right ophth.	2 cm	+	Clipping (2 clips) resection	Excellent	Pregnancy, 4th month at operation
23	S.E.	F	Right carotid	1 cm	–	Clipping (1 clip)	Excellent	
24	M.R.	M	Left carotid (t)	3.5 cm	E	Balloon occl. + trapping	Excellent	Posttraumatic an. transnasal balloon removal 5 steps
25	L.B.	F	Left ophth.	1 cm	+a	Clipping (2 clips)	Excellent	
26	F.H.	F	Left carotid left post. comm.	1 cm	+a	Clipping (1 clip)	Excellent	Both clipped

Table 4. Para- and infraclinoidal aneurysms: Surgical methods ($n = 26$)

Clipping	14
Clipping + Resection	8
Clipping + Resection + Suture	1
Rhaphy (no. 18)	1
Rhaphy + Balloon (no. 16)	1
Balloon + trapping (no. 24)	1

Table 5. Para- and infraclinoidal aneurysms: operative results ($n = 26$)

Excellent	19
Good	2
Poor	–
Death	5

Table 6. Para- and infraclinoidal aneurysms: postoperative mortality ($n = 26$)

Case no.	Cause of postoperative death
6	Myocardial infarction, 6th day
13	Pneumonia, 42nd day
15	Brain edema, 2nd day
17	Epidural hemorrhage, reoperation, 5th day
19	Cerebellar hemorrhage, 21st day

visibility of the adjacent structures and to permit constant control. These approaches make possible the direct surgical management of infraclinoidal aneurysms in the middle and anterior parts of the C3 segment of the internal carotid artery (Dolenc 1983; Johnston 1979; Parkinson 1973; Perneczky et al. 1985, 1988).

References

1. Browder J (1937) Treatment of carotid artery-cavernous sinus fistula. Report of a case. Arch Ophthalmol 18:95–102
2. Dolenc V (1983) Direct microsurgical repair of intracavernous vascular lesions. J Neurosurg 58:824–831
3. Fox JL (1988) Microsurgical treatment of ventral (paraclinoid) internal carotid artery aneurysms. Neurosurg 22:32–39
4. Iwabuchi T, Suzuki Sh, Sobata E (1978) Intracranial direct operation for carotid-ophthalmic aneurysm by unroofing of the optic canal. Acta Neurochir (Wien) 43:163–169
5. Johnston I (1979) Direct surgical treatment of bilateral intracavernous internal carotid artery aneurysms. Case report. J Neurosurg 51:98–102

6. Knosp E, Müller G, Perneczky A (1988) The paraclinoid carotid artery: anatomical aspects of a microneurosurgical approach. J Neurosurg 22:896–902
7. Milenkovic Z, Gopic H, Antovic P, Jovicic V, Petrovic B (1982) Contralateral pterional approach to a carotid-ophthalmic aneurysm ruptured at surgery. J Neurosurg 57:823–825
8. Parkinson D (1965) A surgical approach to the cavernous portion of the carotid artery. Anatomical studies and case report. J Neurosurg 23:474–483
9. Parkinson D (1973) Carotid-cavernous fistula: direct repair with preservation of the carotid artery. Technical note. J Neurosurg 38:99–106
10. Perneczky A, Knosp E, Vorkapic P, Czech TH (1985) Direct surgical approach to infraclinoidal aneurysms. Acta Neurochir (Wien) 76:36–44
11. Perneczky A, Knosp E (1986) The intracavernous connective tissue cover of the internal carotid artery. Anatomy and surgery. In: Scheunemann H, Schürmann K, Helms J (eds) Tumors of the skull base. de Gruyter, Berlin New York, pp 171–177
12. Perneczky A, Knosp E, Czech TH (1987) Para- and infraclinoidal aneurysms. Anatomy, surgical technique and report on 22 cases. In: Dolenc V (ed) The cavernous sinus. Springer, Wien New York, pp 252–271
13. Perneczky A, Knosp E, Matula CH (1988) Cavernous sinus surgery. Approach through the lateral wall. Acta Neurochir 92:76–82
14. Sekhar LN, Schramm VL Jr., Jones NF, Yonas H, Horton J, Latchaw RE, Curtin H (1986) Operative exposure and management of the petrous and upper cervical internal carotid artery. Neurosurgery 6:967–982
15. Umansky F, Nathan H (1982) The lateral wall of the cavernous sinus. With special reference to the nerves related to it. J Neurosurg 56:228–234
16. Vajda J, Juhasz J, Pasztor E, Nyary J (1988) Contralateral approach to bilateral and ophthalmic aneurysms. Neurosurgery 22:662–668

Parasellar Aneurysms: Treatment Options and Results

H. M. Mehdorn, Kiel/FRG, D. Kühne, W.-H. Heienbrok, and
W. Grote, Essen/FRG

Introduction

Parasellar aneurysms include aneurysms of the internal carotid artery (ICA) in its intracavernous and its supraclinoid portion, and those located at the origin of the ophthalmic artery. Because giant aneurysms in the distal portion of the ICA and the A_1 segment of the anterior cerebral artery (ACA) may lead to the same clinical symptoms such as palsy of cranial nerves II, III, IV, V, and VI, and because they may represent the same treatment problems, they are included in this report. Subarachnoid hemorrhage (SAH) is only rarely encountered in the truly parasellar aneurysms, although aneurysms in the cavernous sinus may also have a portion extending intradurally. Treatment consists ideally of clipping the aneurysm to reduce the mass effect, thereby allowing recovery of the cranial nerve function and/or to prevent rebleeding; however, clipping is not always feasible because an aneurysm neck may be absent. Other conventional therapeutic options consist of occluding the ICA by means of a Silverstone clamp without or with extracranial-intracranial bypass (EIAB; Gelber and Sundt 1980) and recently occluding the ICA by means of a detachable balloon (Serbinenko 1974; Berenstein et al. 1984; Raymond and Theron 1986). Also, obliteration of the aneurysm by the balloon technique with placement of a fixed balloon similar to the previously used Fogarty catheter balloon has been reported to be possible in selected cases.

Our experience with these aneurysms is presented here, which has been gained in the management of 45 patients over the past 8 years, 40 of whom have been treated in the last 4 years.

Material and Methods

The 40 patients treated in the last 4 years represent nearly 12% of all aneurysms treated in our institutions over this time, while the 5 patients treated before represent only 3% of the aneurysm patients treated during this period. The age range was 25–77 years, with a female:male ratio of 2.5:1. A summary of aneurysm location and leading symptoms in the patients treated over the last 4 years is given in Table 1. Following the cranial nerve palsy, transient ischemic attacks, and additional SAH were the most frequent symptoms. In one patient, a severe epistaxis following a frontobasal fracture led to angiography revealing a traumatic giant parasellar aneurysm.

Treatment of these aneurysms changed dramatically with the introduction of interventional angiography in our institutions (D. K.). Earlier in the series, patients underwent surgical exploration of the carotid bifurcation and occlusion on the extracranial ICA with a Silverstone clamp or with a Fogarty catheter. In one young patient a Silverstone clamp occlusion of the ICA was not tolerated and an EIAB performed, after which the patient tolerated well ICA occlusion, and optic nerve function returned to normal.

Table 1. Aneurysm location and presenting symptoms

	n
Intracavernous ICA	26
ICA/ophthalmic junction	3
Supraclinoid ICA	7
Anterior communicating artery	2
Posterior communicating artery	1
Nerves II, III, IV, VI palsy	30
Transient ischemic attacks	4
(Additional) SAH	6
Nasal bleeding	1
Headaches	1
Additional seizures	1
Asymptomatic	1

In the last 4 years, interventional angiography techniques were employed in 25 patients. The protocol utilized , with the patient awake and under constant neurological evaluation, consisted of diagnostic angiography, followed by test occlusion of the ICA for 15–20 min (Berenstein et al. 1984) by means of a balloon attached to a coaxial catheter, the catheter being introduced through the ICA under local anesthesia. In a few patients in the later part of the series, collateral circulation was tested using transcranial Doppler sonography (TCD). If TCD showed good collateral circulation, and test occlusion was well tolerated, permanent occlusion of the ICA was achieved with a detachable balloon placed at the site of the aneurysm where it originated from the ICA. Care was taken not to compromise the collateral circulation, particularly anterior and posterior communicating arteries, by inflating the balloon. One patient did not tolerate the test occlusion of the ICA, and an EIAB was performed. Occlusion of the ICA with a Silverstone clamp was performed in another patient, preceded by EIAB following test occlusion of the ICA. In six patients the aneurysms could be selectively occluded by a balloon. In six patients, direct clipping of the aneurysm was possible. No treatment was given to six patients aged over 70 years in whom occlusion of an ICA was thought to carry excessive risk compared to the expected benefit.

Results

In all 19 patients who underwent occlusion of the ICA by means of a balloon, the neurological deficit was reduced significantly or disappeared completely so that visual function of the affected eye returned to useful or normal. The same held true for two patients in whom ICA was occluded by means of a Silverstone clamp and for the six patients in whom the aneurysm could be occluded selectively. Among the seven patients who underwent clipping of the aneurysm, the cranial nerve damage improved in six and remained unchanged in one. In six patients who did not receive any treatment, cranial nerve palsy remained essentially unchanged in five and worsened progressively in one.

It needs to be mentioned that improvement of visual function occurred rapidly following therapy while eye movement improved slowly, mostly within 6 months.

Complications from balloon techniques occurred in two women aged 56 and 60. While a mild hemiparesis in the first patient cleared rapidly, the complication in the second needs to be given in detail. She had presented with visual dysfunction due to a giant supraclinoid aneurysm at the right intracranial carotid bifurcation. Despite an elongated ICA, it was possible to advance a coaxial catheter to the site of the aneurysm and to occlude its neck (Fig. 1 a, b). The balloon was detached without problems, but the patient

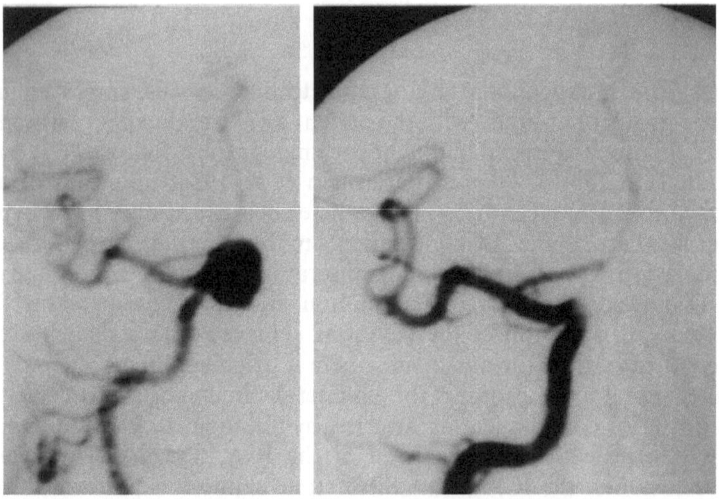

a b

Fig. 1 a–d. A 60-year-old woman presenting with progressive visual field defect and worsening of visual acuity due to a giant aneurysm originating from the supraclinoid internal carotid artery. **a** Preocclusion angiography. **b** Selective balloon occlusions of the aneurysm. **c** Control angiography 4 months later; partial refilling of the aneurysm. **d** Complete occlusion of the aneurysm by ICA balloon occlusion – good collateral circulation via anterior communicating artery

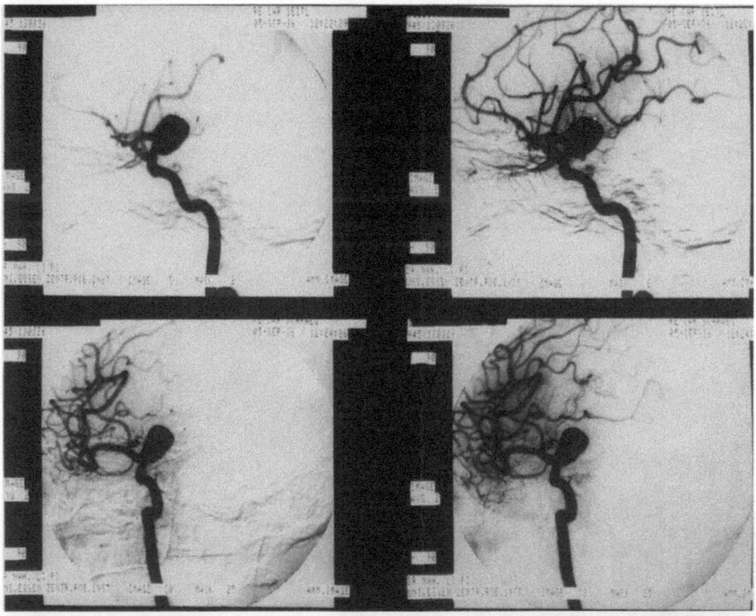

c

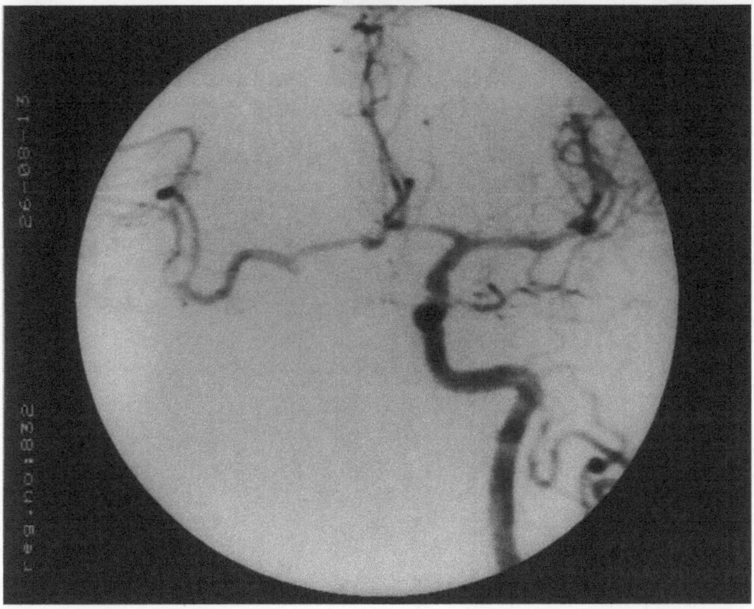

d

Fig. 1c–d

a
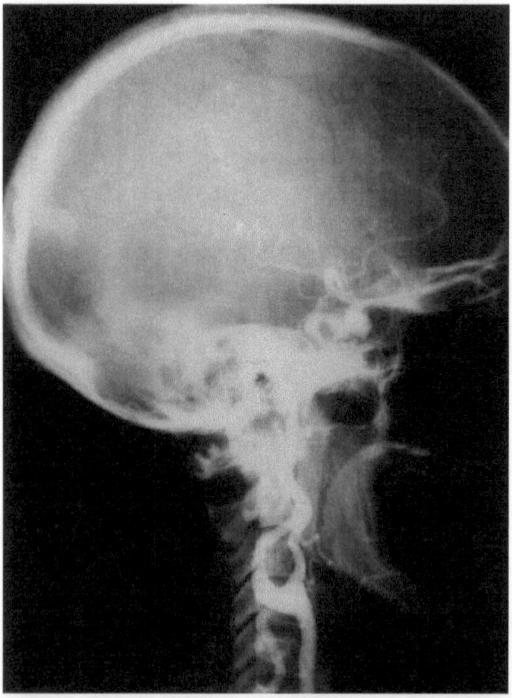

Fig. 2a–d

b
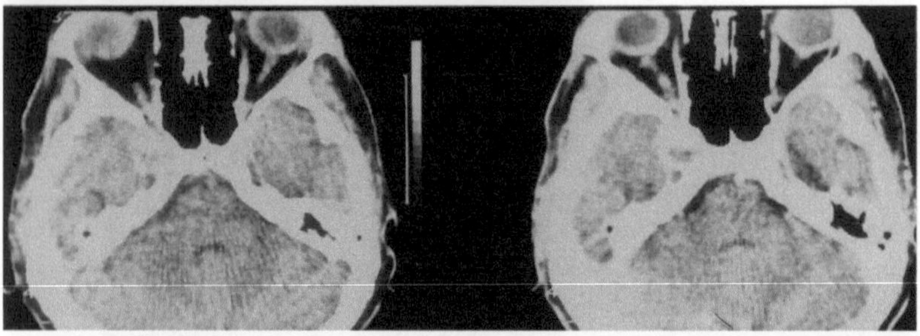

c
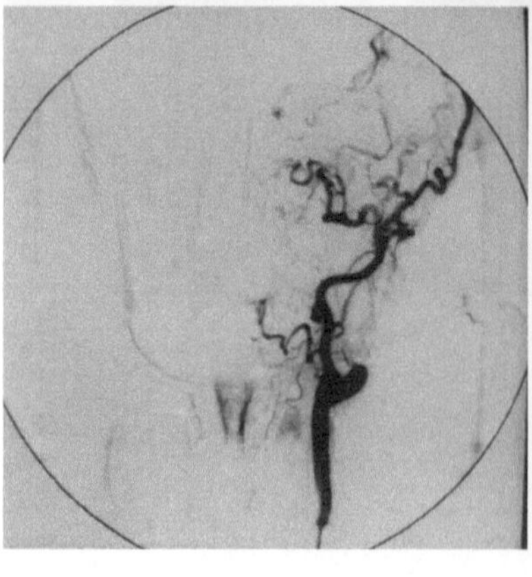

d
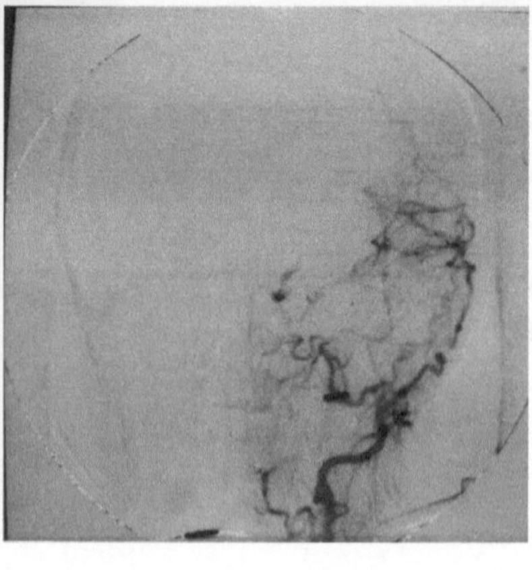

noticed right-sided headaches. Early computed tomography (CT) showed obliteration of the aneurysm, and a later CT scan revealed a right temporal infarction. No new neurological deficit was observed. After rapid improvement of vision and oculomotion, the woman came back 4 months later and complained of the same symptoms as prior to balloon occlusion. Control angiography showed partial filling of the aneurysm (Fig. 1 c), and therefore balloon occlusion of the ICA at the origin of the aneurysm was performed (Fig. 1 d). Following this procedure, cranial nerve function returned to normal.

The other major complication in this series occurred in a 51-year-old woman presenting with left intracavernous aneurysms (Fig. 2 a, b). She did not tolerate digital ICA occlusion, and therefore EIAB was performed together with placement of a Silverstone clamp. Because the bypass functioned well, the ICA was occluded completely the next day, and this was tolerated well (Fig. 2 c, d). While she was still under heparin, she developed an acute right-sided hemiplegia and aphasia which was subsequently shown to be due to a left hemispheric infarction of presumably embolic origin. Retrospective analysis of the postoperative angiogram (Fig. 2 c, d) following complete ICA occlusion showed, besides a well functioning bypass, a tailed contour extending from the left intracranial ICA to the origin of the MCA. It was assumed to represent the negative staining of a tailed thrombus originating from the dead-ended ICA, probably advanced by retrograde flow in the ophthalmic artery.

Discussion

Parasellar aneurysms represent in most surgical series 1.4% (Yasargil 1984) to 3.2% (Sano et al. 1987) of all intracranial aneurysms. Their treatment has been particularly difficult. Direct clipping of the intracavernous ICA aneurysms may be possible with precise knowledge of microsurgical anatomy of the cavernous sinus. Alternatives are occlusion of the ICA either in the neck or at the site of the origin of the aneurysm, without or with adjunctive EIAB. Pendl et al. (1987) experienced 28 patients with intracavernous aneurysms, 13 of which were attacked directly, and 4 of whom died. Sano et al. (1987) reported on 21 infraclinoid ICA aneurysms in a personal series of 660 aneurysms, and clipped 13 of them while 7 were treated by ICA ligation and EIAB. Dolenc et al. (1987) presented 12 aneurysms (out of 15) which could be clipped directly, with preservation of the ICA in 6 (out of 15) and 3 patients dying. Perneczky et al. (1987) reported on 22 aneurysms which they were able to clip. Interestingly, in the latter authors' series, most patients had presented with SAH. Both authors

◄

Fig. 2 a–d. A 51-year-old woman presenting with left nerve III palsy. **a** Initial angiogram, left ICA: intracavernous aneurysm. **b** Initial CT scan without and with contrast: delineating aneurysm. **c/d** Postoperative angiography with good bypass function and filling defect in the distal ICA and proximal middle cerebral artery

reported a mortality rate of about 20%. Direct surgery to intracavernous aneurysms was never attempted in our series, since direct clipping was reserved for purely intradurally located aneurysms, but alternatives were favored in approaching these difficult aneurysms.

Occlusion of the ICA without or with EIAB is a well-established procedure (Gelber and Sundt 1980); ischemic complications may occur immediately following ICA occlusion or later. While cerebral ischemic complication which closely follows ICA occlusion usually is due to poor collateral circulation and can be prevented by performing EIAB, test occlusion of the ICA in the awake patient can obviously not predict the risk of embolic complication subsequent to ICA occlusion, the embolus usually originating from the remaining ICA stump and extending into the middle cerebral artery. This was thought to have occurred in the second complication case presented above.

Balloon occlusion of the ICA at the origin of the aneurysm is a reliable technique which is safe in experienced hands (Berenstein et al. 1984; Raymond and Theron 1986; Fox et al. 1987; Vinuela and Lylyk 1987; Debrun 1987). No mortality and a very low morbidity are also present in our series. Its advantage over the surgical occlusion by being an endovascular technique needs no further discussion; however, it seems interesting that in our experience as well as in that of other authors, the embolic complication is rare. This may be due to the fact that the balloon is placed at the level of the aneurysm and therefore higher up in the ICA than a Silverstone clamp, which leaves the entire ICA as source of emboli.

In order to test the collateral circulation, the test occlusion of the ICA with the patient awake seems to be the most reliable method (Hieshima et al. 1987). Since recently, TCD helps to define collateral pathways and their capacity in order to select patients who might need a bypass, prior to occlusion of the ICA. However, collateral circulatory pathways may also be obstructed temporarily by the aneurysm compressing the communicating arteries, as was the case in one young patient, and in these instances only test occlusion of the ICA at the origin of the aneurysm may give a true picture of collateral capacity.

Occlusion of the aneurysm itself by means of a balloon (Hieshima et al. 1986) is only rarely possible and requires a sufficiently large neck. As occurred in our second complication case, the balloon may loosen for two reasons: weakening of the parent vessel wall and aneurysm neck tissue, or migration or deflation of a detached balloon may permit the aneurysm to dilate again. Therefore, since recently, the balloon is left attached to the coaxial catheter which is left in place, its proximal end being fixed in the subcutaneous layer.

The good clinical outcome in patients treated with endovascular techniques led to the sharp increase of patients presented with these aneurysms to our institutions, and to a percentage of patients with parasellar aneurysms that is higher than in most other purely surgical aneurysm series.

References

Berenstein A, Ransohoff J, Kupersmith M, Falmm E, Graeb D (1984) Transvascular treatment of giant aneurysms of the cavernous and vertebral arteries. Functional investigation and embolization. Surg Neurol 21:3–12

Debrun G, Fox AJ, Drake C, Peerless S, Gervin J, Ferguson G (1981) Giant unclippable aneurysms. Treatment with detachable balloons. AJNR 2:167–173

Debrun GM (1987) Embolization techniques in the treatment of vascular lesions involving the cavernous sinus. In: Dolenc VV (ed) The cavernous sinus. Springer, Wien New York, pp 173–181

Dolenc VV, Cerk M, Sustersic J, Pregelj R, Skrap M (1987) Treatment of intracavernous aneurysms of the ICA and CCFs by direct approach. In: Dolenc VV (ed) The cavernous sinus. Springer, Wien New York, pp 297–310

Fox AJ, Vinuela F, Pelz DM, Peerless SJ, Ferguson GG, Drake CG, Debrun G (1987) Use of detachable balloons for proximal artery occlusion in the treatment of unclippable aneurysms. J Neurosurg 66:40–46

Gelber RB, Sundt TM Jr (1980) Treatment of intracavernous and giant carotid aneurysms by combined internal carotid artery ligation and extra- to intracranial bypass. J Neurosurg 52:1–10

Hieshima GB, Grinnell VS, Mehringer CM (1981) A detachable balloon for therapeutic transcatheter occlusions. Radiology 138:227–228

Hieshima GB, Higashida RT, Halbach VV, Cahan L, Goto K (1986) Intravascular balloon embolization of a carotid-ophthalmic artery aneurysm with preservation of the parent vessel. AJNR 7:916–919

Hieshima GB, Higashida RT, Wapenski J, Halbach VV, Bentson JR (1987) Intravascular balloon embolization of a large mid-basilar artery aneurysm. J Neurosurg 66:124–127

Pendl G, Vorkapic P, Richling B, Koos WT (1987) Strategies in intracavernous saccular aneurysms. In: Dolenc VV (ed) The cavernous sinus. Springer, Wien New York, pp 240–251

Perneczky A, Knosp E, Czech Th (1987) Para- and infraclinoidal aneurysms. Anatomy, surgical technique and report on 22 cases. In: Dolenc VV (ed) The cavernous sinus. Springer, Wien New York, pp 252–271

Raymond J, Theron J (1986) Intracavernous aneurysms: treatment by proximal balloon occlusion of the internal carotid artery. AJNR 7:1087–1092

Sano K, Jain VK, Kato Y, Hoshino M, Shoda M, Tanji H, Kamei Y, Katada K, Kanno T (1987) Treatment of internal carotid infraclinoid aneurysms. In: Dolenc VV (ed) The cavernous sinus. Springer, Wien New York, pp 285–287

Serbinenko FA (1974) Balloon catheterization and occlusion of major cerebral vessels. J Neurosurg 41:125–145

Vinuela F, Lylyk P (1987) Endovascular therapy of vascular lesions of the cavernous sinus. Experience with 129 cases. In: Dolenc VV (ed) The cavernous sinus. Springer, Wien New York, pp 182–197

Yasargil MG (1984) Microneurosurgery II. Thieme, Stuttgart New York, p 37

Direct Operative Approach and Clipping of Intracavernous Nontraumatic Aneurysms

V. LUPRET, D. VIDOVIĆ, and S. VULETIĆ, Zagreb/Yugoslavia

Intracavernous aneurysms constitute a specific entity in today's neurovascular surgery. Only recently the direct operative approach to vascular and tumor pathology has been introduced, and several authors have reported upon successful treatment (Parkinson 1965; Iwabuchi et al. 1978; Dolenc 1985; Hakuba et al. 1987; Perneczky et al. 1987). This paper presents our experience with the direct operative approach and clipping of seven nontraumatic intracavernous aneurysms.

Materials and Methods

The aneurysms of the internal carotid artery that arise between the fibrous ring at the roof of the cavernous sinus and the carotid artery's entrance into the floor of the sinus are called intracavernous. Anatomical studies show that in 15% of cases ophthalmic artery arises from the carotid artery in the cavernous sinus, and thus some aneurysms referred to carotid-ophthalmic should actually be called intracavernous (Ohaegbulam et al. 1987). Our seven patients had aneurysms in the cavernous carotid artery 3 portion of the intracavernous carotid. In cases with this aneurysm location, subarachnoid hemorrhage occurs despite intracavernous location because parts of their sacks sometimes protrude into the subarachnoid space. All patients were operated on via pterional craniotomy. Two different approaches were used for these aneurysms. An ipsilateral pterional approach was used in all aneurysms that had aneurysm sacks oriented laterally. After exposing the optic nerve and the carotid, the anterior clinoid process was drilled away to the extent of approximately 1 cm laterally. The drilling was stopped at the level of the fibrous ring and careful dissection was begun to enter the layer between the carotid wall and the duplicature of the cavernous sinus wall, the space more deeply filled with loose connective tissue. Careful preparation and additional necessary drilling of the bony parts was conducted until the neck was visualized and the appropriate clip positioned. In medially oriented aneurysm an ipsilateral approach is technically less feasible, and so the contralateral modified pterional frontotemporal approach was used. The technique is similar to the ipsilateral approach. The anterior clinoid process is drilled away after circular incision of the dura mater above it between the two optici. After completing the removal of the bone in the extent approximately 1 cm medially to the approached carotid and part of the planum sphenoideum, the layer between the carotid and the cavernous sinus is entered and the neck prepared for clipping. It is important to mention that sometimes, due to the anatomical variations of the sphenoid sinus anatomy, it may be entered and for the prevention of a cerebrospinal fluid leak fibrin glue applied.

Cases and Results

Case 1. A 48-year-old man was admitted to our service after experiencing subarachnoidal hemorrhage 7 days before admission. Neurological examination showed mild right-sided paresis. Computed tomography and cerebral angiography revealed an ischemic area in the left pontine area and an intracavernous aneurysm on the left carotid artery pointing medially. The patient was operated on via a contralateral modified pterional craniotomy and the aneurysm neck occluded. The patient returned to his previous activities with remaining hemiparesis (Fig. 1).

Case 2. A 69-year-old woman was admitted to our service because of vomiting and drowsiness. Upon admission a horizontal nystagmus and paresis of the left oculomotor and facial nerve as well as a positive Babinski sign on the left were recorded. Computed tomography revealed an intracerebral hematoma in the right temporoparietal region and a round hyperdense mass in the right parasellar area which was angiographically found to be a giant intracavernous carotid aneurysm pointing upward. The aneurysm was approached and clipped via the ipsilateral pterional approach. The patient died from pulmonary embolism 4 months after surgery (Fig. 2).

Case 3. A 47-year-old woman presented with subarachnoid hemorrhage and no neurological deficits on admission. Angiographically an intracavernous aneurysm on the right side was visible. The patient was operated via an

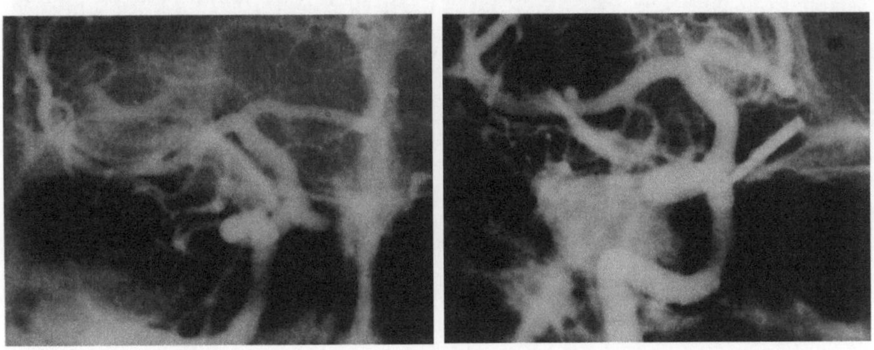

Fig. 1. *Left*, preoperative angiogram; *right*, postoperative angiogram

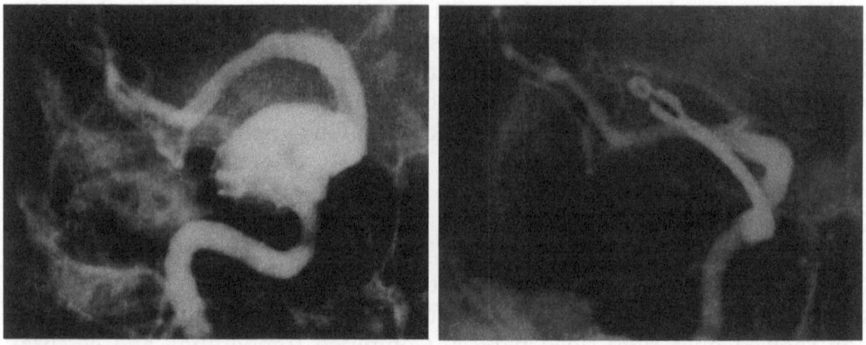

Fig. 2. *Left*, preoperative angiogram; *right*, postoperative angiogram

ipsilateral pterional approach and the aneurysm neck occluded. The postoperative course was uneventful (Fig. 3).

Case 4. A 46-year-old woman presented with subarachnoid hemorrhage. The neurological examination showed no neurological deficits. Cerebral angiography revealed two aneurysms, one at the medial cerebral artery on the left side and the other at the intracavernous carotid on the left side pointing medially. The medial carotid artery aneurysm was clipped via an ipsilateral pterional craniotomy and the cavernous sinus aneurysm via a contralateral modified pterional craniotomy. The patient had transient visual disturbances on the left side that regressed 3 months postoperatively.

Case 5. A 49-year-old woman presented with subarachnoid hemorrhage and otherwise normal neurological findings. The cerebral angiograms showed two aneurysms on the right side, one at the medial cerebral artery and the other at the intracavernous part of the carotid artery pointing medially. The

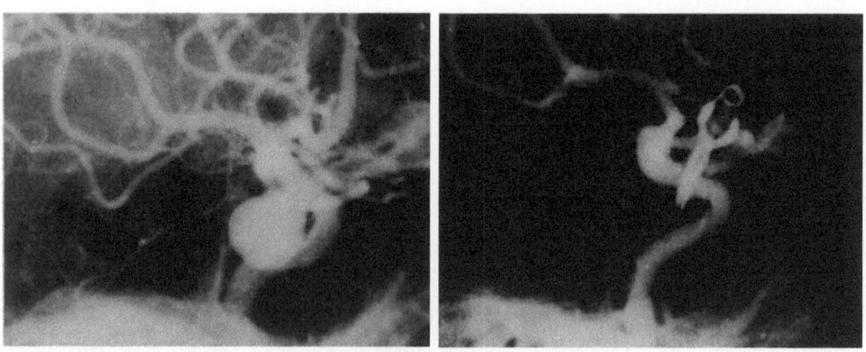

Fig. 3. *Left*, preoperative angiogram; *right*, postoperative angiogram

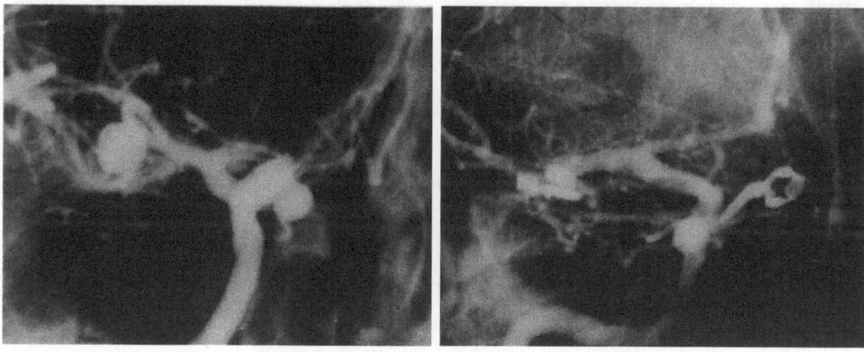

Fig. 4. *Left*, preoperative angiogram; *right*, postoperative angiogram

intracavernous aneurysm was approached via a left modified pterional craniotomy and the medial cerebral artery aneurysm via an ipsilateral pterional craniotomy, and both aneurysms were clipped. The patient had an uneventful recovery (Fig. 4).

Case 6. A 54-year-old man presented to the neurology service with left side hemiparesis and facial and hypoglossal nerve paresis on the left side. Computed tomography showed an ischemic insult in the right medial cerebral artery region. Cerebral angiography showed two aneurysms, one at the carotid-ophthalmic portion of the carotid artery on the left side, and the other at the intracavernous portion of the carotid artery on the right side, pointing medially. Both aneurysms were clipped via a left-sided modified pterional craniotomy (Fig. 5).

Case 7. A 63-year-old patient presented with subarachnoid hemorrhage and mild oculomotor palsy on the right side. Cerebral panangiography showed

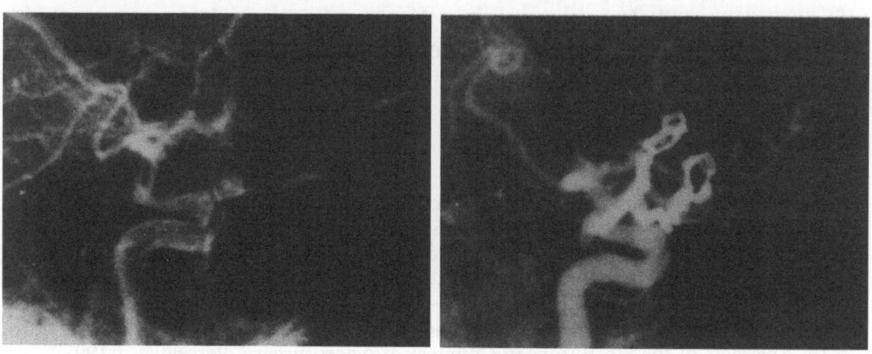

Fig. 5. *Left*, preoperative intracavernous aneurysm angiogram; *right*, postoperative angiogram

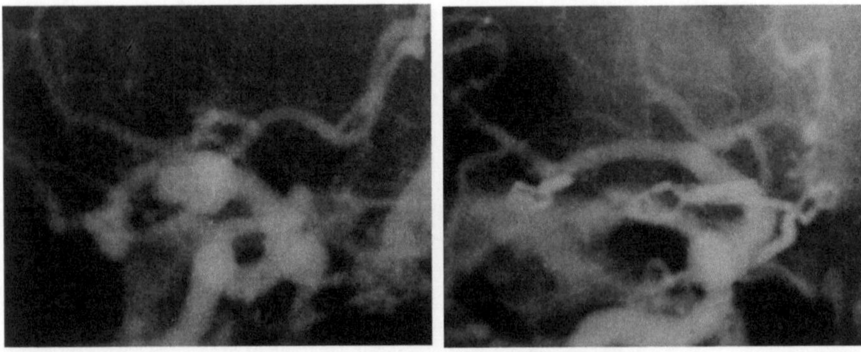

Fig. 6. *Left*, preoperative angiogram; *right*, postoperative angiogram

four aneurysms, all on the right side, one in the cavernous sinus portion of the carotid pointing medially, one at the carotid-ophthalmic portion, one at the posterior communicating artery, and one at the medial cerebral artery. All but the intracavernous aneurysm were clipped via a right pterional craniotomy, and the intracavernous aneurysm was approached and clipped via a contralateral modified pterional craniotomy (Fig. 6).

Discussion

After different approaches for operative management of the intracavernous carotid artery have been advocated (Perneczky et al. 1985; Dolenc et al. 1987; Fukushima 1987; Hakuba 1986), many authors have successfully operated and clipped them. It is our opinion that intracavernous aneurysms should not be treated differently from other intracranial aneurysms. In cases of multiple aneurysms, where the intracavernous sinus aneurysms did not bleed, we decided to treat them by clipping as in other aneurysms; here there exists a risk of rupture as well. It is our opinion that operative surgical treatment with clipping is the treatment of choice in nontraumatic intracavernous aneurysms.

References

Dolenc VV (1985) A combined epi- and subdural direct approach to carotid-ophthalmic artery aneurysms. J Neurosurg 62:667–672

Dolenc VV, Čerk M, Šušteršič J, Pregelj R, Škrap M (1987) Treatment of intracavernous aneurysms of the ICA and CCFs by direct approach. In: Dolenc VV (ed) The cavernous sinus. Springer, Wien New York, pp 297–310

Fukushima T (1987) Direct operative approach for vascular lesions in the cavernous sinus. In: Abstracts of the 8th European Congress of Neurosurgery. Barcelona, p 199

Hakuba A (1986) Surgical approaches to the cavernous sinus: repair of a C-C fistula at the C5 portion of the internal carotid artery. Neurol Surg (Tokyo) 14:601–607

Hakuba A, Matsuoka Y, Suzuku T, Komiyama M, Jin TB, Inoue Y (1987) Direct approaches to vascular lesions in the cavernous sinus via the medial triangle. In: Dolenc VV (ed) The cavernous sinus. Springer, Wien New York, pp 272–284

Iwabuchi T, Suzuki Sh, Sobata E (1978) Intracranial direct operation for carotid-ophthalmic aneurysm by unroofing the optic canal. Acta Neurochir (Wien) 43:163–169

Ohaegbulam S, Dujovny M, Holzer F, Ausman J, Berman K, Diaz F, Mirchandani H (1987) Microsurgical anatomy of the intracavernous carotid arteries and their branches. In: Abstracts of the 8th European Congress of Neurosurgery. Barcelona, p 197

Parkinson D (1965) A surgical approach to the cavernous portion of the carotid artery. Anatomical studies and case report. J Neurosurg 23:474–483

Perneczky A, Knosp E, Vorkapic P, Czech T (1985) Direct surgical approach to infraclinoidal aneurysms. Acta Neurochir (Wien) 76:36–44

Perneczky A, Knosp E, Czech T (1987) Para- and infraclinoidal aneurysms. Anatomy, surgical technique and report on 22 cases. In: Dolenc VV (ed) The cavernous sinus. Springer, Wien New York, pp 252–271

Parasellar Cavernous Angiomas: Report of Two Cases

G. Panzarasa, A. Voci, E. Regalia, E. Geuna, and C. Bellotti, Novara/Italy

Cavernous hemangiomas are relatively rare vascular malformations that may involve any part of the central nervous system but are very unusual in the parasellar region (Namba 1983; Voigt and Yasargil 1976). Because of the complex neurovascular relationships in the cavernous sinus, the surgical approach to parasellar cavernomas presents some serious management problems.

We report two cases, representing 8.6% of 23 intracranial cavernomas histologically verified in the past 10 years.

Case Reports

Case 1. A 61-year-old woman was admitted with a 7-year history of diplopia and a 2-year history of right ptosis, right fronto-orbital headache, and progressive decrease in visual acuity on the right side. Neurological examination revealed blindness and optic atrophy of the right eye, ipsilateral oculomotor palsy with ptosis and right facial hypesthesia. Plain roentgenograms of the skull showed destruction of the petrous bone apex and sellar erosion. Computed tomography (CT) revealed a hyperdense, faintly contrast-enhanced mass in the right middle fossa and parasellar region, eroding the apex of the petrous pyramid (Fig. 1). Cerebral angiography depicted an avascular mass lesion in the right middle fossa with elevation of the internal carotid artery and displacement of adjacent vessels (Fig. 2); no early venous filling or flecked tumor stains were noted. On 11 March 1983 a right frontotemporal craniotomy was performed, under the diagnosis of meningioma. A thinned temporal lobe was retracted, and a smooth, darkred mass was exposed; it was completely enveloped by a capsule that was continuous with and identical to the dura mater of the middle fossa, extending to the sphenoidal ridge anteriorly, the anterior edge of the tentorium posteriorly, and the sella turcica medially. Under the operating microscope the dura was incised, leading to a profuse hemorrhage that was managed by bipolar coagulation and by manual compression with Surgicel. The lesion was completely excised, including the intracavernous sinus part. The cavernous sinus was packed with Surgicel and its lateral dural wall reconstructed and closed by suture. At pathological examination the mass was diagnosed as a cavernoma: it consisted of numerous, closely attached, dilated, thin-walled vascular channels lined with endo-

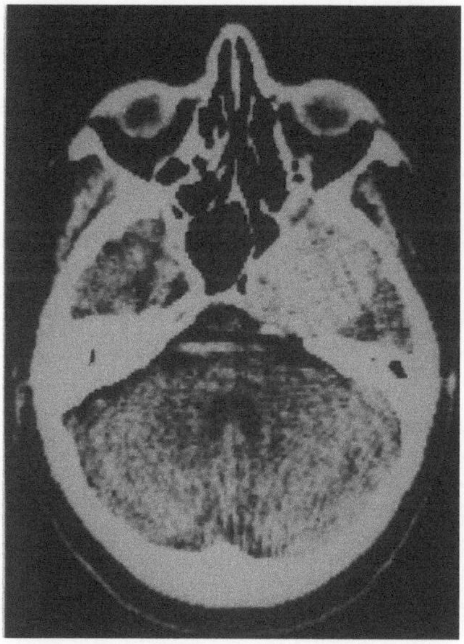

Fig. 1. Enhanced CT scan demonstrating a well-circumscribed, round, high-density mass in the right parasellar region and middle fossa, eroding the apex of the petrous pyramid

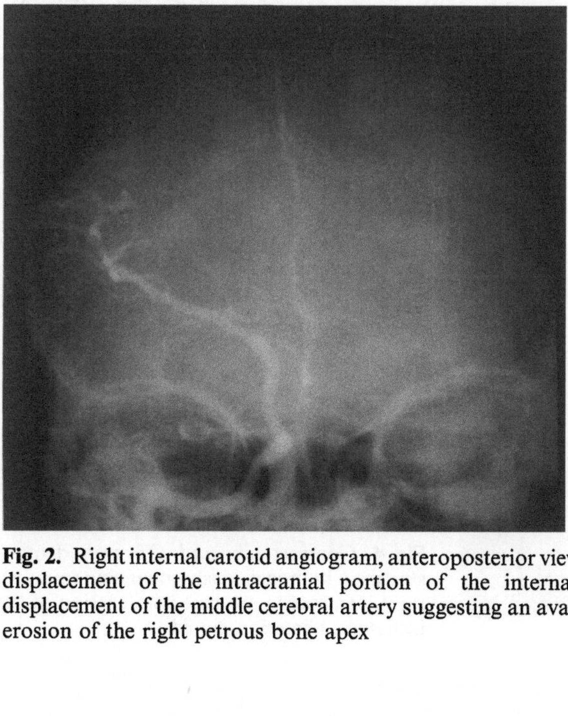

Fig. 2. Right internal carotid angiogram, anteroposterior view, showing elevation and medial displacement of the intracranial portion of the internal carotid artery and upward displacement of the middle cerebral artery suggesting an avascular mass in the middle fossa; erosion of the right petrous bone apex

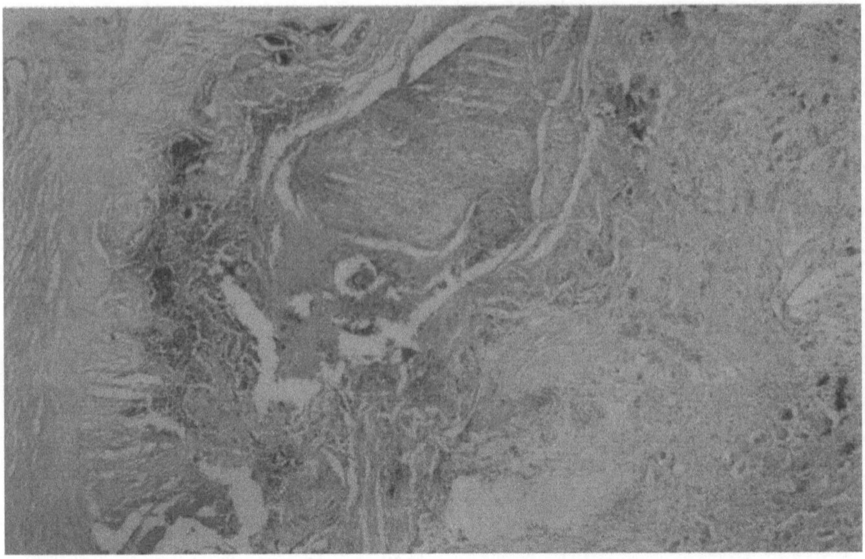

Fig. 3. Photomicrograph of the surgical specimen. *From left to right*, a dural layer with angiomatous lacunae in the innermost part, some trigeminal fibers and ganglion cells, and numerous closely attached thin-walled vascular channels are shown. (H & E stain, × 70)

thelium; no elastic membrane nor intervening neural tissue was evident. A dural layer, showing angiomatous lacunae in the innermost part and some trigeminal ganglion cells from Meckel's cave were unusual findings (Fig. 3). The postoperative course was uneventful; however, total ophthalmoplegia and blindness in the right eye and anesthesia in the second division of the ipsilateral trigeminal nerve persisted. Now, 5 years after operation, no other abnormality is present, and a follow-up CT scan reveals no evidence of recurrence.

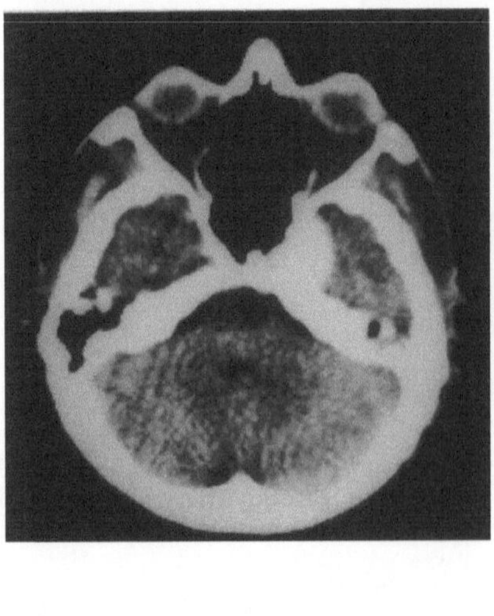

Fig. 4. Contrast-enhanced CT scan demonstrating a homogeneously enhancing roundish right intracavernous mass with erosion of the ipsilateral posterior clinoid

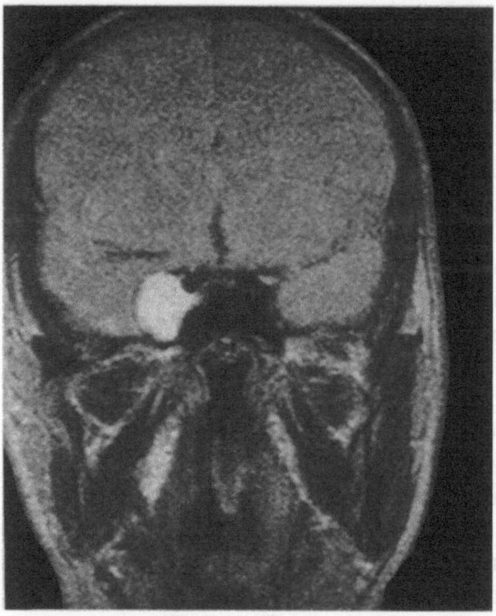

Fig. 5. Coronal magnetic resonance image showing an enlarged right cavernous sinus occupied by hyperintense homogeneous tissue bulging the lateral wall

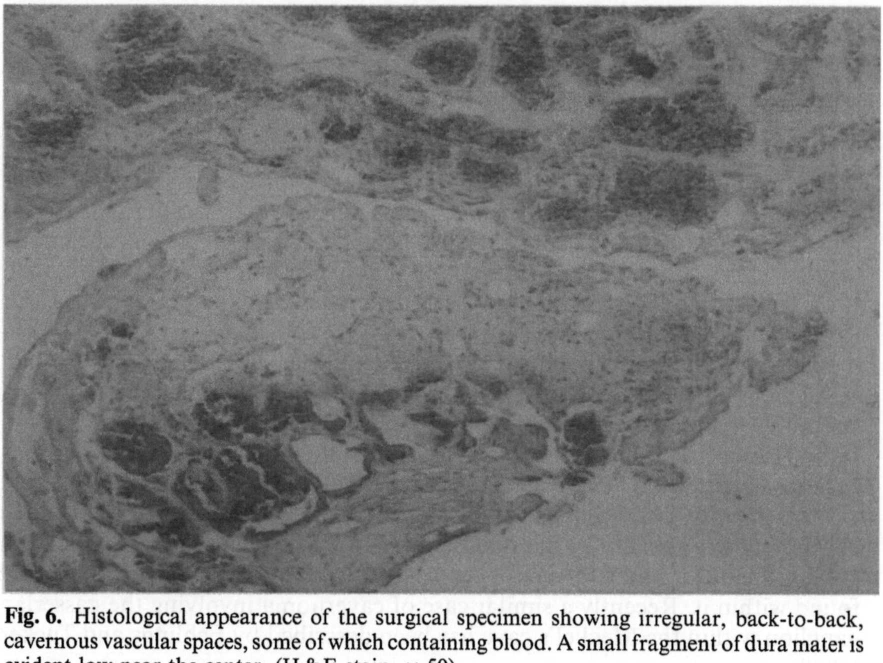

Fig. 6. Histological appearance of the surgical specimen showing irregular, back-to-back, cavernous vascular spaces, some of which containing blood. A small fragment of dura mater is evident low near the center. (H&E stain, × 50)

Case 2. Within 5 years a 49-year-old right-handed man developed two transient episodes of diplopia and right ptosis. On admission, neurological examination was normal. A contrast-enhanced CT scan revealed a hyperdense roundish mass adjacent to the right cavernous sinus and the erosion of the ipsilateral posterior clinoid (Fig. 4). Magnetic resonance imaging (MRI) on the spin-echo pulse sequences showed an enlarged right cavernous sinus occupied by hyperintense homogeneous tissue bulging the lateral wall (Fig. 5). On angiography a slight elevation of the parasellar segment of the right internal carotid artery was noted. On 9 April 1986 the patient underwent a right pterional craniotomy; the temporal lobe was retracted, and with microsurgical technique the internal carotid artery was identified and traced from the apex of the petrous pyramid to its entry into the cavernous sinus. Under a slight intraoperative hypotension the bulged dura was incised within Parkinson's triangle, and a reddish, vascular encapsulated mass appeared. Cranial nerves and internal carotid artery were identified and isolated from the lesion that was shrunk with bipolar coagulation; then piecemeal subtotal removal was performed, leaving a small fragment adherent to the carotid artery. The cavernous sinus was packed with Surgicel and its dural wall closed by suture. Pathologically, the mass was a cavernous hemangioma (Fig. 6). The postoperative course was uneventful, apart from a transient diplopia on the 7th day. A follow-up CT scan performed 2 years later revealed no evidence of a recurrent lesion.

Discussion

Intracranial cavernous angiomas are histologically benign hemartomas representing 5%–13% of central nervous system vascular malformations (Jellinger 1975; McCormick et al. 1968; Russell and Rubinstein 1977; Sarwar and McCormick 1978; Simard et al. 1986). They may involve any part of the brain; most are located in the cerebral hemispheres; the remainder occur in the lateral ventricles, basal ganglia, brainstem and cerebellar hemispheres; occasionally they are located in the dura mater and in the cavernous sinus region (Ito et al. 1977; Kawai et al. 1978; Mori et al. 1980; Namba 1983; Ogasawara et al. 1973; Rosenblum et al. 1986; Voigt and Yasargil 1976).

The parasellar cavernomas may derive from the surrounding intracranial extracerebral tissues, most often arising from the middle fossa and involving structures within the cavernous sinus (extracavernous cavernoma; Mori et al. 1980; Namba 1983) or from the constituents of the sinus itself (intracavernous cavernoma; Ito et al. 1977; Kawai et al. 1978; Rosenblum et al. 1986).

Our former case seemed to originate from the cavernous sinus, then developing extraaxially in the middle fossa, as case 1 referred by Mori et al. (1980); the dura mater formed its capsule, and gasserian ganglion cells were found within it. Recently a similar case of cavernoma involving the gasserian ganglion within the Meckel's cave has been described by Fehlings and Tucker (1988).

The latter case was an effective intracavernous cavernous hemangioma, as the case reported by Rosenblum et al. (1986), i.e., the vascular, encapsulated mass was found within the cavernous sinus, under its dural wall.

Histologically, cavernous hemangiomas are collections of thin-walled vascular channels lined only by endothelium and connective tissue; they have no elastic membrane, no muscle layer, and no intervening neural parenchyma (Giombini and Morello 1978; McCormick 1966; Rosenblum et al. 1986; Russell and Rubinstein 1977; Simard et al. 1986).

Although they are nonneoplastic lesions, they may grow by hemorrhage and by progressive enlargement of the vascular channels causing clinical symptoms due to pressure on neural structures (Rosenblum et al. 1986; Voigt and Yasargil 1976). In the parasellar region they may occur with failing vision, defects in the visual field, double vision, blepharoptosis, facial hypesthesia. Sometimes the pituitary gland is involved. The onset of signs and symptoms is usually insidious; occasionally symptoms are exacerbated during pregnancy; an acute onset may be due to hemorrhage from or into the hemartoma (Kawai et al. 1978; Mori et al. 1980; Namba 1983; Simard et al. 1986; Voigt and Yasargil 1976).

Bony erosions of the sellar and parasellar region are common findings upon X-ray examination of the skull (Ito et al. 1977; Kawai et al. 1978; Mori et al. 1980; Pàsztor et al. 1964; Simard et al. 1986).

CT usually shows a high-dense mass, contrast-enhanced homogeneously; however, the pattern of enhancement is not distinguishable from that of a meningioma (Mori et al. 1980; Numaguchi et al. 1979; Savoiardo et al. 1983; Simard et al. 1986). Therefore, while CT studies may detect a lesion consistent with a cavernous malformation, the CT findings are not specific for this entity.

The sensitivity of MRI increases the probability of detecting a cavernous hemangioma; residual macrophages laden with hemosiderin, related to repeated subclinical hemorrhages or to slow lysis of red blood cells (Bellotti et al. 1985), provide in fact an indelible tissue signature: the decreased signal intensity (SI) on T2-weighted images. On T2-weighted MRI, the combination of a reticulated core of mixed SI with a surrounding rim of decreased SI strongly suggests the diagnosis of cavernous malformation (Rigamonti et al. 1987). In our second case the MRI demonstrated the cavernoma as a hyperintense homogeneous tissue filling the cavernous sinus.

Angiographically parasellar cavernomas are mostly shown as an avascular mass displacing adjacent vessels (due to slow circulation, or previous hemorrhage, or extensive thrombosis), without visualization of tumor stain or feeding arteries and draining veins; often the lesion is fed by the meningohypophyseal trunk (Kawai et al. 1978; Namba 1983; Rigamonti et al. 1987; Voigt and Yasargil 1976).

In any way, the possibility of cavernomas should be kept in mind when investigating a suspected meningioma in the parasellar region, especially in middle-aged women (Mori et al. 1980; Simard et al. 1986). Furthermore, it should be kept in mind that an angiographically avascular mass does not always represent a really avascular lesion; hemorrhage from cavernomas was reported to have been very severe in most cases (Kawai et al. 1978; Mori et al. 1980; Simard et al. 1986).

Total removal of a parasellar cavernous hemangioma may be difficult because of profuse bleeding from the cavernoma itself or from the cavernous sinus and the superior petrosal sinus, and because of the relationships with the cranial nerves. Namba reported in his review (1983) that complete resection was claimed in fewer than half of the cases, with significant operative mortality of about 25%. But at present, the advent of microsurgical techniques has made direct approach and total removal of intra-extracavernous sinus cavernomas feasible, resulting in the resolution of the present complaints, despite the extremely hazardous location (Dolenc 1983; Malis 1976; Malis 1985; Parkinson 1965; Rosenblum et al. 1986; Voigt and Yasargil 1976).

References

Bellotti C, Medina M, Oliveri G et al. (1985) Cystic cavernous angiomas of the posterior fossa. Report of three cases. J Neurosurg 63:797–799

Dolenc V (1983) Direct microsurgical repair of intracavernous vascular lesions. J Neurosurg 58:824–831

Fehlings MG, Tucker WS (1988) Cavernous hemangioma of Meckel's cave. Case report. J Neurosurg 68:645–647

Giombini S, Morello G (1978) Cavernous angiomas of the brain: account of fourteen personal cases and review of the literature. Acta Neurochir (Wien) 40:61–82

Ito J, Takahashi M, Saito A et al. (1977) A case of cavernous hemangioma of the middle fossa with visualization of feeding artery and faint tumor stain in cerebral angiography. Jpn J Clin Radiol 22:339–344

Jellinger K (1975) The morphology of centrally-situated angiomas. In: Pia HW, Gleave JRW, Grote E, Zierski J (eds) Cerebral angiomas: advances in diagnosis and therapy. Springer, Berlin Heidelberg New York, pp 9–20

Kawaguchi S, Ohsawa T (1965) A case of cavernous hemangioma at the base of the skull. Clin Neurol (Tokyo) 5:705–708

Kawai K, Fukui M, Tanaka A et al. (1978) Extracerebral cavernous hemangioma of the middle fossa. Surg Neurol 9:19–25

Malis LI (1976) Tumors of the parasellar region. In: Thompson RA, Green JR (eds) Neoplasia in the central nervous system. Advances in neurology, vol 15. Raven, New York, pp 281–299

Malis LI (1985) Surgical resection of tumors of the skull base. In: Wilkins RH, Rengachary SS (eds) Neurosurgery, vol 1. McGraw-Hill, New York, pp 1011–1021

McCormick WF (1966) The pathology of vascular ("arteriovenous") malformations. J Neurosurg 24:807–816

McCormick WF, Hardman JM, Boulter TR (1968) Vascular malformations ("angiomas") of the brain, with special reference to those occurring in the posterior fossa. J Neurosurg 28:241–251

Mori K, Handa H, Gi H et al. (1980) Cavernomas in the middle fossa. Surg Neurol 14:21–31

Namba S (1983) Extracerebral cavernous hemangioma of the middle cranial fossa. Surg Neurol 19:379–388

Numaguchi Y, Kishikawa T, Fukui M et al. (1979) Prolonged injection angiography for diagnosing intracranial cavernous hemangiomas. Neuroradiol 131:137–138

Ogasawara S, Nakazawa K, Karsura S et al. (1973) Cavernoma on the intracranial base. Report of a case. Brain Nerve 25:735–737

Parkinson D (1965) A surgical approach to the cavernous portion of the carotid artery. Anatomical studies and case report. J Neurosurg 23:474–483

Pàsztor E, Szabò G, Slowik F et al. (1964) Cavernous hemangioma of the base of the skull. Report of a case treated surgically. J Neurosurg 21:582–585
Rigamonti D, Drayer BP, Johnson PC et al. (1987) The MRI appearance of cavernous malformations (angiomas). J Neurosurg 67:518–524
Rosenblum B, Rothman AS, Lanzieri C et al. (1986) A cavernous sinus cavernous hemangioma. Case report. J Neurosurg 65:716–718
Russell DS, Rubinstein LJ (1977) Pathology of tumours of the nervous system. Williams & Wilkins, Baltimore, pp 127–141
Sarwar M, McCormick WF (1978) Intracerebral venous angioma. Case report and review. Arch Neurol 35:323–325
Savoiardo M, Strada L, Passerini A (1983) Intracranial cavernous hemangiomas: neuroradiologic review of 36 operated cases. AJNR 4:945–950
Simard JM, Garcia-Bengochea F, Ballinger WE Jr et al. (1986) Cavernous angiomas: a review of 126 collected and 12 new clinical cases. Neurosurgery 18:162–172
Voigt K, Yasargil MG (1976) Cerebral cavernous hemangiomas or cavernomas. Neurochirurgia 19:59–68

Parasellar Cavernomas Mimicking Meningiomas

J. ZENTNER, Bonn/FRG, G. SCHROTH, E. HEISS, and P. J. EGAN, Tübingen/FRG

Introduction

Cavernomas account for 5%–13% of all intracranial vascular malformations [14]. Usually, they evade direct angiographic visualization [3, 9, 13]. On native computed tomography (CT) they appear as well-demarcated, inhomogenous, hyperdense lesions [4, 15, 18]. Magnetic resonance imaging (MRI) has proven a useful diagnostic tool in these tumors, typically showing a high signal intensity on T1-weighted images with a surrounding irregular, dark rim, more evident on the T2-weighted images [9, 13]. Unusual MRI features of two parasellar cavernomas that we observed, differing from representative findings in these tumors, led us to report on our patients.

Case Reports

Case 1. A 74-year-old woman suffered trigeminal neuralgia on the right side for 2 years. Neurological examination on admission revealed incomplete paresis of the 3rd, 4th, and 6th cranial nerves on the right side as well as mild hemiparesis on the left. Angiographic examination showed a temporal space-occupying lesion with blushlike and partly lacunar contrast enhancement in the venous phase. On the native CT scan, there was a hyperdense, right-sided, temporal lesion involving the cavernous sinus without perifocal edema. The tumor enhanced noticeably with contrast medium (Fig. 1). MRI showed a well-demarcated lesion with a signal intensity similar to that of the surrounding brain on the T1-weighted images. On T2-weighted images the tumor appeared noticeably highlighted showing relaxation times ranging from 140 to 240 ms (mean 200 ms) as calculated from 16 echos of a multispin echo sequence (Fig. 2). Upon operation on 6 June 1985 a well-encapsulated, highly vascularized tumor was found without any signs of previous hemorrhage. The postoperative course was uneventful. Histological examination of the specimen showed a typical cavernoma without signs of an earlier hemorrhage.

Case 2. A 59-year-old woman suffered frontal headaches for 2 months. Neurological examination on admission revealed hypesthesia in the right half of the face and reduced vision on the right. Angiographic examination showed in the late arterial phase a small homogenously vascularized lesion in

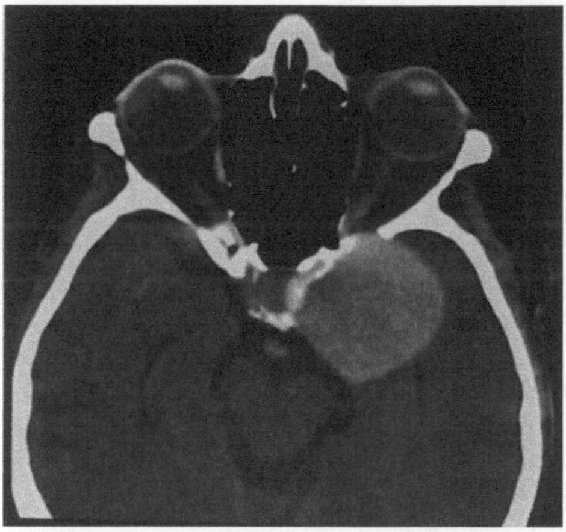

Fig. 1. Axial CT scan of case 1 shows a homogeneously enhancing temporomedial and parasellar space-occupying lesion involving the cavernous sinus

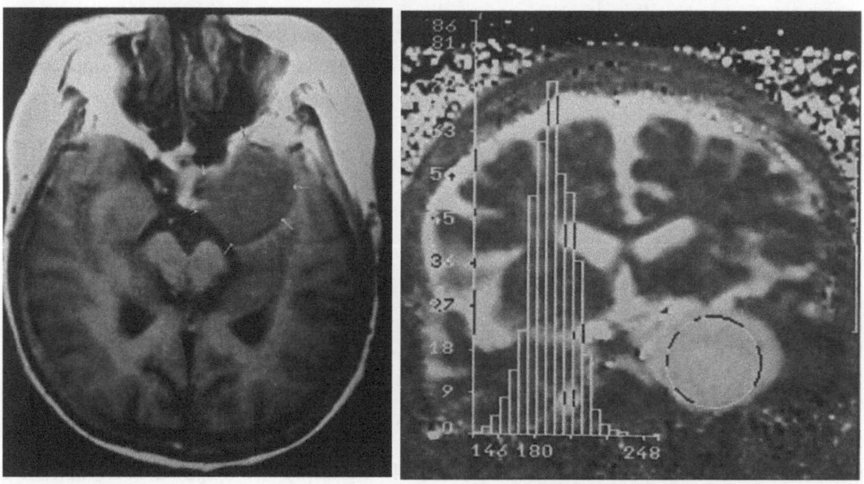

Fig. 2. MRI of case 1 (1.5 T. head coil). T1-weighted spin-echo sequence (SE, 400 ms, 30 ms) demonstrates a low signal intensive lesion (*left side; arrows*). T2-values calculated by linear regression of 16 echos of a CPMG sequence (*right side*) range between 140 and 240 ms

projection on the right cavernous sinus (Fig. 3). CT scan revealed a contrast-enhanced and well-demarcated temporomedial lesion involving the cavernous sinus. On T1-weighted MRI we found a signal intensity similar to that of the surrounding brain (Fig. 4), whereas on T2-weighted MRI the lesion was highlighted, showing relaxation times with mean values of 220 ms as calculated from 16 echos of a multispin echo sequence. Intraoperatively (12 September 1985), the tumor was found to be highly vascularized and well encapsulated.

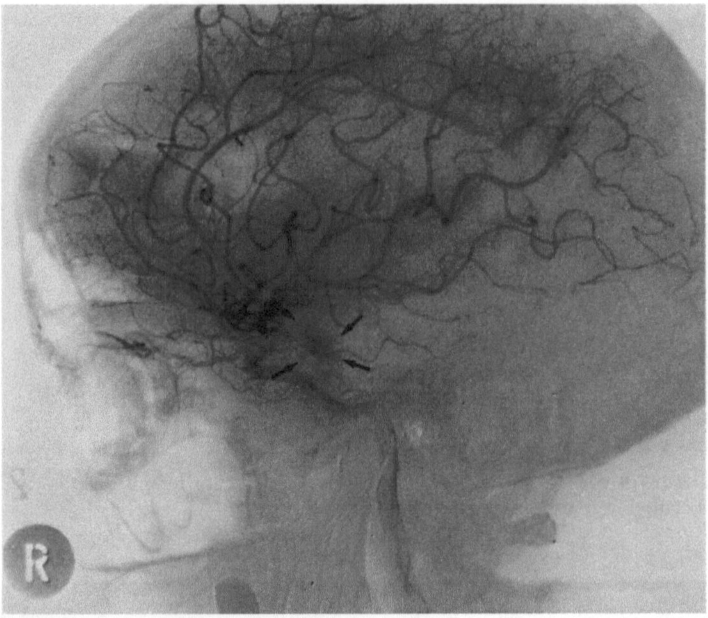

Fig. 3. Internal carotid angiogram of case 2 shows a parasellar lesion which is homogeneously opacified in the late arterial phase (*arrows*)

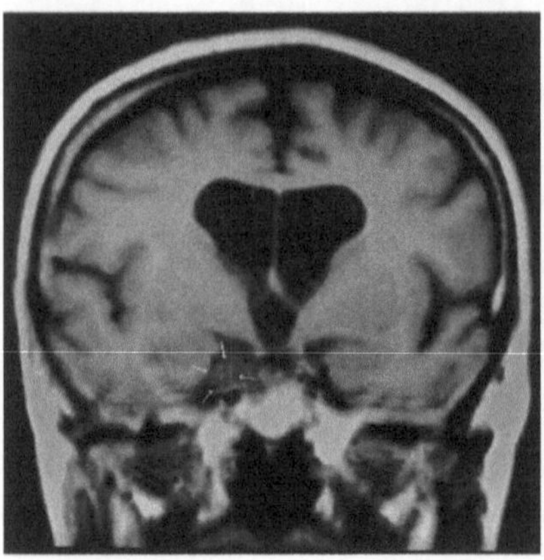

Fig. 4. Coronary T1-weighted MRI of case 2 (SE, 400 ms, 30 ms) reveals a low signal intensive lesion (*arrows*)

The postoperative course was uneventful. Histological examination of the specimen showed a typical cavernoma. There were no signs of an earlier hemorrhage.

Discussion

Cavernomas are known to occur in brain tissue [17, 20], in the ventricles [5, 12], intraspinally [10, 19], in the orbita and eye [6, 11] and in the digestive tract, especially in the liver [1, 2, 7]. Although the natural history of these tumors is still unknown, they frequently become symptomatic by repeated hemorrhage [8]. Consequently, high signal intensity and a surrounding irregular, dark rim representing hemosiderin deposits have been established as typical MRI features of cavernomas [9, 13, 19].

However, the MRI appearance of the cavernomas presented here is quite different from these usual findings. On T1-weighted images signal intensity was similar to that of the adjacent brain tissue. Thus, with additional regard to localization, contrast enhancement in CT scan as well as direct angiographic visualization, the tumors were thought to be meningiomas. On T2-weighted MRI the cavernoma-typical dark rim was lacking. On the other hand, T2-weighted images were highlighted in both cases, showing relaxation times ranging from 140 to 240 ms. This contrasts with our experience with meningiomas, in which T2 relaxation times ranging from 80 to 140 ms were found [16]. Therefore, T2-weighted MRI indicated tumors more vascularized than meningiomas.

In conclusion, the cavernomas presented here are noteworthy both for unusual localization and neuroradiological findings. The atypical MRI features of our cases are thought to be attributable to the lack of prior hemorrhage as confirmed by intraoperative and histopathological findings.

Summary

Two cases of parasellar space-occupying lesions are presented. Both patients were neurosurgically treated under the tentative diagnosis of a meningioma as suggested by neuroradiological findings including CT scan, angiography and T1-weighted MRI. However, both lesions were highlighted on T2-weighted MRI, showing relaxation times of up to 240 ms, thus indicating tumors more vascularized than meningiomas. On operation, highly vascularized and well demarcated tumors were found which were histologically classified to be cavernomas. Neither intraoperatively nor upon histological examination signs of an earlier hemorrhage were found. In conclusion, although MRI features of our cavernomas differ from usual findings in these tumors, T2-weighted images were decisive in both patients with regard to differentiation from meningiomas. It is thought that the unusual MRI findings in our two cases can be explained by the lack of prior hemorrhage.

References

1. Abrams RM, Berenbaum ER, Santos J, Lipson J (1969) Angiographic features of cavernous hemangiomas of the liver. Radiology 92:308–312
2. Barnett PH, Zerhouni EA, White RJ Jr, Segelmann SS (1980) Computed tomography in the diagnosis of cavernous hemangiomas of the liver. AJR 34:439–447
3. Bogren H, Svalander I, Wickbom I (1970) Angiography in intracranial cavernous hemangiomas. Acta Radiol (Diag) 10:81–89

4. Brühlmann Y, Tribolet N de, Bereny J (1985) Les angiomes caverneux intracerebraux. Neurochirurgie 31:271–279
5. Coin CG, Coin JW, Glover MB (1977) Vascular tumors of the choroid plexus: diagnosis by computed tomography. J Comput Assist Tomogr 1:146–148
6. Frenkel M, Russe HP (1967) Retinal teleangiectasia associated with hypogammaglobulinaemia. Am J Ophthalmol 63:215–220
7. Glazer GM, Aisen AM, Francis IR, Gyves JW, Lande J, Adler DD (1985) Hepatic cavernous hemangioma: magnetic resonance imaging. Radiology 155:417–420
8. Giombini S, Morello G (1978) Cavernous angiomas of the brain. Account of fourteen personal cases and review of the literature. Acta Neurochir 40:61–82
9. Gomori JM, Grossmann RI, Goldberg HI, Hackney DB, Zimmermann RA, Bilaniuk LT (1986) Occult cerebral vascular malformations: high-field MR imaging. Radiology 158:707–713
10. Hadlich R (1903) Ein Fall von Tumor cavernous des Rückenmarks mit besonderer Berücksichtigung der neueren Theorien über die Genese des Cavernoms. Virchows Arch 172:429–445
11. Harris GJ, Jakobiec FA (1979) Cavernous hemangioma of the orbit. J Neurosurg 51:219–228
12. Iwasa H, Indei I, Sato F (1983) Intraventricular cavernous hemangioma. Case report. J Neurosurg 59:153–157
13. Lemme-Plaghos L, Kucharczyck W, Brandt-Zawadzki M, Uske A, Edwards M, Norman D, Newton Th (1986) MRI imaging of angiographically occult vascular malformations. AJNR 7:217–222
14. McCormick WF, Schochet SS (1976) Atlas of cerebrovascular disease. Saunders, Philadelphia
15. Savoiardo M, Strada L, Passerini A (1983) Intracranial cavernous hemangiomas: neuroradiologic review of 36 operated cases. AJNR 4:945–950
16. Schroth G, Grodd W, Guhl L, Grauer M, Klose U, Niendorf HP (1987) Magnetic resonance imaging in small lesions of the central nervous system. Improvement by gadolinium-DTPA. Acta Radiol 28:667–672
17. Simard JM, Garcia-Bengochea F, Ballinger WE, Mickle JP, Quisling RG (1986) Cavernous angioma: a review of 126 collected and 12 new clinical cases. J Neurosurg 18:62–172
18. Terao H, Hori T, Matzutani M, Okeda R (1976) Detection of cryptic vascular malformations by computerized tomography. Report of two cases. J Neurosurg 51:546–551
19. Ueda S, Saito A, Inomori S, Kim J (1987) Cavernous angioma of the cauda equina producing subarachnoid hemorrhage. Case report. J Neurosurg 66:134–136
20. Yoshimoto T, Suzuki J (1986) Radical surgery on cavernous angiomas of the brainstem. Surg Neurol 26:72–78

Ophthalmological Manifestations in Cases of Giant Parasellar Aneurysm

U. GIERS, Detmold/FRG and E. SCHÜTTE, ULM/FRG

Introduction

In a normal urban population the incidence of primary subarachnoid hemorrhage is about 16 per 100000. Aneurysms account for 76% of subarachnoid hemorrhage (Pakarinen 1967), therefore the annual incidence of ruptured aneurysm is 1 in 10000. Autopsy figures show that about 1%–2%, i.e., 100–200 in 10000, have an intracranial aneurysm greater than 5 mm in diameter. Thus the likelihood of rupture is 20% in the most likely seen persons over 40 years of age. Yasargil (1984) reports results of 1012 neurosurgically treated cases of intracerebral aneurysm. Of these, 941 had been detected by subarachnoid bleeding, whereas only 71 (7%) were either asymptomatic unruptured aneurysms discovered incidentally during various neuroradiological procedures or had produced focal or generalized perifocal neurological deficits due to their size and localization.

Ophthalmologists are primarily involved when these lesions touch upon areas of ophthalmological interest. This is particularly the case in the parasellar region, of course. Approximately 5.8% of all cases of intracranial aneurysm can be detected prior to rupture with subarachnoid bleeding through neurological dysfunction. Of these 5.8% we are naturally able to notice only those cases that involve recognizable ophthalmological early manifestations. They show a number of characteristics which are of interest because, according to Huber (1986), these are the most important focal neurological signs for aneurysm. These include: the oculomotor (III) nerve, trochlear (IV) nerve, abducens (VI) nerve, visual field loss, visual acuity loss, visual evoked responses, and corneal sensitivity (nerve V). Barr et al. (1971) have presented an overview of the symptoms of intracavernous carotid aneurysm and established that lesions of the third, fourth, and sixth cranial nerves and vision reductions tend to occur more frequently than does involvement of the fifth cranial nerve, since its branches are not generally compressed against bony structures.

Since, according to Brihaye (1979) and Duke-Elder (1971) aneurysm is typically seen in women over the age of 50 years – perhaps with the exception that Miller and Hinkley (1970) see it as being responsible for 12% of perinatal maternal mortality – we, in a military hospital, probably have the best chance of discovering symptoms of an aneurysm prior to rupture in a young man.

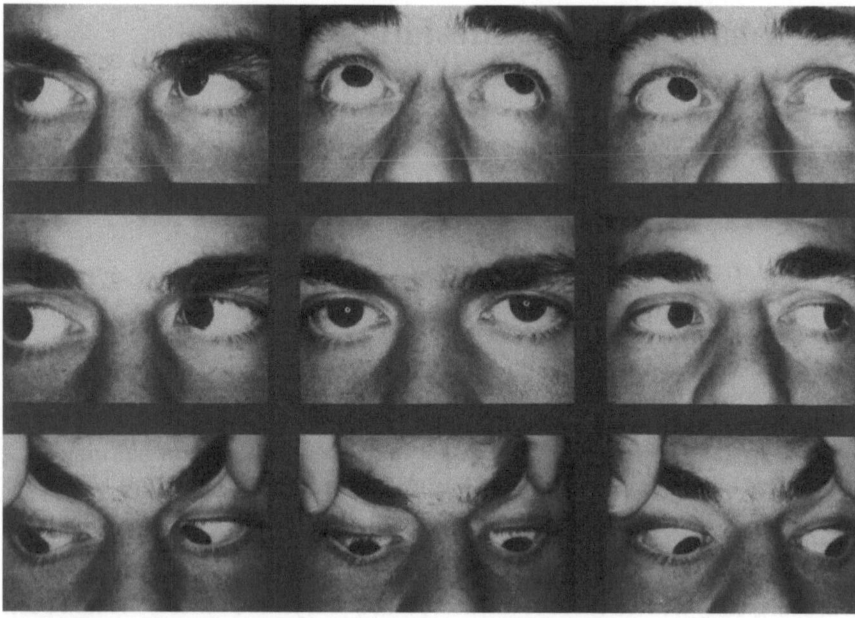

Fig. 1. Examination of the motility-revealed restricted abduction on the right

Case Report

A 21-year-old soldier appeared in our outpatient department and reported suffering from diplopia for 3 days and from diffuse headaches for a number of weeks. The headaches had occurred repeatedly since a car accident in 1983 and had often persisted for 2–3 weeks in the form of a biparietal feeling of pressure. A spontaneous recovery had always occurred after these phases and, this time as well, the headaches had spontaneously disappeared with the occurrence of double vision.

In the course of the subsequent investigations the findings were as follows: Vision was 20/16 on both sides. The front sections of the eyes were not susceptible to stimuli. An examination of motility revealed restricted abduction when looking upward and to the right, to the right and downward to the right (Fig. 1).

Examination on the Hess screen confirmed an impaired function of the lateral rectus muscle with slightly overshooting adduction on the left in the sense of an abducent palsy on the right (Fig. 2). The intraocular pressure was

---

Fig. 2. The Hess screen examination confirmed an abducent palsy on the right with an overshooting adduction on the left

Fig. 3. Nasal lower visual field defect on the right and more or less normal isopters on the left

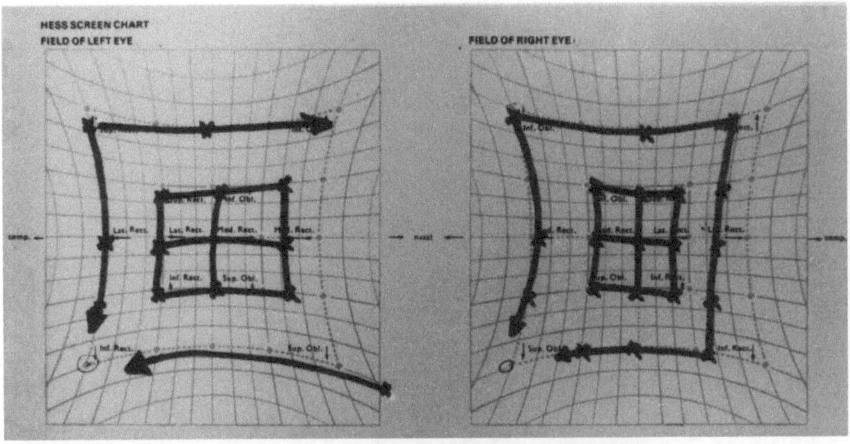
HESS SCREEN CHART
FIELD OF LEFT EYE
FIELD OF RIGHT EYE
temp.
nasal
temp.
Lat. Rect.
Med. Rect.
Inf. Rect.
Sup. Obl.
Inf. Obl.
Sup. Rect.

2

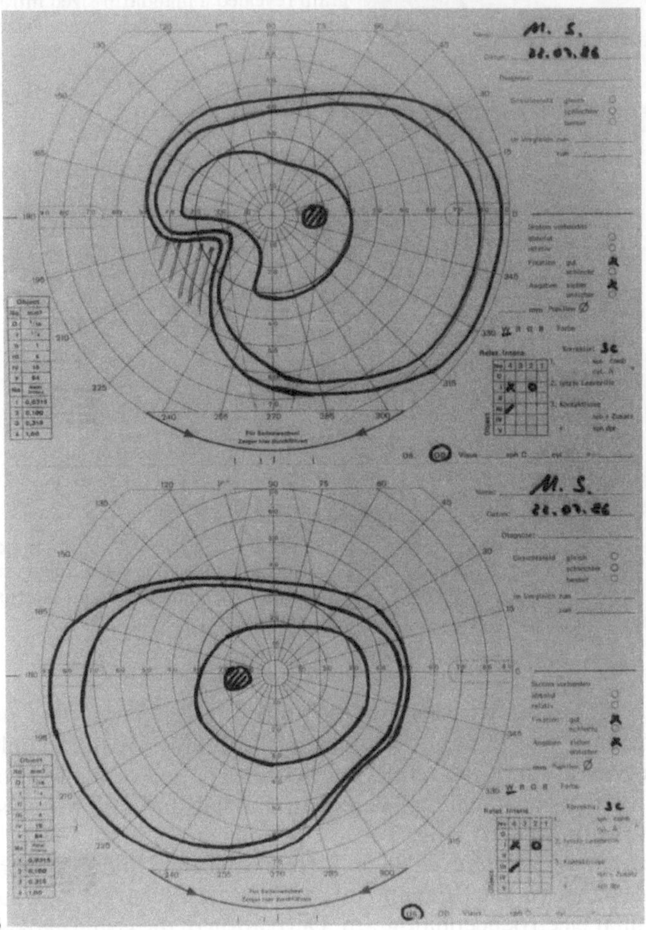
M. S.
M. S.

3

Figs. 2–3

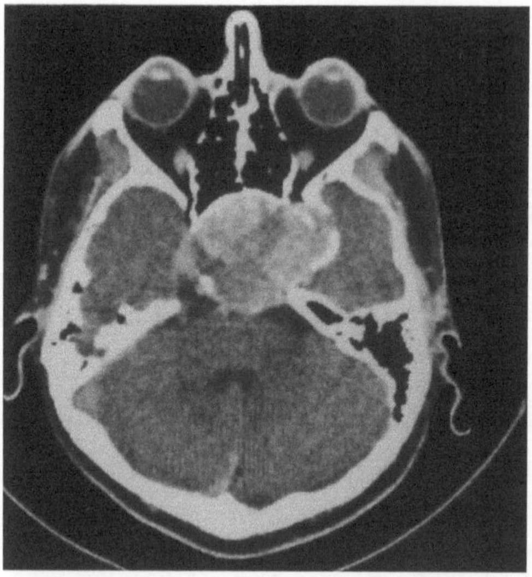

Fig. 4. The cranial computed tomogram revealed a mandarin-sized intra-, supra-, and parasellar tumor sized 45 × 55 mm

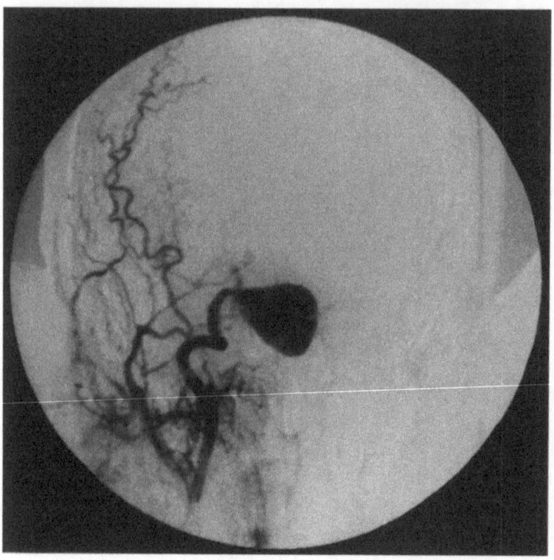

Fig. 5. Digital subtraction angiography showed a giant aneurysm supplied from the right internal carotid artery

15 mmHg on both sides. As we saw on the optic nerve, there were no indications of an increase in intracranial pressure at the fundus. We found only some equatorial degeneration of the retina. In the examination of the field of vision, we found a nasal lower visual field defect on the right and a more or less normal isopters' curve on the left (Fig. 3).

The neurological examination revealed no further abnormal findings; there were no generalized changes in the EEG and no paroxysmal discharges, in what was an overall irregular electroencephalogram. Derivation of the visual evoked potentials showed prolonged latencies in transient visual evoked responses in both eyes with full field stimulus, up to 127 ms on the right and 118.5 ms on the left (standard range up to 112 ms). With a quadrantic stimulus there was a reduced level of reproducibility. These findings would also have been compatible with the presence of a multilocular demyelinating disease.

In the X-ray examination that followed, destruction of the bony sella structure was already visible in the lateral skull radiograph. The cranial computed tomogram revealed a mandarin-sized intra-, supra-, and parasellar tumor (Fig. 4). It should be mentioned at this point that there were no indications of hormonal dysfunction either in the history or clinical examination. Digital subtraction angiography revealed that this was not a tumor but a large aneurysm at the base of the skull supplied from the right internal carotid artery (Fig. 5). We do not wish to report on the neurosurgical treatment that followed, since that would be carrying coals to Newcastle. It should be mentioned merely that the soldier subsequently had the opportunity to become acquainted with a number of the well-known European neurosurgical centers.

Discussion

The present case illustrates that in some cases of aneurysm of the internal carotid artery, with a localization that would lead one to suspect ophthalmological symptoms, such symptoms make the ophthalmologist to be the first port of call. Motility restrictions, such as oculomotor, trochlear, and abducent paralysis, conduction defects of the optic nerve, lesions of the chiasma, and optic radiation with hemiamaurosis and possibly also exophthalmos may occur. Almost those 5% of aneurysms located intracavernous (Jeanmart et al. 1973) may occur by visual field loss. We must consider a wealth of differential diagnoses at this point. Intraocular quadrantic amaurosis may be caused by retinal detachment or other retinal diseases which cause a loss of function in a restricted area. Also glaucoma, retrobulbar neuritis, multiple sclerosis, and neuromyelitis optica may produce unilateral nasal visual field defect. In the parasellar region pituitary adenoma, craniopharyngeoma, suprasellar meningeoma, tumors of the third ventricle, tumors of the cavernous sinus, metastatic tumors and other rare intra-, supra-, and parasellar tumors (chondroma, cholesteatoma, epidermoids, teratoma, parasites) may be responsible for visual field loss. Also inflammatory lesions as chiasmal neuritis, chiasmal arachnoiditis; syphilitic, tuberculous, and actinomycotic basal meningitis should be mentioned in addition to injuries of the chiasmal pathway and manifestations of histiocytotic lesions as histiocytosis X (Leuenberger und Signer 1986).

Nevertheless, observing the development of a scotoma can provide an indication as to the cause of the lesion, since aneurysmatic defunctionalization leads to asymmetrically diverse incongruent hemianopsia that can display pronounced variation in one and the same person during different periods of the disease, while bitemporal hemianopsia, due to a tumor, always shows a regular progression.

References

Barr HWK, Blackwood W, Meadows SP (1971) Intracavernous carotid aneurysms. A clinical pathological report. Brain 94:607–622

Brihaye J (1979) Intracavernous carotid aneurysms. In: Pia HW, Langmaid C, Zierski J (eds) Cerebral aneurysms, advances in diagnosis and therapy. Springer, Berlin Heidelberg New York, pp 67–78

Duke-Elder S (1971) Chiasmal lesions. In: System of ophthalmology, vol XII: Neuroophthalmology. Kimpton, London, pp 299–320

Francois J, Hollwich F (1986) Augenheilkunde in Klinik und Praxis, Bd 3, Teil II: Neuroophthalmologie. Thieme, Stuttgart New York, S 1–459

Huber A (1986) Hirnaneurysmen und Augensymptome. In: Francois J, Hollwich F (Hrsg) Augenheilkunde in Klinik und Praxis, Bd 3, Teil II. Thieme, Stuttgart New York, S 409–426

Jeanmart L, Noterman J, Brihaye J, Batèriaux D (1973) Les anéurysmes de la carotide intracaverneuse. Neurochirurgie 1:61–73

Leuenberger S, Signer E (1986) Histiozytose X bei einem 3 1/2jährigen Mädchen mit unklarem Orbitatumor. Klin Monatsbl Augenheilkd 188:486–487

Miller JH, Hinkley CM (1970) Berry aneurysms in pregnancy: a ten year report. South Med J 63:279–285

Pakarinen S (1967) Incidence, aetiology and prognosis of primary subarachnoidal haemorrhage. A study based on 589 cases diagnosed in a defined urban population during a defined period. Acta Neurol Scand [Suppl] 29:1–128

Pia HW, Langmaid C, Zierski J (1979) Operative treatment. In: Cerebral aneurysms, advances in diagnosis and therapy. Springer, Berlin Heidelberg New York, pp 203–251

Sachsenweger R (1975) Neuroophthalmologie. Thieme, Stuttgart New York

Sundt ThM, Piepgrass DG, Marsh WR, Fode NC (1986) Saphenous vein bypass grafts for giant aneurysms and intracranial occlusive disease. J Neurosurg 65:439–450

Yasargil MG (1984) Clinical considerations, surgery of the intracranial aneurysms and results. In: Yasargil MG (ed) Microneurosurgery, vol II. Thieme, Stuttgart New York, pp 1–55

Extra-Intracranial Anastomosis with Venous Interpositions in Patients with Giant Aneurysms of the Skull Base

H. J. Klein, S. A. Rath, and K. Schmidt, Günzburg/FRG

Introduction

The treatment of giant aneurysms of the skull base continues to challenge neurosurgeons because of their location and size and the common absence of a definitive neck. Drake (1979) reported that 71% of the patients treated had excellent or good outcome, and Koos et al. (1985) differentiated between those patients having been treated directly (63% excellent and good results) and those with indirect procedures (EC/IC bypass and internal carotid artery ligation: 93% good and excellent results). There are numerous reports dealing with a variety of strategies in the treatment of giant aneurysms and aneurysms of the skull base, including auxiliary bypass techniques (Crowell and Jafar 1988; Morgan et al. 1986; Gelber and Sundt 1980; Pendl et al. 1987; Spetzler et al. 1980; Sundt et al. 1986, 1987). Increasingly many authors have shown the value of performing a preliminary superficial temporal artery–middle cerebral artery anastomosis to support the middle cerebral artery circulation before attempting occlusion of the internal carotid artery of middle cerebral arteries.

Perhaps the most clearly defined role of the technique of bypass is in the management of anticipated ischemia resulting from surgical occlusion of the internal carotid or middle cerebral artery in cases of giant aneurysms. The risk of stroke after common carotid ligation, considered the most benign and predictable of the ligation procedures, carries a 15% incidence of early ischemic morbidity (Koos et al. 1985; Schorstein 1946) and 6% incidence of delayed ischemic complications (Black and German 1953). Other authors report incidences of 59% (Locksley 1966) and 22% (Koos et al. 1985) following ligation procedures. The question is whether a conventional anastomosis with superficial temporal artery can immediately provide a sufficient blood supply.

Material and Methods

In the past 3 years five patients out of a total of 95 operated patients with aneurysms were treated for aneurysms of the skull base with a diameter of more than 2.5 cm. All five patients underwent additional bypass procedures. In two of them we performed the final occlusion of the aneurysms only after a conventional anastomosis between the superficial temporal artery and branch of the middle cerebral artery.

Case Reports

Case 1. A 14-year-old girl suffered from a sudden onset of right-sided hemiparesis. Computed tomography (CT) demonstrated a round mass lesion in the Sylvian fissure with considerable surrounding brain edema. Angiography showed a giant aneurysm of the trifurcation of the left middle cerebral artery and an associated aneurysm between the lenticulostriatal arteries and the trifurcation. Two weeks before the direct approach to the aneurysms a superficial temporal artery–middle cerebral artery anastomosis was established. During the second operation we found the two aneurysms to be totally thrombosed while the first angiography demonstrated aneurysms 2.5×2.7 and 0.8×1.0 cm in size. A partial occlusion was seen also in our second case. This tendency to aneurysmal thrombosis after EC/IC bypass procedure is in congruence with the findings of Morgan et al. (1986). These authors combined a venous interposition bypass with an internal carotid artery ligation without direct approach to the aneurysmatic complex. In our opinion this procedure is limited in cases with an aneurysmatically induced space-occupying process able to cause visual disorders (cases 2–5 in our series) and is hazardous in giant aneurysms or aneurysms of the skull base with recent subarachnoid hemorrhage.

Case 2. In a 30-year-old woman suffering from visual symptoms caused by a giant aneurysm of the anterior communicating artery we performed the final occlusion of the aneurysm only after a bilateral conventional anastomosis (in two operations) and a clamping procedure of the left internal carotid artery at the neck, so that four procedures were necessary. The clinical outcome was uneventful. The postoperative observation period is now 25 months. Both anastomoses are patent and the actual Doppler sonography shows an elevated diastolic baseline at the side of previous aneurysm and internal carotid artery occlusion which is distinctly different from the lower diastolic value of an unanastomosed superficial temporal artery. The Doppler sonography at the noncompromised hemisphere reveals no increased diastolic plateau, and this patent anastomosis without a diastolic plateau in the Doppler signal over time had an anterograde flow near the brain surface and thus did not provide true cerebral perfusion. In contrast, the anterograde flow through the anastomosis could be followed to depth of several centimeters at the side of internal carotid artery occlusion.

Our first approach to aneurysms with a venous interposition was a short vein graft between the proximal superficial temporal artery and a cortical branch of the MCA in a 44-year-old man with a giant aneurysm of the left internal carotid artery. We have used this method since October 1986, and up to now 31 patients have received an anastomosis via venous transplant (long saphenous vein). In 28 patients, this was performed as revascularizing surgery for 25 carotid occlusions, one medial cerebral artery occlusion, one extensive, high-grade stenosis at the base of the skull. Only those patients with vascular lesions and corresponding symptoms were operated on, when they also showed a decreased autoregulatory reserve in transcranial Doppler sonography under CO_2 respiration with a reduced increase in flow velocity (Widder et al. 1987).

Case 3. A 44-year-old man suffered from an acute subarachnoid hemorrhage and a visual disturbance on the left eye due to a giant aneurysm of the left internal carotid artery representing a diameter of 29 mm. A short vein bypass between the proximal superficial temporal artery and a proximal branch of the medial cerebral artery was performed 12 days prior to the aneurysm operation. After these two operations the later course was uneventful, and the angiogram showed a complete filling of the left MCA region via the short vein graft (Figs. 1, 2). In this operation the surgery required lowering the arterial blood

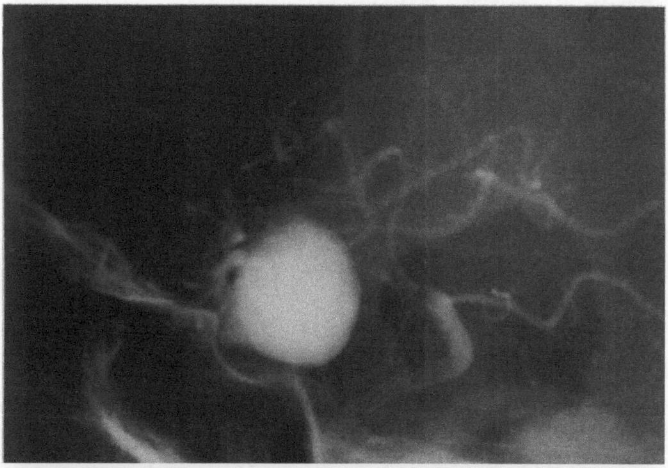

1

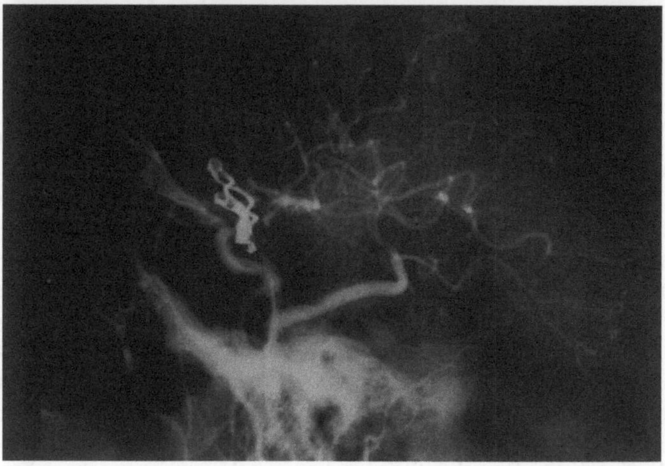

2

Fig. 1. Case 3: giant aneurysm of the left carotid artery. In the first operation a strategic short vein bypass was performed

Fig. 2. Case 3: complete filling of the left MCA region via the short vein graft. The carotis interna aneurysm and the internal carotid artery, distal to the ophthalmic artery, are occluded

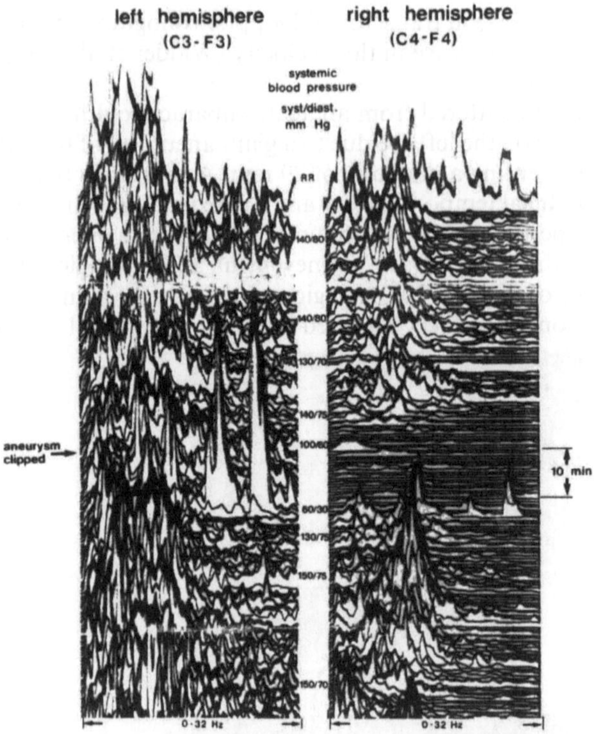

Fig. 3. Case 3: continuous EEG recording in compressed spectral array mode. The hemisphere with additional bypass blood support shows a nonaffected EEG whereas the right hemisphere demonstrates burst suppression EEG after induced hypotension. Two weeks after EC/IC with venous interposition

pressure to 60/30 mmHg because the clips repeatedly snapped off from the internal carotid artery aneurysm. During this period, EEG frequency analysis and compressed spectral array monitoring showed a total zero line over the right nonaffected hemisphere while electrical brain activity was not reduced for the left side. We attributed this to the high volume of flow through the venous bypass with simultaneous maintenance of the flow in the aneurysmatic vessel and the increased perfusion capacity compared to the healthy hemisphere (Fig. 3). During the postoperative course up to now we have observed a progressive transformation of the Doppler signal of the bypass from an initial externa type (stand-by type) into an interna type (demand-type; Fig. 4). Our next step was – and this is our current concept – to perform the bypass operation with a long vein graft and definite aneurysm closure in one step. Due to pathophysiological reasons and a review of literature concerning bypass surgery in giant aneurysms and aneurysms of the skull base we first establish the venous bypass between the external carotid artery and a M2 vessel to obtain

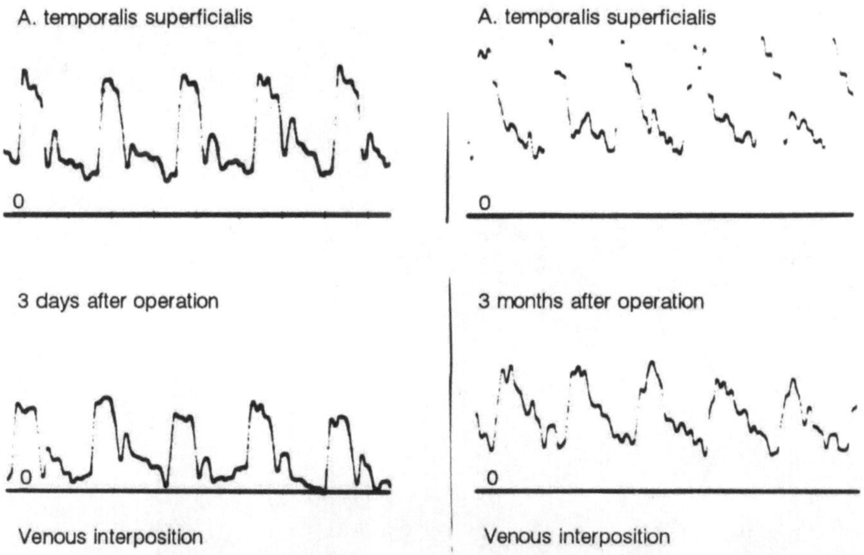

Fig. 4. Case 3: transformation of the Doppler signal from an externa type (stand-by type) into an interna type (demand-type) during the postoperative course

an immediate blood supply of approximately 100 ml/min via the graft (Sundt et al. 1987) and then perform the direct approach to the aneurysm.

Case 4. In a 45-year-old man suffering from a gradual visual loss and a temporal visual field defect we found a very hard and partially arteriosclerotic aneurysm of the right internal carotid artery. The basal side of the aneurysm (25 × 27 mm in diameter) was intimately related to the dura of the skull base causing trigeminal pain in the first division. To provide an immediate high flow the vein graft was interposed between the external carotid artery and one M2 segment just beside the aneurysm. During the operation a definitive control of the arteriosclerotic and aneurysmatic complex was only possible with clipping both aneurysm and the internal carotid artery distal to the ophthalmic artery and proximal to the posterior communicating artery. Postoperative angiography demonstrated good flow through the saphenous vein graft, which extended from the external carotid artery in the neck to the M2 segment of the middle cerebral artery, filling the whole right hemisphere (Figs. 5, 6).

Case 5. A 59-year-old woman, operated for giant aneurysm of the left internal carotid artery just distal to the ophthalmic artery with bypass procedure and clipping of the aneurysm under occlusion of ICA, suffered a delayed and progressive stroke with brachiofacial right-sided incomplete hemiparesis on the second postoperative day. Simultaneously, SEP leads taken from the right hemisphere showed a significant prolongation and reduction of

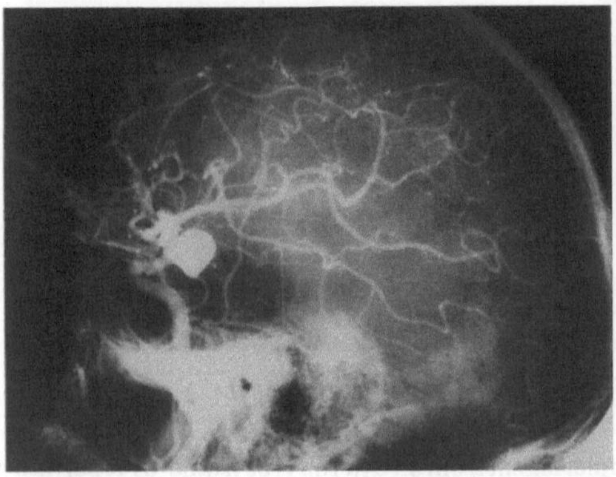

5

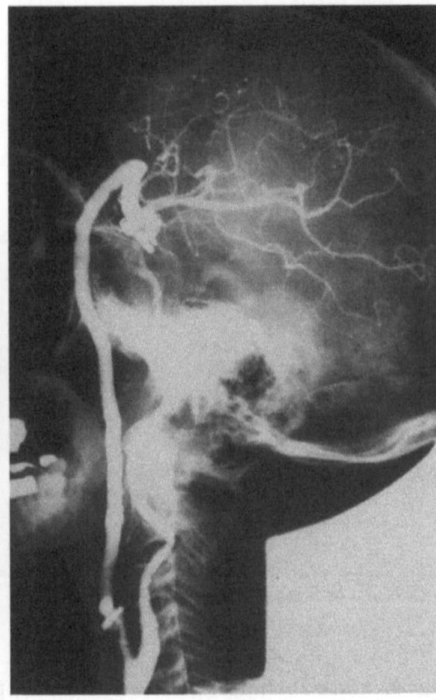

6

Fig. 5. Case 4: giant aneurysm of the right carotid artery, partially arteriosclerotic altered

Fig. 6. Case 4: postoperative angiogram showing the extent of filling from the vein bypass after clipping of an internal carotid artery aneurysm with occlusion of the vessel. The whole operative procedure was performed in one session

the N20/P25 complex with a loss of the later components, and postoperative angiogram revealed an occlusion of the EC/IC bypass after primary patency. There was a slight recovery of the hemiparesis after 2 months so that the patient was in an ambulatory state and independent.

Conclusion

We think there is physiological, clinical, and radiological evidence that the venous interposition may be superior to the superficial temporal artery – middle cerebral artery bypass for more effective and immediate protection against ischemia (Gelber and Sundt 1980; Spetzler et al. 1980; Sundt et al. 1987). This high level of blood supply is made possible by the large diameter of the vein as well as by the option of connecting the vein at a very proximal position both in the donor as well as the recipient organ, so that the length of the grafts also permits an anastomosis in the region of the proximal Sylvian fissure. With this operative procedure, which starts with the establishment of a high-volume bypass, the clipping and resection of the aneurysm and eventual occlusion of the intracranial carotid artery is possible in one step without additional ischemic risks. Some giant aneurysms and particularly aneurysms of the skull base are so intimately related to parent arteries and perforating arteries or so inaccessibly situated as to render direct approach extremely dangerous and often impossible. The anastomosis may provide collateral circulation to essential vascular territories beyond these aneurysms so that they can be obliterated more safely even if it is necessary to occlude a parent artery. The limitation of the venous interposition are those areas of the brain which are supported by the anterior perforating arteries and the lenticulostriatal arteries eventually being in a direct and fatal neighborhood to the aneurysmatic complex (Yasargil 1984).

References

Black SPW, German WJ (1953) The treatment of internal carotid artery aneurysms by proximal arterial ligation. A follow up study. J Neurosurg 10:590–601

Crowell RM, Jafar JJ (1988) Direct brain revascularization. In: Schmidek HH, Sweet WH (eds) Operative neurosurgical techniques, vol 1. Grune & Stratton, Orlando, pp 783–795

Drake CG (1979) Giant intracranial aneurysms: experience with surgical treatment in 174 cases. Clin Neurosurg 26:12–95

Gelber BR, Sundt TM Jr (1980) Treatment of intracavernous and giant carotid aneurysms by combined internal carotid ligation and extra to intracranial bypass. J Neurosurg 52:1–10

Koos WT, Perneczky A, Schuster H, Vorkapic P (1985) Zerebrale Riesenaneurysmen. In: Schürmann K (Hrsg) Der zerebrale Notfall. Urban & Schwarzenberg, München, S 163–176

Locksley HB (1966) Report on the cooperative study of intracranial aneurysms and subarachnoid hemorrhage. J Neurosurg 25, Section V Part II

Morgan M, Besser M, Dorsch N, Segelov J (1986) Treatment of intracranial aneurysms by combined proximal ligation and extracranial-intracranial bypass with vein graft. Surg Neurol 26:85–91

Pendl G, Vorkapic P, Richling B, Koos WT (1987) Strategies in intracavernous saccular aneurysms. In: Dolenc VV (ed) The cavernous sinus. Springer, Wien New York, pp 240–251

Schorstein J (1946) Carotid ligation in saccular intracranial aneurysms. Br J Surg 28:50–70

Spetzler RF, Schuster H, Roski RA (1980) Elective extracranial-intracranial arterial bypass in the treatment of inoperable giant aneurysms of the internal carotid artery. J Neurosurg 53:22–27

Sundt TM Jr, Piepgras DG, Marsh WR, Fode NC (1986) Saphenous vein bypass grafts for giant aneurysms and intracranial occlusive disease. J Neurosurg 65:439–450

Sundt TM Jr, Piepgras DG, Marsh WR, Fode NC (1987) Bypass vein grafts for giant aneurysms and severe intracranial occlusive disease in the anterior and posterior circulation. In: Sundt TM Jr (ed) Occlusive cerebrovascular disease. Saunders, Philadelphia, pp 439–464

Widder B, Paulat K, Hackspacher J, Kornhuber HH, Mayr E (1987) Selektive Indikationsstellung zur Karotis- und EC-IC-Bypasschirurgie mit Hilfe des CO_2-Testes. In: Widder B (Hrsg) Transkranielle Dopplersonographie bei zerebrovaskulären Erkrankungen. Springer, Berlin Heidelberg New York Tokyo, S 133–137

Yasargil MG (1984) Special operative problems. In: Yasargil MG (ed) Microneurosurgery, vol I. Thieme, Stuttgart, pp 268–269

Embolization of Four Cases of Carotid-Cavernous Sinus Fistulas by Retrograde Catheterism of the Superior Orbital Vein

J. P. Caron, Y. Raulo, A. Gaston, Creteil/France,
P. Lasjaunias, Le Kremlin-Bicetre/France, and J. Courtheoux, Caen/France

Introduction

According to Dolenc [9] Mont as early as 1935 advocated in cases of carotid-cavernous sinus fistulas (CCSF) with anterior drainage through the superior orbital vein (SOV) the resection and ligature of this vein. He suggested scouring the lumen with a small tube brush soaked in 50% glucose. Also, as described by Dolenc, Brainerd in 1953 attempted to cure an "erectile tumor of the orbit" by retrograde injection of lactate of iron. Dandy et al. in 1937 [3] and 1941 [4] expressed the opinion that the SOV route should not be used for treating CCSF. In 1975 Peterson et al. [11] used a percutaneous approach for passing through the SOV to treat successfully CCSF. In 1981 Debrun et al. [6] maintained that the endovenous route was an excellent way to treat CCSF drained anteriorly through the SOV; however, their first percutaneous approaches failed, and in 1983 [7] Debrun wrote that "retrograde navigation through the SOV was virtually impossible in cases of CCSF with totally or mainly anterior drainage through the ophthalmic veins." But they were recently successful [8], in two cases using a surgical approach. In fact, Uflacker et al. [13] in 1986 published the most important article on retrograde catheterism through the SOV for treating CCSF drained anteriorly, presenting three successful cases out of five by a surgical approach "deep in the orbit." Since this article, very few cases of CCSF embolization through the SOV have been published. Here we present four such cases, coming from three different French neuroradiological and neurosurgical teams.

Case Reports

Case 1 (Prof. Lasjaunias). A 64-year-old woman presented 15 days after a car accident with a right pulsating exophthalmos, conjunctival edema, and orbital pains. Angiography revealed a CCSF. Obliteration of the CCSF was first attempted by endoarterial route, but this failed. One month later, the exophthalmos had increased, orbital pains remained, and visual acuity was slightly impaired. A percutaneous approach of the CCSF by retrograde catheterism of the facial vein and SOV was then decided. When the catheter reached the third portion of the SOV, the rupture of this vein resulted in a substantial increase in the exophthalmos, with eyelid swelling. Intermittent internal carotid artery (ICA) compression, repeated over 45 min, stopped the exophthalmos increase. An eyelid hematoma appeared the following hours. Three months later, the exophthalmos had disappeared, and visual acuity was normal. Control angiography revealed total cure of the fistula.

Cases 2 and 3 (Dr. Courtheoux). A 61-year-old woman presented with a bilateral spontaneous CCSF, with increasing ophthalmologic symptoms: red-eye syndrome, ocular pressure tension up to 32, slight exophthalmos, and orbital pains. Angiograms showed on each side a CCSF vascularized by the ICA meningo-hypophyseal trunk and drained through the SOV. Occlusion of the meningo-hypophyseal trunk by endoarterial route being too hazardous, retrograde embolization through the SOV was decided. After orbitonasal skin incision, the right SOV was exposed and dissected. The SOV was too narrow to be catheterized with a releasable balloon catheter, and two small Gianturco steel coils were pushed with a 5-F catheter into the SOV and released in its medial part. Within 3 weeks, right ocular pressure tension became normal; conjunctival congestion had largely decreased; the slight proptosis had disappeared. At control angiography, the fistula remained visible, but the SOV was thrombosed.

Two months later, the left SOV was approached with the same surgical procedure, a very thin catheter was introduced into the cavernous sinus, and

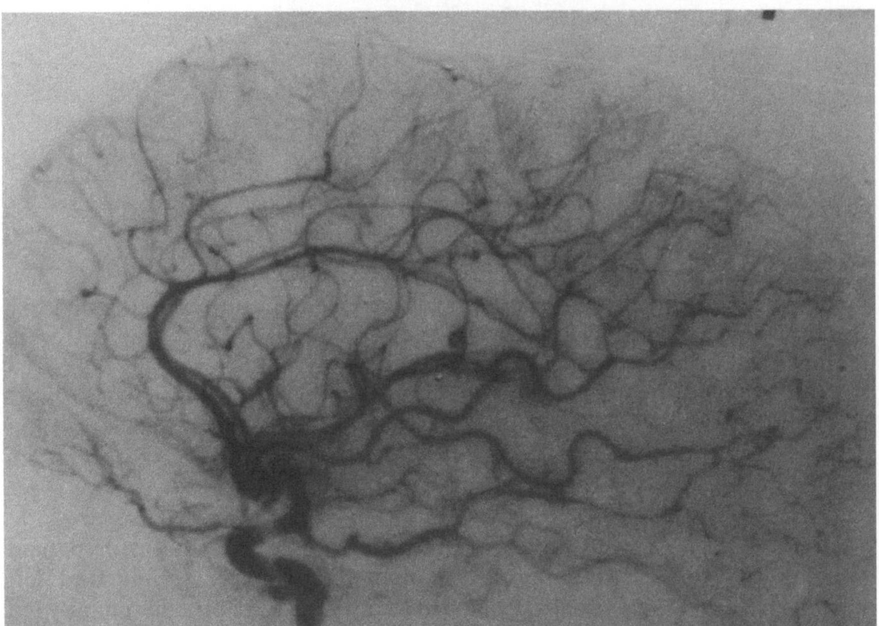

Fig. 1. Case 3: left ICA angiogram before embolization

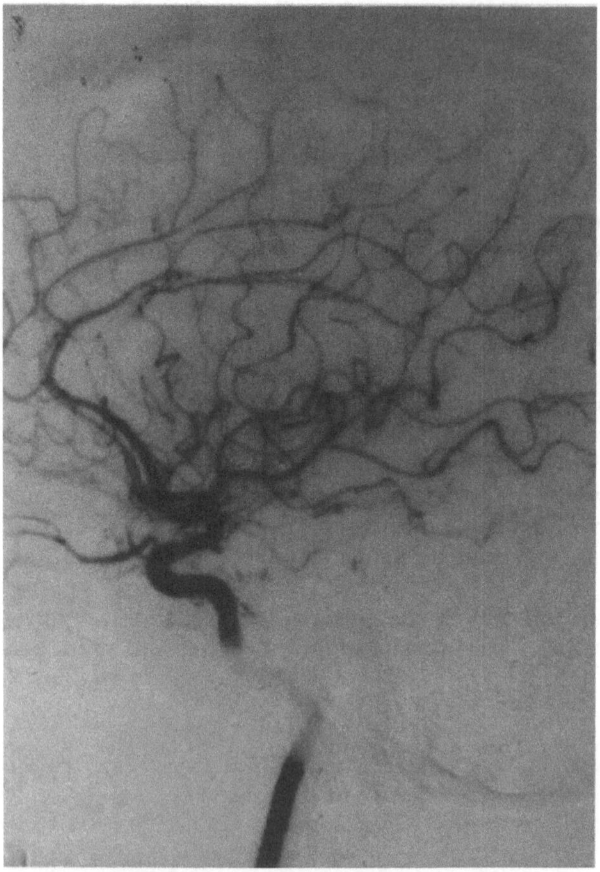

Fig. 2. Case 3: left ICA control angiogram after embolization with sclerosing agent, CCSF has disappeared

0.5 cm^3 tetradecylsulfate (a sclerosing agent) was injected into it. The immediately postoperative period was marked by an important inflammatory reaction, with eyelid edema, chemosis, and orbital pains. Eye movements were limited in all directions, visual acuity decreased to 4/10. Corticoids improved the situation within 3 weeks. Pain and chemosis slowly disappeared. Within 2 months, there was no more proptosis; eye movements were normal, and visual acuity had returned to 10/10. At control angiography, SOV was thrombosed, and the fistula was cured (Figs. 1, 2).

Case 4 (Profs. Caron, Raulo, Gaston). In 1965 a 1-year-old male patient suffered a severe right temporal trauma, with transient coma and left hemiplegia. Soon thereafter he developed a complex CCSF, predominantly fed by the right ICA. A direct surgical approach to repair the intracavernous ICA was unsuccessfully attempted in 1966, and resulted in ligature of ICA in its intracavernous portion and in the neck. In 1975, for a subjective and objective temporal bruit, headaches, right eye proptosis, and periorbital excessive venous dilatation, bilateral embolizations of both external carotid artery branches feeding the fistula were made by Prof. Picard (Nancy). A temporary decrease in proptosis and of periorbital venous circulation was obtained, as well as relief of temporal bruit. In October 1986, the patient came to our department for recurrent hedaches and right orbitotemporal subjective and objective bruit. Venous periorbital circulation was excessive, right eye visual acuity was 1/20. Ocular pressure tension was at the superior normal limit. Proptosis of the right eye was 6 mm. Supraselective angiograms showed that the CCSF was still vascularized by different branches of the right external carotid artery and by the intracavernous portion of the right ICA, which was itself revascularized by a branch of the pharyngeal ascending artery, by a meningeal branch of the internal maxillary artery, and by meningeal branches of the left ICA (Fig. 3). Drainage was assured essentially by the enlarged and tortuous SOV. Direct feeding branches of the external carotid artery were first embolized, except for the ascending pharyngeal artery one anastomosed with a meningeal branch of the right vertebral artery. Ophthalmologic symptoms remained unchanged. Therefore, retrograde embolization through the SOV was decided. After surgical removal of the orbital superior margin and orbital roof, the SOV was dissected deep into the orbit (Fig. 4). A 5-F catheter was introduced into the SOV, and a releasable balloon was pushed into it; however,

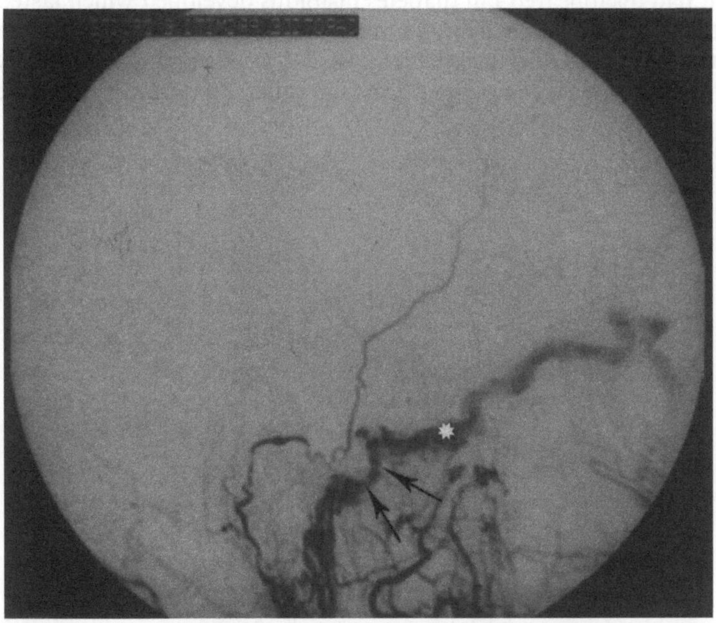

Fig. 3. Case 4: right primitive carotid angiogram before embolization. *Arrows* indicate the revascularization of the intracavernous portion of ICA. *White star* indicates fistula drained by SOV

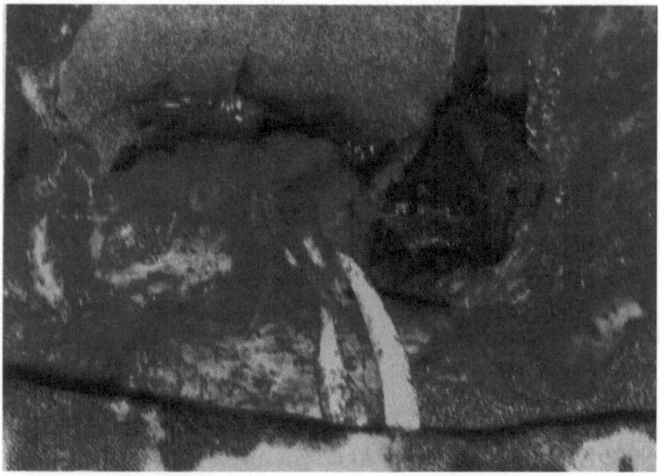

Fig. 4. Case 4: surgical exposure of the dilated and tortuous SOV, deep into the orbit

probably because of the deep bending of the vein or its narrowing at its connection with the cavernous sinus (as assessed by Mullan), it could not enter the cavernous sinus. After a second unsuccessful attempt with a smaller catheter, two inflated balloons were released into the SOV, quite close to but not into the cavernous sinus. The immediate postoperative period was eventful: bruit had disappeared but diabetes insipidus developed which lasted 1 month; on the 2nd postoperative day an important orbital ecchymotic edema extended to the right face and neck which lasted 2 weeks and was probably secondary to the rupture of some other drainage veins. Five months after the

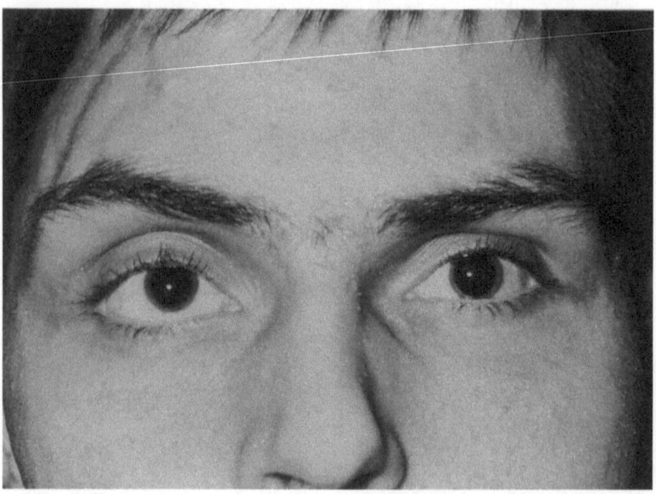

Fig. 5. Case 4: patients photograph, 5 months after embolization

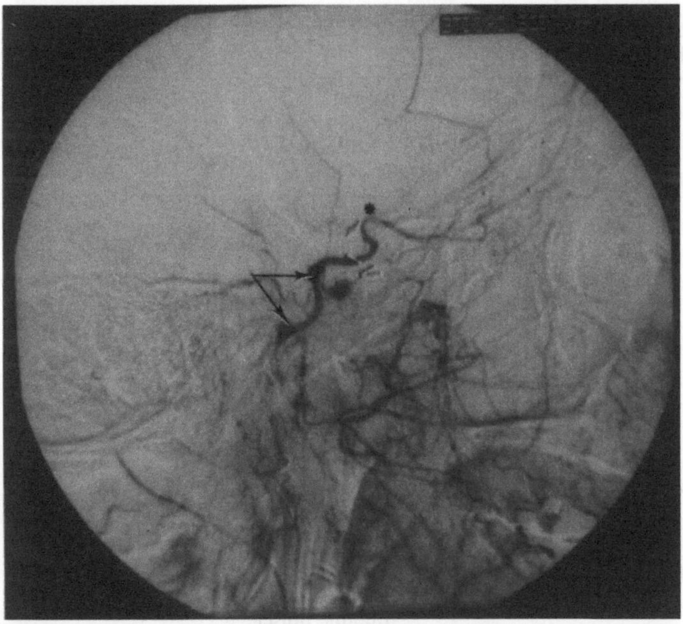

Fig. 6. Case 4: after embolization, the fistula remains. SOV is still visible, but its caliber has become normal

operation, bruit and headaches were absent. Periorbital venous circulation was normal. Visual acuity had improved to 1/10. Ocular pressure tension was normal at 10. Proptosis was reduced to 1 millimeter. Movements of the eye were normal (Fig. 5). At control angiography, the fistula was still present, fed only by the branch of the ascending pharyngeal artery. The SOV was still partly permeable but very narrowed (Fig. 6).

Discussion

Materials used by different authors to obtain thrombosis of the fistula or of the SOV are quite different; Tress et al. [12] and Courtheoux et al. used Gianturco steels coils; Uflacker et al. [13], steel spring coils; Debrun et al. [8], Caron et al. releasable balloons, which do not enter the cavernous sinus very easily; a sclerosing agent (tetradecylsulfate) was used by Courtheoux once but with a severe inflammatory reaction; isobutylcyanoacrylate has also been used but is considered by many authors as dangerous, due to its possible propagation into the ICA and thrombosis of it. Many authors feel that very thin catheters, such as Tracker's, should now facilitate entering the cavernous sinus. Table 1 shows that the surgical approach is the safer way to enter the catheter into the cavernous sinus or in the deeper portion of the SOV; except for Peterson [11], no author [10, 6, 7] was successful using a percutaneous approach.

Table 1. Percutaneous (P) or surgical (S) approach: Location of vein catheterism, entering the cavernous sinus or deep portion of SOV

Year	Authors	*n*	Approach	Location of vein catheterism	Entering the cavernous sinus or deep portion of SOV
1970	Peterson	3	P	Facial vein	+
1979	Mullan	2	P	Facial vein	–
1981	Debrun	2	P	Facial or angular vein	–
1983	Tress	1	S	SOV deep in the orbit	+
1986	Uflacker	5	S	SOV in the orbit	3+ 2–
1986	Lasjaunias	1	P	Facial vein	–
1987	Courtheoux et al.	2	S	SOV behind trochlea of superior oblique muscle	2+
1988	Debrun et al.	2	S	SOV behind trochlea of superior oblique muscle	2+
1988	Caron, Raulo, Gaston et al.	1	S	SOV deep in the orbit	+
	Total	19	8 percutaneous approaches		3+(Peterson) 5–
			11 surgical approaches		9+ 2–

Entering into the cavernous sinus is the ideal goal, which permits the radical cure of the CCSF (as Peterson [11], Tress [12], Uflacker [13], Debrun [8], and Courtheoux did). But in some cases (as Caron's and Courtheoux's on the right side) thrombosis of the deeper part of the SOV, very close to its connection to the cavernous sinus, is sufficient for relief of the clinical ophthalmologic symptoms; no symptom arises from the remaining fistula.

Rupture of the SOV is the main complication of this procedure. Tress [12] mentions such a complication, due to a perioperative venous control angiogram; he treated this complication by an anterior chamber paracentesis and surgical incision of the inferior orbital septum. A similar rupture occurred in Lasjaunias's case before any embolization while pushing the catheter in the SOV. The treatment was a repeated ICA compression over 45 min, which indeed cured the fistula. The same complication (or rupture of other draining veins) happened in Caron's case, which was treated by slightly compressive dressing. But in no case did this complication alter the long-term postoperative result.

Conclusions

On the basis of the literature and our own experience, we consider the following observations as important:

- One must not forget that some CCSF (more frequently spontaneous ones) never need any treatment; some cure spontaneously; others are cured after carotid angiogram (Voigt [15], Bennett [2] or repeated intermittent carotid compression (Stolpmann, Lasjaunias).
- As a consequence, only cases with severe and increasing symptoms (decrease of visual acuity; increased ocular pressure tension; excessive pain or bruit, important exophthalmos) require an active treatment.
- Retrograde embolization through the SOV must not be used before all possible endoarterial embolizations with conservation of the ICA flow have been attempted.
- This retrograde intravenous route is to be reserved for CCSF with exclusively anterior drainage, and preferentially for slow flow CCSF [1].
- Depending on the danger of rupture of the SOV, this retrograde route (as Uflacker et al. [13] have already observed) is recommended when the SOV has had sufficient time to become arterialized.
- Very thin recent catheters (such as Tracker's) should facilitate the approach to the fistula itself.
- Given these conditions, retrograde embolization through the SOV is a valuable procedure for the treatment of CCSF drained anteriorly through the SOV.

Summary

Four cases of carotid-cavernous sinus fistulas have been treated by retrograde catheterism of the superior ophthalmic vein (SOV) by three different medical teams. Reasons for treatment were bruit and headache (one case) or ophthalmic symptoms (three cases).

Different techniques used for catheterism of the SOV are exposed and discussed, as well as materials used for embolization. From this limited experience, this method seems to bring good results in ophthalmologic symptoms, through reduction or suppression of the SOV drainage; but getting into the cavernous sinus itself to obliterate the fistula is hazardous, because of bending and narrowing of the SOV at its connection with the cavernous sinus itself.

It seems safer to wait until the walls of the SOV have had time to acquire a certain thickness.

References

1. Barrow DL, Spector RH, Braun EF, Landman JA, Tindall SC, Tindall GT (1985) Classification and treatment of spontaneous carotid-cavernous sinus fistulas. J Neurosurg 62:248–256
2. Bennett DR, Van Dyk HJL, Davis DO (1973) Carotid-cavernous sinus fistula closure following angiography. JAMA 224:1637–1639
3. Dandy WE (1937) Carotid-cavernous aneurysms (pulsating exophthalmos). Zentralbl Neurochir 2:77–113, 165–206
4. Dandy WE, Follis RH Jr (1941) On the pathology of carotid cavernous aneurysms (pulsating exophthalmos). Am J Ophthalmol 24:365–385
5. Debrun GM, Lacour P, Caron JP et al. (1975) Experimental approach to treatment of carotid cavernous fistulas with an inflatable and isolated balloon. Neuroradiology 9:9–12
6. Debrun G, Lacour P, Vinuela F, Fox A, Drake CG, Caron JP (1981) Treatment of 54 traumatic carotid-cavernous fistulas. J Neurosurg 55:678–692
7. Debrun G (1983) Treatment of traumatic carotid-cavernous fistula using detachable balloon catheters. AJNR 4:355–356
8. Debrun G, Vinuela F, Fox AJ, Davis KR, Ahn HS (1988) Indications for treatment and classification of 132 carotid-cavernous fistulas. Neurosurgery 22:285–289
9. Dolenc VV (1987) The cavernous sinus. A multidisciplinary approach to vascular and tumorous lesions. Springer, Wien New York
10. Mullan S (1974) Experience with surgical thrombosis of intracranial berry aneurysms and carotidcavernous fistulas. J Neurosurg 41:657–670
11. Peterson E, Valberg JD, Ventureyra E (1975) Percutaneous treatment of carotid cavernous fistula with preservation of the carotid artery. Presented at the Canadian Meeting of Neurological Sciences, June 19, 1975
12. Tress BM, Thompson KR, Klug GL, Mee RRB, Crawford B (1983) Management of carotid-cavernous fistulas by surgery combined with interventional radiology. J Neurosurg 159:1076–1081
13. Uflacker R, Lima S, Ribas GC, Piske RL (1986) Carotid-cavernous fistulas: embolization through the superior ophthalmic vein approach. Radiology 159:275–179
14. Vinuela F, Lylyk P (1987) Endovascular therapy of vascular lesions of the cavernous sinus. Experience with 129 cases. In: Dolenc VV (ed) The cavernous sinus. A multidisciplinary approach to vascular and tumorous lesions. Springer, Wien New York
15. Voigt K, Sauer M, Dichgans J (1971) Spontaneous occlusion of a bilateral carotid-cavernous fistula studied by serial angiography. Neuroradiology 2:207–211

Subject Index

MIX
Papier aus verantwortungsvollen Quellen
Paper from responsible sources
FSC® C105338

If you have any concerns about our products,
you can contact us on
ProductSafety@springernature.com

In case Publisher is established outside the EU,
the EU authorized representative is:
Springer Nature Customer Service Center GmbH
Europaplatz 3, 69115 Heidelberg, Germany

Printed by Libri Plureos GmbH
in Hamburg, Germany